Hazards and Treatment of Wastewater Contaminants

Hazards and Treatment of Wastewater Contaminants

Dipika Jaspal
Arti Malviya

WILEY

Published by John Wiley & Sons, Inc., Hoboken, New Jersey.

Library of Congress Cataloging-in-Publication Data:

Names: Jaspal, Dipika author | Malviya, Arti author | John Wiley & Sons publisher
Title: Hazards and treatment of wastewater contaminants / by Dipika Jaspal, Arti Malviya.
Description: Hoboken, New Jersey : Wiley, [2026] | Includes index.
Identifiers: LCCN 2025034107 | ISBN 9781394217342 hardback | ISBN 9781394217359 adobe pdf | ISBN 9781394217366 epub | ISBN 9781394217373 ebook
Subjects: LCSH: Water—Pollution | Water—Purification | Sewage—Purification
Classification: LCC TD420.J37 2025
LC record available at https://lccn.loc.gov/2025034107

Cover Design: Wiley
Cover Image(s): © hero mujahid/Shutterstock

Contents

Preface *xv*

1 Introduction *1*
1.1 Water Pollution: A Global Concern – Facts and Figures *2*
1.1.1 Sustainable Development Goals *2*
1.1.2 Climate Change and Water Stress *3*
1.1.3 Wastewater and Its Types *5*
1.1.3.1 Storm Water/Surface Runoff *6*
1.1.3.2 Domestic or Municipal Wastewater *6*
1.1.3.3 Agricultural Wastewater *7*
1.1.3.4 Industrial Wastewater *7*
1.2 Wastewater Characteristics *8*
1.2.1 Color *8*
1.2.2 Odor *8*
1.2.3 Turbidity *9*
1.2.4 Nutrients *10*
1.2.5 Surfactants *10*
1.3 Hazards of Wastewater and the Need for Treatment *11*
1.3.1 Aquatic Ecosystem *11*
1.3.2 Humans *11*
1.4 Need to Treat Wastewater *12*
1.5 Scope and Objectives of the Book *12*
References *13*

2 Overview of Contaminants *17*
2.1 Contaminants *17*
2.2 Harm Posed by Contaminants *18*
2.3 Sewage Wastewater *19*
2.4 Nonsewage Wastewater *20*
2.4.1 Industrial Contaminants *20*
2.4.2 Agricultural Runoffs *21*
2.4.3 Oil Spill *22*
2.4.4 Radioactive Contaminants *23*
2.4.5 Microbial Contaminants *25*

2.4.6 Emerging Contaminants 25
2.4.6.1 Microplastics 26
2.4.6.2 Surfactants 27
2.4.6.3 Pharmaceuticals 27
2.5 Detecting Contaminants 28
2.5.1 Physical Methods 28
2.5.2 Chemical Methods 29
2.5.3 Biological Methods 29
2.5.4 Electrochemical Methods 29
2.5.5 Advanced Analytical Methods 29
2.6 Removal of Wastewater Contaminants 29
2.7 Incidents of Contamination: Case Studies 30
2.7.1 Flint, Michigan Water Crisis (2014) 30
2.7.2 Oil Spill in the Gulf of Mexico (2010) 30
2.7.3 West Virginia Chemical Spill – Charleston (2014) 30
2.7.4 Contamination of the River Ganga in India 31
2.7.5 Minamata Bay Disaster (1932–1968) 31
2.8 Successful Elimination of Contaminants 31
2.8.1 Tuas Water Reclamation Plant (Singapore) 31
2.8.2 Wastewater Reclamation (Taiwan) 32
2.8.3 Advanced Oxidation Processes (AOPs) (Japan) 32
2.8.4 Constructed Wetlands (Sweden) 32
2.8.5 Bioreactors for Wastewater Treatment (Saudi Arabia) 33
2.9 Important Considerations 33
2.9.1 Education and Public Awareness 33
2.9.2 Prevention of Pollution at Source and Best Practices 33
2.9.3 Wastewater Management 34
2.9.4 Use of Technology 34
2.9.5 Research and Development 34
2.9.6 Polluted Water Body Restoration 34
2.9.7 Policies and Collaboration 35
2.10 Challenges 36
References 36

3 Sewage Wastewater: Sources, Characteristics, and Treatment 43
3.1 Characteristics of Sewage Wastewater 45
3.1.1 Sewage Physical Characteristics 45
3.1.2 Chemical Features of Sewage 46
3.1.3 Biological Features of Sewage 48
3.2 Treatment of Sewage Wastewater 49
3.2.1 Preliminary Treatment 50
3.2.1.1 Screening 50
3.2.1.2 Grit Chamber 51
3.2.1.3 Detritus Tank 51
3.2.1.4 Skimming Tanks 51
3.2.2 Primary Treatment of Sewage Wastewater 52
3.2.2.1 Sedimentation 52
3.2.2.2 Sedimentation with Coagulation 53

3.2.3 Secondary Treatment of Sewage Wastewater *54*
3.2.3.1 Aerobic Biological Treatment *54*
3.2.3.2 Anaerobic Treatment Method *62*
3.2.3.3 Other Biological Methods *62*
3.2.4 Tertiary Treatment of Sewage Wastewater *67*
3.3 Case Studies *73*
3.3.1 Sewage Treatment Plant: A Case Study of the Delawas District, Jaipur, India *73*
3.3.2 Sewage Treatment Plant: A Case Study of the Okhla District, New Delhi, India *74*
References *74*

4 Industrial Contaminants *77*
4.1 Industrial Wastewater *77*
4.2 Wastewater Contaminants from Different Industries *78*
4.2.1 Mining Industry *78*
4.2.2 Agricultural Industry *78*
4.2.3 Iron and Steel Industry *79*
4.2.4 Nuclear Industry *79*
4.2.5 Pulp, Paper, and Textile Industry *79*
4.2.6 Food Industry *80*
4.2.7 Organic Chemicals Industry *81*
4.2.8 Recreation Industry *82*
4.3 Categories of Industrial Contaminants *82*
4.4 Monitoring and Detection of Industrial Contaminants *83*
4.5 Transport and Fate of Contaminants in the Environment *84*
4.5.1 Fate of the Contaminants *85*
4.6 Toxicity of Industrial Wastewater *86*
4.6.1 Impact on Waterbodies *86*
4.6.2 Impact on Soil and Groundwater *87*
4.6.3 Impact on Human Health *87*
4.6.4 Impact on Biodiversity and Ecosystem *87*
4.6.5 Impact of Some Specific Contaminants *87*
4.7 Treatment Technologies *88*
4.7.1 Adsorption *89*
4.7.2 Membrane Separation and Technologies *89*
4.7.3 Electro-technologies *91*
4.7.4 Advanced Oxidation Processes *91*
4.7.5 Integrated Techniques *91*
4.7.6 Constructed Wetlands and Integrated Wetlands *93*
4.8 Case Studies of Industrial Contamination of Water Bodies *95*
4.8.1 Groundwater Pollution in Southeast Vietnam *95*
4.8.2 The Elk River Chemical Spill, West Virginia *95*
4.8.3 Minamata Disease (Kumamoto Prefecture) *96*
4.8.4 Kala Shah Kaku Industrial Complex, Pakistan *96*
4.9 Industry Best Practices *96*
4.9.1 Zero Liquid Discharge *97*
4.9.2 Advanced Treatment Technologies for Wastewater *97*

4.9.3 Adoption of Green Alternatives *97*
4.9.4 Regulatory Compliance *98*
4.9.5 Industrial Wastes and Wastewater Through the Circular Economy *98*
4.10 Challenges and Research Directions *99*
References *99*

5 Heavy Metals in Wastewater – Toxicity Assessment and Remediation *113*
5.1 Sources of Heavy Metal Pollution *114*
5.1.1 Point Sources *114*
5.1.2 Nonpoint or Diffuse Sources *114*
5.2 Toxicity Assessment *116*
5.2.1 Effect on Aquatic Organisms *117*
5.2.2 Effects on Plants *118*
5.2.3 Effect on Human Health *118*
5.3 Treatment Methods *123*
5.4 Novel Methods for Aqueous Heavy Metal Remediation *125*
5.4.1 Adsorption by Novel Adsorbents *125*
5.4.2 Phytoremediation *125*
5.4.3 Photocatalysis *128*
5.4.4 Electrochemical Method *132*
5.4.5 Nanotechnology for Metal Removal *133*
5.5 Permissible Limits of Heavy Metals in Water Bodies *137*
5.6 Case Studies *140*
5.6.1 Kutubdia Channel in Bangladesh: Contamination of Water and Sediments by Metals *140*
5.6.2 Heavy Metal Contamination of River Cauvery, Tamil Nadu, India *141*
References *142*

6 Dye Wastewater Characteristics and Treatment *153*
6.1 Classification of Dyes *156*
6.2 Toxicity Assessment of Dyes *161*
6.2.1 Toxicity in the Aqueous Environment *162*
6.2.2 Toxicity Effects on the Living Organisms *163*
6.3 Treatment of Dye Wastewater *165*
6.3.1 Physical Treatment *165*
6.3.2 Chemical Treatment *166*
6.3.2.1 Ozonation *168*
6.3.2.2 Wet Air Oxidation *169*
6.3.2.3 Fenton Process *170*
6.3.2.4 Photodegradation *170*
6.3.3 Biological Treatment *171*
6.3.4 Novel/Hybrid Treatment Methods *175*
6.4 Legislative Guidelines for Dye-loaded Wastewater *180*
6.5 Case Studies *181*
6.5.1 Textile Wastewater Treatment Plant – Arequipa, Peru *181*
6.5.2 Efficiency Analysis of Textile Wastewater Treatment Plant of Foshan, China *183*
References *184*

7 Contamination by Oil Spills *191*
7.1 Characteristics of Oils *191*
7.2 Important Oil Properties *192*
7.3 Oil Behavior in the Aquatic Environment *193*
7.4 Impacts of Oil Spills *195*
7.4.1 Impacts on the Environment *196*
7.4.1.1 Marine Ecosystems *196*
7.4.1.2 Coastal Ecosystems *196*
7.4.1.3 Contamination of Water Bodies *197*
7.4.2 Impacts on Humans *197*
7.4.3 Economic Impacts *197*
7.5 Detection of Oil Spills and Monitoring *198*
7.5.1 Remote Sensing *198*
7.5.2 Machine Learning and Deep Learning *199*
7.5.3 Laser Fluorosensors *200*
7.5.4 Ground Penetrating Radars (GPR) *200*
7.5.5 Recent Techniques *200*
7.6 Technologies and Materials for Oil Spill Cleanup *200*
7.6.1 Some Important Studies *201*
7.6.2 Innovative Techniques *201*
7.6.3 Integration of Methods *202*
7.6.4 Materials for Oil Spill Cleanup *202*
7.6.4.1 Nanomaterials and Nanotechnology *204*
7.7 Prevention *205*
7.8 Important Factors in the Cleanup Activity *206*
7.8.1 Environmental Factors *206*
7.8.2 Nature of the Oils *206*
7.8.3 Technological Factors *207*
7.8.4 Organizational Factors *207*
7.8.5 Socioeconomic and Political Factors *207*
7.9 Treatment of Oil Spills *207*
7.9.1 Use of Skimmers *207*
7.9.2 Oil-absorbent Pads *208*
7.9.3 Burning In Situ (Thermal Remediation) *208*
7.9.4 Sorbents *209*
7.9.5 Hot Water and High Pressure Washing *211*
7.9.6 Oil Spill Dispersants *211*
7.9.7 Oil Boom *212*
7.9.8 Manual Labor *213*
7.9.9 Bioremediation *213*
7.10 Some Major Oil Spill Incidents *214*
7.10.1 Deepwater Horizon Oil Spill *214*
7.10.2 Exxon Valdez Oil Spill *215*
7.10.3 Prestige Oil Spill *215*
7.10.4 Montara Oil Spill *215*
7.11 Disposal and Treatment of Wastes Generated *216*
7.12 Strategies for Prevention and Mitigation *219*

7.13 Legislation and Role of Authorities in the Cleanup of Oil Spills *219*
7.13.1 The US Regulations *220*
7.13.2 The UK Regulations *220*
References *221*

8 Pesticide Elimination from Hydro Environment *231*
8.1 Classification of Pesticides *232*
8.2 Scenario of Global Pesticide Consumption *237*
8.3 Toxicity Assessment of Pesticides *238*
8.4 Treatment of Pesticide Wastewater *241*
8.4.1 Physical Treatment Methods *243*
8.4.1.1 Membrane Filtration *243*
8.4.1.2 Plasma Process *243*
8.4.1.3 Ultrasonication *243*
8.4.1.4 Adsorption *244*
8.4.1.5 Evaporation Beds *247*
8.4.2 Chemical Treatment Method *247*
8.4.2.1 Iron-enhanced Sand Filters *247*
8.4.2.2 Chlorination *247*
8.4.2.3 Advanced Oxidation Process *248*
8.4.2.4 Photochemical Degradation *249*
8.4.2.5 Ozonation *249*
8.4.2.6 Fenton Process *250*
8.4.2.7 Electrocoagulation *250*
8.4.3 Biological Treatment Methods *252*
8.4.3.1 Pesticide Degradation Using Ligninolytic Fungi *252*
8.4.3.2 Pesticide Degradation Using Bioaugmented Activated Sludge *253*
8.4.3.3 Pesticide Degradation Using Microbial Consortium *253*
8.4.3.4 Pressurized Activated Sludge for Pesticide Elimination *253*
8.4.3.5 Combined Aerobic and Anaerobic Biological Treatment *253*
8.4.3.6 Membrane Bioreactor for Pesticide Removal *254*
8.4.4 Hybrid Technologies *257*
8.5 Legislative Framework of Pesticides *259*
8.6 Case Studies *263*
8.6.1 Removal of Pentachlorophenol from Wastewater *263*
8.6.2 Pesticide Removal from Secondary Wastewater of Waste Treatment Plant, Tunisia *265*
References *266*

9 Radioactive Contaminants *271*
9.1 Detection of Radioactive Contaminants *271*
9.1.1 Gamma Ray Spectrometry *271*
9.1.2 Liquid Scintillation Counting (LSC) *273*
9.1.3 Mass Spectrometry *273*
9.1.4 Alpha Spectrometry *275*
9.1.5 Beta Counting *276*
9.2 Methods of Treatment *276*
9.2.1 Chemical Precipitation and Coagulation *277*

9.2.2 Membrane Technique *279*
9.2.3 Sorption *280*
9.2.4 Photo-induced/AOP Processes *281*
9.2.5 Bioremediation *284*
9.2.6 Electrokinetic Remediation (EKR) and Ion Exchange *287*
9.2.7 Evaporative Method *289*
9.2.8 Integrated Methods *290*
9.3 Case Studies on Radioactive Contaminants *293*
9.3.1 Hanford Site, Washington, USA *293*
9.3.2 Fukushima Daiichi Nuclear Disaster, Japan *294*
9.3.3 Mayak Facility and the Techa River, Russia *295*
9.3.4 Chernobyl Disaster, Soviet Union *296*
9.4 Technologies and Materials Studied for Radioactive Contaminants *296*
9.4.1 Novel Technologies *297*
9.4.2 Materials for Radionuclide Wastewater Remediation *298*
9.5 Role of Regulatory Authorities *300*
9.6 Challenges and Future Needs *301*
References *302*

10 Microbes as Contaminants *317*
10.1 Treatment Methods *320*
10.1.1 Primary Sedimentation *321*
10.1.2 Coagulation-flocculation *322*
10.1.3 Filtration *323*
10.1.4 Conventional Disinfection *323*
10.1.5 Chlorination *324*
10.1.6 Ultraviolet (UV) Light *324*
10.1.7 Ozonation *325*
10.1.8 Waste Stabilization Ponds *326*
10.1.9 Constructed Wetlands *326*
10.1.10 Novel Methods *328*
10.1.10.1 Carbon Nanotubes *328*
10.1.10.2 Membrane Bioreactor *329*
10.1.10.3 Biofilters *330*
10.1.10.4 Advanced Oxidation Process *332*
References *332*

11 Pharmaceutics as Emerging Contaminants in Wastewater: Hazards and Removal Technologies *339*
11.1 Sources of Pharmaceutical Waste *339*
11.1.1 Pharmaceutical Manufacturing Facilities *340*
11.1.2 Healthcare Facilities *340*
11.1.3 Personal Care Products *340*
11.1.4 Veterinary Institutions *341*
11.1.5 Wastewater Treatment Plants *341*
11.1.6 Landfills *341*
11.1.7 Other Sources *341*
11.2 Types of Pharmaceutical Waste *341*

11.3 Toxicity Assessment *344*
11.4 Treatment Methods *350*
11.4.1 Physical Treatment Methods *350*
11.4.1.1 Autoclaving *350*
11.4.1.2 Precipitation *350*
11.4.1.3 Ion Exchange *350*
11.4.1.4 Coagulation *351*
11.4.1.5 Floatation Process *351*
11.4.1.6 Solvent Extraction Process *352*
11.4.1.7 Evaporation *352*
11.4.2 Physio-chemical Methods *353*
11.4.2.1 Advanced Oxidation Process *353*
11.4.2.2 Photocatalytic Degradation *353*
11.4.2.3 Electrochemical Removal *354*
11.4.2.4 Ozonation *354*
11.4.2.5 Adsorption *355*
11.4.2.6 Metal–Organic Frameworks *357*
11.4.3 Biological Treatment *358*
11.4.3.1 Constructed Wetlands *358*
11.4.3.2 Anaerobic Treatment *359*
11.4.3.3 Activated Sludge Process *359*
11.4.4 Novel Technologies *360*
11.4.4.1 Membrane Separation Technology *360*
11.4.4.2 Membrane Bioreactor System *362*
11.5 Legislative Framework of Pharmaceuticals *363*
11.6 Case Studies *365*
11.6.1 Sewerage Treatment Plant, Ghana *365*
11.6.2 Pharmaceutical Wastewater Treatment Using High-rate Algal Pond *366*
References *367*

12 Surfactants: Behavior and Remediation Process *381*
12.1 Sources of Surfactants *382*
12.2 Types of Surfactants *383*
12.3 Toxicity Assessment of Surfactants *385*
12.3.1 Surfactants and the Abiotic Environment *388*
12.3.2 Surfactants in the Biotic Environment *389*
12.4 Analytical Techniques for Surfactants *394*
12.5 Treatment Techniques *394*
12.5.1 Physical Treatment *394*
12.5.2 Chemical Treatment *400*
12.5.3 Biological Treatment *404*
12.5.4 Integrated Treatment Methods *407*
12.6 Legislative Framework for Surfactant Discharge *409*
12.7 Case Studies *410*
12.7.1 Surfactant-loaded Wastewater – Lion Corporation, Thailand *410*
12.7.2 River Tigris, Iraq – A Surfactant Loaded Freshwater Source *410*
References *411*

13 Fertilizers and Nutrients *415*
13.1 Fertilizers in Agriculture *415*
13.1.1 Types of Fertilizers *417*
13.1.1.1 Inorganic Fertilizers *417*
13.1.1.2 Organic Fertilizers *418*
13.1.1.3 Slow-release Fertilizers *419*
13.1.1.4 Micronutrient Fertilizers *420*
13.2 Hazards of Fertilizers in Water/Wastewater *420*
13.2.1 Pathways of Contamination by Fertilizers *420*
13.2.1.1 Leaching *420*
13.2.1.2 Surface Runoff *420*
13.2.1.3 Agricultural Drainage *421*
13.2.1.4 Acid Rain Due to Deposition in the Atmosphere *421*
13.2.2 Impact on the Water Quality *422*
13.2.2.1 Eutrophication *422*
13.2.2.2 Mechanism of Eutrophication *423*
13.3 Consequences of Eutrophication and Nutrients in Water *423*
13.4 Control of Eutrophication and Nutrients in Water *426*
13.4.1 Regulating Nutrient Inlet in the Water Bodies *426*
13.4.2 Improving Methods Adopted in Wastewater Treatment Plants *426*
13.4.3 Management of Agricultural Practices *426*
13.4.4 Public Contribution *426*
13.4.5 Regulatory Authorities *427*
13.4.6 Effluent Diversion *427*
13.4.7 Phosphate Striping and Buffer Striping *427*
13.4.8 Treatment Using Constructed Wetlands *427*
13.4.9 Other Methods *428*
13.4.10 Ecological Approaches *428*
13.5 The USEPA for Nutrient Pollution *430*
13.6 Food and Agriculture Organization of the United Nations *430*
13.7 Policies and Schemes for the Use of Fertilizers *431*
13.7.1 The EU Nitrates Directive (1991) *431*
13.7.2 The US Clean Water Act (1972) *431*
13.7.3 New Soil Health Card Scheme (2015) *432*
13.8 Case Studies on Fertilizer Contamination *432*
13.8.1 Dead Zone – Gulf of Mexico *432*
13.8.2 Punjab – Groundwater Nitrate Pollution *433*
13.8.3 China – Eutrophication of Lake Water *433*
13.9 Efforts on Successful Remediation *434*
13.9.1 Strategies Adopted in Denmark *434*
13.9.2 Initiatives for Organic Farming in Europe *435*
13.9.3 Wetland Restoration Techniques *435*
13.10 Management of Fertilizer Use *436*
13.11 Technologies and Methods for Nutrient Removal *436*
13.12 Recent Studies on Nutrient Removal *439*
References *442*

14 Microplastics *453*
14.1 Challenges in the Recycling of Plastics *454*
14.2 Plastics to Microplastics *456*
14.3 Toxicity Assessment of MPs *457*
14.4 Legislation for MP Management *461*
14.5 Treatment of Microplastic-containing Wastewater *465*
14.5.1 Primary Treatment *466*
14.5.2 Secondary Treatment *466*
14.5.3 Tertiary Treatment *467*
14.5.4 Newer Treatment Methods *467*
14.5.4.1 Biomass-based Treatment *467*
14.5.4.2 Membrane Technology *469*
14.5.4.3 Advanced Oxidation Process *471*
14.5.4.4 Electrocoagulation *472*
14.5.4.5 Photocatalysis *475*
14.5.4.6 Plasma Technology *476*
14.5.4.7 Phytoremediation *477*
14.5.4.8 Biotechnological Methods *478*
14.6 Case Studies *481*
14.6.1 MP Removal Efficiency of Wastewater Treatment Plants in Zhengzhou, China *481*
14.6.2 Groundwater Contamination by MPs *482*
References *483*

15 Wastewater Treatment Technologies: Performance and Mechanism *495*
15.1 Mechanisms of Different Processes *495*
15.1.1 Adsorption *495*
15.1.2 Ion Exchange *503*
15.1.3 Membrane Filtration *504*
15.1.4 Coagulation *506*
15.1.5 Advanced Oxidation Processes *507*
15.1.6 Fenton Process *508*
15.1.7 Electro-Fenton Process *508*
15.1.8 Ozone Treatment *508*
15.1.9 Electrochemical Oxidation *509*
15.1.10 Photocatalytic Degradation *511*
15.1.11 Sonochemical Method *515*
15.1.12 Biologically Assisted Degradation *516*
15.1.13 Membrane Bioreactor *518*
15.1.14 Constructed Wetlands *519*
15.1.15 Phytoremediation *521*
References *524*

Conclusions and Recommendations *531*
Index *535*

Preface

The book *Hazards and Treatment of Wastewater Contaminants* presents the gist of the extensive research that has been carried out on several wastewater contaminants, which pose a threat to the environment and living organisms. Owing to the presence of these contaminants, the quality of water in the water bodies deteriorates. Thus, there is an urgent need for a proper understanding of the nature of these contaminants, to adopt methodologies for their treatment, and to save the environment.

The contents of the book have been designed to help academicians, researchers, wastewater treatment professionals, and practitioners working in this domain to get more information on the different categories of possible contaminants in wastewater and several treatment methods used for their removal.

The entire study has been presented in 15 chapters, which deal with a wide range of information on hazardous wastewater contaminants. The book is focused on different categories of contaminants in the aqueous environment, their toxicity, and methods for their treatment and removal. Emphasis has been given to the studies incorporating novel methods, and the integration of different methods and materials for the treatment of these contaminants. The conclusions and recommendations chapter incorporates the significant outcomes.

Providing all necessary information on wastewater contaminants in one place would help as a ready reference for researchers, which has been the motive of the work. We are very optimistic that our efforts will be valuable, and the researchers will find the content informative and beneficial.

Our gratitude to the Almighty for the opportunity and the countless blessings given to us in our project journey. We express our sincere gratitude to those who have directly or indirectly helped us in the shaping of the book and have supported our endeavors. We hope that this book will be a meaningful small contribution to the global efforts toward wastewater treatment and environmental sustainability.

Dr. Dipika Jaspal
Dr. Arti Malviya
(Authors)

1

Introduction

Water is an important life-sustaining commodity. It is fundamental for the ecosystem, agriculture, industries, and living organisms. The availability of clean water for drinking and several other purposes is, however, declining due to mounting industrialization, population growth, land development, urbanization, and pollution. The uncertainties expressed due to climatic changes further pose a threat to the availability of usable water for almost 80% of the world's population [1, 2]. As per a report published by UNESCO, approximately 26% of the population globally has no access to safe drinking water, which poses a threat to livelihood. Also, only 10% of the wastewater is being treated for reuse and making it safe.

Although water is being replenished every year naturally through the hydrological cycle, due to the continuous increase in population and industrialization, there has been a competition for pure and clean water. The management of water demand without any change in the supply will be the key to effective water management. Industrialization is a vital element in the growth of the economy. However, industrialization has been one of the prominent and instrumental factors contributing to the deterioration of the available water bodies. Industries such as the paper industry, textile, iron and steel, tannery, and distillery are the prominent contributors. Although the discharge limits of different components in wastewater are in place, several industries tend to directly discharge their wastes into the water streams and water bodies without treatment. Several components, ranging from colorants to heavy metals, are being let into the water bodies [3]. Some constituents in the industrial effluents can be easily treated with the conventional treatment facilities, but some components, such as oils, fats, heavy metals, etc., need special treatment [4].

1.1 Water Pollution: A Global Concern – Facts and Figures

There has been deterioration in the quality of water available in the water streams and water bodies, due to the influx of chemicals and contaminants from the industries, agriculture, domestic, and commercial places. The consumption of unsafe drinking water leads to the death of almost 829000 people every year. Several diseases, such as diarrhea, cholera, gastrointestinal diseases, etc., are caused [5]. In addition to the mentioned several skin diseases, allergies [6] and even cancer [7] are also witnessed after coming in contact with polluted and contaminated water. Polluted water not only has harmful implications for human life, but the aquatic flora and fauna are also equally affected by its impacts. The oxygen level in the water bodies reduces substantially, posing a threat to the different varieties of fish in the water, degrading their ability to reproduce. Further, the pollutants harm the gills, epithelial layer, reduce metabolism, and cause several diseases [8]. Some toxic substances, such as mercury, can be converted into more toxic forms, such as methyl mercury, due to the action of the microbes in the water bodies. Some aquatic organisms can further amplify the concentration of several toxicants without any harm to them. Many histological changes which appear in the bodies of fish can be correlated directly to the level of contaminants in the water bodies, thus serving as biomarkers of pollution [9]. Another harmful effect of water pollution is the retarding light penetration, limiting the photosynthetic activity, leading to an improper growth of the aquatic flora [10].

Considering the adverse effects of water pollution, there has been a shift in the focus of different policies in the current scenario. Focus is now also being laid on providing clean water by setting goals and adopting different measures. Water quality problems are generally unaddressed, especially in countries of Asia and Africa, owing to poor wastewater management. According to a report by the United Nations, in case we do not work toward controlling water quality, this would pose a risk to the well-being of around 4.8 billion people around the globe by 2030 [11].

1.1.1 Sustainable Development Goals

Unavailability of clean drinking water is a concern across the globe, and measures are to be taken to make clean water available to the population, ensuring enhanced human health. One of the main agendas brought to the forefront for international action in the 2030 Sustainable Development Goals (SDGs) is water quality. The demand for safe and clean water is increasing in domestic, industrial, and agricultural sectors due to an increase in the population. Setting international goals in the SDGs is a clear indication of the pressing need for water quality management. Access to safe and clean drinking water and sanitation has been identified as a human right by the United Nations General Assembly in 2010. Goal 6 in the SDGs is dedicated to clean water and sanitation. The target points under goal 6 [11] are:

6.1 Having affordable and safe drinking water for all by 2030, and having universal and equitable access.
6.2 To have adequate and equitable hygiene and sanitation for all, ending open defecation, with special attention to females and those in vulnerable situations, by 2030.
6.3 Reducing pollution and minimizing release of hazardous materials and chemicals. It aims to increase recycling and safe reuse of wastewater globally by 2030.
6.4 Increasing water-use efficiency by 2030 across all sectors, making sure that the withdrawals are sustainable, addressing the scarcity of freshwater, and reducing the number of affected people from water scarcity.

As per the data by the United Nations, UN water, the level of global water stress in the last year 2020 is shown in Figure 1.1.

The region-wise water stress is depicted in Figure 1.2.

6.5 Implementation of water resources management at all levels, by 2030, incorporating appropriate trans boundary cooperation by 2030.
6.6 Restoring and protecting water-related ecosystems, by 2020, including forests, mountains, rivers, wetlands, lakes, and aquifers.
6.7 To expand international cooperation for developing countries, along with capacity-building support, sanitation, and water-related programs and activities by 2030, including water efficiency, wastewater treatment, water harvesting, desalination, reuse, and recycling technologies.
6.8 Strengthening and supporting the participation of local communities in the improvement of sanitation and water management.

1.1.2 Climate Change and Water Stress

The changes in the overall climatic patterns have substantial implications for the water resources. The rate of water evaporation, uncertainty in rainfall, scarcity of rainfall and its excess, increased temperature of water bodies affecting aquatic life, along with the physical and economic consequences, cannot be ignored. The United Nations experts mention that there is a 20% drop in the water resources which are renewable for every 1 °C increase in the average temperature of the globe. This intensifies the problem of water crises in the regions which are already under stress. For these regions under water stress, agriculture further becomes a concern due to unpredictable climate and rainfall.

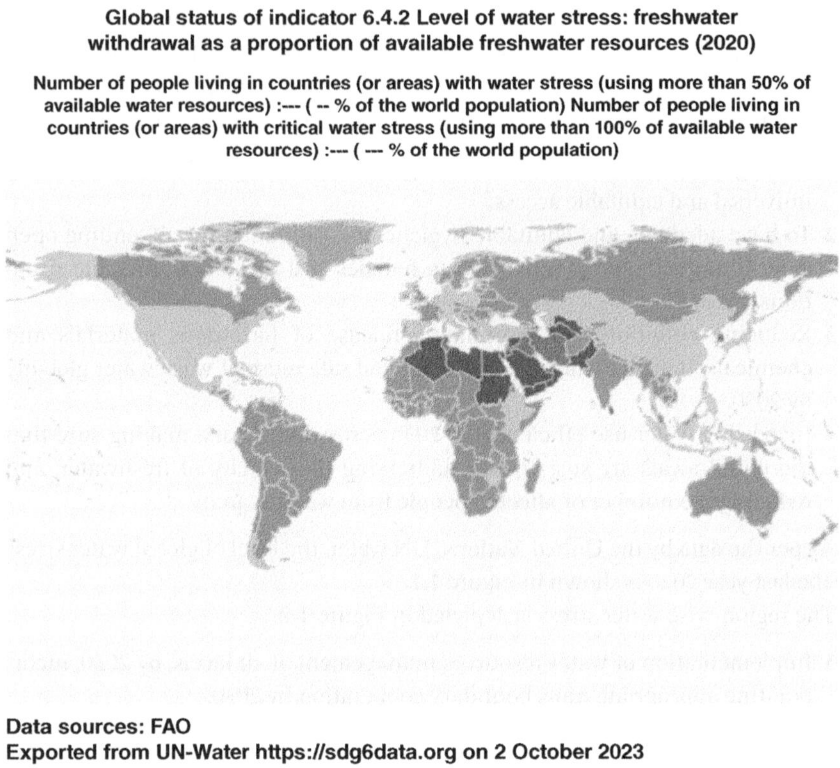

Figure 1.1 Regions of water stress globally. Adapted from UN-Water https://sdg6data.org.

In these regions, unpredictable and uncertain changes in the climate also lead to floods in certain regions, which sweep off the beneficial topsoil and components, which can clog the water treatment facilities. Due to increased global temperature, soil moisture decreases, which is necessary for plant growth and controlling droughts. The winter precipitation also increases. Further, an increased water temperature affects the decomposition of the organic matter, thereby affecting the concentration of dissolved oxygen in the water bodies. The quality of water in the water bodies is thus affected, inhibiting its use and posing a greater pressure on the existing and available clean water sources. There is an increase in the sea level, which leads to an intrusion of salt into the groundwater, thereby affecting the freshwater sources affecting coastal water supply. Events of high-intensity rainfall also adversely affect the water quality, as these lead to increased runoff and erosion.

It thus becomes crucial to protect existing water resources as well as to follow a wastewater management approach, by collecting, treating, and reusing

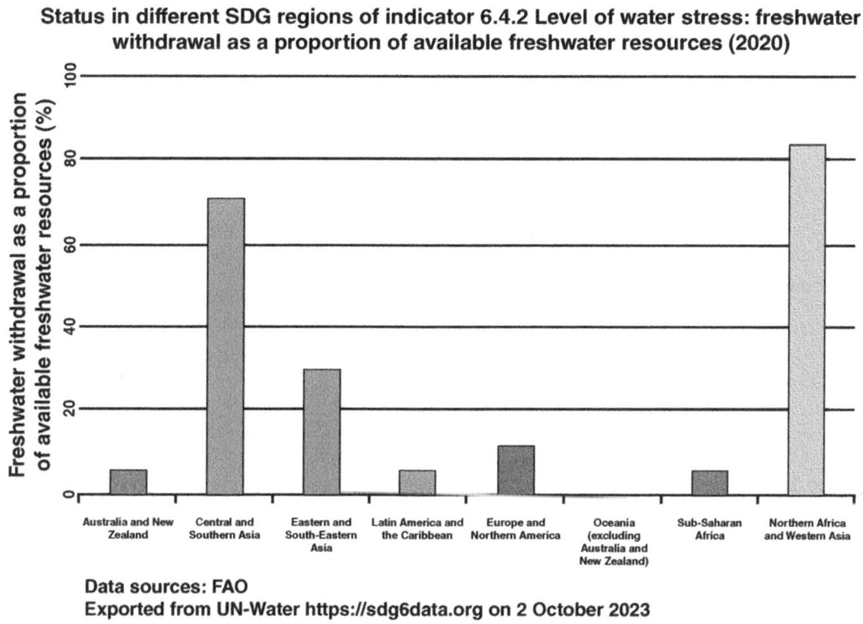

Figure 1.2 Regions of maximum freshwater withdrawal in proportion to the available water resources. Adapted from UN-Water https://sdg6data.org.

wastewater. An effective wastewater management system aims at treating and managing wastewater for nonpotable utilization. It also intends to combat pollution and damage to the environment, sustenance of the water table and water bodies.

1.1.3 Wastewater and Its Types

Wastewater can be, in general, be classified into two main types: nonsewage and sewage wastewater. Water discharged from domestic and household activities is placed under sewage water. It is the wastewater coming from the sanitary facilities, houses, educational institutes, hotels, hospitals, public places, etc. It mainly comprises of urine and feces. On the other hand, the wastewater generated from all the nondomestic or commercial places and water processing activities carried out in the industries is nonsewage wastewater. It is devoid of any human waste. Rainwater, stormwater, floodwater water, etc., are also categorized under nonsewage water.

To have a better understanding of the composition of the wastewater and for effective water management, wastewater can also be divided into various types based on their source of generation [12] as shown in Figure 1.3.

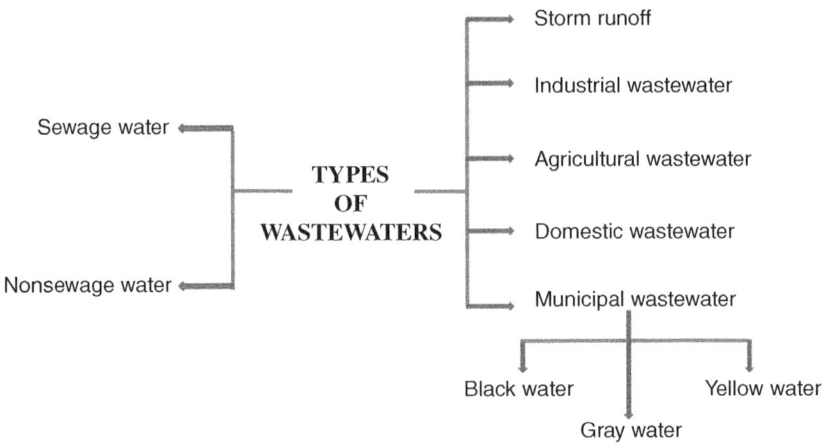

Figure 1.3 Types of wastewater.

1.1.3.1 Storm Water/Surface Runoff

This is the runoff water from rain or floods not seeped into the ground. The runoff water flows directly into water bodies or is directed via proper channels into drains reaching water sources. Water gets polluted during its journey from the land surface to a water source. This includes:

1. Atmospheric runoff due to rain or meltwater
2. Infiltration runoff in which the surface water seeps into the sewers through some cracks and becomes part of wastewater
3. Drainage water
4. Washing wastewater produced due to the cleaning or watering activities of greenery, roads, garages, public places, gardens, etc.

1.1.3.2 Domestic or Municipal Wastewater

This is the water produced from day-to-day activities of human beings. It is collected from the discharge points of houses, restaurants, offices, shopping malls, and multiplexes usually in urban areas. Domestic or municipal water can be of different categories depending upon the outlets from where it is released (Figure 1.4) [13–16].

1. Black water is the most contaminated of all categories. It refers to the aqueous waste generated from sanitary facilities, washrooms, kitchens, etc. It contains urine, feces, toilet paper, superfluous food particles, and other household chemicals used as cleaning agents. Various components in this

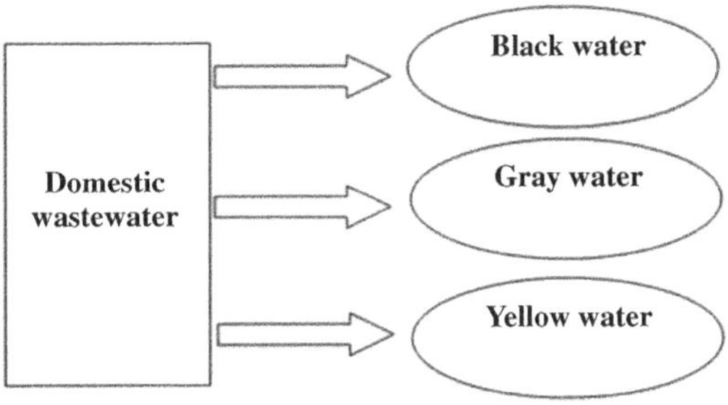

Figure 1.4 Categories of domestic wastewater.

category are water from the toilets, basins, cleaning chemicals, soaps, urine, fecal matter, etc., making this category of wastewater the most hazardous of all categories. The water is extremely contaminated with disease-causing pathogens.

2. Gray water or sullage refers to the black water not having fecal matter and food waste. It is linked with the residual water having soap, shampoo, laundry detergent and toothpaste from sinks, bathrooms, industrial units, etc. Gray water is relatively less hazardous and does not contain fecal matter or urine. It contains water from basins, sinks, and kitchens and contains soaps, detergents, and cleaners. This type of water is also contaminated, but is not loaded with pathogenic species.
3. Water comprising of mainly urine is yellow water. Yellow water is generally deprived of other contaminants.

1.1.3.3 Agricultural Wastewater

The water emerging from the agricultural practices is termed agricultural runoff. It is also referred to as irrigation tailwater. Such wastewater is contaminated with sediments, nutrients, fertilizers, pesticides, and animal waste, which finally reach the nearby water bodies.

1.1.3.4 Industrial Wastewater

Water discharged as effluent due to various industrial activities is termed industrial wastewater. The composition of this wastewater is variable as it depends on the types of processing involved. Effluent from the laundry industry, textiles, dairy, sugar manufacturing, leather industry, pharmaceutics, food and beverage production, mining industry, petrochemical, and power plants all fall under this category. This wastewater requires specific treatment methods to make it fit for discharge into the environment.

1.2 Wastewater Characteristics

Some of the important wastewater characteristics, such as color, odor, and turbidity, are affected due to the presence of specific contaminants.

1.2.1 Color

Several factors, such as dyes, heavy metals, soil, etc., lead to the appearance of color in the water bodies. The presence of this color affects the aesthetics and the aquatic life. Table 1.1 gives the different colors imparted to wastewater due to the presence of different contaminants.

Methods which can be used for color removal are depicted in Figure 1.5.

The different methods for the removal of color have been discussed at length in the subsequent chapters.

1.2.2 Odor

Odors are indicative of the presence of contaminants on a large scale or further degradation of wastewater characteristics. Odor is associated with the presence of degrading organic substances. Generally, foul odor is due to improper treatment of wastewater by the treatment plants. Odor from the wastewater bodies makes it difficult to work in the vicinity, creating a poor environment. Several methods for the removal [21] of foul odor are depicted in Figure 1.6.

Table 1.1 Contaminants and colors imparted by them [16–18].

S. No.	Components	Color imparted	References
1.	Minerals	Yellow	[17]
2.	Algal growth along with nutrients and fertilizers	Dark green or brown	[17]
3.	Iron	Brown or red	[18]
4.	Manganese	Black/brown	[18]
5.	Organic matter	Yellow	[19]
6.	Fabric dyes, inks from printers	Blue, green, violet, red	[19]
7.	Oil or diesel	Orange, brown	[19]
8.	Detergents	White	[19]
9.	Copper	Blue/green	[20]

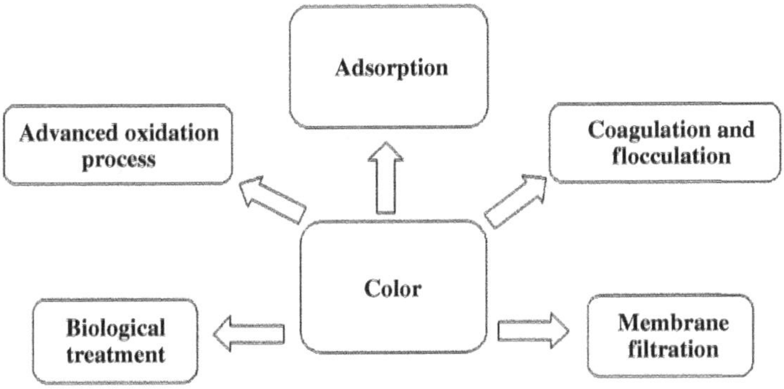

Figure 1.5 Methods for color removal.

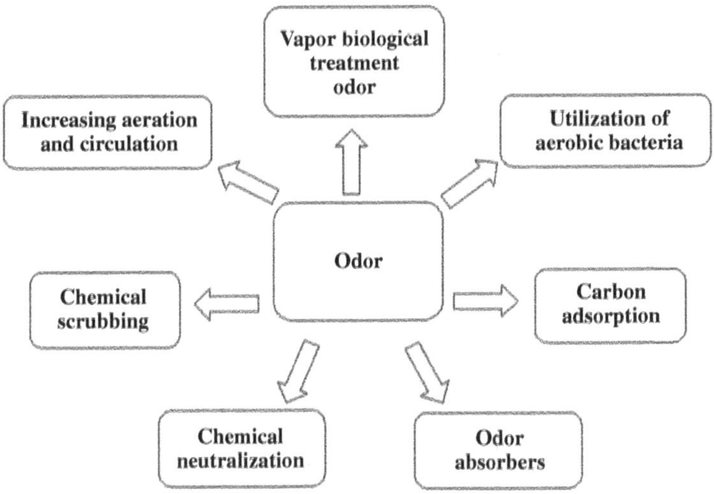

Figure 1.6 Methods for odor removal.

1.2.3 Turbidity

Turbidity is one of the important preliminary aspects indicating pollution in the water streams or the presence of contaminants. Turbidity or cloudiness is usually due to the presence of fine soil particles. These fine particles can also be of decaying organic matter or biological solids. Turbidity usually provides shelter to different microbes, which can thrive and grow by attaching themselves to these suspended particles [22]. Turbidity in water samples can be treated or removed by adopting the methods as shown in Figure 1.7.

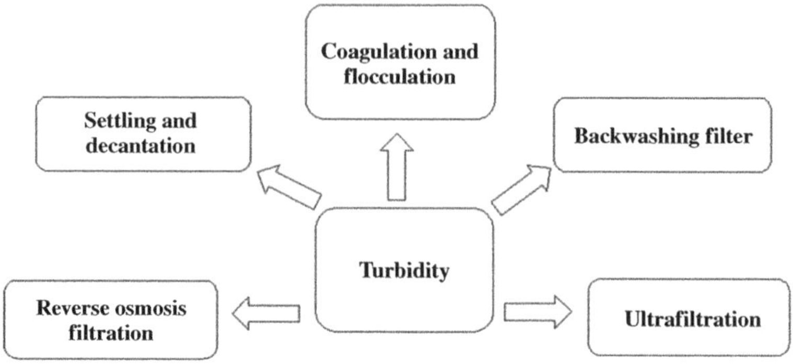

Figure 1.7 Methods for turbidity removal.

1.2.4 Nutrients

Nutrients can be present in wastewater in inorganic or organic form. Several nutrients are present in wastewater along with other organic and inorganic substances and particles. These nutrients are drawn into the water bodies from the local sources, leachate from landfills, and agricultural runoffs. Some examples of important nutrients that can be present in wastewater are nitrogen and phosphorus [23]. Nitrogen is a nutrient important for building of cell wall [24]. Phosphorus can be present in organic and inorganic forms. Orthophosphate is the available bioreactive form, which can be utilized by the aquatic beings [25].

Generally, there is a minimal concentration of nutrients which are present in the water bodies. An increase and a high concentration of nutrients in wastewater can lead to eutrophication, death of aquatic species, loss of habitat, and a rise in turbidity [26].

1.2.5 Surfactants

The appearance of foam on the surface of water bodies or water streams can be due to the presence of chemicals known as surfactants. Surfactants are another category of contaminants contributing to water pollution and leading to adverse impacts on aquatic beings. Surfactants make their way into the water streams through the use of detergents, soaps, personal care products, and wastewater treatment plants [27]. Surfactants in higher concentrations can cause hazardous implications to the aquatic and terrestrial ecosystems [28, 29]. In addition to the toxic effects, surfactants also enhance the dissolution capacity of several other contaminants in water [30]. Several methods can be adopted for the elimination of surfactants as shown in Figure 1.8 [31].

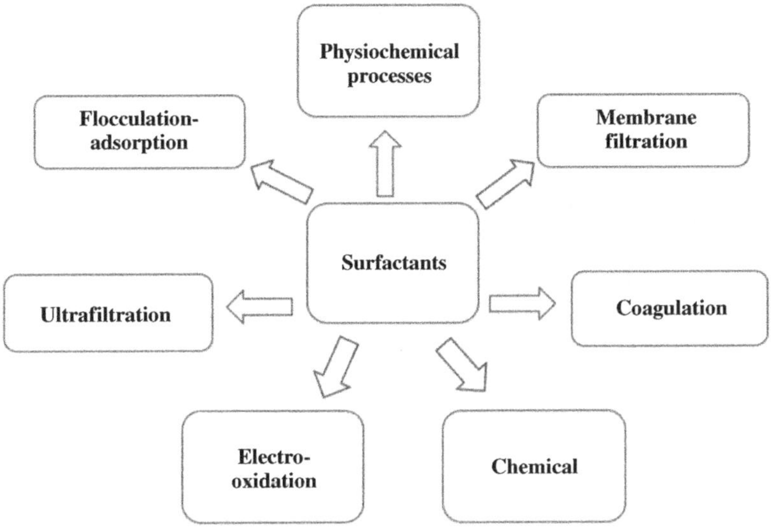

Figure 1.8 Methods for surfactant removal.

1.3 Hazards of Wastewater and the Need for Treatment

Wastewater containing several contaminants suspended, dissolved, and free-floating leads to a deterioration in aesthetics and can pose several impacts and threats to aquatic and human beings. The effects on the photosynthetic activity, metabolism of aquatic creatures, and diseases in humans are the most prominent implications posed by the contaminants in wastewater. Several effects of wastewater and the different contaminants are given in the following sections.

1.3.1 Aquatic Ecosystem

The contaminants in wastewater can lead to direct harm to the aquatic creatures. Effect on the food chain, reproductive activity and metabolism, damage to the gills, ingestion, choking, and death of the organisms could be the main harm to the aquatic beings. Another important effect is the depletion of the oxygen level, damage to the coral reefs, and growth of algae, fungi, and weeds [32].

1.3.2 Humans

Wastewater containing contaminants, when it comes in contact with the human body or is ingested, leads to various health-related issues such as skin irritation and several waterborne diseases [33]. Diarrhea, gastrointestinal diseases, malnutrition, etc., are caused due to the consumption of unsafe and polluted water. In addition, exposure to contaminants like metals, dyes, and surfactants can lead to poisoning, organ disorders, and growth-related issues [34].

1.4 Need to Treat Wastewater

Water contamination is a global problem. Wastewater treatment is a crucial area, due to the harmful effects caused by the pollutants on the water bodies and living organisms. The intensity of contamination and deterioration varies from region to region. In the global context, there has been a rapid deterioration of the quality of water in the rivers of Asia, Africa, South America, etc. One of the direct and obvious reasons is the population density and rapid industrialization.

Continuous degradation of water quality, improper management, and lack of proper treatment of wastewater pose a threat to the availability of clean water, leading to a crisis of usable water for the increasing population [35]. The need for proper treatment methods for contaminant elimination is challenging, due to the lack of strict regulations and strategies for varied types of contaminants. There is a mounting threat to the water bodies with the rising pollution and influx of contaminants in the water bodies.

Considering the reduction in the availability of clean water and the hazardous effects of several contaminants on the aquatic beings, humans, and the ecosystem, the treatment of wastewater for the reduction or elimination of contaminants, and possible reuse becomes very essential. Sustainable development targets universal access to secure and affordable water. This is in harmony with the universal right to water and sanitation, where everyone has the right to physically access clean water for personal and domestic use. Also, with the climatic changes, with the increase in population, the reuse and recycling of water are the aspects of utmost importance. Improved water supply and proper water resources management can help boost the nation's economic growth.

1.5 Scope and Objectives of the Book

This book dives deep into the several categories of contaminants in wastewater, associated hazards, and the different treatment methods adopted for their eradication. All major and minor contaminants, ranging from sewage to microplastics and microorganisms, have been discussed at length. The details of these contaminants, their hazardous traits, several methods used for their treatment, and several studies, which have been carried out for the elimination, form an integral part of all the chapters. Several case studies related to the contaminants have also been incorporated, giving a glance at the present status of the wastewater treatment facilities and their performance. The sources, toxic effects, and treatment of various emerging contaminants have been discussed. The book has been written in a manner to address each contaminant at length as a dedicated chapter, helping the readers to clearly understand the impact and the methodology of treatment for every contaminant. Rules and regulations are mandatory to be formulated to provide a framework for the prevention and remediation of water pollution. The legislation related to the discharge of each type of contaminant has been included in the book to recognize the role of regulatory authorities in mitigating water contamination.

The main objective of this book is to provide a comprehensive content on various wastewater contaminants in all possible aspects, which will help understand wastewater characteristics, contaminant toxicity, and measures to be adopted for their mitigation.

References

1 Sophocleous, M. (2004). Global and regional water availability and demand: prospects for the future. Natural Resources Research, Volume 13, pp. 61–75. https://doi.org/10.1023/B:NARR.0000032644.16734.f5.

2 Oki, T., and Kanae, S. (2006). Global hydrological cycles and world water resources. Science, Volume 313, Issue 5790, pp. 1068–1072. https://doi.org/10.1126/science.1128845.

3 Rajput, R.S., Pandey, S., and Bhadauria, S. (2017). Status of water pollution in relation to industrialization in Rajasthan. Reviews on Environmental Health, Volume 32, Issue 3, pp. 245–252. https://doi.org/10.1515/reveh-2016-0069.

4 Shah, M.P. (2017). Wastewater pollution. Journal of Applied Biotechnology and Bioengineering, Volume 3, Issue 1, pp. 262–263. https://doi.org/10.15406/jabb.2017.03.00054.

5 Li, L., Haoran, Y., and Xiaocang, X. (2022). Effects of water pollution on human health and disease heterogeneity: a review. Frontiers in Environmental Science, Volume 10, p. 880246. https://doi.org/10.3389/fenvs.2022.880246.

6 Hanif, M., Miah, R., Islam, M., et al. (2020). Impact of Kapotaksha river water pollution on human health and environment. Progressive Agriculture, Volume 31, Issue 1, pp. 1–9. https://doi.org/10.3329/pa.v31i1.48300.

7 Marmot, M., Atinmo, T., Byers, T., et al. (2007). Food, nutrition, physical activity, and the prevention of cancer: a global perspective. Nutrition Bulletin, Volume 30, Issue 2, pp. 168–172. https://doi.org/10.1111/j.1467-3010.2005.00470.x.

8 Malik, D.S., Sharma, A.K., Sharma, A.K., et al. (2020). A review on impact of water pollution on freshwater fish species and their aquatic environment. In: Kumar V, Kamboj N, Payum T, et al. (eds) Advances in environmental pollution management: wastewater impacts and treatment technologies, Volume 1. pp. 10–28. https://doi.org/10.26832/aesa-2020-aepm-02.

9 Grueso-Gilaberth, R.N., Jaramillo-Timarán, K.S., Ospina-Pérez, E.M., et al. (2020). Histological description and histopathology in *Polypedilum* sp. (Diptera: Chironomidae): a potential biomarker for the impact of mining on tributaries. Annals of the Entomological Society of America, Volume 113, Issue 5, pp. 359–372. https://doi.org/10.1093/aesa/saaa008.

10 Hakeem, K.R., Bhat, R.A., and Qadri, H. (2020). Concerns and threats of contamination on aquatic ecosystems. In: Hakeem K, Bhat R, Qadri H (eds) Bioremediation and biotechnology. Springer, Cham, pp. 1–26. https://doi.org/10.1007/978-3-030-35691-0_1.

11 Progress on ambient water quality, mid-term status of SDG indicator 6.3.2, and acceleration needs, with a special, focus on health, 2024, United Nations, UN Water, 2024. https://www.unwater.org/publications/progress-ambient-water-quality-2024-update (Last accessed 25.08.2024).

12 Ahmed, J., Thakur, A., and Goyal, A. (2021). Industrial wastewater and its toxic effects. In: Shah MP (ed) Biological treatment of industrial wastewater. The Royal Society of Chemistry, pp. 1–14. https://doi.org/10.1039/9781839165399-FP005.

13 https://books.rsc.org/books/edited-volume/937/chapter/741366/Industrial-Wastewater-and-Its-Toxic-Effects (Last accessed 25.08.24).

14 Yin, H., Xie, M., Zhang, L., et al. (2019). Identification of sewage markers to indicate sources of contamination: low cost options for misconnected non-stormwater source tracking in stormwater systems. Science of the Total Environment, Volume 648, pp. 125–134. https://doi.org/10.1016/j.scitotenv.2018.07.448.

15 Al-Jayyousi, O.R. (2003). Greywater reuse: towards sustainable water management. Desalination, Volume 156, Issues 1–3, pp. 181–192. https://doi.org/10.1016/S0011-9164(03)00340-0.

16 Friedler, E., Butler, D., and Alfiya, Y. (2013). Wastewater composition. In: Larsen T, Udert K, Lienert J (eds) Source separation and decentralization for wastewater management. IWA Publishing, p. 241. https://doi.org/10.2166/9781780401072.

17 https://www.netsolwater.com/what-is-yellow-wastewater-and-how-to-manage.php?blog=5683&srsltid=AfmBOoo9VgPO_UL7trEIAqalia4IY_azOBuQqkYAeLDLXUfxAv_vHFrc (Last accessed 06.11.24).

18 Clean Water Team. Color of water fact sheet, State Water Resources Control Board Fact Sheet 3.1.5.9. https://www.waterboards.ca.gov/water_issues/programs/swamp/docs/cwt/guidance/3159.pdf (Last accessed 19.9.25).

19 https://www.austintexas.gov/sites/default/files/files/Watershed/Austin-Guides/Common-Water-Quality-Complaints-Color-Guide.pdf (Last accessed 06.11.24).

20 https://www.filterwater.com/t-color-in-water.aspx (Last accessed 06.11.24).

21 https://chemtech-us.com/all-about-odor-control-in-wastewater-treatment/ (Last accessed 08.11.24).

22 https://complete-water.com/resources/how-to-treat-turbidity-in-water (Last accessed 10.11.24).

23 Rout, P.R., Shahid, M.K., Dash, R.R., et al. (2021). Nutrient removal from domestic wastewater: a comprehensive review on conventional and advanced technologies. Journal of Environmental Management, Volume 296, p. 113246. https://doi.org/10.1016/j.jenvman.2021.113246.

24 Głazowska, S., Baldwin, L., Mravecet. J., et al. (2019). The source of inorganic nitrogen has distinct effects on cell wall composition in *Brachypodium distachyon*. Journal of Experimental Botany, Volume 70, Issue 21, pp. 6461–6473. https://doi.org/10.1093/jxb/erz388.

25 https://www.ljea.org/2018-state-lake-james-watershed/chemical-analysis-water-quality/nutrients/ (Last accessed 13.11.24).

26 https://niwa.co.nz/freshwater/kaitiaki-tools/what-impacts-interest-you/nutrient-overloading/causes-nutrient-overloading/nutrients-and-wastewater (Last accessed 10.11.24).

27 Kulkarni, D., and Jaspal, D. (2023). Surfactants in waste water: development, current status and associated challenges. Materials Today Proceedings. https://doi.org/10.1016/j.matpr.2023.11.022.

28 Petrie, B., Barden, R., and Kasprzyk-Hordern, B. (2015). A review on emerging contaminants in wastewaters and the environment: current knowledge, understudied areas and recommendations for future monitoring. Water Research, Volume 72, pp. 3–27. https://doi.org/10.1016/j.watres.2014.08.053.

29 Collivignarelli, M., Miino, M.C., Baldi, M., et al. (2019). Removal of non-ionic and anionic surfactants from real laundry wastewater by means of a full-scale treatment system. Process Safety and Environmental Protection, Volume 132, pp. 105–115. https://doi.org/10.1016/j.psep.2019.10.022.

30 Badmus, S.O., Amusa, H.K., Oyehan, T.A., et al. (2021). Environmental risks and toxicity of surfactants: overview of analysis, assessment, and remediation techniques. Environmental Science and Pollution Research, Volume 28, pp. 62085–62104. https://doi.org/10.1007/s11356-021-16483-w.

31 Jena, G., Dutta, K., and Daverey, A. (2023). Surfactants in water and wastewater (greywater): environmental toxicity and treatment options. Chemosphere, Volume 341, pp. 140082. https://doi.org/10.1016/j.chemosphere.2023.140082.

32 https://www.aquaread.com/blog/impact-of-water-pollution-on-aquatic-life/ (Last accessed 21.11.24).

33 Wu, B. (2020). Human health hazards of wastewater. In: Ren H, Zhang X (eds) High-risk pollutants in wastewater. Elsevier, pp. 125–139. https://doi.org/10.1016/B978-0-12-816448-8.00006-X.
34 Lin, L., Yang, H., and Xu, X. (2022). Effects of water pollution on human health and disease heterogeneity: a review. Frontiers in Environmental Science, Volume 10. https://doi.org/10.3389/fenvs.2022.880246.
35 Plessis, A. (2022). Persistent degradation: global water quality challenges and required actions. One Earth, Volume 5, Issue 2, pp. 129–13. https://doi.org/10.1016/j.oneear.2022.01.005.

2

Overview of Contaminants

Considering several concerns arising due to a reduction in the available fresh water sources and water stress, it becomes important to discuss the various aspects that predominantly giving rise to this situation. The deterioration in the water quality poses a threat to human beings and aquatic life, hampers the ecosystem functioning, and affects the socioeconomic growth. Out of the total wastewater generation, 80% of the domestic and industrial wastewater is let into the water streams without any treatment. Point and nonpoint sources of water pollution have been instrumental in the deterioration of water quality around the globe. Improper sanitation of several regions, effluent from the industries, agricultural runoffs, chemicals, and emerging contaminants has played a vital role in affecting the water quality. The lower- and middle-income countries face the maximum exposure to the pollutants [1].

2.1 Contaminants

The substances degrading the quality of water are termed as contaminants. Contaminants can belong to several categories. Figure 2.1 shows the different possible broad categories in which the contaminants can be categorized.

In the physical category of contaminants, suspended contaminants such as dirt and sand contribute, whereas turbidity occurs in water due to the presence of very fine particles of these contaminants. Chemical contaminants include contaminants such as mercury, lead, and chromium as inorganic contaminants and nutrients such as nitrogen and phosphorus, acids, and alkalis. Organic contaminants can be pharmaceuticals, pesticides, solvents, and emerging contaminants such as microplastics or nanomaterials. In addition to the inorganic and organic

Hazards and Treatment of Wastewater Contaminants, First Edition. Dipika Jaspal and Arti Malviya. Published 2026 by John Wiley & Sons, Inc.

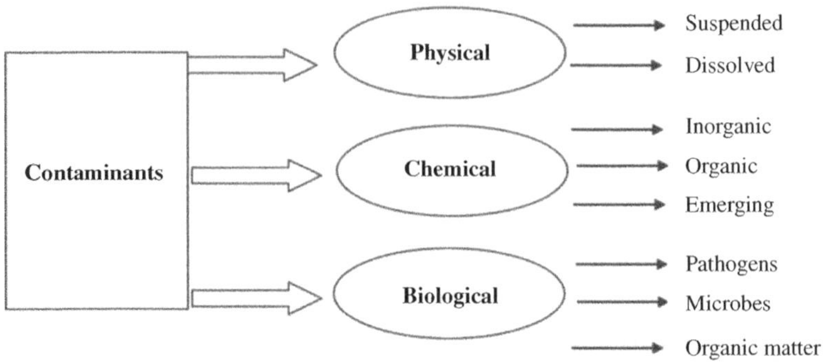

Figure 2.1 Categories of contaminants.

contaminants, biological organisms and pathogens, along with organic matter, can also pollute the water bodies.

2.2 Harm Posed by Contaminants

The presence of contaminants in wastewater raises an alarm due to the following reasons:

- **Risk to Human and Aquatic Life:** Several harmful effects include the occurrence of several diseases in humans and aquatic organisms, in addition to the effects on the aquatic flora.
- **Effect on the Environment:** Contaminants lead to eutrophication, leading to a reduction in the oxygen level in water, impacting the aquatic beings. Destruction of habitats can also result in an increase in the contaminant levels. These also alter the aquatic ecosystem and alter several biogeochemical cycles. Also, effects on biodiversity take place.
- **Effect on Aesthetics:** One of the most visible effects of the presence of contaminants is on aesthetics. Some contaminants impart colored appearances to the water bodies, having a direct impact on the photosynthetic activity.
- **Economic Implications:** Several economic implications are also related to the presence of contaminants in wastewater, which involve expenditures on health-related issues, contamination of soil due to the use of polluted water. Effect on the cost incurred from fisheries and the decline in agriculture results. Many funds are invested on cleaning the waterbodies and the growth of weeds due to excessive nutrients.

Several important contributors to water pollution and those adding to the pool of contaminants have been discussed in the subsequent sections.

2.3 Sewage Wastewater

Sewage wastewater is the liquid waste of washrooms, sanitary facilities, kitchens, generated from domestic sources, industrial units, and commercial places, or from storm runoff. It includes black, gray, and yellow wastewater. Sewage is an issue to be addressed with priority, as a voluminous amount of the water can be reclaimed and reused by adopting a proper sewage treatment technique.

Sewage water contains 99% water, and the rest is solid matter of organic and inorganic nature. The concentration of sewage is usually measured in terms of chemical oxygen demand (COD), biological oxygen demand (BOD), and suspended solids. Table 2.1 shows the typical composition of the domestic sewage water.

Owing to the global issues related to poor sanitation conditions at different places, sewage affects the ecosystem. The concentration level of different nutrients gets elevated, including heavy metals, pathogens, nutrients, and endocrine disruptors, which causes significant harm to the aesthetics and poses a threat to the inhabitants. Considering the volume of sewage being generated with the increasing population and existing sanitation conditions, greater efforts in this direction are needed [3]. Approximately 4.5 billion people live without proper

Table 2.1 Composition of untreated sewage water.

		Concentration (mg/L)		
S. No.	**Characteristics**	**Weak**	**Medium**	**Strong**
1.	Fixed solids	145	300	525
2.	Total dissolved solids (TDS)	250	500	850
3.	Volatile solid	105	200	325
4.	Total solids	350	720	1200
5.	Suspended solids	20	10	5
6.	Total organic carbon	35	15	10
7.	BOD	400	200	100
8.	Total nitrogen	90	50	20
9.	COD	1000	500	250
10.	Phosphorus	15	8	4
11.	Alkalinity	200	100	50
12.	Chlorides	100	50	30

Source: Adapted from [2].

sanitation [4], which also results in widespread contamination of surface water [5]. The production of sewage water is related to the water consumption. Factors like climate, water pressure at the supply end, water cost, population size, water availability, and water supply systems influence the sewage generation. The quality of sewage water is assessed by parameters like solids, presence of nitrogen, phosphorus, and indicators of fecal contamination and organic matter. Bacteria, viruses, protozoa, and helminthes are the main micropollutants present in sewage water. The indicators of sewage pollution are the fecal viruses [6].

2.4 Nonsewage Wastewater

Wastewater devoid of any human waste is nonsewage wastewater. The composition and the contaminants present in nonsewage wastewater depend upon the source of its origin. Thus, it can be well understood by categorizing it into below mentioned types:

2.4.1 Industrial Contaminants

Industrial wastewater refers to the water with suspended and dissolved industrial substances that comes out of the industrial outlets. With the rise in population, there has been rapid industrialization to cater to the large population. Water from industries and industrial areas is let into the available water sources either untreated or without proper treatment. This poses a concern as people have to rely on these water bodies due to the decline in clean water sources. Further, the use of this water for irrigation also leads to subsequent complications. The chemicals and contaminants tend to accumulate, adversely affecting the aquatic life. Contaminants such as dyes, bleaching agents, oils, chemicals, etc., are majorly the available substances, adversely affecting the quality of water [7, 8]. Different types of contaminants emerge from different industrial sectors such as laundries, the beverage industry, oil refineries, textile industries, etc. The contaminants may include a variety of components, ranging from dyes, heavy metals, pesticides, pharmaceuticals, oils, etc. The composition of the effluent coming from industries varies due to divergence in the processes, water is subjected to, and the materials being processed. In general, industrial wastewater is characterized by turbidity, odor, color, temperature, COD, BOD, nitrogen, phosphorus, pH, alkalinity, etc. Heavy metal and dye-loaded effluent is carcinogenic due to its non-biodegradable nature. This, if present in drinking water, can lead to critical health issues in the living organisms. The reason is easy adsorption of heavy metals on the suspended particles present in the water, which settle down to the riverbed and are slowly released, posing a threat to the ecosystem.

Table 2.2 Different industrial contaminants.

S. No.	Industry	Contaminants	References
1.	Textile	Bleaching agents, heavy metals, surfactants, alkaline bases, sodium silicate	[9]
2.	Paper	Phenolic compounds, dyes, adsorbable organic halogens, organic materials, biocides, resins, acids, tannins, sterols, lignin-derived compounds	[10–12]
3.	Sugar industry	Pb^{2+}, Fe^{3+}, and Mn^{2+}	[13]
4.	Leather	Lime sludge, salts, acids, and sulfides	[14]
5.	Paint manufacturing	Zinc, chromium, volatile organic compounds (VOCs), lead	[15]
6.	Pharmaceutical industries	Drugs, caffeine, naproxen, diclofenac	[16]
7.	Agriculture	Insecticides, pesticides, fertilizers	[17, 18]
8.	Metal	Cadmium, chromium, lead, arsenic, and mercury	[19]
9.	Plastic	High-density polyethylene, polyvinyl chloride, polypropylene, low-density polyethylene, styrofoam, or polystyrene, miscellaneous plastics	[20]
10.	Petrochemicals	Ammonia, sulfide, phenols, benzo, and other hydrocarbons, oil, phenols, suspended solids, metals, ammonia, dissolved minerals	[21]

Table 2.2 describes the different contaminants emerging from different industries.

2.4.2 Agricultural Runoffs

Agricultural pollution is the pollution due to food crops, livestock, biofuel crops, and animal feed. Agricultural pollution has taken over as a major factor degrading the coastal and inland waters, primarily due to eutrophication. The most common contaminant has been identified as nitrate [22]. Almost 38% of the water bodies in the European Union are witnessing pressure due to agricultural runoffs. The main source of pollution in the United States of America is agriculture (US EPA, 2016). In countries like China also a large share of water pollution is due to agriculture (FAO, 2013). In the European Union, around 38% of the water bodies are under a threat of agricultural pollution (WWAP, 2015) [23], as around 70% of

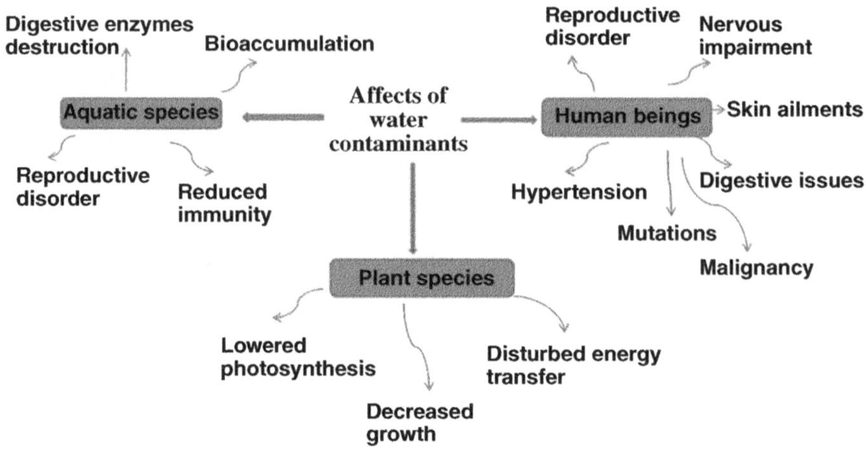

Figure 2.2 Effects of water contaminants.

water in the world is used for agriculture. Cows, hens, pigs, etc. poop has to be disposed at suitable places. This waste is not treated like human waste, where the sewage is sent to the treatment plants. This is instead spread in the agricultural farms to dispose which often acts as manure, but is beyond the absorption capacity of the soil. Along with the rainwater runoff, all these nutrients are let into the water bodies, leading to eutrophication. Nitrogen-based fertilizers are the major category of nutrients which have been used over the past century. These nitrogen-based fertilizers make their way into the nearby rivers, lakes, and ponds. In addition to eutrophication, the growth of the algae is also enhanced due to the loads of nutrients passing into the water bodies. The following figure summarizes the impact of contaminants in general on human health and the ecosystem (Figure 2.2).

2.4.3 Oil Spill

It is an intentional or unintentional release of oil into the marine ecosystem. Oil spills pose a significant threat to the marine ecosystem. Oil spills are a result of offshore explorations, tank and pipeline ruptures, and transportation activities. The problems associated with oil spills get aggravated if these spills reach the shoreline, which is the area of maximum active underwater biological activities. Depending upon the volume of the activity, spills could be small, medium, or large, ranging from 7 to greater than 700 tons [23]. Chemical changes, such as degradation and physical transformation, take place when oil enters the aquatic system. The damage caused by an oil spill depends on the location of the spill, the

Table 2.3 Categories and effects of oil spills.

S. No.	Type	Type of oil	Characteristics	Toxicity
1.	Class A	Jet fuel, crude oil, gasoline	Most toxic, flammable property, light, mixed easily with water	Ingested, inhaled, and absorbed by wildlife, posing a danger
2.	Class B	Kerosene	Highly inflammable and is difficult to clean	Less toxic than class A
3.	Class C	Crude oil	Heavy and highly sticky, unlike Class B oils, they do not dilute in water but are costly to clean up.	It can cause a high level of contamination and can stick to animals.
4.	Class D	Heavy crude oil	Harden when heated and solidify	Forms a layer over the organisms
5.	No petroleum oil spills	Vegetable oils from food industries and factories	Degrades water quality	Leads to suffocation and dehydration of wildlife

type of oil, weather conditions, and season. The oil floating on the surface tends to spread quickly with wind, tides, and weather conditions, blocking the oxygen connect of the aquatic life, obstructing the photosynthetic activity, forming a coat over the feathers of birds, etc. [24]. The aquatic beings can be affected by these oil spills internally and externally. The organisms living in the shoreline areas are affected by the oil floating on the surface. When this oil sinks to the bottom of the sea, it causes harm to the fish in the water. Owing to the physical, chemical, and biological changes, the characteristics of the oils can change and can prove to be even more harmful. The emulsion created by water and oil can persist for years. Hunting and fishing can be further affected and can have adverse impacts on the economy and the livelihood of people who rely on these activities. The important sources of drinking water can also be affected. The different types of oil spills and their effects have been summarized in Table 2.3 [25, 26].

2.4.4 Radioactive Contaminants

More and more countries are utilizing nuclear technology for the generation of radioactive substances, which are used for a variety of applications. A large amount of radioactive wastewater is generated by the nuclear reactors, hospitals, research units, radiopharmacy units, and fuel processing plants. Radionuclides have been causing substantial concerns over the past few decades owing to the pollution caused and the impact on public health. These require proper stabilization techniques and solidification to properly dispose off the radionuclide, which can penetrate the aquatic environments, causing potential risk to human beings and aquatic fauna. Therefore, the elimination of the nuclear waste from the wastewater becomes crucial [27]. Table 2.4 shows the different categories of radioactive contaminants, depending upon the extent of radioactivity shown by them [28].

Table 2.5 gives some categories of contaminants and their sources [27].

Uranium (U (VI)), strontium (Sr (II)), thorium (Th (IV)), cesium (Cs (I)), etc., are some of the radioactive elements which are highly toxic and hazardous [29]. Nuclear radiation from radioactive substances can lead to harmful effects on the lungs, skin, bone marrow, genetic effects, hereditary disorders, severe mental illness, and even death [30, 31].

Table 2.4 Categories of radioactive contaminants.

S. No.	Category	Activity	Sources
1.	Low-level waste (LLW)	Only 1% of the radioactivity, out of 90% of the volume of radioactive waste	Industries, hospitals, nuclear fuel cycle
2.	Intermediate-level waste (ILW)	7% of the radioactivity, out of 4% of the volume of radioactive waste	Chemical sludge, ion-exchange resins, radiation therapy, and contaminated materials from reactor decommissioning
3.	High-level waste (HLW)	95% of the radioactivity, out of 3% of the volume of radioactive waste	Uranium fuel burning in a nuclear reactor

Table 2.5 Various radioactive contaminants in wastewater and their sources.

S. No.	Radioactive contaminants	Sources
1.	^{14}C, ^{3}H, U (^{233}U, ^{234}U, ^{235}U, ^{238}U), Th (^{228}Th, ^{232}Th)	Nuclear Power Plant
2.	^{14}C, ^{3}H, ^{32}P, ^{35}S, ^{125}I, ^{99}Tc-m, ^{131}I, ^{85}Sr	Radiolabels and radiopharmaceuticals/ medical diagnosis and treatment centers
3.	^{90}Sr, ^{137}Cs	Nuclear reactors

2.4.5 Microbial Contaminants

Rapid industrialization and an increase in population have led to the generation of a large volume of wastewater, where more than 75% of the water with several contaminants is discharged. In addition to several contaminants, pathogenic microorganisms are an important category of contaminants. Several microbes, such as bacteria, fungi, algae, viruses, and protozoa, are introduced into the water streams and water supply due to poor sanitary conditions and unhealthy practices. Pathogenic infestation takes place due to fecal disposal, industrial runoff, manure, domestic water outlets, bathing activities, etc. [32]. Out of the mentioned sources, fecal matter is the main inlet source of microbes in wastewater [33].

Microbes can cause serious threats to human life and functioning. Therefore, their detection and treatment become inevitable [34]. For portable water, there are processes such as chlorination and the use of ultraviolet radiation, which are carried out for disinfection, but the process might be complicated for wastewater. If the turbidity of drinking water is ≥ 1 nephelometric turbidity units (NTU), disinfection is not efficient. A higher concentration of particles is usually present in treated wastewater. Waste treated wastewater typically has higher concentrations of particles in the range of 1–10 NTU. Several techniques are being used for the detection of microbes in wastewater. The simplest of all techniques is the culture technique.

2.4.6 Emerging Contaminants

Emerging contaminants naturally exist, or are man-made chemicals, or microbes, which have the potential to enter the environment and cause adverse ecological impacts and health issues. These are persistent contaminants and are thus recognized as the contaminants with "emerging concern" or "chemicals of emerging concern" [35]. Recently impacts of emerging contaminants have been reported in water bodies worldwide. These contaminants could be present in surface and underground water sources. They are usually not the main targets of the research community and regulatory authorities [36]. Such contaminants are generally unmonitored but can enter the environment, causing adverse effects on living organisms. The most commonly quoted emerging contaminants are personal care products, pharmaceuticals, surfactants, pesticides, nanomaterials, plasticizers, and fire retardants [37, 38]. These pollutants are not newly discovered, but are already well-known contaminants, recently identified with their new harmful effects and mode of action. These are those pollutants that have invaded the living organism through an alternate route. Emerging contaminants are biologically active and can bioaccumulate. They are persistent and recalcitrant. Different categories of emerging contaminants are mentioned in Table 2.6.

Table 2.6 Different categories of emerging contaminants.

S. No.	Category	Types
1.	Pharmaceuticals	Drugs
2.	Personal care products	Disinfectants, cosmetics, surfactants, food additives, and domestic biocides
3.	Industrial chemicals	Flame retardants, polychlorinated biphenyl, antimicrobial substances, pesticides, and food additives
4.	Disinfection byproducts	Halonitromethanes, nitrosamine, haloacetonitriles, haloacetic acids, trihalomethanes
5.	Algal toxins	Microcystin, cyanotoxins
6.	Biocides and their metabolites	Pesticides
7.	Devices for bioterrorism and disruption	Biological weapons

Emerging contaminants can lead to several hazardous implications, including endocrine disruption in organisms. These are the compounds affecting the normal functioning of the hormones. These are capable of mimicking hormones, modifying hormonal action, and also disrupting the hormone production [39]. These usually enter the water streams through wastewater treatment plants, sewer flow, hospitals, and direct discharge of untreated wastewater from where these tend to enter the water sources. The concentration of these contaminants present in the environment is very less (in μg/L or ng/L,) but the paucity of their concentration levels changing after entry into the water bodies is high [40–43]. It is imperative to regulate emerging contaminants in water bodies to mitigate their ill effects. Several technologies have been developed to eradicate the emerging contaminants from the aqueous environment, which include physical, chemical, and biological methods.

2.4.6.1 Microplastics

Microplastics are fragmented plastic particles having size less than 5 mm and have been identified as one of the emerging contaminants in the water bodies. These diminish the aesthetic beauty of the water ecosystem and poses long term health issues in living organisms. The global escalation in the microplastic pollution alarms the need for effective strategies and technology for pollution mitigation.

Microplastics can appear in spherical to elongated fiber-like shapes in water. The main contributors of microplastics in the water bodies include wastewater treatment facilities, the fishing sector, plastic manufacturing sectors, and runoff from dumping sites and ineffective sewage or garbage disposal. Plastic debris is a mixture of various types and shapes of plastic, ranging from micro to macro size. The presence and transportation of plastics in aqueous environments are influenced by factors like size, shape, composition and hydrodynamics, water chemistry, and interactions. Microplastics have been found to cause physiological and biological effects in living organisms. These also act as a vector for other contaminants. They have been reported to host antibiotic-resistant bacteria and genes that pose a significant impact on the health of living organisms. They have been reported to disturb the feeding capacity of marine organisms. They can inhibit the growth of vegetation and can enter the human system via ingestion, inhalation, or contact. Recent studies have suggested that microplastics should be considered as contaminants of immediate concern [44].

2.4.6.2 Surfactants

Surfactants are surface-active agents with the presence of hydrophobic and hydrophilic groups, and the ability to reduce interfacial energy. Surfactants can be nonionic, anionic, amphoteric, or cationic. The application and function of surfactants depend on the structure of surfactants. Their easy fabrication and large applicability escalate the chances of their presence in the water bodies, discharged from facilities employing surfactants.

The major contributors of the surfactants are industrial and municipal wastewater, rainwater runoff, household discharges, and discharges from treatment facilities. Though natural microbial action degrades the surfactants to some extent, they may result in the generation of toxic and bioaccumulative products. Surfactants at very low concentrations can cause harmful and deadly effects on aquatic organisms. Surfactants have different toxicological consequences that depend on their structure, characteristics, concentration, and exposure time [45].

2.4.6.3 Pharmaceuticals

Pharmaceutically active compounds are one of the major types of emerging contaminants that can cause deadly impacts on living organisms, even at very low concentrations. With the increase in the requirement of pharmaceuticals, including personal care products, there has been a sharp increase in these contaminants in water. These mainly arise from the drug formulation industries and are biologically active compounds. These are used to prevent, cure, or treat various ailments, infections, and diseases. These include antibiotics, analgesics, anti-inflammatory, and antidepressant drugs. Personal care products include lotions, hair dyes, lipsticks, bath soaps, fragrances, sunscreens, creams, used to improve the aesthetics and quality of daily life. There has been an immense growth in the production of these substances. These are persistent in aqueous media and are recalcitrant. The effluents having such compounds lead to genotoxic and ecotoxicological impacts on living organisms. Thus, effective treatment of pharmaceuticals and cosmetic-loaded wastewaters is essential before discharging them to the water environment [46].

The toxicity assessment and treatment of the above-mentioned contaminants are discussed in detail in the subsequent chapters.

2.5 Detecting Contaminants

Several methods can be used for the detection of the presence of contaminants in wastewater. Figure 2.3 shows the different methods by which contaminants can be successfully identified and detected.

2.5.1 Physical Methods

These methods rely on the physical appearance of the samples and mainly include turbidity assessment, measurement of color, and conductivity, which can be affected by the presence of different salts and contaminants.

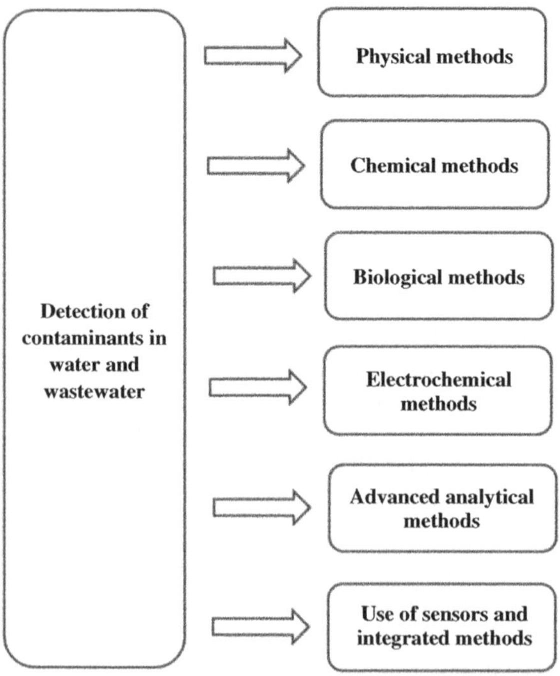

Figure 2.3 Methods for detection of contaminants in wastewater.

2.5.2 Chemical Methods

Measurement of BOD and COD are direct indicator of the pollution of the water streams. Contaminants such as metals can be detected by several spectrophotometric and chromatographic methods.

2.5.3 Biological Methods

Methods such as microbial testing methods are helpful for the detection of contaminants, such as *Coliform* bacterial testing, total plate count, and membrane filtration. Biosensors, such as bioluminescent assays and toxicity analysis on organisms due to several contaminants, are some of the methods.

2.5.4 Electrochemical Methods

The use of ion-selective electrodes can help identify the presence of contaminants such as heavy metals, ammonia, etc. In addition to this, electrochemical sensors can detect the presence of chlorides and sulfates. Conductivity meters are also used for contaminant measurements.

2.5.5 Advanced Analytical Methods

Advanced analytical methods such as mass spectrometry have been used for the detection of trace quantities of pesticides. X-ray fluorescence is also one such technique used for the detection of some contaminants, such as metals, etc.

For monitoring large water bodies, satellite imaging is also being used. Real-time monitoring of contaminants has also been possible with the technique of applying sensor-based monitoring.

2.6 Removal of Wastewater Contaminants

Several methods are employed for the removal of contaminants, which include physical, chemical, biological, and advanced methods. In addition to all these, there is an integration of methods also carried out for contaminant elimination. Some physical methods involve the use of filtration, flocculation, and coagulation. The use of certain chemicals is done in the chemical treatment of wastewater for contaminant eradication. The biological methods include the use of bioreactors, activated sludge, etc., and the use of membrane filtration, advanced oxidation processes are some examples of the advanced methods being used. The methods used for specific categories of contaminants have been discussed in subsequent chapters.

2.7 Incidents of Contamination: Case Studies

Contaminants of different categories are responsible for the degradation of the water quality. There have been different cases worldwide where large-scale destruction and harm to living beings and the ecosystem have been caused. Some such examples to discuss the impact of toxicants have been discussed below.

2.7.1 Flint, Michigan Water Crisis (2014)

There was a switch of the water supply from Detroit's system to the Flint River. There were continuous complaints raised mentioning the complications and the significant changes in smell, taste, and color, in addition to the various health effects, which were overlooked by the government [47]. In a study carried out by Virginia Tech, samples from several hundreds of houses were collected and tested. The results indicated the presence of elevated levels of lead.

Impact: Lead contamination and poisoning affected the population in terms of retarded cognitive ability, impact in learning ability, behavioral issues, hampering the metal and social well-being.

2.7.2 Oil Spill in the Gulf of Mexico (2010)

It was one of the major disasters in the world. Deepwater Horizon, an oil drilling rig operative in the Macondo Prospect in the Gulf of Mexico, exploded and sank. Almost four million barrels of oil were released into the water for several days before it could be controlled [48].

Impact: A report by the Marine Mammal Commission [49] mentioned the long-term effects of the incident. Damage to the natural resource was a very evident impact. Increase in the fish sores or lesions, and destruction of fish on a large scale, spread of carcinogenic polycyclic aromatic hydrocarbons, impact on birds, and ground water contamination, etc. The long-term impacts still persist [50].

2.7.3 West Virginia Chemical Spill – Charleston (2014)

Gallons of chemicals (almost 10000 gallons) used for coal processing leaked into Elk River (West Virginia). The primary chemical was 4-methylcyclohexanemethanol (MCHM) in addition to some other chemicals such as propylene glycol phenyl ether (PPH) and dipropylene glycol phenyl ether [51].

Impact: Elka River, being a municipal source of water and supplying water to almost 30000 people, the leak led to the disruption of the supply of clean water for a large population and impact on the natural ecosystem. In addition to this people reported several health impacts such as vomiting, diarrhea, itching etc. [52].

2.7.4 Contamination of the River Ganga in India

The River Ganga has now become one of the most polluted rivers in the world [53]. Although it is a sacred river, it has undergone massive pollution. People release the ashes after cremations, human bodies are burnt and dumped into the river due to spiritual significance, flowers, sewage, agricultural runoffs, industrial effluents [54], etc., are all dumped into the Ganga, deteriorating the quality of water, increasing the concentration of heavy metals, mercury, etc. A report by CPCB (Central Pollution Control Board), Ministry of Environment and Forests, Government of India [55] mentions several studies that have been carried out to analyze the extent of pollution over the years, which indicates the deteriorating quality.

Impact: The impacts of the pollution of the Ganges have led to severe impacts on humans as well as wildlife. Consumption of the water from the Ganga has been responsible for waterborne illness such as hepatitis, cholera, dysentery, etc. Reduction in the population of freshwater dolphins and turtles has been a direct impact of pollution on the Ganges [56].

2.7.5 Minamata Bay Disaster (1932–1968)

A plastic and petrochemical company Chisso Corporation released almost 27 tons of mercury into Minamata bay. This led to a widespread impact of lead poisoning on the fish and humans who consumed them. A study in 2001 suggested that almost 2 million people ingested mercury.

Impact: Hundreds of people died after consuming mercury-contaminated local fish. Permanent disabilities, blindness, insomnia, headaches, tremors, brain damage, and several sensory disorders occurred [57, 58].

2.8 Successful Elimination of Contaminants

Several technologies have been successfully applied for the removal of contaminants from waterbodies or wastewater. Some of the successful projects have been mentioned below:

2.8.1 Tuas Water Reclamation Plant (Singapore)

This plant will be treating domestic and industrial wastewater in an energy-efficient manner. This plant will be proving its worth in the removal of contaminants from wastewater using membrane bioreactors with the combination of membrane filtration and biological treatment. The treated water has been successfully used for reuse and as portable water [59]. The domestic wastewater will be purified for further use, and the industrial wastewater will be treated to meet the international standards of discharge into the sea [60].

2.8.2 Wastewater Reclamation (Taiwan)

In order to cater to the arising water shortage in Taiwan the reclaimed water from industries and municipal water treatment plants has been an important source [61]. The wastewater samples are pretreated and exposed to a variety of treatment techniques including reverse osmosis, ion exchange, use of activated carbon etc. Some setups in Taiwan have successfully treated wastewater with the use of a combination of several techniques.

Some examples are Innolux Corporation, which has treated wastewater with many contaminants such as organics, inorganics, etc., enabling reclamation, reuse, and recovery of chemicals [62]. Treatment methods such as coagulation, flocculation, the use of activated carbon, reverse osmosis, etc., are being used for water treatment. Another example is the Fengshan reclaimed water treatment plant. Preliminary treatment processes, along with ultrafiltration and reverse osmosis, effectively and efficiently treat the wastewater [63].

2.8.3 Advanced Oxidation Processes (AOPs) (Japan)

There has been a widespread application of AOPs in Japan for the treatment of wastewater [64]. Yokohama treats the municipal wastewater using ozonation in combination with the use of activated carbon for the elimination of contaminants such as pharmaceuticals, organics, and organic pollutants present in trace quantities, which are difficult to remove by conventional methods [65]. The contaminated river water is also treated using ozonation. Kunijima water purification facility in the city of Osaka involves the use of ozonation and granular activated carbon for the removal of odor and contaminants such as trihalomethane and microorganisms [66].

2.8.4 Constructed Wetlands (Sweden)

Free Water Surface Constructed Wetlands (FWS) in Sweden treat the wastewater from the wastewater treatment plants. The main aim of these wetlands has been the removal of nitrogen [67, 68]. These have also been used to remove nitrogen from the agricultural catchments [69]. A report [70] in 2019 mentions the use of FWS for the treatment of pharmaceutical effluent. Constructed wetlands showed a good potential for the treatment of active pharmaceutical ingredients (APIs).

2.8.5 Bioreactors for Wastewater Treatment (Saudi Arabia)

Bio reactors have proven their efficiency in treating several wastewater contaminants. Membrane bioreactors (MBRs), moving bed biofilm reactors (MBBRs) are some in this category. The MBBR technology has been successfully implemented in Saudi Arabia [71, 72]. The MBBR system has proven its efficiency in the effective treatment of COD, BOD, and total suspended solids (TSS).

2.9 Important Considerations

With the increase in population and rapid industrialization there has been an alarming rise in the level of contaminants being let into the water bodies. The scarcity of available usable water adds to the deteriorating situation. To manage the scenario in addition to the involvement of regulatory bodies a joint effort involving the community also becomes important.

2.9.1 Education and Public Awareness

The general public should be made aware of the several consequences of contaminating the water bodies and the importance of maintaining the cleanliness of the water bodies. Campaigns educating on the different methods of safe disposal of contaminants, wastes, and chemicals should be conducted. Methods of segregation of wastes should be highlighted to tackle the hazardous and toxic contaminants.

2.9.2 Prevention of Pollution at Source and Best Practices

Another important aspect is to stop the pollution and spread of contaminants at the source. This would involve regulating the emergence and spread of contaminants from various industrial and domestic sources. Various methods which can be adopted are imposing strict regulations and pollution control strategies for discharge of chemicals, heavy metals, etc., into the water bodies. Implementation of stringent industry guidelines for waste disposal will help in reduction of the impact of contaminants.

Considering the arising scarcity of clean water technologies for the reuse and recycling of wastewater, and encouraging zero liquid discharge will be helpful. Agricultural runoffs also play a vital role in water contamination, and hence the adoption of sustainable practices in agriculture, such as regulated pesticide and fertilizer use and organic farming, will be some best practices which can be followed.

2.9.3 Wastewater Management

Effective and efficient wastewater treatment and management are the keys to combating the spread of contaminants through the wastewater treatment plants (WWTP) before these are discharged into the water streams. Decentralized setups should be preferred over centralized systems for load reduction and efficient treatment. The existing plants for wastewater treatment should be monitored thoroughly for operations and upgraded with the latest and impactful technologies for a growing population and increasing demands.

The management of stormwater and preventing it from mixing with sewage and entering the water bodies should be avoided. Keeping separate systems for the collection of treatment of storm water and sewer can help prevent the contamination of lakes and rivers. Reduction in runoff of the rainwater leading to the inlet of several contaminants in the water streams by the construction of permeable pavements, rain gardens, etc., can help reduce contamination.

2.9.4 Use of Technology

The treatment of various contaminants can be done using different worth adopting technologies that have proven their worth. Several impactful technologies are available, which can be chosen depending upon the possible conditions for operation and the type of contaminants to be treated. Various methods and advanced wastewater treatment technologies are now available, which can be implemented in different sectors.

Some of the latest and upcoming technologies are nanotechnology, involving the use of nanomaterials for the treatment of chemicals, pesticides, pharmaceuticals, etc. Research is being carried out using artificial intelligence, remote sensing, etc., to predict and monitor the spread of various contaminants and track their discharge. Solitary techniques are now been combined with other technologies known as integrated technologies for wastewater treatment.

2.9.5 Research and Development

The growing variety of emerging contaminants makes it essential that research is carried out for better and efficient treatment of different categories to reduce the threat posed by them. Carrying out studies on effective amalgamation of different techniques for remediation and innovative methods of wastewater treatment and eradication of contaminants is continuously required.

2.9.6 Polluted Water Body Restoration

Efforts in restoring the available water bodies also become essential due to the growing water shortage, owing to the growing population. The use of different water purifying bacteria and plants will help remove contaminants to some extent. Also, the removal of contaminants present in sediments will be helpful.

2.9.7 Policies and Collaboration

Effective measures to control pollution and contaminants should be taken up by the industries, communities, and the regulatory authorities, considering the regional and national policies for waste disposal, wastewater treatment, and contaminant discharge norms. Efforts in the direction of international collaboration for new technologies and innovative solutions for the treatment of contaminants and wastewater management should be accelerated. This will help in cross-border efforts and will not be confined to the national level.

There are several regulatory authorities and environment protection agencies which provide regulations for the safe discharge of contaminants in wastewater emerging from several sources. Although these guidelines vary from region to region, yet these provide an understanding of the several permissible limits of contaminants in wastewater being discharged into the water sources.

The safe discharge of the contaminants into the environment is governed by:

1. **Requirements of Permits:** Obtaining necessary permits is a must for ensuring the water quality and health requirements.
2. **Effluent Limits:** The limits of several contaminants are defined based on the region and the source from where wastewater is emerging. The limits of important parameters such as COD, BOD, TSS, and nutrients are defined by the regulatory agencies.
3. **Reporting and Monitoring:** These are essential to ensure that the industrial setups/discharging sources are generating contaminants within permissible limits by continuous monitoring and reporting to the regulatory authorities.
4. **Treatment Technologies:** It is necessary that the different industrial sectors adopt treatment technologies for regulated emergence of contaminants.
5. **Best Practices and Continuous Improvement:** A continuous pursuit for continuous improvement and best practices be adopted for minimal discharge of contaminants into the environment.

Some prominent national and international agencies that provide these regulations are as mentioned below:

1. United States Environmental Protection Agency (EPA) – United States of America
2. European Union Water Framework Directive – European Union
3. Central Pollution Control Board (CPCB) – India
4. World Health Organization (WHO)

To protect the environment and to monitor the discharge limits, the Environment Protection Rules [73] were enacted in India in 1986, in response to the 1972 United Nations Conference on the Human Environment at Stockholm. The objectives of the act include protection and improvement of the environment, prevention of hazards to living species and property, and maintenance of a harmonious relationship between humans and the environment. Below are the key features of these rules:

1. Discharge standards [74] are mentioned for several pollutants to ensure the safe discharge of the contaminants which are released into the water bodies.
2. Necessary permissions are provided to authorize the operations that might lead to environmental pollution.
3. Enforcement of compliance for discharge standards, failing which penalties can be enforced, or the setup can be ordered to cease operations.
4. Allows the stakeholders to voice their concerns during consultations.
5. Imposing fines and penalties for violation of rules.

India enacted "The Water Act, 1974" and "Water Rules, 1975" in response to the critical problems of the quality and quantity of water. The act is concerned with restoring the wholesomeness of water. Schedule VI (Rule 3A) mentions the discharge limits of various contaminants in inland surface water, public sewers, irrigation land, and marine coastal areas. The limits have been defined for various parameters such as pH, temperature, COD, BOD, etc., and different categories of contaminants such as heavy metals.

2.10 Challenges

Although several technologies have made the removal of contaminants easy, yet there could be several challenges in the treatment, which could arise due to a matrix and diversity of contaminants being present, newer categories of contaminants known as emerging contaminants, challenges in applicability due to large-scale contamination, and the cost of treatment. The variability of composition depending on different conditions is also an important aspect. High energy considerations and sludge production in some processes also needs to be considered very carefully.

References

1 Hutton, M., and Shafahi, M. (2019). Water pollution caused by leather Industry: a review. Proceedings of the ASME 2019 13th International Conference on Energy Sustainability collocated with the ASME 2019 Heat

Transfer Summer Conference. ASME 2019 13th International Conference on Energy Sustainability. Bellevue, Washington, USA. https://doi.org/10.1115/ES2019-3949.

2 Fahad, A., Radin, M., Mohamed, S., et al. (2019). Wastewater and its treatment techniques: an ample review. Indian Journal of Science and Technology, Volume 12, Issue 25, pp. 1–13. https://doi.org/10.17485/ijst/2019/v12i25/146059.

3 Wear, S.L., Acuna, V., McDonald, R., et al. (2021). Sewage pollution, declining ecosystem health, and cross-sector collaboration. Biological Conservation, Volume 255, p. 109010. https://doi.org/10.1016/j.biocon.2021.109010.

4 World Health Organization (2015). Progress on Sanitation and Drinking-water – 2015 Update and MDG Assessment. World Health Organization and UNICEF, WHO Press, Geneva, Switzerland, https://www.unicef.org/publications/index_82419.html (cited 9 June 2023).

5 Kiulia, N.M., Hofstra, N., Vermeulen, L.C., et al. (2015). Global occurrence and emission of rotaviruses to surface waters. Pathogens, Volume 4, Issue 2, pp. 229–255. https://doi.org/10.3390/pathogens4020229.

6 Toribio-Avedillo, D., Anicet R., Blanch, M.M., et al. (2021). Bacteriophages as fecal pollution indicators. Viruses, Volume 13, Issue 6, p. 1089. https://doi.org/10.3390/v13061089.

7 Global health, how water pollution in India kills millions. https://www.borgenmagazine.com/water-pollution-in-india/ (Last accessed 23.07.2023).

8 Azimi, A., Azari, A., Rezakazemi, M., et al. (2017). Removal of heavy metals from industrial wastewaters: a review. ChemBioEng Reviews, Volume 4, Issue 1, pp. 37–59. https://doi.org/10.1002/cben.201600010.

9 Fu, F., and Wang, Q. (2011). Removal of heavy metal ions from wastewaters: a review. Journal of Environmental Management, Volume 92, Issue 3, pp. 407–418. https://doi.org/10.1016/j.jenvman.2010.11.011.

10 Wang, Z., Xue, M., Huang, K., et al. (2011). Textile dyeing waste water treatment. In: Hauser PJ (ed). Advances in treating textile effluent, IntechOpen, pp. 91–116. https://doi.org/10.5772/22670.

11 Buyukkamaci, N., and Koken, E. (2010). Economic evaluation of alternative wastewater treatment plant options for pulp and paper industry. Science of the Total Environment, Volume 408, Issue 24, pp. 6070–6078. https://doi.org/10.1016/j.scitotenv.2010.08.045.

12 Rizwan, K., Rasheed, T., Bilal, M., et al. (2022). Pulp and paper industry-based pollutants, and their adverse impacts. In: Bhat R, Kumar A, Nguyen TA, et al. (eds). Micro and nano technologies, nanotechnology in paper and wood engineering, Elsevier, pp. 143–160. https://doi.org/10.1016/B978-0-323-85835-9.00005-2.

13 Lindholm-Lehto, P.C., Knuutinen, J.S., Ahkola, H.S., et al. (2015). Refractory organic pollutants and toxicity in pulp and paper mill wastewaters.

Environmental Science and Pollution research, Volume 22, pp. 6473–6499. https://doi.org/10.1007/s11356-015-4163-x.
14 Rahim, M.A., and Mostafa, M.G. (2021). Impact of sugar mills effluent on environment around mills area. AIMS Environmental Science, Volume 8, Issue 1, pp. 86–99. https://doi.org/10.3934/environsci.2021006.
15 Krithika, D., and Philip, L. (2016). Treatment of wastewater from water-based paint industries using submerged attached growth reactor. International Biodeterioration and Biodegradation, Volume 107, pp. 31–41. https://doi.org/10.1016/j.ibiod.2015.10.017.
16 Ranjan, N., Singh, P.K., and Maurya, N.S. (2022). Pharmaceuticals in water as emerging pollutants for river health: a critical review under Indian conditions. Ecotoxicology and Environmental Safety, Volume 247, pp. 114220. https://doi.org/10.1016/j.ecoenv.2022.114220.
17 Aktar, W., Sengupta, D., and Chowdhury, A. (2009). Impact of pesticides use in agriculture: their benefits and hazards. Interdisciplinary Toxicology, Volume 2, Issue 1, pp. 1–12. https://doi.org/10.2478/v10102-009-0001-7.
18 https://19january2017snapshot.epa.gov/nps/nonpoint-source-agriculture_.html (Last accessed 23.9.24).
19 Preeyaporn, K., and Young, R.S. (2011). Advances in carcinogenic metal toxicity and potential molecular markers. International Journal of Molecular sciences, Volume 12, Issue 12, pp. 9576–9595. https://doi.org/10.3390/ijms12129576.
20 http://swachhbharaturban.gov.in/writereaddata/SBM%20Plastic%20Waste%20Book.pdf (Last accessed 23.09.24).
21 Santo, C.E., Vilar, V.J.P., Bhatnagar, A., et al. (2013). Biological treatment by activated sludge of petroleum refinery wastewaters. Desalination and Water Treatment, Volume 51, Issues 34–36, pp. 6641–6654. https://doi.org/10.1080/19443994.2013.792141.
22 Rusănescu, C.O., Rusănescu, M., and Constantin, G.A. (2022). Wastewater management in agriculture. Water, Volume 14, Issue 21, p. 3351. https://doi.org/10.3390/w14213351.
23 https://www.undp.org/sites/g/files/zskgke326/files/migration/sa/232273E.pdf (Last accessed 20.02.24).
24 Chughtai, R., and Asif, Z. (2021). Study fate of pollutants due to oil spill in sea water through multimedia environmental modeling. International Journal of Environmental Science and Technology, Volume 18, pp. 761–770. https://doi.org/10.1007/s13762-020-02849-z.
25 Asif, Z., Chen, Z., An, C., et al. (2022). Environmental impacts and challenges associated with oil spills on shorelines. Journal of Marine Science and Engineering, Volume 10, Issue 6, p. 762. https://doi.org/10.3390/jmse10060762.
26 https://opgplus.com/types-of-oil-spills/ (Last accessed 23.09.24).

27 https://www.safewater.org/fact-sheets-1/2017/1/23/oil-spills (Last accessed 23.09.24).
28 https://world-nuclear.org/information-library/nuclear-fuel-cycle/nuclear-waste/radioactive-waste-management (Last accessed 23.09.24).
29 https://www.downtoearth.org.in/water/radioactive-pollution-in-water-a-global-concern-for-human-health-80637 (Last accessed 23.09.24).
30 Changlun, C. (2019). Emerging natural and tailored nanomaterials for radioactive waste treatment and environmental remediation. In: Chen C (ed) Interface science and technology, Volume 29, Elsevier Science, pp. 2–337.
31 International Atomic Energy Agency (2006). Chernobyl's legacy: health, environmental and socio-economic impacts and recommendations to the Governments of Belarus, the Russian Federation and Ukraine, 2nd ed. International Atomic Energy Agency, Vienna, p. 55.
32 Ma, H., Shen, M., Tong, Y., et al. (2023). Radioactive wastewater treatment technologies: a review. Molecules, Volume 28, Issue 4, p. 1935. https://doi.org/10.3390/molecules28041935.
33 Bonetta, S., Pignata, C., Gasparro, E., et al. (2022). Impact of wastewater treatment plants on microbiological contamination for evaluating the risks of wastewater reuse. Environmental Sciences Europe, Volume 34. https://doi.org/10.1186/s12302-022-00597-0.
34 Sharma, M., Agarwal, S., Agarwal, R.M., et al. (2023). Recent advances in microbial engineering approaches for wastewater treatment: a review. Bioengineered, Volume 14, Issue 1. https://doi.org/10.1080/21655979.2023.2184518.
35 https://www.ncbi.nlm.nih.gov/pmc/articles/PMC7126130/ (Last accessed 23.09.24).
36 Rosenfeld, P.E., and Feng, L.G.H. (2011). Chapter 16: Emerging contaminants. In: Rosenfeld PE, Feng LGH (eds) Risks of hazardous wastes. William Andrew Publishing, pp. 215–222. https://doi.org/10.1016/B978-1-4377-7842-7.00016-7.
37 Dulio, V., Van Bavel, B., Brorström-Lundén, E., et al. (2018). Emerging pollutants in the EU: 10 years of NORMAN in support of environmental policies and regulations. Environmental Sciences Europe, Volume 30. https://doi.org/10.1186/s12302-018-0135-3.
38 Cabeza, Y., Candela, L., Ronen, D., et al. (2012). Monitoring the occurrence of emerging contaminants in treated wastewater and groundwater between 2008 and 2010. The Baix Llobregat (Barcelona, Spain). Journal of Hazardous Materials, Volumes 239–240, pp. 32–39. https://doi.org/10.1016/j.jhazmat.2012.07.032.
39 Keerthanan, S., Jayasinghe, C., Biswas, J.K., et al. (2020). Pharmaceutical and personal care products (PPCPs) in the environment: plant uptake, translocation, bioaccumulation, and human health risks. Critical Reviews in Science and

Technology, Volume 51, Issue 12, pp. 1221–1258. https://doi.org/10.1080/10643389.2020.1753634.

40 Kumar, R., Qureshi, M., Vishwakarma, D.K., et al. (2022). A review on emerging water contaminants and the application of sustainable removal technologies. Case Studies in Chemical and Environmental Engineering, Volume 6, p. 100219. https://doi.org/10.1016/j.cscee.2022.100219.

41 Abdulrazaq, Y., Abdulsalam, A., Rotimi, L., et al. (2020). Classification, potential routes and risk of emerging pollutants/contaminant. In: Nuro A (ed) Emerging contaminants. Intech Open. https://doi.org/10.5772/intechopen.94447.

42 Dey, S., Bano, F., and Malik, A. (2019). Pharmaceuticals and personal care product (PPCP) contamination—a global discharge inventory. In: Prasad MNV, Vithanage M, Kapley A (eds). Pharmaceuticals and personal care products: waste management and treatment technology, Butterworth-Heinemann, pp. 1–26. https://doi.org/10.1016/B978-0-12-816189-0.00001-9.

43 https://www.epa.gov/chemicals-under-tsca (Last accessed 20.11.24).

44 Stapleton, M.J., and Hai, F.I. (2023). Microplastics as an emerging contaminant of concern to our environment: a brief overview of the sources and implications. Bioengineered, Volume 14, Issue 1, p. 2244754. https://doi.org/10.1080/21655979.2023.2244754.

45 Quesada, H.B., Baptista, A.T.A., Cusioli, L.F., et al. (2019). Surface water pollution by pharmaceuticals and an alternative of removal by low-cost adsorbents: a review. Chemosphere, Volume 222, pp. 766–780. https://doi.org/10.1016/j.chemosphere.2019.02.009.

46 Samal, K., Mahapatra, S., and Ali, M.H. (2022). Pharmaceutical wastewater as emerging contaminants (EC): treatment technologies, impact on environment and human health, Energy Nexus, Volume 6, p. 100076. https://doi.org/10.1016/j.nexus.2022.100076.

47 https://www.nrdc.org/stories/flint-water-crisis-everything-you-need-know#summary (Last accessed 31.01.25).

48 https://www.epa.gov/enforcement/deepwater-horizon-bp-gulf-mexico-oil-spill (Last accessed 31.01.25).

49 https://www.mmc.gov/wp-content/uploads/longterm_effects_bp_oilspil.pdf (Last accessed 31.01.25).

50 https://www.natlenvtrainers.com/blog/article/the-environmental-impact-of-the-deepwater-horizon-oil-spill (Last accessed 31.01.25).

51 https://ntp.niehs.nih.gov/whatwestudy/topics/wvspill (Last accessed 31.01.25).

52 https://dhhr.wv.gov/news/chemical-spill/documents/elkrivermedicalrecordsummary.pdf (Last accessed 31.01.25).

53 https://www.seasmartschool.com/blog/2022/2/17/12-most-polluted-rivers-in-the-world (Last accessed 31.01.25).

54 Dwivedi, S., Mishra, S., and Tripathi, R.D. (2018) Ganga water pollution: a potential health threat to inhabitants of Ganga basin. Environment International, Volume 117, Issue 2, pp. 327–338. https://doi.org/10.1016/j.envint.2018.05.015.

55 https://cpcb.nic.in/wqm/pollution-assessment-ganga-2013.pdf (Last accessed 31.01.25).

56 https://unacademy.com/content/neet-ug/study-material/chemistry/pollution-of-the-ganges/ (Last accessed 31.01.25).

57 Curry, J.M. (2006). Japan remembers Minamata. The Lancet Summit, Volume 367, Issue 9505, pp. 99–100. https://doi.org/10.1016/S0140-6736(06)67944-0.

58 Kessler, R. (2013). The Minamata convention on mercury: a first step toward protecting future generations. Environmental Health Perspectives, Volume 121, Issue 10, pp. A304–A309. https://doi.org/10.1289/ehp.121-A304.

59 https://www.straitstimes.com/singapore/tuas-water-reclamation-plant-on-track-for-2026-opening-despite-rising-costs (Last accessed 31.01.25).

60 https://www.pub.gov.sg/Professionals/Requirements/Used-Water/DTSS/Tuas-Water-Reclamation-Plant (Last accessed 31.01.25).

61 Cheng, H.H., Yu, W.S., Tenget, S.C., et al. (2013). Reclaimed water in Taiwan: current status and future prospects. Sustainable Environment Research, Volume 33. https://doi.org/10.1186/s42834-023-00177-8.

62 https://www.innolux.com/Uploads/CsrReport/Innolux%202021ESG%20Report-EN_f_189497.pdf (Last accessed 31.01.25).

63 WRB. (2022). Progress Report on Reclaimed Water from Municipal Wastewater Treatment Plants in Kaohsiung City. Water Resources Bureau, Kaohsiung City Government, Kaohsiung.

64 Kashiwaya, M., and Annaka, T. (1980). Current research on advanced wastewater treatment in Japan. Journal (Water Pollution Control Federation), Volume 52, Issue 5, pp. 1008–1012. https://www.jstor.org/stable/25040826.

65 https://panorama.solutions/sites/default/files/12_japan_project_brief_yokohamasewage.pdf (Last accessed 31.01.25).

66 https://www.kubota.com/corporatehistory/ourchallenges/japan-water-01/1/ (Last accessed 30.01.25).

67 Andersson, J.L., Bastviken, S.K., and Tonderski K.S. (2005). Free water surface wetlands for wastewater treatment in Sweden: nitrogen and phosphorus removal. Water Science and Technology, Volume 51, Issue 9, pp. 39–46. https://doi.org/10.2166/wst.2005.0283.

68 Ge, C. (2019). The seasonal dynamics of removal of nitrogen in free water surface constructed wetlands in cold climate. Halmstad university, school of business, engineering and science. Master's Thesis Program in Applied Environmental Science.

69 Tonderski, K.S., Arheimer, B., and Pers, C.B. (2005). Modeling the impact of potential wetlands on phosphorus retention in a Swedish catchment. AMBIO: A Journal of the Human Environment, Volume 34, Issue 7, pp. 544–551. https://doi.org/10.1579/0044-7447-34.7.544.

70 CCB (2019). Report - Removal efficiency of pharmaceuticals in constructed wetlands in Sweden. Uppsala, Sweden.

71 https://enviromatch.com/containerized-wastewater-treatment-t-mbbr-system-enviromatch-saudiarabia// (Last accessed 30.01.25).

72 https://emveestech.com/waste-water-treatment-in-riyadh// (Last accessed 30.01.25).

73 https://cpcb.nic.in/upload/home/epa/THE%20ENVIRONMENT.pdf (Last accessed 28.09.24).

74 https://www.cpcb.nic.in/GeneralStandards.pdf (Last accessed 12.01.25).

3

Sewage Wastewater: Sources, Characteristics, and Treatment

Sewage wastewater or municipal wastewater is the liquid discharged from domestic sources like kitchens, washrooms, sanitary facilities, or is generated from commercial institutions. Sewage wastewater comprises of a black, gray, or yellow coloration. It generally constitutes 99.9% water and the remaining organic and inorganic solids, either in a suspended, dissolved, or colloidal state. The generation of the sewage is influenced by various parameters like climate, water availability, water pressure, population dependent on water, and water supply systems. The quality assessment of sewage is done by analyzing its solid content, presence of elements like nitrogen and phosphorus, indicators suggesting the type of contamination, and total organic matter. The main micro pollutants contaminating water are bacteria, viruses, protozoa, and helminthes, along with inorganic and organic load. Their invasion and escalation in the water bodies can lead to several damaging effects both to the environment as well as living species. The harmful effects on living systems range from physical ailments to genetic alterations. Bacterial infections from water sources can cause typhoid, cholera, and tuberculosis, while protozoan invasion can lead to dysentery. Sewage wastewater is directly linked with life-threatening diseases like pneumonia, encephalitis, meningitis, hepatitis, paralysis, and liver and intestinal damage. The negative impacts of these microorganisms severely affect human survival when present in water bodies in large concentrations. Owing to the existence of a great number of microbial species, organic impurities, and inorganic contaminants in sewage water, it becomes important to assess the characteristics of sewage wastewater and adopt suitable treatment methods to eradicate the harmful effects associated with it.

Oxygen is needed to support the biological degradation of certain contaminants. Sewage wastewater or municipal wastewater is loaded with oxygen-demanding substances. Organic matter and ammonia are usually associated with sewage water. These substances are degraded into simpler forms by the consumption of oxygen, but this decreases the level of oxygen essential for the sustenance of aquatic life. Infectious microbes or pathogens may seep into the ground through sewage from cities. Humans and other living organisms may come in contact with these either by drinking contaminated water or through activities such as swimming, fishing, etc. [1].

Sewage is associated with a large amount of nitrogen, carbon, and phosphorus. Conventional procedures do not eliminate them, which overstimulates water plants and may cause interference in water treatment systems, develop disagreeable taste and odor in drinking water, or can result in algal blooms, followed by eutrophication. Sewage water is also associated with various inorganic and synthetic organic chemicals, which are toxic to fish, aquatic life, and humans, even at very low concentrations [2].

Since sewage water gets contaminated each day, assessment of sewage wastewater quality is of utmost importance. Its analysis is crucial for the well-being of the public as well as the environment. Waste management always succeeds wastewater characterization. Thus, sewage water must be analyzed for its characteristics to evaluate the quality of sewage in terms of the variety and amount of contaminants present in it. The analysis of sewage wastewater is essential because:

1. The analysis helps to understand the quality of the sewage in terms of types of impurities and their relative strength.
2. The analysis indicates the possible hazards associated with the sewage water and pollution potency.
3. It helps to draw conclusions on the kind of treatment facilities to be acquired and also to decide about the type of process to be followed for decontamination.
4. It helps to choose the size of the operational units and requirement strategies, both for chemicals as well as for equipment.
5. Analysis of wastewater is usually followed both before and after subjecting the wastewater to various treatment processes.
6. The water quality assessment after the treatment helps to ascertain whether the required standards of sewage discharge have been attained.
7. It helps to understand the feasibility of the process and the efficiency of the treatment strategies and methodologies employed for treating sewage wastewater.
8. It forms an integral part of the wastewater management procedures and is employed as a tool for assessing the quality of the effluent obtained after treatment, in addition to the fulfillment of discharge standards.

Based on characteristics of sewage, various treatment methodologies can be adopted, which decide the extent of treatment and evaluation of waste disposal. The characteristics of sewage wastewaters vary from place to place, as these depend upon the lifestyle of the society living in the region. Amount of daily water usage, type and nature of water supply, quantity of water supply, and sewerage system conditions are some of the deciding factors of the quality of sewage wastewater generated. The analysis of sewage water comprises investigating the physical, chemical, and biological traits of sewage water.

3.1 Characteristics of Sewage Wastewater

3.1.1 Sewage Physical Characteristics

Physical traits of sewage wastewater include:

1. **Color and Odor:** The color and odor tell about the longevity and strength of the sewage. Both of these give information regarding the state of sewage water. These also indicate the freshness of the sewage wastewater. Fresh sewage water is usually cloudy in appearance, which becomes dark in color, with the emerging foul and pungent smell of hydrogen sulfide, as it ages. The analysis of color and odor is a preliminary examination indicating the biochemical modifications occurring in the sewage wastewater. The temperature of the sewage water is crucial in bringing about the change in color and odor of the sewage water. Introduction of other contaminants intensifies the color and the odor of the sewage wastewater.
2. **Temperature:** On average, the temperature of the sewage wastewater is around 20 °C. At this temperature, about two to six hours is sufficient to convert the fresh sewage into stale or septic sewage water. Temperature is assessed to ascertain the availability of oxygen in wastewater and the efficiency of sedimentation and biological treatment systems. Sewage water usually has a higher temperature as it contains processed water from various facilities. It influences the solubility of oxygen, aeration capacity, biological activity, and efficiency of the sedimentation process.
3. **Turbidity:** The load in sewage wastewater increases due to the presence of fecal matter, organic waste, oils, greases, soaps, paper, etc. The suspended and colloidal solids existing in wastewater introduce turbidity in sewage water, which can be detected using a turbidometer.
4. **Solids:** Solids present in sewage wastewater are highly putrescible. Also, they require careful disposal strategies. The total solid matter present is known as "total solids." It may be either suspended, dissolved, colloidal, or settleable solids. The solids can be categorized as mentioned in Table 3.1.

The different types of solid categories that can be present in sewage wastewater have been described below:

1. **Total Solids:** Total solids in mg/L are determined by evaporating the given volume of sewage sample at 103–105 °C and weighing the dry residue left.

Table 3.1 Types of solids.

Type of solids	Particle size	Nature of solids
Suspended solids	Solids >1.0 μm	Suspended solids remain floating in water.
Colloidal solids	Solids in the range from 1.0 to 10^{-3} μm	These are the finely divided suspensions present in water.
Dissolved solids	Solids <10^{-3} μm	These are present in dissolved form in sewage.
Settleable solids	Solids >10 μm	These solids settle down when sewage water is kept undisturbed for some time.

2. **Suspended Solids:** These are nonfilterable. Suspended solids (in mg/L) are determined by filtering the sewage water through a 1 μm filter unit and weighing the dry residue left.
3. **Filterable Solids:** Filterable solids can be calculated by the difference between the suspended solids and total solids. These also include the dissolved and colloidal solids.
4. **Organic and Inorganic Solids:** In order to examine the amount of organic and inorganic solids, the dry total solid is burnt and ignited at 550 °C in a muffle furnace, for 12–20 minutes, followed by weighing. The weighed solid represents fixed or inorganic solids. The decrease in the weight due to ignition represents the organic solid fraction of the total solids.
5. **Settleable Solids:** Settleable solids (in mg/L) can be estimated by placing sewage in an especially designed Imhoff tank (conical container), undisturbed for two hours. The quantity of the settleable solids is determined after decantation of the settled solids, which can be dried and weighed.

Information about solids in wastewater is essential as the equipment size is decided based on the quality of the inorganic and organic settleable solids. In addition to this, the design of biological treatment units and disposal management can be done with proper information about the solid content in wastewater.

3.1.2 Chemical Features of Sewage

The chemical characteristics of the sewage wastewater can be ascertained based on the following parameters [3]:

1. **pH:** The hydrogen ion concentration is a valuable variable in the sewage water analysis. The pH of sewage wastewater usually varies in the range 5.5–8.0. Acidity and alkalinity of wastewater are important, as the efficiency

of various treatment methods depends upon them. The acidity or alkalinity of wastewater helps in deciding the type of reagents and amount of chemicals essential to carry out the decontamination and wastewater treatment. Fresh sewage is generally alkaline, while due to bacterial action and the formation of acids, the pH value drops. The knowledge of pH is thus indicative of the chemical characteristics and changes taking place in the sewage wastewater samples.

2. **Chlorides:** The strength of sewage is determined by the presence of chlorides. The presence of a large amount of chloride content indicates infiltration of saline water into the sewage system. Chloride presence also specifies the discharge of the chloride-bearing waste from various industries into wastewater. An average sewage flow of 150 liters per capita per day results in 150 mg/L of chloride content from the normal water supply.
3. **Nitrogen:** The sources of nitrogen in sewage water are proteins, amines, amino acids, and urea. In sewage water, nitrogen can be present as ammonia, nitrites, nitrates, organic nitrogen, and albuminoidal nitrogen, and indicates various stages of decomposition of nitrogen-containing compounds. Ammonia, nitrogenous substances like ammonium hydroxide and ammonium salts, indicate an early decomposition stage of sewage. Nitrites are formed by the partial oxidation of nitrogenous wastes, while the formation of nitrates indicates complete oxidation and stability of wastewater. The untreated sewage has a nitrogen concentration of 20–50 mg/L, which is measured as TKN (Total Kjeldahl Nitrogen). TKN measures the content of organic nitrogen and ammonia. Phosphorus in raw sewage is found as phosphates, which are in the range of 5–10 mg/L.
4. **Hydrogen Sulfide:** Anaerobic decomposition of waste in sewage water results in the formation of hydrogen sulfide gas. The presence of hydrogen sulfide is not only offensive but also has a corrosive effect on the concrete structures of sewage treatment units.
5. **Fats or Oils or Greases:** Domestic sewage water, water from kitchens, restaurants, and garages, contains oils and greases. Their presence can decrease the efficiency of the treatment process. These are among the primary contaminants to be removed from sewage water before subjecting it to other treatment processes.
6. **Dissolved Oxygen (DO):** Dissolved oxygen content is vital for aquatic organisms. The treated sewage wastewater must have at least 4 mg/L dissolved oxygen before it is discharged into the water bodies. Dissolved oxygen is determined by iodometric or Winkler's method. It is an oxidation-reduction process carried out chemically to liberate iodine, which is equivalent to the oxygen present in water. Dissolved oxygen measurements are vital for aerobic treatment processes, intended to purify sewage wastewater.

7. **Biological Oxygen Demand (BOD):** Wastewater treatment involves the biological degradation of the organic load present in sewage. The biologically inactive matter present in sewage cannot be oxidized by bacteria. The oxygen amount required for the aerobic decomposition of biologically active organics present in the sewage is called as biological oxygen demand. This directly indicates the amount of organics associated with sewage wastewater. The oxidation process depends upon various factors, including the amount and type of organic load, temperature, types of microbes involved, and the presence of oxygen. BOD is also an indicator of organic pollution and helps to choose the required treatment and disposal method during sewage treatment. Aerobic decomposition proceeds in two stages. In the first stage, demand for oxygen arises due to the conversion of all carbonaceous matter to carbon dioxide and water. It is also termed as carbonaceous demand. In the second stage of oxygen demand, oxidation of nitrogenous matter takes place over a prolonged period. As the complete degradation takes an indefinite time, thus, practically a reaction period of five days at 20 °C is considered for expressing the BOD. The BOD of sewage can be determined as per Equation 3.1:

$$\text{BOD}_5 \text{ at } 20\,°\text{C (mg/L)} = \frac{(D_1 - D_2)}{f} \tag{3.1}$$

   where
   D_1 is the dissolved oxygen of diluted sewage before incubation
   D_2 is the dissolved oxygen after five days incubation at 20 °C
   f is the dilution factor, and is the ratio of volume of diluted sample to volume of undiluted sample
8. **Chemical Oxygen Demand (COD):** This gives an information about the oxygen requirement for chemical oxidation of organic matter, whether biologically active or inactive. In this regard, the most common oxidizing agent used is potassium dichromate or potassium permanganate. This analysis of sewage wastewater is a rapid test and gives an information about the cumulative oxygen demand. Raw sewage has about 100–400 mg/L, of BOD while, COD is between 200 and 700 mg/L.
9. **Relative Stability:** It is the ratio of dissolved oxygen available in sewage or waste, to the total oxygen needed to completely satisfy the carbonaceous stage of BOD. It gives a rough idea about the quality of the sewage.

3.1.3 Biological Features of Sewage

The biological characteristics of sewage include the presence of living species and several microbial populations like viruses, bacteria, protozoa, rotifers, fungi, etc. Few of these microbes are beneficial in the wastewater treatment plants, as they contribute in the stabilization of the organic matter. Total bacterial count is sometimes determined to know about the presence of toxic industrial waste which affects the performance of biological treatment of wastewater. Also, analysis of pathogenic microbes can help to design proper treatment methods. Extensive knowledge is required to analyze and recognize the amount and type of microbes present in the sewage wastewaters. The presence of organic load in sewage makes it highly putrescible and biodegradable. Sewage has intestinal bacteria like *Fecal coliforms*, *Clostridium perfringens*, *Fecal streptococci*, and many other pathogenic microbes responsible for causing health related risks like dysentery, typhoid and diarrhea. Protozoa like *Giardia*, *Balantidium coli*, fungi like *Fusarium*, *Sporotrichum* are also associated with sewage. The presence of few microbes is required or is beneficial at the later stages of biological treatment of sewage wastewater. A detailed description on the effects and removal of pathogenic organism from wastewaters is discussed in another chapter.

3.2 Treatment of Sewage Wastewater

Sewage treatment is a multistage process aiming at the restoration of wastewater. Sewage treatment is required to make water free from contaminants, which are offensive and are of toxic nature. This is not only needed for safe disposal of water but is an important part in view of reuse and recycling of water. Usually for sewage treatment though a variety of combination of techniques are possible (Figure 3.1), yet standard procedures are usually followed.

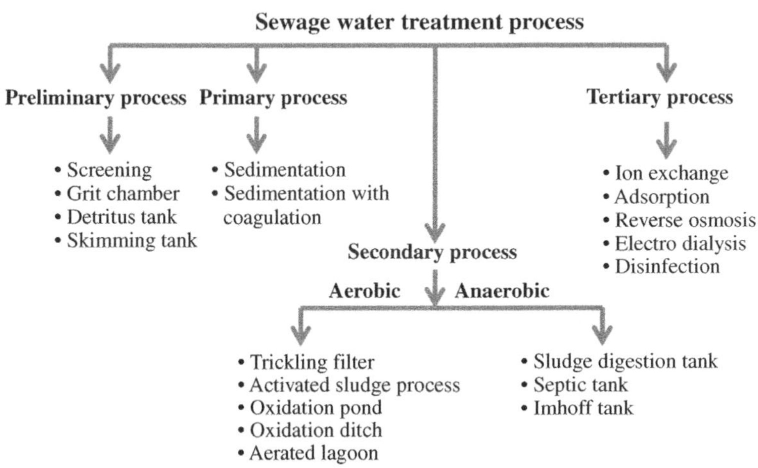

Figure 3.1 Sewage wastewater treatment.

The sewage treatment methods are mainly classified as below:

1. **Preliminary Treatment:** The first step involved in sewage wastewater treatment is the preliminary treatment. It is carried out for the removal of floating material, leaves, rags, papers, dead species, plastics, cans, and heavy settleable materials, including oily impurities. The process reduces the BOD to 30%.
2. **Primary Treatment:** Primary treatment involves the removal of the solids present in the sewage water. It makes use of the sedimentation process wherein the sewage is stored in specially designed basins, allowing the settling of suspended particles. The supernatant liquid and sediment sludge are directed to subsequent treatment units for further purification.
3. **Secondary Treatment:** The treatment involves the use of microbes to remove finely suspended and leftover solids from sewage, either aerobically or anaerobically. It is also called as biological treatment process. The process mainly converts the complex organics into simpler products. The biological unit that is used to carry out the treatment process of sewage wastewater is termed an aerobic biological unit and includes filters, aerated tanks, oxidation ponds, etc. The process also covers the anaerobic way of decomposition, where anaerobic biological units are employed. It reduces the BOD to about 95% and the bacterial load to 98%.
4. **Tertiary Treatment:** Tertiary or final treatment is an additional and advanced treatment executed to make the water fit for reuse in various activities [4]. The final treatment is usually performed to make water fit for drinking or to ensure its suitability, before disposal into any water body. Thus, the sewage treatment is confined to the biological treatment process only.

3.2.1 Preliminary Treatment

3.2.1.1 Screening

This is the preliminary step for removing larger contaminants like leaves, rags, paper, wood, dead organisms, etc., from water, using screens or bars (drum screen, bar screen, cutting screens, band screens).

Screening is essential to be carried out, as a preliminary step, to avoid any damage to pumps and equipment parts, and to avoid interference in the efficiency of other successive treatment processes. Screens consist of parallel bars or rods, wire meshes, or perforated plates of variable pore sizes and shapes. Usually, rectangular-shaped medium screens are used in sewage treatment units. The screens are placed at an inclined position, at an angle of 30–60°, in the direction of flow of sewage. This helps proper and convenient collection of debris and also keeps the screen free from clogging. Also, the slope of the screen decreases the flow velocity of the incoming sewage water and makes screening more effective. The design velocity is very crucial at this stage [5].

3.2.1.2 Grit Chamber

The sewage water which is usually contaminated with soil, grit, or road debris, is allowed to pass through a channel in a grit chamber at a specific velocity (velocity grit chamber), thereby settling down the inorganic substances. Grit chamber intends to remove heavy and hard inorganics from sewage water, in order to avoid wearing out of pumps and pipes due to grit and sandy materials. The presence of these can interfere with the digestion of sewage and can reduce the capacity of the digestion tank, the manure value, and the fuel value of the digested sludge. The removal of these materials at this stage is thus essential before sewage water is pumped or sent to the sedimentation tanks. The flow velocity is usually in the range of 0.5–0.9 m/s, which helps to avoid forced screening through the bars, and is neither too slow to let the grit settle down. The grits are periodically removed and dumped on wasteland for disposal (Figure 3.2).

3.2.1.3 Detritus Tank

In addition to the grit chamber, tanks known as detritus tanks are employed, where grit and fine sand particles, inorganics, and some organics are eliminated from sewage water. This differs from a grit chamber in terms of a slightly longer detention period and small horizontal flow speed.

3.2.1.4 Skimming Tanks

In addition to the grit chamber, skimming tanks are employed to get rid of any oily or greasy materials associated with water, especially from kitchen and restaurant outlets.

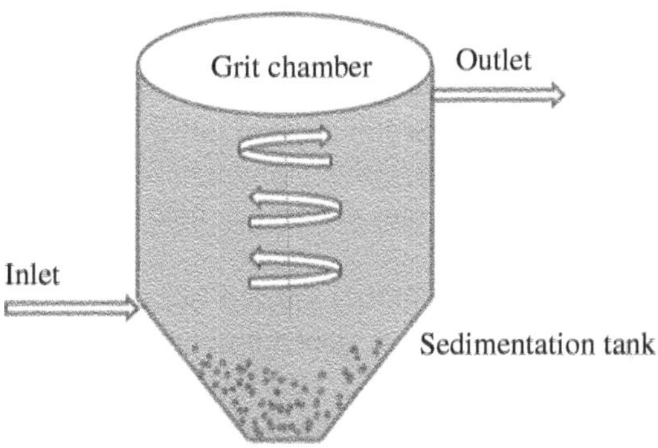

Figure 3.2 Grit chamber.

Sewage from kitchens and restaurants, garages, and oil refineries includes fats, fatty acids, oils, waxes, etc. They interfere with the working of treatment units and may form odorous scum, both in operational units and at discharge points. The presence of such substances inhibits the microbial growth required for aerobic decomposition and interferes with the activated sludge process. Such materials are thus removed using skimming tanks. These are narrow, three-compartment rectangular tanks through which the compressed air can be introduced upwards. The principle involved in the process is the coagulation and congealing of the greasy and oily material present in the sewage. Manual or mechanically driven removal of the rising material can be ensured from the stilling compartments [6]. This step is usually followed by the sedimentation process.

3.2.2 Primary Treatment of Sewage Wastewater

3.2.2.1 Sedimentation

When the impurities in the sewage are too fine or too heavy to be removed by the preliminary operations, sedimentation tanks are employed. The process is executed to:

1. Remove all the settleable organics present in the sewage, accounting for its BOD.
2. Prevent any excessive load on the subsequent treatment units and clogging of filter units.
3. To avoid any type of interference caused by the suspended impurities to the aeration tendency in biological treatment units.
4. The process essentially increases the performance of the downstream processes.
5. It aids in preremoval of hazardous suspended contents of sewage.
6. The goal of the sedimentation process is to remove any type of sludge accumulation when sewage is discharged into water bodies.

Sedimentation tanks are the units where the sewage water is kept undisturbed in large settling basins or primary settling tanks. Sedimentation tanks can either be batch type or continuous flow type. These can be either circular or rectangular in design. All types of sedimentation tanks have four zones, i.e., inlet zone, settling zone, outlet zone, and sludge zone.

In sewage treatment plants, sedimentation is mostly carried out twice, viz, primary sedimentation and secondary sedimentation, depending upon whether it is performed before the secondary process or after the biological process. Sedimentation can either be simple or chemically aided sedimentation. The principle involved in the sedimentation process is the difference in the specific gravity of the solids present in the sewage water and the liquid. Solids being denser settle down by gravity, under undisturbed conditions. The tank in which the flow of sewage water is stopped is called a sedimentation tank, and the time required for the settlement of the solids present in the sewage water is called detention time or detention period. In this step, most of the organics and fine suspensions are allowed to settle down under gravity in tanks, by retaining sewage wastewater in them for about four to twelve hours. The process ensures 60% removal of suspended solids, while only 40% of BOD can be eliminated. The sedimentation process depends upon the viscosity of wastewater, or the temperature at which the treatment process is being carried out. The velocity of wastewater flow also governs the precipitation rate of the suspended impurities associated with the sewage wastewater. The size, shape, and specific gravity of the components of sewage water also affect the rate of settlement.

3.2.2.2 Sedimentation with Coagulation

Sewage water is composed of various colloidal and finely dispersed solids that cannot be eliminated by the plain sedimentation process. They are thus required to be accumulated as settleable particles by making use of certain chemicals known as coagulants. This type of treatment is called sedimentation with coagulation, chemical-aided sedimentation, or chemical precipitation. It mainly comprises of three operational units [7] which are:

1. **Chemical Coagulation Unit:** These units involve the process of coagulation. The process involves feeding the appropriate nature and amount of chemicals known as coagulants into the tank for coagulation to occur. These, when added to water, form insoluble gelatinous precipitates, which absorb and entangle the non-settleable fine solids. Coagulation is the process in which the finely suspended charged particles are destabilized by use of chemicals like ferrous sulfate, aluminum sulfate, polyelectrolytes, alum, sodium aluminates, lime, and soda, etc. Iron salts like ferrous sulfate, ferric sulfate, and ferric chloride are among the chief coagulants employed in sewage wastewater treatment. This unit is also equipped with a stirrer for efficient mixing of the sewage with the chemicals.
2. **Flocculation Unit:** This unit provides the agglomerated particles sufficient time to get compacted into settleable solids, termed as flocs. Slow and gentle mixing taking place in the flocculators in the flocculation unit brings about the necessary changes.
3. **Sedimentation Unit:** This is the last section of the coagulation cum sedimentation tank, wherein the flocculated matter settles down, to give clear liquid. The chemically aided process is not implemented in modern treatment plants, as secondary treatment units facilitate the functions performed by the coagulants and are quite efficient and effective. Also, the chemicals at this

stage may destroy the effective microbes needed for sludge digestion, and may increase the total dissolved solid content of the sewage, including the cost of treatment. But, chemical treatment has an advantage over the simple sedimentation process, in terms of greater BOD removal, less space requirement, and efficient removal of phosphates.

3.2.3 Secondary Treatment of Sewage Wastewater

3.2.3.1 Aerobic Biological Treatment

Sewage wastewater after the primary treatment is subjected to a biological treatment process, which mainly involves the oxidation of the suspended impurities [8]. Common approaches for oxidation are trickling filters, aeration tanks, oxidation ponds, and the activated sludge method [9]. The most effective biological process for the removal of dissolved organics is the aerobic process, as it is fast as well as results in the production of inoffensive by-products like carbon dioxide and water (Equation 3.2).

$$\text{Organic matter} + \text{Bacteria} + \text{Oxygen} \rightarrow \text{Biomass} + CO_2/H_2O/NH_3 \qquad (3.2)$$

The biological treatment process is of two types, attached growth system and suspended growth system. Table 3.2 illustrates the major features of the two processes.

The process of separating suspended and dissolved impurities from the sewage by passing through a porous medium is called filtration. The media includes silica, anthracite coal, garnet sand, metal fabric, granular carbon, etc. Below is the sequence of different processes carried out when sewage is filtered through the media:

1. **Mechanical Straining:** It refers to the lodging and removal of large particles that have a size larger than the filter media pore size. The particles are mainly arrested in the upper few centimeters of the filter media.
2. **Sedimentation:** The pores of the filter media act as small sedimentation tanks. Here, the smaller particles settle down or get adsorbed and cling to the media grain surface, due to the force of attraction between the media and the sewage particles.
3. **Biological Action:** The organics associated with sewage along with the microbial species like bacteria, form a layer or film over the filter media. The layer is called "Schmutzdecke," which is the home for microorganisms. This layer is a biological layer, with trapped matter, on the surface of the slow sand filter. The organic matter present in sewage acts as the food for the microbial population surviving on the media. The bacterial species carry out the biological degradation of the organics present in the sewage water.

Table 3.2 Types of biological treatment processes.

Biological treatment process	
Attached growth system	**Suspended growth system**
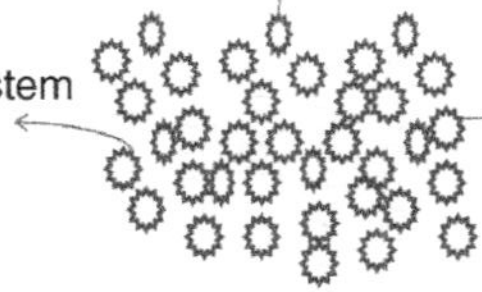	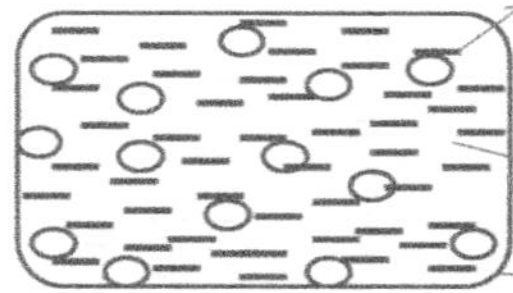
In the attached growth system microbial population remains attached to the support	In a suspended growth system microbial population is suspended in the liquid medium
Inert supports like sand or gravel are used	The sewage wastewater is the only liquid medium, without any solid support
It includes trickling filters	It includes activated sludge process, oxidation ponds, lagoon etc.
It constituents a heterogeneous system	It constitutes a homogenous system
The method is less effective	It is more effective for pathogenic removal
More stable system	It is less stable as is subjected to shock loading
Biofilms develop on the media which eventually falls off after thickening	Microorganisms are retained in the flocs forms and can be retrieved for further treatment

Source: Adapted from [10].

4. **Electrolytic Action:** Charges on the filter media and sewage particles are neutralized by each other. The charged particles are thus removed from sewage due to neutralization.

3.2.3.1.1 Trickling Filters

Trickling filters are commonly used in biological treatment units. These are also called sprinkling or percolating filters. They can be high-rate filters, low-rate filters, intermediate-rate filters, or roughing filters, depending upon the organic load and hydraulic parameters. The trickling filter system is an aerobic sewage wastewater treatment method in which the microbes attached to the filter media remove the complex organics from sewage wastewater [11]. The basic components of a trickling filter system include:

1. A filter medium over which the microbial species grows, using the organic matter present in the wastewater.
2. A concrete enclosure for the filter media (circular, rectangular or square shaped).
3. A sewage wastewater distribution system (fixed or moving nozzle) along with a provision for artificial or natural ventilation.
4. A system for the removal and disposal of sludge produced during the treatment process.

Trickling filters are made up of bed of stones, brick pieces, gravel, slag, sand, and a wide range of plastic, synthetic materials and impervious rocks with proper channels for air entry. Sewage is allowed to distribute and percolate from the top of the filter. The high specific surface area of the media materials provides a large space for the biofilm formation. The filtration of sewage from the bed involves the consumption of organics from sewage wastewater by the slimy microbial population flourishing on the impervious bed (Figure 3.3).

The biological film usually is of 0.1–0.2 mm thickness and has both aerobic and anaerobic layers. The aerobic bacteria present on the upper portion of the slime layer utilize oxygen for the degradation of organics present in sewage and generate water and carbon dioxide. With time, the microbial population of aerobic bacteria, facultative bacteria, anaerobic bacteria, ciliates, protozoans, and micro fauna increases in the slime layer, thereby increasing its thickness. When the sewage water trickles down, the microbial species present in the lower zone of the slime layer do not get a sufficient amount of organics and oxygen, and they starve. This makes the microbial species consume their own cytoplasm matter, and they are removed from the slime layer. This process is called sloughing of the filter. The sloughed off film and the treated sewage water are collected through an underdrainage system and are directed to a subsequent clarification tank.

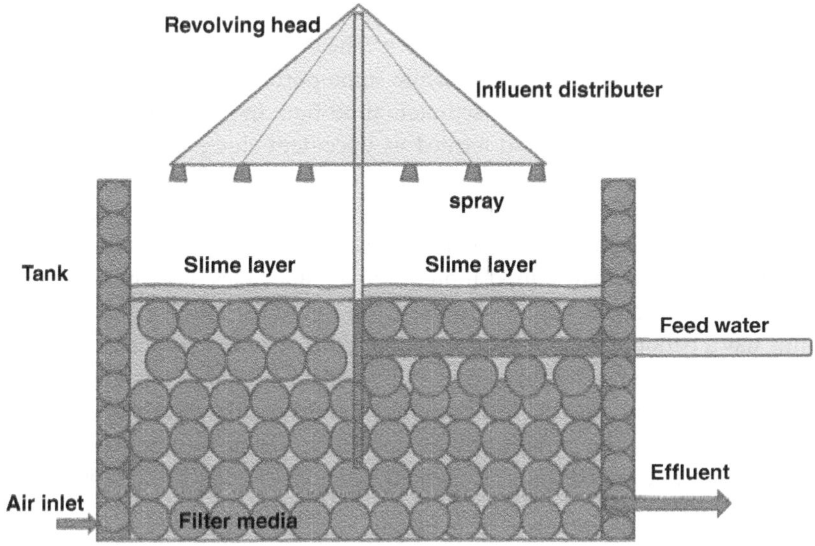

Figure 3.3 Trickling filter.

The efficiency of the trickling filter is estimated in terms of the reduction of BOD. The trickling filter unit reduces the BOD by 95% and 80% of the suspended impurities. It is possible to achieve the 20 : 30 effluent standard (BOD less than 20 mg/L : less than 30 mg/L of suspended solids), along with lowered nitrification, and quick recovery to changes in the influent BOD. It is associated with the problems of biofilms in addition to sedimentation of particles within the distributor arms, large area requirement, and fly and odor nuisance. The efficiency of the trickling filter can be enhanced by recirculation system. It is the method in which the effluent from the trickling filter is redirected to the filter unit, which increases the contact time of the waste with the microbes. This also ensures continuous and uniform sloughing of the growth. It also aids in ventilation in the system [12].

Rotating biological contractors are another type of attached growth systems working on the principle the same as that of trickling filters, except the involvement of rotating media and stationary sewage water.

3.2.3.1.2 Activated Sludge Process

Activated sludge process is an aerobic biological treatment process in which the suspended microorganisms decompose the sewage organics. In the activated sludge method, active microbes in high concentrations (2000–8000 mg/L, bacteria, protozoa, and fungi) are added to the sewage along with aeration and agitation, due to which the oxidation of the organic solids happens. The process is carried out in an aeration tank where the active biomass (bacteria and protozoa) metabolizes the organics by using dissolved oxygen. The organic matter partly gets oxidized into carbon dioxide and water, while some portion of it forms new cells. The new cell formed settles in the sedimentation tank in subsequent processes. The settled biological floc is also termed as flocculent sludge. A small portion of the flocculant sludge or activated sludge is redirected to the aeration tank for remixing with the influent. The main function performed by the recycled sludge is to grow the microbial population which can accelerate the organic removal process or BOD. A complete decomposition of the organic takes place in suspended form. This ensures better contact between the microbial species and organics and thus has a better efficiency than trickling filters. The key processes involved are adsorption, carbonaceous oxidation and nitrification due to assimilative and endogenous microbial growth. The coagulated and flocculated matter settles to the bottom which can be periodically removed. The liquid so produced is clear with no odor. The activated sludge treatment assembly (Figure 3.4) consists of:

1. Aeration tank with microbial suspension
2. Activated sludge recirculation system
3. Aeration system facility
4. Sludge disposal facility and
5. Secondary sedimentation tank (clarifier)

The activated sludge processes can be of the following types:

1. **Conventional Activated Sludge Process:** This is one of the most widely employed processes for the biological treatment of sewage wastewater. The process begins with the entry of the air into an aeration tank loaded with sewage. The supply of oxygen is maintained either by mechanical or diffused aeration. This is followed by a settling tank (clarification tank) where the biological floc settles down and clear supernatant liquid is formed. The

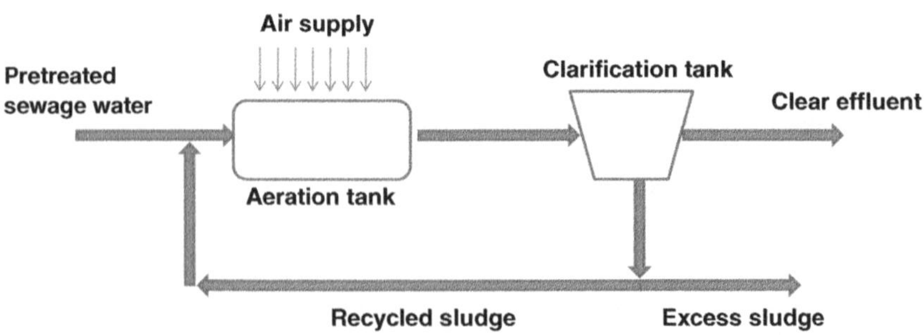

Figure 3.4 Activated sludge process.

excess sludge is either directed to ultimate disposal, or is reused in aeration tank as returned sludge. This reentry of the sludge ensures availability of the microbial population for degradation purpose. The process involves oxidation of the carbonaceous matter and nitrogen containing substances. The main factors influencing the activated sludge process are the loading rate of sewage, flow pattern and mixing regime. The conventional method helps to reduce the BOD level to about 85–92%.

2. **Modified Activated Sludge Process:** In order to meet the specific treatment objectives several modifications have been done in the activated sludge process. The basis of modification depends on the aeration time, way of introducing the sewage in treatment tanks, degree of mixing, and concentration of active biomass. A few modified activated sludge processes are given below:
   a. **Step Aeration:** It involves the addition of settled sewage at various locations along the aeration tank length, and the addition of sludge at the entry of the aeration tank (Figure 3.5). This modified activated sludge process helps to generate more uniform BOD throughout the aeration unit.
   b. **Tapered Aeration:** This is a modified form of treatment in order to achieve an optimum utilization of air. In this, compressed air is supplied near the inlet of the aeration tank and then gradually reduced towards the end of the tank (Figure 3.6). The pressure of the introduced air is slightly higher at the inlet.
   c. **Contact Stabilization:** In contact stabilization activated sludge process re-aeration of the returned activated sludge from the secondary clarifier is carried out. It involves the use of an additional re-aeration tank in which the effluent from the secondary sedimentation tank is re-aerated for three to six hours and then directed to the contact aeration tank for efficient mixing. This is usually employed for sewage with a high organic

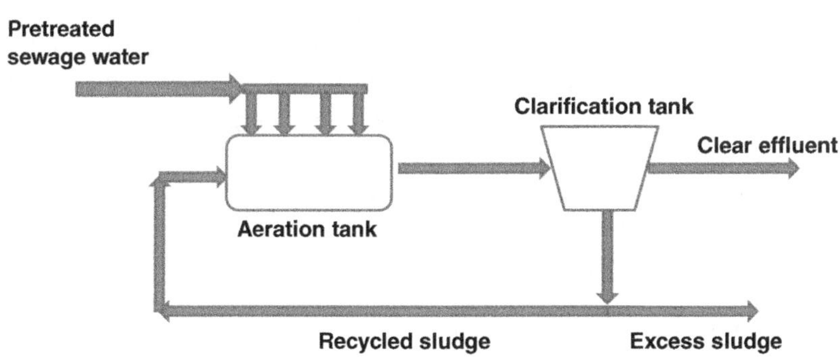

Figure 3.5 Step aeration process.

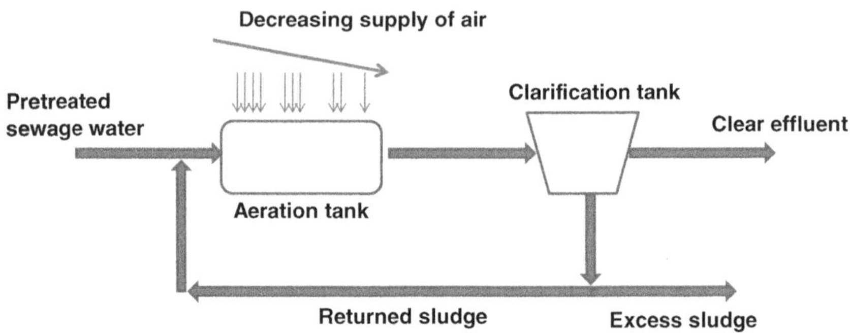

Figure 3.6 Tapered activated sludge process.

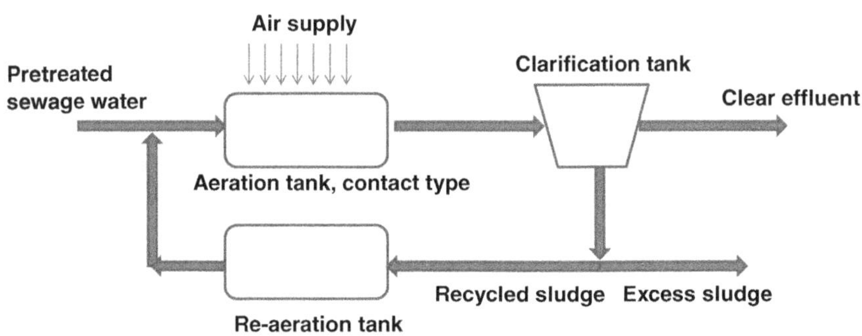

Figure 3.7 Contact stabilization activated sludge process.

load. It is seen as an efficient method for the removal of colloidal and suspended matter (Figure 3.7).

d. **Complete Mixing or High-rate Aeration Process:** It involves the complete mixing of sewage either by distributing the sewage and return sludge from one side of the tank, and withdrawing the aerated sludge from the opposite side, or by using mechanical aerators at the center of the tanks. In this method, the food-to-microorganism ratio is high, along with high mixed liquor suspension sludge. Though it suffers from the disadvantage of difficult sludge settlement, it is one of the widely accepted activated sludge processes. This method reflects an increased operational stability and treatment efficiency, even for toxic biodegradable pollutants.
e. **Extended Aeration:** This is similar to the complete mixing aeration process, except that the primary sedimentation tank is omitted. This scheme can be followed for sewage water treatment of small communities with

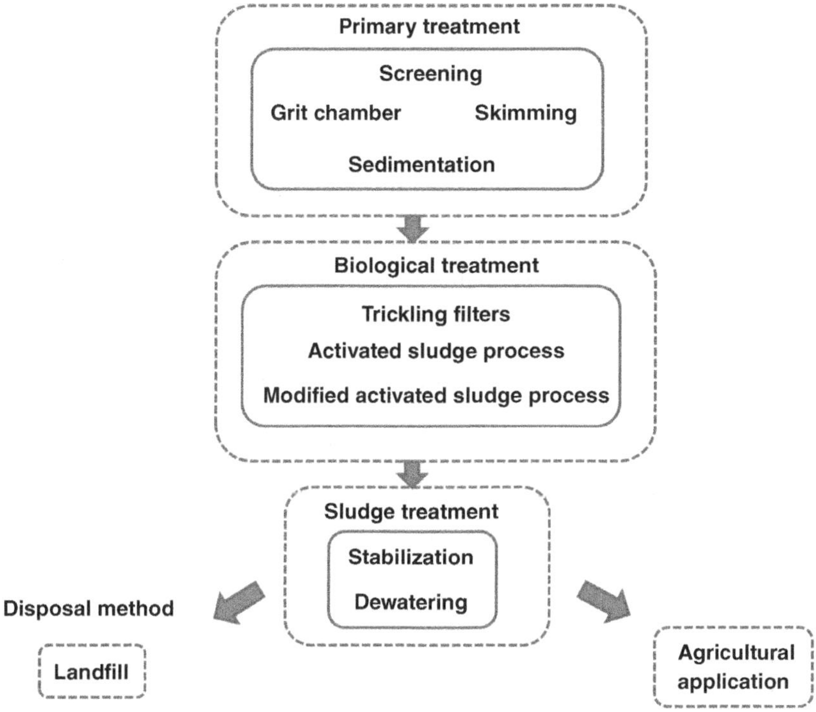

Figure 3.8 Biological treatment process.

flow rates below 0.1 million gallons per day. The process is characterized by low organic load, greater aeration time, high mixed liquor suspended solids, and low food to microorganism ratio. Because of the long retention time, mixed liquor suspended solids are efficiently stabilized. The sludge production in this type of modified activated sludge process is less and does not require post-digestion treatment. The BOD removal efficiency is up to 95–98%.

Activated sludge process is comparatively better and efficient than other physical and chemical treatment methods in terms of cost, secondary pollutant formation, and large-scale implementation. Figure 3.8 illustrates a complete flow sheet of various processes employed in the sewage treatment.

The biomass collected from either the trickling filters or the activated sludge process represents a good volume of organic load, and its removal is done in secondary sedimentation units. The sludge obtained is secondary sedimentation and differs from the primary sedimentation tank as it is a fully aerated sludge. This sludge is thus called activated sludge. It is innocuous and loaded with microbial biomass, nitrogen, and phosphorus, which can act as excellent manure on dehydration. Sludge is then subjected to various treatment units to stabilize the organic matter and reduce the volume of the waste generated.

3.2.3.2 Anaerobic Treatment Method

This method is carried out to stabilize the organic load obtained from the secondary sedimentation tank. Sewage sludge is a semi-liquid material with high water content. It is subjected to a sludge digestion tank to lessen its putrescibility, pathogenic contents, and to enhance its dewatering characteristics. Sludge is then stabilized by anaerobic decomposition in sludge digestion tanks. The process involves the conversion of the organics into methane (biogas), carbon dioxide, and water, and the volume of sludge is reduced to a considerable extent. In anaerobic decomposition, the hydrolysis of water takes place, followed by combination with carbon, nitrogen, and sulfur. The end products formed in the process are methane, hydrogen, ammonia, carbon dioxide, hydrogen sulfide, organic acids, etc. The anaerobic treatment involves the breaking down of large organic compounds into simple amino acids, fatty acids, sugar, and glycerol. These substances are then absorbed by the anaerobic bacteria, and simple inorganic substances are released as byproducts, along with sludge. The sludge obtained can either be used as fertilizer or landfilled. The selection of the sludge treatment process depends on factors like site-specific sludge composition, cost involved, and intended fate of treated sludge [13].

For the treatment of sludge, anaerobic digestion is the most preferred method (Figure 3.9). The anaerobic process produces valuable fuel gas like methane along with carbon dioxide. The remaining solid sludge is used either as fertilizer or, in a few cases, also dumped on land.

3.2.3.3 Other Biological Methods

3.2.3.3.1 Oxidation Pond

Oxidation ponds or stabilization ponds are specially designed shallow earthen ponds to treat domestic sewage water [14]. Oxidation ponds can also be used for the degradation of industrial wastewater by providing a proper retention time of 5–50 days. These are constructed partly in excavation and partly in embankment. The pond harnesses the symbiotic interaction of the bacteria and algae flourishing in the pond to stabilize the organic matter present in the sewage water in the presence of sunlight. The sewage organics act as food for the bacteria, which consume or degrade the organic matter by oxidizing it to carbon dioxide, water, and nitrates. The algae symbiotically utilize these products for their proliferation, growth, and to generate oxygen via the process of photosynthesis. It helps mitigate the inorganic components comprising nitrogen and phosphorus. The

Figure 3.9 Anaerobic digester with discharge pipe feeding into a wetland: Jarratt Dairy in Los Lunas, New Mexico. *Source*: United States Department of Agriculture/Wikimedia Commons/Public Domain.

oxygen so produced is taken up by the bacteria for aerobic decomposition of the waste matter present in sewage. In this way, the symbiotic interaction of the microbial species helps in treating the wastewater. The efficiency of the oxidation pond community is based on the organic load present in the sewage. Oxidation ponds are classified into various types depending upon the type of contaminants they are dealing with, the mechanism involved during the purification process, available land area, climatic conditions, and design of the artificially constructed ponds. Oxidation ponds can be:

1. **Aerobic Oxidation Ponds:** These are the stabilization ponds that decompose the organics aerobically. These stabilize the liquid contaminants in the presence of the appropriate amount of sunlight. These are shallow ponds that allow the availability of oxygen through wind action and photosynthesis.
2. **Anaerobic Ponds:** These types of ponds are employed as a pretreatment process for high loads of industrial wastewater and municipal sewage effluents. The depth of the pond is between 2.5 and 5 m, which provides anaerobic conditions for organic degradation. A detention period of about 50 days results in a reduction of 85% of the BOD. But, the process suffers from the disadvantage of a bad odor.
3. **Facultative Ponds:** The sewage water is mostly subjected to the facultative pond for clarification. These earthen ponds have both aerobic and anaerobic zones present, and thus offer better efficiency. The term oxidation pond refers

to facultative ponds. The upper region of these ponds has aerobic conditions wherein the aerobic bacteria absorbs oxygen and degrades the organic matter suspended in the sewage. The symbiotically associated algal species ensure the availability of oxygen by performing the photosynthesis process. The sewage sludge or the settleable particles produced during the aerobic decomposition settle to the bottom, where anaerobic decomposition takes place, and volatile products like organic acids, alcohols, hydrogen sulfide, ammonia, and methane are produced, which move to the top. As these reach the top aerobic zone, they are aerobically decomposed, reducing the odor nuisance. The treated sewage water is discharged through the outlet system. The sludge produced from the oxidation pond acts as valuable manure in the agricultural fields.

The treatment of sewage water by stabilization is influence by various factors like:

a. **Air:** The availability of oxygen is one of the crucial factors governing the aerobic decomposition of the biodegradable pollutants.
b. **Sunlight:** Algal growth is fostered by the consumption of inorganics in the presence of sunlight.
c. **Microorganisms:** Diverse arrays of microbial species like bacteria, algae, fungi, protozoans, insects, larvae, crustaceans, viruses, and nematodes contribute to transforming the complex organics into simpler forms. The symbiotic association between bacteria and algae is the chief mechanism operative in the process. Major species of bacteria dominating the facultative pond include *Achromobacter*, *Thiospirillum proteus*, *Pseudomonas*, and *Rhodothece*. Prominent algal genera thriving in the pond are *Scenedesmus*, *Chlorella*, *Microcystis*, and *Euglena*.
d. **Sewage Water Quality:** The sewage water quality dictates the method adopted for treatment. Sewage water with a high organic load can be treated efficiently by stabilization ponds.

3.2.3.3.2 Oxidation Ditch

Oxidation ditches are the modified activated sludge process units which can be installed for sewage treatment of wastewater of small communities, institutions, isolated installations, and less dense population [15]. This method is profitable and has been successfully used for small land areas where large-sized oxidation ponds cannot be developed. This sewage treatment system evolved in the Netherlands, with a full-scale plant developed in Holland. Currently, more than 9000 municipal oxidation ditches are installed and operational in the United States. Oxidation ditches provide greater retention time to sewage wastewater for eliminating biodegradable matter, in which the complex organics are converted into settleable solids and stable inorganic compounds. These treatment

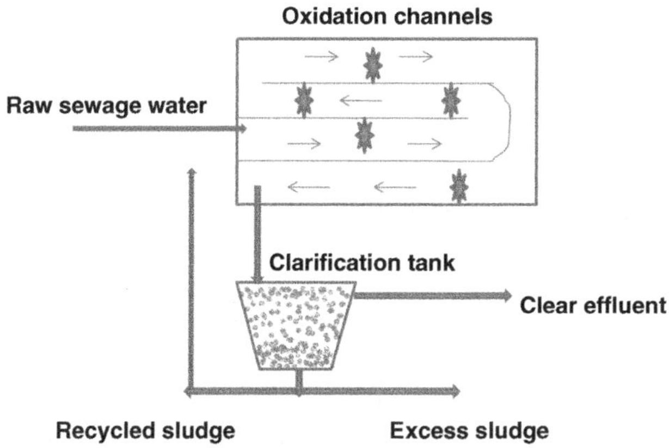

Figure 3.10 A typical oxidation ditch.

units do not require any primary sedimentation setup. The raw sewage is directly fed into long concrete channels or ditches, known as aeration units. These are usually 150–1000 m long, 1–5 m wide, and 1–5 m deep. Each ditch channel is provided with vertical or horizontally mounted rotors, which help in agitation and circulation of sewage, thereby aerating it. A typical oxidation ditch system (Figure 3.10) consists of:

1. Multichannel configuration having an oval or horseshoe shaped basin
2. Screening device (bar screens), and
3. Horizontal or vertical surface aerators or rotors (rotating biological contactors)

The sewage velocity is maintained more than 30 cm/s to keep the solid content in suspension, due to which the greater contact time is provided with the microbial population. The treated sewage is then directed to a separate sedimentation tank where the sludge settles down. The settled sludge is directed to the anaerobic sludge treatment facility. A small quantity of sludge is continuously redirected to the oxidation ditches to maintain good amount of biomass for aerobic decomposition or digestion. A mixture of incoming sewage and returned sludge is termed as mixed liquor. The overall pollutant confiscation efficiency of the unit for suspended solids and BOD observed is 95% and 98%, respectively.

3.2.3.3.3 Constructed Wetlands

Constructed wetlands (Figure 3.11) are artificially created wetlands for treating not only sewage water but also gray water, storm water, and industrial waste effluent. This treatment technique uses natural biogeochemical mechanism and physical transformations to remove organics and nutrients form sewage water,

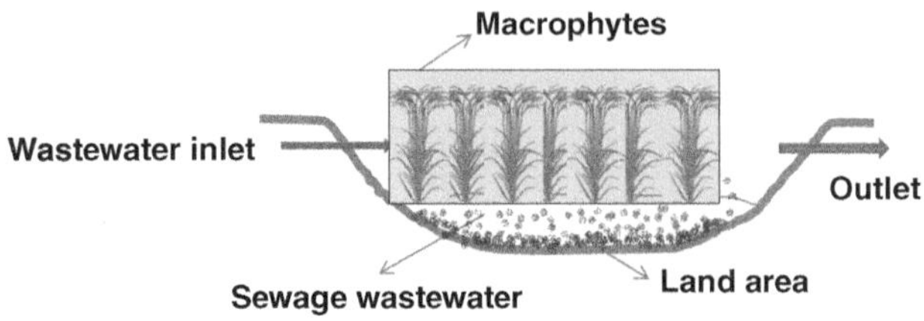

Figure 3.11 A typical constructed wetland.

while simultaneously providing safe environment in a controlled manner. Free water surface systems could be of several categories such as floating-leaved, vegetation-free floating, emergent and submerged, depending upon the type of macrophytes involved in remediation. Several such different systems have been used in different countries. The first free water surface constructed wetland was established in the Netherlands in 1967 [16, 17]. The major components of constructed wetland are:

1. Influent wastewater (sewage or industrial effluent)
2. Substrate layer consisting of gravels
3. Vegetation (macrophytes)
4. Microorganism (array of bacteria, protozoans, algal species)

The decontamination technique involves vegetation, soil and assemblies of diverse microorganism to improve water quality. The contaminant removal is influenced by the climatic conditions and wastewater composition.

The removal of contaminants from the wastewater occurs by a number of physical chemical and biological transformations. Filtration and adsorption are the main operating mechanisms along with microbial action. Both aerobic and anaerobic microorganisms help with the decomposition of complex organics. The nitrogen content present in the wastewater is usually removed by nitrification process and ammonia volatilization. The wetland algal species maintains appropriate medium for nitrification and ammonification process to take place. Phosphorus is precipitated along with iron and calcium compounds present in the medium. The impurities from the sewage are taken up by the macrophytes, but are stored in them temporarily only, as these return back to the water as the plant dies. Constructed wetlands are in operational worldwide including North America, Australia, and Europe. Presently, hybrid constructed wetland are in use, which are especially designed to eliminate complete nitrogen and phosphorous present in the sewage. The hybrid constructed wetlands can be employed for treating not only sewage wastewater but also wastewater generating from compost leaching, textile effluents, pharmaceuticals, pesticide manufacturing plants, winery industry, or fish aquaculture. The energy input in constructed wetland is nominal as they are self-sustaining systems. Also, the operational and maintenance cost is also less in comparison to the other physicochemical treatment methods. Constructed wetland help eliminate the pollutants present in the sewage wastewater and also aid other ecological processes like carbon sequestration, storm water filtration, increased greenery and a diversified habitat. The only limitation is that they have a large area requirement.

3.2.4 Tertiary Treatment of Sewage Wastewater

After the secondary treatment, liquid effluent and solid sludge is obtained. Tertiary treatment is carried out to eliminate non-biodegradable pollutants. Tertiary treatment is very crucial for elimination of highly toxic pollutants. It is one of the most advanced and rigorous methods of treatment. It is followed after primary and secondary wastewater treatments, which aim at obtaining water up to the level of safe discharge into the environment. When the objective of the treatment is to make water reusable for water-intensive applications and drinking water, tertiary treatment is executed. Tertiary sewage water treatment tends to a level of water purification that makes water safe for reuse in various applications, as well as for human consumption.

In tertiary treatment, the removal of phosphorus and nitrogen is carried out to increase the quality of effluents before they are discharged into the receiving waters. It mainly comprises of chemical methods and also includes physical methods. Tertiary treatment is also termed as an advanced treatment process. Treatment options depend upon the characteristics of effluents received after the secondary treatment process and end use [18].

The tertiary treatment is carried out to obtain clean water, safe for discharge in the ecosystem or for reuse. This advanced treatment process is quite use-specific. The sewage water treatment is usually confined to the secondary treatment method. The tertiary treatment is adopted when:

1. The quality of the effluent by the secondary treatment does not sufficiently meet the discharge standards.
2. For efficient removal of nitrogen and phosphorus to avoid eutrophication.
3. When the discharge point of the treatment plant is near the water body, in proximity to the population.

Tertiary treatment may be carried out using:

1. **Disinfection:** Disinfection is the process which involves adding up of oxidants to the effluent from secondary treatment units to kill bacteria and viruses, and to eliminate parasites like *Giardia* and *Cryptosporidium*, which

can cause very serious illnesses. The water disinfectant should be effective in killing the microbes potentially in the contact time available, and should be easy to handle and safe to use. This is performed to protect the downstream users from waterborne diseases. The disinfection process mainly occurs by following steps:

a. Microbial cell wall damage
b. Alternation of the permeability of cell
c. The change in the nature of the protoplasm of the cell
d. Inactivation of the enzyme activity responsible for the metabolic functions of the microbes.

Disinfection of water can be carried out either by physical methods or by chemical ways. The most common way of disinfecting water using chemicals includes chlorination and ozonation. The chlorination process involves the addition of chlorine or its compounds to water in order to oxidize the impurities present in water. Chlorine can be applied either in liquid (sodium hypochlorite) or in solid form (calcium hypochlorite). When chlorine is added to water it forms hypochlorous acid or hypochlorite ions, which act as disinfecting agents.

Nowadays, the use of chloramines is increasing for disinfecting the public water supply. This method of disinfection is known as chloramination. It is much more effective than chlorine and can help protect the quality of water throughout the distribution system. Ozone is a powerful oxidant and a virucide. It is more effective against bacteria than chlorine and requires only a short contact time for efficient action (10 to 30 minutes). Unlike chlorination, ozonation is not associated with the generation of any harmful residues. But, due to high capital cost and maintenance requirements, ozonation is employed sparingly. A typical ozone disinfection system consists of ozone feed preparation, ozone generation, contact with contaminants, and ozone destruction.

2. **De-chlorination:** The chlorinated water contains excess chlorine, which must be removed as it is very harmful to living organisms. It also reduces the water quality when present in large concentrations. Thus, de-chlorination is done by adding sodium bisulfite to water. Chlorine ions are removed, and treated water can then be considered safe for its release into the aquatic system or for reuse.
3. **Coagulation-flocculation:** Chemical coagulation and flocculation are adopted with highly contaminated sewage water, where the secondary treatment method does not give water of the required standard. This technique involves the use of chemicals called coagulants, which help in the neutralization of the charges on the fine suspended impurities, including microbial population, and then agglomeration of these to form heavier flocs,

which can easily be settled down. The choice of coagulant depends upon the nature of the suspended solids to be removed, the conditions of water to be treated, design parameters, and the cost involved in the treatment. Inorganic coagulants like aluminum sulfate, sodium aluminates, and ferrous sulfate are the most commonly used inorganic salts in the treatment. These salts, when added to water, get hydrolyzed and neutralize the suspended particles. The precipitates so formed polymerize to form microfloc and finally heavier floc, which facilitates the settlement of impurities.

4. **Adsorption:** It is the process of transfer of a molecule from the solution onto a solid surface. Adsorption is a surface phenomenon. This method is adopted as a tertiary purification process to eliminate the non-biodegradable contaminants from wastewater by allowing the water to pass through a bed of adsorbent. The impurities are adsorbed on the surface of the adsorbent and are removed. Adsorbents are characterized by the presence of a large surface area, which helps retain the impurities. Some commonly used adsorbents in water treatment are activated carbon, alumina, bone char, siliceous material, clays, etc.
5. **Ion Exchange:** The method of ion exchange is very important in wastewater treatment, as it contributes to improving the water quality and also provides water as per standards, needed for various industries and the domestic sector, for reuse. This process is an extension of adsorption. In this process, the wastewater is made to pass through the bed of ion exchange resins to eliminate the ions of choice from the wastewater. The ion exchange resins are high molecular weight polymeric substances (polyacrylates and polystyrenes) that have the characteristic of attracting the impurity ions associated with the wastewater and exchanging them with the desirable ions, effectively purifying the water.
6. **Membrane Separation:** Wastewater loaded with dissolved impurities is filtered through a semipermeable membrane under pressure. Membrane separation includes microfiltration, ultrafiltration, nanofiltration, forward osmosis, and reverse osmosis.

The other method employed for water purification includes ultraviolet radiation, electro dialysis, thermal reduction, and air stripping. The type of tertiary treatment process used for water treatment depends upon the characteristics of the effluent from previous treatment units.

Table 3.3 summarizes the pros and cons of various biological methods employed to treat sewage wastewater for its safe disposal or reuse.

Sewage treatment demands proper understanding and execution of various treatment processes as per the composition of sewage wastewater. The discussed methods are the conventional ways for treating wastewater to achieve

Table 3.3 Biological techniques involved in sewage wastewater treatment.

Technique	Key features	Advantages	Disadvantages
Trickling filters	1. A trickling filter is an attached-growth type of biological treatment process 2. The efficiency of the attached growth type is less than suspended growth system 3. It is a type of fixed-bed biological reactor	1. High loading rate 2. The land requirement is comparatively less than the suspended growth system 3. Less filter media needed 4. No skilled supervision 5. The system is simple in operation 6. It ensures 75% BOD removal and 80% suspended solids removal 7. It is a self-cleaning device 8. It has fewer requirements for mechanical equipment 9. The power requirement is quite low 10. Has an efficient nitrification unit	1. It is associated with a high cost 2. The head loss of the system is high 3. A primary sedimentation unit is necessary to be present before feeding the effluent to the filters 4. The velocity of distributor arms should be maintained at less than 1.2 m/s 5. Fly, snail, vector, and odor nuisance is possible 6. Requires regular attention by the supervisor 7. Possible accumulation of excessive biomass, which reduces the performance of the filter 8. It leads to huge sludge discharge 9. The filters may have a high rate of clogging
Activated sludge process	1. It is a type of suspended-growth biological treatment 2. It proves to be better and more efficient than trickling filters	1. The process results in clear effluent 2. The system has no fly or odor nuisance 3. Removal of BOD and suspended solids is greater than 90% 4. The cost involved in installation is low 5. Sludge obtained in the process has fertilizer value	1. Operational cost associated with the system is high 2. It is sensitive to changes in the influent 3. Sludge produced from the system needs proper disposal 4. Constant supervision is needed 5. The quantity of returned sludge has to be maintained 6. Energy-intensive

Stabilization pond or oxidation ponds, or lagoons	1. Aerobic pond 2. Anaerobic pond 3. Facultative pond	1. It has a low installation and operational cost 2. The system can be easily modified 3. There is no requirement for skilled supervision 4. Effective for pathogenic removal and is natural energy efficient system	1. More land needed 2. Nuisance due to flies and odor 3. Not suitable for toxic contaminants 4. The effectiveness is influenced by climatic conditions 5. It has limited efficiency during cold climatic conditions
Oxidation ditch	1. These are constructed for the sewage treatment of small communities 2. It is a type of extended aeration type	1. There is no need for primary sedimentation unit 2. Aeration of the tank is ensured mechanically 3. It results in about 98% BOD removal 4. Removal of suspended matter is greater than 95% 5. The process produces less sludge in comparison to another biological process 6. Energy-efficient execution	1. Power consumption is more 2. Can be used for low volume of sewage only 3. Applicable only for sewage treatment for small communities 4. These are expensive to be maintained

(Continued)

Table 3.3 (Continued)

Technique	Key features	Advantages	Disadvantages
Constructed wetlands	1. Sewage flows through artificial vegetated pond systems 2. Treatment by combined action of soil matrix and soil root interface	1. No oxygenation required 2. Less maintenance and operational cost 3. It can be established harmoniously into the landscape 4. It is an environmentally sensitive approach 5. It is a multipurpose treatment method 6. Less cost intensive than other methods as is self-sustaining system	1. Requires large land areas 2. Unpleasant odor can be associated due to presence of specific type of contaminants in wastewater 3. It may not efficiently treat urban wastewater having toxic components 4. The anaerobic decomposition of pollutants releases around 25% of the Earth's atmospheric methane one of the most potent greenhouse gas

the traditional aim of treatment. With growing population, the load on the sewage treatment systems is also increasing. Thus, it is imperative that a constant modernizing in technologies and infrastructure should be simultaneously done to meet the demand of growing population. Newer techniques can be incorporated to achieve and improve the water treatment goals. Monitoring of the effluent at regular intervals can ensure better results and ensure complete reclamation and reuse of wastewater. Public awareness, willingness and partnership of public and private sectors can help facilitate the expansion of robust sewage wastewater treatment systems, to achieve sustainable development goals.

3.3 Case Studies

3.3.1 Sewage Treatment Plant: A Case Study of the Delawas District, Jaipur, India

The demand of sewage water treatment capacity is challenging to achieve due to proliferation in population, especially in the urban regions. As per the report form Centre for Science and Environment (2016) about three-fourths of urban sewage remains untreated and is directly discharged into the water bodies.

The first commissioning of sewage wastewater treatment was done in Delawas, Pratap Nagar, Jaipur, India, in 2006. Since then, it is treating 125 million liters of sewage per day. It is known as the best plant for its chemical free treatment and no pump utilization, for bringing wastewater from 25 km sewerage. It is a popular treatment plant with energy saving installation. The different components of the sewage wastewater treatment plant were screening, grit chamber, sludge treatment tanks and effluent discharge. The effluent characteristics emerging from sewage treatment plant were in agreement with the regulatory norms. The treatment plant produced biogas and digested sludge for utilization as manure. The generated biogas could fulfill 75–80% energy requirement for operation and maintenance. The plant did not incorporated tertiary treatment units. But, with expanding population the gap between the treatment capacity and demand of clear water widened, as the characteristics of the influent changed considerably. A successfully operating sewage treatment plant turned into a failure. It was required to understand the modifications required in the plant to meet with the changing load. Operational and monetary constraints were also a challenge for the treatment plant.

Hence, it is imperative to tackle these challenges to guarantee that the plant can operate effectively and meet the mounting requirements for sewage wastewater treatment [19].

3.3.2 Sewage Treatment Plant: A Case Study of the Okhla District, New Delhi, India

The Okhla sewage wastewater treatment plant is located in the Okhla industrial area in New Delhi, India. The plant works with the aim of decontaminating and purifying the sewage water for its safe disposal in the Yamuna River. The plant has been commissioned to tackle 60 million liters of wastewater daily using the activated sludge process. The sewage plant covers three main stages. First, wastewater purification and monitoring, second, the disposal of sludge produced, and finally, the disposal of the treated wastewater. The disposal site for the effluent from sewage treatment plant is the Yamuna River. The treated water is disposed off after a check for the discharge norms. The maintenance of discharge quality standards is needed to avoid any adverse impact on the ecosystem of the river Yamuna. Continuous increment in the population and sewage load requires additional investment both in infrastructure and technology to ensure effective operation of Okhla treatment plant.

With the collaboration of government agencies and organizations in the private sector, it is important to keep on improving the water quality in the Yamuna River, and to shield the well-being of the communities relying on it [20].

References

1 https://www3.epa.gov/npdes/pubs/primer.pdf (Last accessed 23.09.24).
2 Koul, B., Yadav, D., Singh, S., et al. (2022). Insights into the domestic wastewater treatment (DWWT) regimes: a review. Water, Volume 14, Issue 21, p. 3542. https://doi.org/10.3390/w14213542.
3 Use of reclaimed water and sludge in food crop production. (1996). National academies of sciences, engineering, and medicine. The National academies press, Washington, DC. https://doi.org/10.17226/5175.
4 Widyarani, Wulan, D.R., Hamidah, U., et al. (2022). Domestic wastewater in Indonesia: generation, characteristics and treatment. Environmental Science and Pollution Research International, Volume 29, Issue 22, pp. 32397–32414. https://doi.org/10.1007/s11356-022-19057-6.
5 Fytili, D., and Zabaniotou, A. (2008). Utilization of sewage sludge in EU application of old and new methods-a review. Renewable and Sustainable Energy Reviews, Volume 12, Issue 1, pp. 116–140. https://doi.org/10.1016/j.rser.2006.05.014.
6 Boudjabi, S., Ababsa, N., and Chenchouni, H. (2023). Sewage and sewage treatment. In: Brinkmann R (ed). The Palgrave handbook of global sustainability, Palgrave Macmillan, Cham. https://doi.org/10.1007/978-3-031-01949-4_50.

7 Kesari, K.K., Soni, R., Jamal, Q.M.S., et al. (2021). Wastewater treatment and reuse: a review of its applications and health implications. Water, Air, and Soil Pollution, Volume 232, Issue 5, pp. 208. https://doi.org/10.1007/s11270-021-05154-8.
8 Deswal, A., and Deswal, S. (2021). A basic course in environmental studies. Dhanpat Rai and Company.
9 Ghaitidak, D.M., and Yadav, K.D. (2013). Characteristics and treatment of greywater-a review. Environmental Science and Pollution Research, Volume 20, pp. 2795–2809. https://doi.org/10.1007/s11356-013-1533-0.
10 Harahap, J., Gunawan, T., Suprayogi, S., et al. (2021). A review: domestic wastewater management system in Indonesia. IOP Conference Series: Earth and Environmental Science. https://doi.org/10.1088/1755-1315/739/1/012031.
11 Weber Jr., W.J. (1972). Physicochemical processes for water quality control, 1e, John Wiley and Sons, New York, pp. 533–596.
12 https://www3.epa.gov/npdes/pubs/trickling_filter.pdf (Last accessed 23.09.24).
13 Zinicovscaia, I. (2016). Conventional methods of wastewater treatment. In: Zinicovscaia I, Cepoi L (eds). Cyanobacteria bioremediation wastewaters, Springer, Cham, Switzerland, pp. 17–25. https://link.springer.com/chapter/10.1007/978-3-319-26751-7_3.
14 Ireland, D. (1998). Sewage treatment: general introduction. Encyclopedia of hydrology and lakes. Encyclopedia of earth science, Springer, Dordrecht. https://doi.org/10.1007/1-4020-4497-6_203.
15 https://www3.epa.gov/npdes/pubs/oxidation_ditch.pdf (Last accessed 23.09.24)
16 Vymazal, J. (2010). Constructed wetlands for wastewater treatment. Water, Volume 2, Issue 3, pp. 530–549. https://doi.org/10.3390/w2030530.
17 Hassan, I., Chowdhury, S.R., Prihartato, P.K., et al. (2021). Wastewater treatment using constructed wetland: current trends and future potential. Processes, Volume 9, Issue 11, p. 1917. https://doi.org/10.3390/pr9111917.
18 Zagklis, D.P., and Bampos, G. (2022). Tertiary wastewater treatment technologies: a review of technical, economic, and life cycle aspects. Processes, Volume 10, Issue 11, p. 2304. https://doi.org/10.3390/pr10112304.
19 Rukshar, Vyas, A.D., and Bhatnagar, N. (2025). Feasibility of sewage treatment plant with changing influent characteristics and quantity-a case study of Delawas STP at Jaipur city. In: Roshan Dash R, Mohapatro S, Behera M (eds). Pollution control for clean environment—Volume 1. ICPCCE 2023. Lecture notes in civil engineering, Volume 415, Springer, Singapore. https://doi.org/10.1007/978-981-97-7842-3_13.
20 Khan, M.S., Saifi, S., and Kacker, S. (2023). Design of sewage treatment plant: a case study of Okhla district, New Delhi. International Journal of Scientific Research in Engineering and Management. Volume 7, Issue 6, pp. 1–6. https://doi.org/10.55041/IJSREM22636.

4

Industrial Contaminants

The escalating population is making industrialization and urbanization inevitable. Industrialization has led to several developments but has given rise to several challenges. The mounting number of industries has led to water, air, and land pollution, posing harm to humans and aquatic life. Industrialization has also resulted in widespread deforestation, adversely affecting the climate. Increased industrial setups have led to issues such as increased waste generation, in addition to the pollution, being a matter of concern. Water bodies have been always found to be one of the easiest sites for dumping off the processed water from these industries. Water, after serving the specific purpose in the industries, is let out in the form of wastewater. It contains several contaminants and is known as polluted industrial wastewater.

This polluted water with hazardous pollutants should not be discharged into the water bodies without proper treatment. All the contaminant concentrations should lie well within the permissible limits of wastewater discharge. Several industries are still not adhering to the legislation laid down by the governmental agencies for the wastewater discharge. One of the important parameters high for industrial wastewaters is chemical oxygen demand (COD). This is an indicator of the oxygen requirement for the digestion of chemicals present in industrial wastewater. Many times, the COD of the industrial wastewater is very high, leading to oxygen depletion in the water bodies.

4.1 Industrial Wastewater

Different industries contributing to the pollution of water bodies have been portrayed in Figure 4.1 [1]. Contaminated water from these industries, when it enters the aquatic environment, leads to several adverse impacts, depending on the nature of the contaminants.

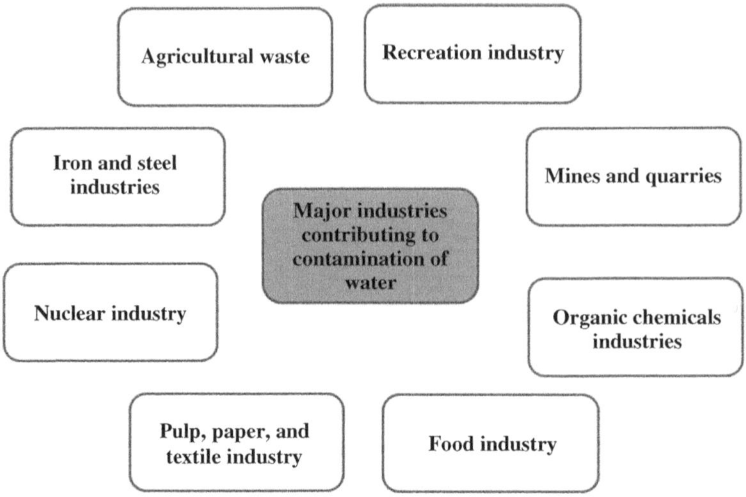

Figure 4.1 Major industrial contributors of wastewater contaminants.

It is very important to treat the industrial wastewater before it is discharged into the water bodies, as it bears a high pollutant load and poses substantial impacts on humans and the aquatic ecosystem [2].

4.2 Wastewater Contaminants from Different Industries

4.2.1 Mining Industry

Acid mine drainage (AMD) takes place when sulfur minerals in the rocks interact with oxygen and water, making sulfuric acid. The acid formed is capable of leaching heavy metals such as chromium, cadmium, copper, zinc, arsenic, etc., into the water bodies [3, 4]. Heavy metals are also let into the water bodies through the mine drainage. Suspended sediments and solids are responsible for turbidity and reduced light penetration in the water bodies.

4.2.2 Agricultural Industry

Wastewater from agriculture is generated from livestock farming, use of fertilizers, pesticides, agricultural products, and irrigation. Wastewater from agricultural activities contaminates the quality of groundwater and affects human health. Contaminants such as pesticides, herbicides, and fertilizers present in wastewater are predominantly used in the fertilizer industry and tend to percolate into the water bodies through runoffs. Apart from the contaminants mentioned, the agriculture industry is an important contributor of nutrients in the water bodies and is responsible for the deterioration of water quality in the lakes and ponds. In the United States, in 94% of water samples, at least one pesticide was found in the tested sample [5].

Toxic chemicals from the agricultural industry have been contributing to water pollution for the last many decades [6]. An example worth mentioning is dichloro diphenyl trichloroethylene (DDT), which is a chemical extensively used as an insecticide. This chemical gets adsorbed well on the soil and eventually gets washed off into the nearby rivers and lakes [7, 8]. This chemical is capable of affecting the phytoplanktons by obstructing the passage of sunlight and can enter the food chain [9, 10].

4.2.3 Iron and Steel Industry

An important contributor to water pollution is the wastewater from the iron and steel industry. Several processes of this industry include operations of the blast furnace, coke production, sintering, etc., which generate an array of contaminants such as suspended solids, inorganic and organic compounds, oils, toxicants, heavy metals, etc.

Several contaminants from this industry are generated, such as phenols, cyanides, and isocyanides, heavy metals, ammonia, nitrogen, etc., which are very toxic, carcinogenic, and tend to accumulate [11].

4.2.4 Nuclear Industry

The major applications of the nuclear industry include nuclear medicine, reprocessing facilities, reactors, etc. Nuclear reactors are cooled with the use of water, generating a large amount of liquid waste. Liquid wastewater contains large amounts of radioactive and nonradioactive wastes, posing threats to the environment and humans [12]. This wastewater is capable of causing biological mutations. Approximately 400000 tons of used fuel is discharged all over the globe [13].

4.2.5 Pulp, Paper, and Textile Industry

The pulp and paper industry generates several pollutants, which pollute the waterbodies, in addition to greenhouse gases (GHGs). Several impacts of the pollutants from these industries include health problems, impact on the food chain, and harm to aquatic life. Organochlorine compounds are formed, known as adsorbable organic halides (AOX), during the bleaching process, which are released into the waterbodies, causing pollution [14]. These are responsible for several ailments such as skin diseases, respiratory diseases, reproductive damages, and mutations in addition to chronic disorders [15].

It has been reported that the effluent and sludge from the pulp and paper industries has a very high biological oxygen demand (BOD), COD, and besides has high amounts of suspended solids, derivatives of lignin, tannin, AOX, fatty acids, resin acids, etc. [16]. In addition to these gaseous pollutants, for example, methyl mercaptan, hydrogen sulfides, sodium sulfide, other inorganic compounds such as magnesium, nickel, copper, and iron, also emerge, which are responsible for adversely affecting the marine organisms. Hexadecanoic acids, trimethylsilyl ether, β-sitosterol 1-tetradecane, methanol, benzoic acid, alkylated phenols, tetrachlorocatechol, trichlorocatechol, 2-methoxy phenol, decalone, abietic acid, chloroguaiacols, etc., have been reported to cause endocrine disruption in the aquatic organisms. Pentachlorophenol (PCP) emerging from the pulp and paper industries is considered to be one of the most toxic pollutants, which is extremely harmful for aquatic organisms and humans [17]. It results in carcinogenicity, mutagenicity, and is considered to be very toxic. Many endocrine-disrupting chemicals are also prominent in the wastewater from the pulp and paper industry.

4.2.6 Food Industry

A high pollution load results from the food industry on the water bodies. The contaminants from the food industry include fats, nutrients, organic matter, chemicals, and proteins. Different types of food industries generate varied types of water pollutants and contaminants (Table 4.1).

Table 4.1 Types of water contaminants emerging from the food industry with their toxicity.

Type of industry	Pollutant	Toxic effects	References
Milk processing and dairy	*E. coli* and other pathogens, detergents, phosphorus, fats	Eutrophication, oxygen depletion, foodborne illness	[18]
Meat processing	Nitrogen compounds, fats, blood, and pathogens	Hazardous of food safety, high BOD and COD	[19]
Sugar refining	Oil and grease, suspended solids, metals, and metalloids	Increased COD, clogging of water streams	[20]
Beverage industry	Fluorides, microplastics and heavy metals, organics, suspended solids	Bioaccumulation, fluorosis	[21]
Seafood processing	Organic matter, salts, detergents, pathogens, nitrogen	Increase in salinity, increased oxygen demand	[22]

4.2.7 Organic Chemicals Industry

Industries predominantly dealing with organic chemicals and responsible for organic pollutants are mentioned in Figure 4.2.

Industries involved in the production and use of organic chemicals are also substantial contributors to water pollution. Petrochemical industries are one of the vital contributors to water pollution. The contaminants include oil, soaps, aromatic and polycyclic hydrocarbons, phenols, hydrocarbon derivatives, surface active agents, sulfides, etc. [23, 24]. These harmful chemicals are capable of affecting the skin, liver, and can cause cancer, hormonal disorders, etc. [25].

With the revolutionization in plastic and resin manufacturing industries, there have been serious concerns related to their contribution to pollution. Millions of gallons of wastewater are dumped into the United States waterways due to plastic industries containing contaminating pollutants such as phosphorus, nitrogen, dioxins, 1,4-dioxane, along with other harmful pollutants. Plastic nurdles (tiny plastic pellets), which are used for the manufacture of plastic goods, release plastic particles and microplastics into the water streams [26]. Resin manufacturing

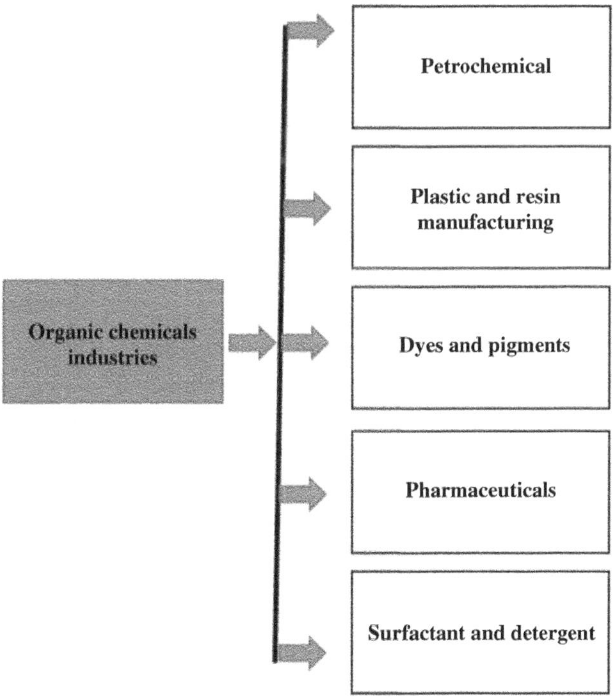

Figure 4.2 Organic chemicals industries.

industries produce phenolic compounds [27, 28], releasing phenol, which does not degrade readily, causing pollution.

The dye and pigment industry employs tons of water for the processing of the products, releasing large volumes of wastewater with dyes and colorants, as contaminants, which are eventually released into the water bodies. Dyes are used in various sectors such as cosmetics, textiles, and pharmaceuticals. Dyes constitute complex structures, making them non-biodegradable in the environment [29]. Dyes and pigments in water pose severe toxicity, mutagenicity, and carcinogenicity [30].

Pharmaceuticals are another category of contaminants which is are responsible for water pollution, as many contaminants such as undigested pharmaceuticals, antibiotics, hormones, etc., are hazardous for the aquatic organisms [31, 32].

With the inevitable growing utilization of soaps, detergents, wetting agents, emulsifiers, etc., are being released in large quantities into the water bodies, affecting the aquatic species and human health [33].

4.2.8 Recreation Industry

Though not very visible but the recreation industry also plays a significant role in the pollution of water bodies. Several activities, such as marine activities and boating, result in the introduction of oils and greases, antifouling paints, copper, zinc [34], sunscreens, fragrances, and preservatives. Adding to this, the tourism activity, microbial contamination owing to improper sanitation, along with garbage and plastic waste, is being dumped into the water bodies [35].

4.3 Categories of Industrial Contaminants

The contaminants from the industry are classified according to their nature as shown in Figure 4.3.

Organic contaminants primarily are the outcomes of the textile, petrochemical, chemical, and pharmaceutical industries.

Based on the category mentioned above, organic pollutants are volatile organic compounds (VOCs) such as xylene, benzene, toluene, originating from the petrochemical refining industries, phenolic compounds

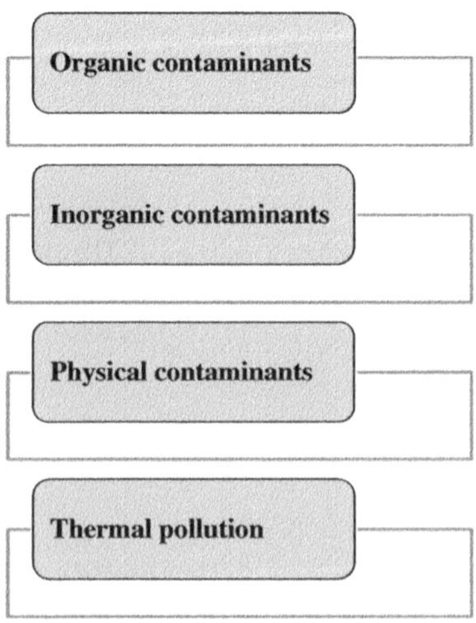

Figure 4.3 Broad categories of industrial contaminants.

from the resin manufacturing and oil refining industries [36]. Persistent_organic pollutants such as dioxins are a result of the pesticide manufacturing industries. Crude oil and pesticides are the contaminants from the petroleum refinery industries and agricultural runoffs. Inorganic pollutants such as heavy metals, lead, cadmium, mercury, nutrients such as nitrogen and phosphorus, cyanides, salts, acids, and alkalis [37] are other important contaminants. Physical contaminants such as sand and silt from the mining and construction industry, oil and grease from the food industry, and suspended solids belong to this category [38, 39]. Waterbodies usually maintain a steady temperature. Several industrial processes release effluents at elevated temperatures into the water bodies. This leads to a change in the ambient water quality and serves as an indirect pollutant. It tends to affect the aquatic ecosystem. The oxygen holding capacity of warm water is less than cool water, which results in oxygen depletion, creating an alarming situation for aquatic flora and fauna [40].

4.4 Monitoring and Detection of Industrial Contaminants

It is important to detect and monitor industrial contaminants to ensure the safety of water bodies and public health. A wide range of detection and monitoring techniques have been used for industrial contaminants, which have also been discussed, specific contaminant-wise, in subsequent chapters.

As per the classification mentioned above, the detection and monitoring techniques have been tabulated (Table 4.2).

Table 4.2 Detection and monitoring techniques for contaminants.

Technique	Key features	References
LC-MS (liquid chromatography coupled with mass spectrometry) GC-MS (gas chromatography coupled with mass spectrometry)	Quantitative and qualitative analysis, the capacity for the separation of contaminants	[41]
Surface-enhanced Raman spectroscopy	Removal efficiency is high for anti-inflammatory and nonsteroidal drugs, proved to be a low-cost reusable method	[42]
Fluorescence spectroscopy	High selectivity and sensitivity	[43]
Biosensors	Rapid and sensitive detection, possibility of onsite monitoring	[44]

(Continued)

Table 4.2 (Continued)

Technique	Key features	References
Microfluidic devices	Cost-effective, rapid response, easy method	[45]
Surface-enhanced Raman spectroscopy	Environment-friendly and cost-effective	[42]
Atomic absorption spectroscopy (AAS)	Portable, low-cost cost and simple	[46]
Electrochemical sensors	Simple process, economical, and portable	[47]
Colorimetric sensors	Simple method, sensitive and reproducible	[48, 49]
Passive samplers (e.g., DGT)	Can be used for a complex matrix of contaminants, with specificity	[50]
Turbidity sensors (Nephelometry)	Feasibility for large-scale operations, real-time monitoring possible	[51, 52]
Oil content analyzer (infrared-based)	Sensitive and rapid detection and monitoring	[53]
Particle imaging/scanning	Detection of an array of contaminants is possible	[54]
Gravimetric analysis	High accuracy, cost-effective, can be used for a variety of contaminants	[55]
Thermal infrared (IR) remote sensing	Help in the accurate determination of contaminants along with other techniques	[56]
IoT sensors	Real-time monitoring with early warnings	[57]
Fiber optic temperature sensors	High sensitivity, small, resistant to electromagnetic interference	[58]

4.5 Transport and Fate of Contaminants in the Environment

Contaminants once in the environment move from one point to the other under the influence of several factors and tend to either remain in the aquatic environment or get transformed into newer components. When they make their way into the water bodies, they either migrate in the surface water or enter the groundwater.

Contaminants often make their way into the surface water through leaching, runoffs, and industrial discharges. The factors that affect their movement into the water bodies are the flow of water, turbulence, solubility, and their interactions with the sediments [59]. For the passage of contaminants to groundwater, the main responsible factor is the percolation through soil. Factors affecting the migration of these contaminants to the underground water are the permeability and the porosity of soil, adsorption and desorption processes, and chemical degradation [60]. The chemical structure of the contaminant, along with the presence of organic matter and clay influences the percolation.

4.5.1 Fate of the Contaminants

During the passage of these contaminants into the surface water bodies or underground water, the important changes possible are shown in Figure 4.4 [61].

Understanding the manner in which the contaminants move in water helps to ascertain the extent to which the spread would take place, creating an impact, aiding better management of contaminants. With the help of several mathematical models, the concentrations of these contaminants with time and space can be calculated. Different aspects, such as water flow, turbulence, velocity, and mixing tendency, help in ascertaining how the contaminants move and create an impact.

The contaminants are spread over long distances by bulk fluid motion in advection, whereas in the case of dispersion, the variations in the velocity due to turbulence help the spread. Diffusion of contaminants can also take place when there is a movement from an area of a higher concentration to a lower concentration by molecular motions, which are random [61]. Sorption refers to the contaminant interaction with solids, such as soil. This usually deals with the subsurface behavior of the contaminants [62]. It's a combination of adsorption and absorption, which are greater concentration of the contaminant on the

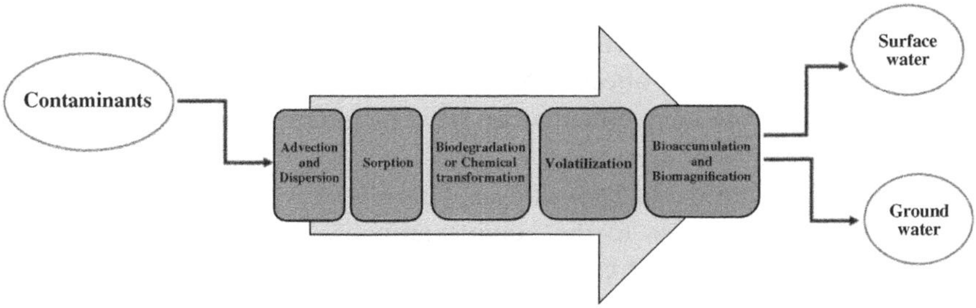

Figure 4.4 Fate of contaminants from industries.

surface, and the penetration of the contaminant into the bulk of the soil particles, respectively. The physical and chemical characteristics of the contaminants, fluid media, and the surface of the solid, soil traits are some vital aspects deciding the sorption process taking place. Some contaminants during the passage through soil or water tend to biodegrade. The extent to which biodegradation takes place is site-specific. Some of the contaminants undergo chemical transformations, generally photochemical reactions. Volatilization of the contaminants depends upon pH and temperature. A greater evaporation of the contaminants takes place in warmer environments [63]. Bioaccumulation is defined as a process by which the accumulation of contaminants takes place in the bodies of organisms, biologically, over a period of time. This accumulation in the body of the organisms takes place when the ingestion is more than the rate of metabolic excretion, for example, industrial contaminants such as pesticides and heavy metals. Bioaccumulation depends upon the chemical form and the solubility of the contaminant [64]. Biomagnification is a process in which the concentration of the contaminants inside an organism becomes high [65]. This accumulation and concentration increase at successive trophic levels of the food chain. The contaminants are present in small amounts at lower levels but have significant impacts on the organisms at the higher trophic levels.

4.6 Toxicity of Industrial Wastewater

Several effects of the industrial-specific contaminants have been discussed at length chapterwise. However, the overall impact has been presented in this section.

4.6.1 Impact on Waterbodies

Harmful industrial contaminants are led into the waterways. Effects of industrial contaminants on the water bodies can be detrimental to the aquatic ecosystem and humans. Contaminant inlet at the onset can make the water turbid and can visibly impact the aesthetics. The light penetration and the photosynthetic activity of the plants are adversely affected. Not only is the appearance of the waterbodies is altered, but the contaminants can have long-term effects on the aquatic organisms.

Hazardous contaminants such as dyes, pharmaceuticals, heavy metals, etc., can cause serious threats to the aquatic animals, ranging from temporary disorders to mutagenic and carcinogenic effects. Nutrient excess can lead to increased growth of algae and weeds in the waterbodies, further aggravating the problem.

4.6.2 Impact on Soil and Groundwater

Industrial contaminants can also percolate to the underground water table through the soil, thereby adversely impacting the groundwater. Industrial activities, for example, mining, rapidly lead to the deterioration of the underground water quality [66]. Activities and sources which lead to pollution of underground water sources are undetected leakage from industrial collectors and sewers, underground storage tanks, areas of bulk chemical storage, waste injection tanks, liquid effluent, inappropriate disposal practice, and leachate from solid waste dumping sites, and catastrophic discharges [67].

4.6.3 Impact on Human Health

Pollution of water sources adversely affects human health substantially. Right from physical external contact with polluted water to the consumption of water from contaminated sources, an array of health complications has been well reported in literature, which are discussed in subsequent chapters. Contaminants such as heavy metals, acids, alkalis, surfactants, bacteria, etc., pose several health threats, with minor ailments from skin irritation and allergies to health complications such as cancer.

4.6.4 Impact on Biodiversity and Ecosystem

Every specific living species present in the water bodies is essential for a proper balance of the ecosystem. Industrial contaminants tend to bioaccumulate and biomagnify with time, affecting the entire ecosystem. Industrial wastewaters, which are let into the waterbodies, tend to elevate the temperature of the water, decreasing the dissolved oxygen level, which is necessary to be present in the water bodies for a healthy life of aquatic organisms, as well as for the hydrosphere. An oxygen reduction destroys the natural cleaning tendency of the water bodies, hampering the ecosystem. Industrial effluents with high temperature, when plunged into waterbodies, also make water warmer for several species to thrive in.

Degradation of the habitats, loss of species, and biological changes, etc., result from the passage of contaminants in the water bodies [68]. Chemicals such as insecticides and pesticides get accumulated in the food chain, causing severe impacts.

4.6.5 Impact of Some Specific Contaminants

Below are a few negative effects due to the presence of certain contaminants that are let into the water bodies from various industries (Table 4.3).

Table 4.3 Toxicological impacts of industrial contaminants.

S. No.	Industrial contaminants	Hazardous effects	References
1.	Textile contaminants like bleaching agents, surfactants, alkaline bases, colorants, resins, sterols, lignin-derived compounds	Allergic reactions, turbidity, effect on photosynthetic activity of aquatic plants, carcinogenicity, mutagenicity, irritation to skin, respiratory disorders	[69–71]
2.	Metals like lead and manganese are involved in the sugar industry, paint, and fabrication industry	Low oxygen is harmful to aquatic life, asthma, allergies, and eczemas in children, reproductive damage, carcinogenic, teratogenic, causes oxidative stress, nervous system impairments, organ damage, and affects growth and development	[72–74]
3.	Drugs like carbamazepine, diclofenac	Aquatic flora and fauna are at high risk	[75]
4.	Agricultural chemical agents like pesticides, fertilizers, and larvicides	Increased nitrogen and phosphorus concentration	[76, 77]
5.	Lime, salts, acids, and sulfides	High water salinity	[78]
6.	Polymeric contaminants, microplastics, and nanoplastics	Ingestion and suffocation, deaths in birds, mammals, and fishes, cancer, heart damage, effects on hormones	[79]
7.	Hydrocarbons, oil, phenols, tannins, resins, and chlorinated organic compounds	Oxygen level depletion, toxic to both human and aquatic species, pneumonia, respiratory illnesses, rhinitis, tracheitis, mutagenicity, genotoxicity	[80, 81]

4.7 Treatment Technologies

Depending on the nature of contaminants present in wastewater, different strategies and processes need to be adopted for treatment. In general, the processes adopted are:

1. Removal of solid waste
2. Removal of oil and grease
3. Treatment of organics
4. Treatment of acids and alkalis

Depending upon the category of contaminants and the industries from which wastewater is emerging, several technologies are used to treat water and wastewater [82, 83]. Some of them are mentioned below.

4.7.1 Adsorption

In this method, the contaminants adhere onto the substrate (adsorbent) by physical or chemical bond and get eliminated from the wastewater sample. Adsorption could be physical or chemical, depending upon the layers deposited and the type of bonds formed [84, 85]. Different adsorbent materials [86] can either be used raw or after treatment with some chemicals or activation. Several adsorbents used for the treatment of various industrial adsorbents are given in Table 4.4.

4.7.2 Membrane Separation and Technologies

Membrane techniques have been widely used for the removal of industrial contaminants due to their cost-effectiveness, selectivity, and ease in scaling up [103]. Several processes have been explored and are being used, such as

Table 4.4 Different sorbent materials for industrial contaminants.

S. No.	Industries	Adsorbents used	References
1.	Dye manufacturing	Activated carbon, activated carbon from waste materials, bio-sorbents, natural and synthetic materials, waste materials from agriculture and industries, hybrid nanomaterials, composites and nanocomposites, magnetic nanocomposite materials, polymers	[87]
2.	Textile	Activated carbon, zeolite, clay, natural biosorbents	[88, 89]
3.	Pharmaceutical	Activated carbon, nanocomposites and magnetic materials, chitin, lignin	[90, 91]
4.	Agriculture	Metal–organic frameworks, biochar, nanomaterials, polymers, silica-based materials, and clays	[92–94]
5.	Paper and pulp	Composites, bottom ash, activated carbon	[95–97]
6.	Paint manufacturing	Activated carbon from biosorbents	[98]
7.	Petrochemical	Biomass-derived products, anthracite	[99, 100]
8.	Leather	Adsorbent prepared from biomass, nanoparticles	[101, 102]

ultrafiltration (UF), microfiltration (MF), nanofiltration (NF), and reverse osmosis (RO) [104]. Though these methods have several advantages, the drawback associated is fouling [105], where unwanted substances tend to get deposited on the pores of the membranes [106]. Membrane fouling depends upon the type of water being treated, the conditions of operation, and the kind of membrane used [107].

Technologies such as ultrafiltration remove the contaminants depending upon the particle sizes in the range of 10^{-3}–1 microns and show limitations in terms of decontamination [108]. Microfiltration, on the other hand, separates the contaminant particles by physical screening under low pressure, for particles ranging from 0.1 to 10 microns. Processes such as nanofiltration (retaining diameter 10^{-3}–10^{-2}) [109, 110] and reverse osmosis (retaining diameter 10^{-4}–10^{-3}) [111, 112] are high-pressure processes. The advantages and disadvantages of different membrane technologies are mentioned in Table 4.5.

Table 4.5 Membrane technologies: advantages and disadvantages.

S. No.	Technology	Solutes filtered	Advantages	Disadvantages	References
1.	Microfiltration	Grease, oil, fat, colloids, organics, and micro particles	Large pore sizes enable efficient removal of contaminants at low pressure	Unable to treat color, odor, and bad taste causing substances, fouling	[114]
2.	Ultrafiltration	Oils, organics, pigments and proteins, sugar, and microplastics	Low-pressure treatment	Membrane fouling	[115]
3.	Reverse osmosis	Monovalent ions and all contaminants	Less maintenance, removal of a variety of contaminants, cost effective	Wastage of water, which can be utilized	[116, 117]
4.	Nanofiltration	Divalent anions and cations, sodium chloride, lactose and sucrose, sulfates, and pigments	High efficiency, reduction of COD, and can work at low pressure	Significantly affected by feed characteristics, porosity, morphology, and charge of contaminants	[118]

Source: Adapted from [113].

4.7.3 Electro-technologies

Technologies such as electro flotation (EF), electrocoagulation (EC), electro oxidation/reduction, and electrodeposition have been used for the treatment of industrial contaminants [119]. EF has been found to be fruitful in the removal of grease and oil, colloidal particles, and organic pollutants in industrial wastewater. In this method, the pollutants and contaminants are made to float on the water surface by electrolysis. In the technique of electrocoagulation, coagulants are generated in situ by using iron or aluminum electrodes. Metal ions are liberated at the anode, and at the cathode, hydrogen is released. This helps the flocculated contaminant particles to float. In the method of electrochemical oxidation, strong organic compounds are decomposed [120]. This process can take place directly or indirectly. In the direct method, a direct switch of electrons to the reducer from the anode takes place, whereas in the indirect method, an in situ electro-generation of the oxidizing species occurs at the surface of the anode. In electrochemical reduction, reduction takes place by the electrons provided exter nally [121]. Electrodeposition is a method to recover contaminants such as heavy metals from wastewater [120].

4.7.4 Advanced Oxidation Processes

Many of the conventional technologies fail to treat the wastewater completely. Advanced oxidation processes (AOPs) help in the efficient removal of contaminants from several industries such as paper and pulp, textile, petrochemicals, refineries, and pharmaceuticals [121]. These technologies offer several advantages as they can be carried out at mild temperatures and atmospheric pressure. AOPs substantially reduce the contaminants and allow the treated water to be used for further conventional processes. Some of these processes are photocatalysis, the Fenton process, ozonation, and electrochemical processes. In the AOPs, highly reactive hydroxyl radicals (·OH) are generated in sufficient quantities, helping in the removal of organic substances and several contaminants, thereby increasing the biodegradability of wastewater for effective treatment [122]. These methods help the industries to meet the stringent regulations related to the contaminant outlet from the industries. Table 4.6 portrays the advantages and disadvantages of different AOPs for industrial wastewater.

4.7.5 Integrated Techniques

In addition to the solitary treatment techniques, focus has now shifted to the use of integrated technologies, which have added advantages over the individual techniques. The treatment of wastewater sometimes becomes challenging due to

Table 4.6 AOPs: advantages and disadvantages.

S. No.	AOP	Mechanism	Advantages	Disadvantages	References
1.	UV/H_2O_2	UV light is used for the breakdown of contaminants. Hydrogen peroxide and ozone can be used for the effective treatment, along with UV.	Efficient for a variety of contaminants, with no residues in treated water	High energy requirement due to UV. The cost of hydrogen peroxide and ozone is high	[122]
2.	Ozone-based processes	Process of making and collapsing of microbubbles and cavities in a time frame of milliseconds. Highly reactive free radicals are formed, increasing the surface area for solid catalysts, also magnifying mass and heat transfer rates.	Cost efficiency of operation, no unnecessary secondary contaminants	High energy requirement	[123, 124]
3.	Photocatalysis	Oxidation reactions are initiated in the presence of photo catalysts. Light-activated catalysts generate hydroxyl radicals.	Harnessing the potential of solar light, contaminants are broken down into water, carbon dioxide, and inorganic minerals	Limitations due to light scattering, catalyst fouling, and the problem of separation of fine catalyst from the treated water	[125]
4.	Hydrodynamic cavitation	Involves growth, nucleation, and the implosion of cavities filled with vapor or gas in a short period (usually milliseconds)	High mass transfer capacity and efficiency, low operating cost	Several factors govern the efficiency, e.g., operational parameters, inductor characteristics	[126]
5.	Electrochemical AOPs	The technique of generating highly reactive hydroxyl radicals converts contaminants into carbon dioxide and water by the transfer of electrons.	Helpful for degradation of soluble or insoluble dyes, reduction of COD, presence of salt in wastewater has no effect	High cost, dependence on the nature of the contaminant to be treated	[127, 128]
6.	Fenton-based processes	Peroxides and iron ions react, generating reactive oxygen species helpful in oxidizing inorganic and organic contaminants.	Inexpensive reagent, simple setup, promising results for contaminants like polypropylene, fast reaction, easy generation of hydroxyl radicals	Necessary acidic conditions to be persistent for effectiveness, sludge generation	[129]

the complexity of wastewater generated from different industries [130]. Also, to comply with the stringent regulations, it becomes imperative to look for innovative methods for efficient and economical industrial wastewater treatment. Several physicochemical methods, such as AOPs, ion exchange, membrane filtration, and coagulation/flocculation, are very sensitive and efficient; however, they require a large amount of chemicals and are not cost-effective, and lead to the production of secondary pollutants and sludge. Solitary methods sometimes fail to eradicate the different pollutants, thereby posing a threat to the water quality [131]. Physical, chemical, and biological methods of wastewater treatment can be combined as a wholesome approach towards wastewater treatment.

Several novel methods and techniques, such as artificial intelligence, can be used for integration with other techniques [132]. A combination of solitary electrocoagulation (EC) and combined anodic oxidation (AO), electro-Fenton (EF), and peroxi-coagulation (PC) has been tested [133–135]. A combination of membrane separation and physicochemical processes has been tested for soft drink industries [136], giving remarkable results in terms of COD reduction. An integration of agents such as ozone and hydrogen peroxide, followed by ultrasonication, has also been tested for polymer and dye industry wastewater [137]. It was observed that the COD removal reduced from 6026 to 2026 ppm within 60 minutes of processing time, and with a combination of zone and hydrogen peroxide, followed by ultrasonication. Photocatalytic methods integrated with membrane treatment have been used for the eradication of emerging contaminants [138]. For fertilizer industry wastewater, a combination of adsorption proved to be fruitful in reducing approximately 85% of COD and ammoniacal nitrogen [139]. Several researchers are exploring different treatment technologies that can cater to a wider matrix of pollutants.

An amalgamation of treatment technologies has not only led to complete treatment of wastewater contaminants but also have moved to the concept of zero liquid discharge (ZLD), supporting the circular economy.

4.7.6 Constructed Wetlands and Integrated Wetlands

Constructed wetlands (CWs) are the engineered, low-cost, and sustainable setups which are designed involving vegetation, soil, adsorbents, and microbial assemblies, etc., to effectively treat industrial wastewater. For more than 50 years, constructed wetlands have been used successfully for the treatment of industrial wastewater. Constructed wetlands have been successfully employed for industries such as meat processing, petrochemical, pulp and paper, dairy, wine, and textile industries. The processes majorly operative in these wetlands are sorption, photodegradation, sedimentation, precipitation, evapotranspiration, volatilization, plant uptake, diffusion, and degradation by microbes. Surface and subsurface flow constructed wetlands of the type horizontal and vertical flow, in addition to several hybrid constructed wetlands, have been reported and explored. Vertical subsurface flow systems (VF), surface flow systems (FWS), and horizontal subsurface flow systems (HSSF) are widely used and are sometimes combined, known as hybrid systems, depending upon the type of contaminants to be eliminated [140].

FWS are generally shallow beds ranging from 20 to 40 cm with vegetation planted. As the wastewater passes through the planted beds, the contaminants get eliminated by processes such as sorption, biological degradation, precipitation, oxidation, sedimentation, and reduction. Though these are low-cost systems, but have low removal rates. Also, these can lead to bad odors and require large surfaces. The dimension of HSSF is 30–60 cm deep, filled with media such as gravel, supporting vegetation, and helping the formation of biofilms. These are planted with vegetation, and the water column remains 5–10 cm below the surface. Wastewater passes horizontally through several aerobic and anaerobic zones. The biggest advantage of these types of wetlands is the high removal efficiency and less surface requirement. Although there could be a risk of pore clogging in a fine-gravel bed. Vertical flow systems are 60–100 cm deep. The wastewater is fed from the top of the wetland, which passes through the different layers. The performance efficiency is good, however demands high inlet loads [141]. Constructed wetlands have shown remarkable results in terms of the treatment of wastewater from several industries (Table 4.7).

Table 4.7 Constructed wetlands for several important industries.

S. No.	Industry	Contaminants treated	COD (%)	BOD (%)	TSS (%)
1.	Textile	Sulfates and colorants	70	–	88
2.	Paper and pulp	Resins, fatty acids, and phenols	–	27–90	55–68
3.	Refinery	Mineral oils, phenol	54–93	50–98	44–95
4.	Tannery	Total Kjeldahl nitrogen, heavy metals, and turbidity	59–65	48–73	65–72
5.	Dairy	Total Kjeldahl nitrogen	92–97	34–98	62–96
6.	Food	Total Kjeldahl nitrogen, phosphates, nitrates	11–92	89–94	81
7.	Winery	Total Kjeldahl nitrogen, phenol, tannin	49–99	92	30–97
8.	Steel	Heavy metals	50–77	–	–

A variety of techniques are being explored, along with the integration of techniques to treat industrial wastewater. New technologies, along with an integration of artificial intelligence, are also being investigated for efficient contaminant removal from industrial wastewater.

4.8 Case Studies of Industrial Contamination of Water Bodies

4.8.1 Groundwater Pollution in Southeast Vietnam

The groundwater pollution of Ho Chi Minh City and Tay Ninh was studied and reported by a group of researchers [142].

In Tay Ninh, a province in southeast Vietnam, the development of industrial crop processing industries has taken place. The wastewater produced from these industries is about 4–15 cubic meters (m^3) every ton of the industrial raw material processed. These factories are far from the water distribution networks and mostly employ underground water. Pond system is usually used without impermeable liners for the treatment of wastewater with high organic content. This has resulted in infiltration and subsequent degradation in the quality of groundwater in terms of color and odor, with increased levels of ammonia and Total Organic Carbon (TOC).

An open dumping site, Dong Thanh in Ho Chi Minh City, used to have a maximum capacity of solid waste of 4000 tons, generating a large quantity of leachate. Owing to a high dumping of 6.5 million tons of solid waste, it was closed in 2003. The leachate from earthen pots without liners deteriorated the quality of water in the shallow wells of households, with nitrogen, in addition to high BOD and COD of the water streams. At present, due to cost considerations, there are no proper plans, as mentioned in the study, which makes it necessary to have impermeable liners for the stabilization ponds [142].

4.8.2 The Elk River Chemical Spill, West Virginia

The Elk River, the longest river in West Virginia, was responsible for the spread of a chemical known as 4-methylcyclohexanemethanol, a coal cleaning chemical, from a leak through a rusted containment tank of Freedom Industries at the bank. The event occurred on 9 January 2014, affecting the drinking water of almost 300000 West Virginians [143].

Although no fatalities were reported, but exposed people experienced skiing irritation, nausea, vomiting, and rashes, etc. In addition to this, the event led to several people without portable water for weeks.

4.8.3 Minamata Disease (Kumamoto Prefecture)

The discharge of Methylmercury (MeHg) took place from a chemical plant (Chisso Co. Ltd.) in May 1956, Minamata City, Japan. A high level of mercury contamination, around 5.61–35.7 ppm, was reported in the marine products. People who ingested shellfish and fish reported ailments such as dysarthria, auditory disturbances, ataxia, sensory disturbances, tremors, and constriction of the visual field [144]. With the decline in serious cases, there was a gradual increase in the patients with chronic illnesses. Thousands of cases of chronic illness have been recognized.

A significant amount of wastewater was being discharged into the Minamata Bay from 1932 to 1958. The factory was shut down in 1968 by that time it had converted to a dead bay. Till date, studies on the Minamata disease are still on, with the reported cases of 12890 people in the study. Patients suffer from severe pain, paralysis, convulsions, blindness, and deafness, mental health, and growth problems [145].

The victim compensation for the affected population and bay clean up amounts to about 2.829 billion USD.

4.8.4 Kala Shah Kaku Industrial Complex, Pakistan

Studies have been reported on the contamination of groundwater by Kala Shah Kaku, an industrial area in Punjab [146]. This region is dominated by industries such as paper and pulp, tanneries, steel, textile, refining industries, petrochemicals, food processing industries, chemical industries, etc. The generated wastewater is let into the Deg Nalla, which eventually enters the River Ravi. Ground water samples were analyzed, where it was observed that the hardness, turbidity, potassium level, total dissolved solids, and conductivity were high and outside the range, as prescribed by the World Health Organization (WHO). Moreover, the industrial discharges are mixed with municipal water, making it unfit for irrigation [147]. The release of this contaminated discharge has been a matter of concern.

Other studies [148] of this region reported a high total coliform count and *Escherichia coli* for the majority of the water samples. This was a result of the discharge of highly polluted wastewater from the industries in this region. Several heavy metals were also found in several water samples, such as Cd, Mn, Zn, Fe, Ni, Cr, Pb, and Cu.

4.9 Industry Best Practices

These are the different remarkable methods and measures that have been adopted by the industries. Some of the general ones have been mentioned below.

4.9.1 Zero Liquid Discharge

As several industrial contaminants are let into the water streams from various industrial technologies, such as ZLD, are being adopted to regulate contamination and efficient use of water. This technology eliminates the liquid waste that is generated, thereby maximizing the efficiency of water usage. ZLD often involves a high energy usage and cost [148]. Despite the mentioned, ZLD is gaining attention globally, making its use necessary due to the arising scarcity of water and the expenditure involved in the disposal of wastewater [149]. ZLD is fast being adopted from developed nations to developing nations such as India.

In the previous years in the ZLD thermal method, evaporation of brine was carried out in a brine concentrator, a brine crystallizer, and an evaporation pond. The distillate is reused, whereas the residue is used for the recovery of salts or sent to landfill [150]. In the current scenario, technologies such as membrane technology, reverse osmosis, etc., are being adopted. The impacts of ZLD on the environment include GHG emissions, for which modified ZLD technologies are being developed.

4.9.2 Advanced Treatment Technologies for Wastewater

Understanding the hazardous effects of water pollution and the need to treat and remove the contaminants, it becomes essential to adopt and develop advanced methods of treatment for wastewater from several industrial setups. Details of different methods that are being used have been dealt with and discussed in different chapters of this book elaborately.

Emphasis is also being given to innovative methods and materials for the contaminant treatment. In addition to this, due emphasis is being given to innovations in the field with the use of the latest techniques, which also have the potential for integration with other techniques for enhanced efficiency of contaminant removal.

4.9.3 Adoption of Green Alternatives

Adoption of greener materials has been emphasized for combating the production of harmful and toxic contaminants in the water streams. The products being used for several processes, the solvents used, and the operation of different processes is being carried out and tried in greener ways. Green chemistry leads to a transformation of several industrial practices into sustainable services.

In terms of the environmental impact, the traditional manufacturing processes lead to the generation of toxic contaminants with a high consumption of energy. Green chemistry, on the other hand, reduces the toxic impacts, improving the conditions, resulting in minimal environmental degradation. Green chemistry and greener ways focus more on the use of environment-friendly materials, leading to enhanced sustainability. Several energy-saving innovations result in less consumption of resources, proving to be environmentally friendly. Some key principles of green chemistry include the prevention of wastes, improved energy efficiency, design of safer chemicals, use of renewable feed stalks, and atom economy, dealing with the maximum utilization of reactants in the formation of products [150].

Despite several advantages in the use of greener approaches, there could be possible limitations such as economic feasibility, lack of support by the regulatory authorities, and technological limitations. These application challenges for the greener approach come with the possible solutions with provisions of subsidies and incentives, establishment of stronger regulatory frameworks, and investments in research and development [151].

4.9.4 Regulatory Compliance

Having a regulatory framework well in place for the emergence of all the emitted contaminants from several industries is a must. With the changing times and several emerging contaminants, a modification in the policies and framework becomes a matter of utmost importance. The use of risk assessment tools and their updation, depending upon the emerging risks is extremely beneficial for gaining control over the emergence of industrial contaminants.

As per the European Union's Industrial Emissions Directive (IED), the use of Best Available Techniques (BAT) becomes a mandate for the minimization of emissions of contaminants in wastewater discharges [151]. BAT are the most environmentally friendly and economical technique for several industrial sectors, as agreed in the "Sevilla process" in the European Commission's Joint Research Centre, in Seville, Spain. In this, the prominent areas of issues have been identified requiring attention, such as heavy metals, chemicals, etc., arising from and linked to industrial pollution. In the United States of America, the Environmental Protection Agency provides the effluent discharge guidelines to several industries [152]. Regular monitoring and reporting of the standards is an essential component.

4.9.5 Industrial Wastes and Wastewater Through the Circular Economy

In this concept, waste being generated is minimized, making the best of the resources. The key aspects of circular economy are resource recovery, water reuse, minimization of waste, and a closed-loop approach where the recovered sources are again reused in the industries [153, 154]. Wastes emerging from industrial wastewater, such as heavy metals, nutrients, etc., are recovered and reused [155].

4.10 Challenges and Research Directions

There are significant challenges in the emergence of industrial contaminants in the water streams due to the diversity of the contaminants. Certain contaminants of emerging nature, such as microplastics, pharmaceuticals, etc., resist the conventional treatment methods, eventually impacting humans and aquatic flora and fauna.

Advanced technologies are generally energy-intensive, costly, and with limitations of scalability. This emphasizes the need for research and innovation to come up with several feasible, cost-effective, and environmentally friendly alternatives. More freedom is required to be given to research through academic institutions. In addition to this integration of technologies is being explored to serve the purpose of the treatment of wastewater contaminants. One such example is Europe, where there has been much focus on the integration of novel technologies with the existing WWTPs (wastewater treatment plants) [156].

References

1 https://npolluted.com/industries-that-contribute-most-to-water-pollution/ (Last accessed 19.06.25).
2 Mekuria, D.M., Kassegne, A.B., and Asfaw, S.L. (2021). Assessing pollution profiles along Little Akaki River receiving municipal and industrial wastewaters, central Ethiopia: implications for environmental and public health safety. Heliyon, Volume 7, Issue 7, p. e07526. https://doi.org/10.1016/j.heliyon.2021.e07526.
3 Johnson, D.B., and Hallberg, K.B. (2005). Acid mine drainage remediation options: a review. Science of the Total Environment, Volume 338, Issues 1–2, pp. 3–14. https://doi.org/10.1016/j.scitotenv.2004.09.002.
4 Simate, G.S., and Ndlovu, S. (2014). Acid mine drainage: challenges and opportunities. Journal of Environmental Chemical Engineering, Volume 2, Issue 3, pp. 1785–1803. https://doi.org/10.1016/j.jece.2014.07.021.
5 Hamilton, P.A., Miller, T.L., and Myers, D.N. (2004). Water Quality in the Nation's Streams and Aquifers Overview of Selected Findings, 1991–2001. Circular 1265. U.S. Dept. of the Interior, U.S. Geological Survey, Reston, VA, https://doi.org/10.3133/cir1265.
6 United Nations Environment Program (UNEP) (2008). Draft Final Review of Scientific Information on Lead. Europe, 211.

7 Aqeel, M., Jamil, M., and Yusoff, I. (2014). Soil contamination, risk assessment and remediation. In: Hernandez-Soriano MC (ed). Environmental risk assessment of soil contamination, InTech. https://doi.org/10.5772/57287.

8 Tillett, T. (2005). Promoting health in Texas colonias. Environmental Health Perspectives, Volume 113, Issue 7, pp. 452–55.

9 Kesic, R., Elliott, J.E., and Fremlin, K.M., et al. (2021). Continuing persistence and biomagnification of DDT and metabolites in northern temperate fruit orchard avian food chains. Environmental Toxicology and Chemistry, Volume 40, Issue 12, pp. 3379–3391. https://doi.org/10.1002/etc.5220.

10 Niu, S., Xia, Y., Yang, C., et al. (2023). Impacts of the steel industry on sediment pollution by heavy metals in urban water system. Environmental Pollution, Volume 335, p. 122364. https://doi.org/10.1016/j.envpol.2023.122364.

11 Kadadou, D., Said, E.A., and Ajaj, R., et al. (2023). Research advances in nuclear wastewater treatment using conventional and hybrid technologies: towards sustainable wastewater reuse and recovery. Journal of Water Process Engineering, Volume 52, p. 103604. https://doi.org/10.1016/j.jwpe.2023.103604.

12 https://world-nuclear.org/information-library/nuclear-fuel-cycle/fuel-recycling/processing-of-used-nuclear-fuel (Last accessed 03.05.25).

13 Mandeep, Gupta, G.K., and Liu, H., et al. (2019). Pulp and paper industry–based pollutants, their health hazards and environmental risks. Current Opinion in Environmental Science and Health, Volume 12, pp. 48–56. https://doi.org/10.1016/j.coesh.2019.09.010.

14 Yankovych, H., Vaclavikova, M., and Melnyk, I. (2023). A review on absorbable organic halogens treatment technologies: approaches and application. Sustainability, Volume 15, Issue 12, p. 9601. https://doi.org/10.3390/su15129601.

15 Ali, M., and Sreekrishnan, T. (2001). Aquatic toxicity from pulp and paper mill effluents: a review. Advances in Environmental Research, Volume 5, Issue 2, pp. 175–196. https://doi.org/10.1016/S1093-0191(00)00055-1.

16 Singh, A.K., and Chandra, R. (2019). Pollutants released from the pulp paper industry: aquatic toxicity and their health hazards. Aquatic Toxicology, Volume 211, pp. 202–216, https://doi.org/10.1016/j.aquatox.2019.04.007.

17 Seiler, J.P. (1991). Pentachlorophenol. Mutation Research, Volume 257, Issue 1, pp. 27–47. https://doi.org/10.1016/0165-1110(91)90018-q.

18 https://organicabiotech.com/dairy-industry-causes-water-pollution/ (Last accessed 05.05.25).

19 Water Pollution from Slaughter Houses, Three Quarters of U.S. Meat Processing Plants that Discharge into Waterways Violated their Permits, 2016–2018, Report-Environmental Integrity Project (EIP) October 11, 2018 (Last accessed 05.05.25).

20 Singh, P., Verma, A., Pratibha, et al. (2025). A review on the environmental impact and management of sugar mill effluent through phytoremediation.

Environmental Quality Management, Volume 34, Issue 3, p. e70027. https://doi.org/10.1002/tqem.70027.

21 Falletti, L., Conte, L., and Zaggia, A., et al. (2015). Food industry wastewater treatment plant based on flotation and MBBR. Modern Environmental Science and Engineering, Volume 1, Issue 2, pp. 94–98. https://doi.org/10.15341/mese(2333-2581)/02.01.2015/006.

22 Venugopal, V., and Sasidharan, A. (2021). Seafood industry effluents: environmental hazards, treatment and resource recovery. Journal of Environmental Chemical Engineering, Volume 9, Issue 2, p. 104758. https://doi.org/10.1016/j.jece.2020.104758.

23 Sharma, A., Sharma, P., Sharma, A., et al. (2017). Hazardous effects of petrochemical industries: a review. Recent Advances in Petrochemical Science, Volume 3, Issue 2, p. 555607.

24 Israel, A.U., Obot, I.B., Umoren, S.A., et al. (2007). Effluents and solid waste analysis in a petrochemical company- a case study of Eleme petrochemical company Ltd, Port Harcourt, Nigeria. Journal of Chemistry, Volume 5, Issue 1, pp. 74–80. https://doi.org/10.1155/2008/805957.

25 Lin, M.C., Yu, H.S., Tsai, S.S., et al. (2001). Adverse pregnancy outcome in a petrochemical polluted area in Taiwan. Journal of Toxicology and Environmental Health. Part A, Volume 63, Issue 8, pp. 565–574. https://doi.org/10.1080/152873901316857743.

26 https://environmentalintegrity.org/news/report-documents-epas-failure-to-regulate-wastewater-from-plastics-plants/ (Last accessed 05.05.25).

27 Jayapal, M., Vijayakumar, M., and Jagadeesan, H. (2024). Aerobic treatment processes for resin industry wastewater. In: Arun Ramnath R, Sanjay MR, Sergey M, et al. (eds). Woodhead publishing series in composites science and engineering, biocomposites - bio-based fibers and polymers from renewable resources, Woodhead Publishing, pp. 295–317. https://doi.org/10.1016/B978-0-323-97282-6.00005-.

28 Patel, K.K.S., and Mohanty, B. (2023). Treatment of wastewater containing hazardous pollutant-phenol generating from resin manufacturing industry by adsorption. International Journal of Research in Engineering and Science (IJRES), Volume 11, Issue 3, pp. 26–32.

29 Maheshwari, K., Agrawal, M., and Gupta, A.B. (2021). Dye pollution in water and wastewater. In: Muthu SS, Khadir A (eds). Novel materials for dye-containing wastewater treatment. sustainable textiles: production, processing, manufacturing & chemistry. Springer, Singapore. https://doi.org/10.1007/978-981-16-2892-4_1.

30 Lacasse, K., and Baumann, W. (2004). Textile chemicals-environmental data and facts Autoren, Springer Verlag Berlin Heidelberg, Verlag.

31 Hejna, M., Kapuścińska, D., and Aksmann, A. (2022). Pharmaceuticals in the aquatic environment: a review on eco-toxicology and the remediation potential of algae. International Journal of Environmental Research and Public Health, Volume 19, Issue 13, p. 7717. https://doi.org/10.3390/ijerph19137717.

32 Estrada-Almeida, A.G., Castrejón-Godínez, M.L., Mussali-Galante, P., et al. (2024). Pharmaceutical pollutants: ecotoxicological impacts and the use of agro-industrial waste for their removal from aquatic environments. Journal of Xenobiotics, Volume 1, Issue 4, pp. 1465–1518. https://doi.org/10.3390/jox14040082.

33 Arora, J., Ranjan, A., and Chauhan, A. et al. (2022). Surfactant pollution, an emerging threat to ecosystem: approaches for effective bacterial degradation. Journal of Applied Microbiology, Volume 133, pp. 1229–1244. https://doi.org/10.1111/jam.15631.

34 https://eponline.com/Articles/2017/03/27/The-Environmental-Impacts-of-Boating.aspx (Last accessed 05.05.25).

35 Whitman, R.L., and Nevers, M.B. (2003). Foreshore sand as a source of *Escherichia coli* in nearshore water of a Lake Michigan Beach. Applied and Environmental Microbiology, Volume 69, Issue 9. https://doi.org/10.1128/AEM.69.9.5555-5562.2003.

36 Kato, S., and Kansha, Y. (2024). Comprehensive review of industrial wastewater treatment techniques. Environmental Science and Pollution Research, Volume 31, pp. 51064–51097.

37 Ahmed, J., Thakur, A., and Goyal, A. (2021). Biological treatment of industrial wastewater. In: Shah MP (ed) The royal society of chemistry. Royal Society of Chemistry, pp. 1–14. https://doi.org/10.1039/9781839165399.

38 Vysokomornaya, O.V., Kurilenko, E.Y., and Shcherbinina, A.A. (2015). Major contaminants in industrial and domestic wastewater. MATEC Web of Conferences, Volume 23, p. 01041.

39 https://www.lgsonic.com/water-contamination/ (Last accessed 05.05.25).

40 Vallero, D.A. (2019). Chapter 20: Thermal pollution. In: Letcher TM, Vallero DA (eds) Waste, 2e. Academic Press, pp. 381–404. https://doi.org/10.1016/B978-0-12-815060-3.00020-7.

41 Kadadou, D., Tizani, L., Alsafar, H., et al. (2024). Analytical methods for determining environmental contaminants of concern in water and wastewater. Methods X, Volume 12, p. 102582. https://doi.org/10.1016/j.mex.2024.102582.

42 Terry, L.R., Sanders, Potoff, R.H., et al. (2022). Applications of surface-enhanced Raman spectroscopy in environmental detection. Analytical Science Advances, Volume 3, Issue 3-4, pp. 113–145. https://doi.org/10.1002/ansa.202200003.

43 Jing, Z.-B., Wang, W.L., Nong, Y.J., et al. (2023). Fluorescence analysis for water characterization: measurement processes, influencing factors, and data analysis. Water Reuse, Volume 13, Issue 1, pp. 33–50. https://doi.org/10.2166/wrd.2022.065.

44 Fdez-Sanromán, A., Bernárdez-Rodas, N., Rosales, E., et al. (2025). Biosensor technologies for water quality: detection of emerging contaminants and pathogens. Biosensors, Volume 15, Issue 3, p. 189. https://doi.org/10.3390/bios15030189.

45 Aryal, P., Hefner, C., Martinez, B., et al. (2024). Microfluidics in environmental analysis: advancements, challenges, and future prospects for rapid and efficient monitoring. Lab Chip, Volume 24, pp. 1175–1206. https://doi.org/10.1039/D3LC00871A.

46 Abdelmonem, B.H., Kamal, L.T., Elbaz, R.M., et al. (2025). From contamination to detection: the growing threat of heavy metals. Heliyon, Volume 11, Issue 1, e41713. https://doi.org/10.1016/j.heliyon.2025.e41713.

47 Olayiwola Idris, A., Orimolade, B., Dennany, L., et al. (2023). A review on monitoring of organic pollutants in wastewater using electrochemical approach. Electrocatalysis, Volume 14, pp. 659–687.

48 Willian, N., Pardi, H., and Fitriyah, D. (2023). Colorimetric detection of mercury using silver nanoparticles: a review. BIO Web of Conferences, Volume 70, p. 01008. https://doi.org/10.1051/bioconf/20237001008.

49 De Matteis, V., Cascione, M., Fella, G., et al. (2020). Colorimetric paper-based device for hazardous compounds detection in air and water: a proof of concept. Sensors, Volume 20, Issue 19, p. 5502. https://doi.org/10.3390/s20195502.

50 Senila, M. (2023). Metal and metalloid monitoring in water by passive sampling – a review. Reviews in Analytical Chemistry, Volume 42, Issue 1, p. 20230065. https://doi.org/10.1515/revac-2023-0065.

51 Vu, C.T., Zahrani, A.A., and Duan, L. (2023). A glass-fiber-optic turbidity sensor for real-time In situ water quality monitoring. Sensors, Volume 23, Issue 16, p. 7271. https://doi.org/10.3390/s23167271.

52 Rocher, J., Jimenez, J.M., and Tomas, J. (2023). Low-cost turbidity sensor to determine eutrophication in water bodies. Sensors, Volume 23, Issue 8, p. 3913. https://doi.org/10.3390/s23083913.

53 Pahlavan, R., Pahlavan, N., Tabrizchi, M., et al. (2012). In situ trace analysis of oil in water with mid-infrared fiberoptic chemical sensors. Analytical Chemistry, Volume 84, Issue 3, pp. 1274–1280. https://doi.org/10.1021/ac201664p.

54 Koppanen, M., Kesti, T., Kokko, M., et al. (2022). An online flow-imaging particle counter and conventional water quality sensors detect drinking water contamination in the presence of normal water quality fluctuations. Water Research, Volume 213, p. 118149. https://doi.org/10.1016/j.watres.2022.118149.

55 Can, N., and Altindal, A. (2017). Gravimetric detection of pesticide in water samples. 2nd International Conference on Civil and Environmental Engineering (Cappadocia-2017), Nevşehir, Turkey, p. 3430.

56 Yang, J., Hu, Y., Zhanget, J., et al. (2023). Identification of marine oil spill pollution using hyperspectral combined with thermal infrared remote sensing, Frontiers in Marine Science, Volume 10, p. 1135356. https://doi.org/10.3389/fmars.2023.1135356.

57 Tang, J., Lin, H., and Tian, Q. (2024). Design and realization of a water quality monitoring system based on the Internet of Things. Water Practice and Technology, Volume 19, Issue 9, pp. 3538–3554. https://doi.org/10.2166/wpt.2024.222.

58 Raju, B., Kumar, R., and Dhanalakshmi, S. (2021). Review of fiber optical sensors and its importance in sewer corrosion factor analysis. Chemosensors, Volume 9, Issue 6, p. 118. https://doi.org/10.3390/chemosensors9060118.

59 Moschino, V., Delaney, E.L., and Ros, D. (2012). Assessing the significance of Ruditapes philippinarum as a sentinel for sediment pollution: bioaccumulation and biomarker responses. Environmental Pollution, Volume 171, pp. 52–60. https://doi.org/10.1016/j.envpol.2012.07.024.

60 Hassan, M., Khan, M.J., and Ali, S.S. (2021). Ground water quality assessment in industrial zones of central Punjab, Pakistan. Pakistan Journal of Science, Volume 73, Issue 1, pp. 177–185.

61 https://library.fiveable.me/environmental-chemistry-ii/unit-6/advection-dispersion-diffusion-aquatic-systems/study-guide/m1DvxH4SD3VX2Ph5 (Last accessed 13.05.25).

62 Piwoni, M.D., and Keeley, J.W. (1990). Basic concepts of contaminant sorption at hazardous waste sites (Ground water issue), EPA.

63 https://www.atsdr.cdc.gov/pha-guidance/conducting_scientific_evaluations/exposure_pathways/environmental_fate_and_transport.html (Last accessed 13.05.25).

64 https://www.gy4es.org/post/bioaccumulation-causes-effects-solutions (Last accessed 13.05.25).

65 Drouillard, K.G. (2008). Biomagnification, In: Jørgensen SE, Fath BD (eds). Encyclopedia of ecology. Academic Press, pp. 441–448. https://doi.org/10.1016/B978-008045405-4.00377-3.

66 British Geological Survey. Groundwater Fact Sheet: The Impact of Industrial Activity. https://nora.nerc.ac.uk/id/eprint/528008.

67 Morris, B.L., Lawrence, A.R., Chilton, P.J., et al. (2003). Groundwater and its susceptibility to degradation: a global assessment of the problems and options for management. Early Warning and Assessment Report Series, RS, 03-3. United Nations Environment Programme, Nairobi, Kenya.

68 https://krakensense.com/blog/water-pollution-aquatic-biodiversity (Last accessed 14.05.25).
69 Fu, F., and Wang, Q. (2011). Removal of heavy metal ions from wastewaters: a review. Journal of Environmental Management, Volume 92, pp. 407–418.
70 Wang, Z., Xue, M., Huang, K., et al. (2011). Textile dyeing wastewater treatment. Advances in Treating Textile Effluent, Volume 5, pp. 91–116.
71 Buyukkamaci, N., and Koken, E. (2010). Economic evaluation of alternative wastewater treatment plant options for pulp and paper industry. Science of the Total Environment, Volume 408, pp. 6070–6078.
72 Lindholm-Lehto, P.C., Knuutinen, J S., Ahkola, H.S., et al. (2015). Refractory organic pollutants and toxicity in pulp and paper mill wastewaters. Environment and Pollution Research, Volume 22, pp. 6473–6499.
73 Koedrith, P., and Seo, Y.R. (2011). Advances in carcinogenic metal toxicity and potential molecular markers. International Journal of Molecular Science, Volume 12, Issue 12, pp. 9576–9595.
74 Krithika, D., and Philip, L. (2016). Treatment of wastewater from water-based paint industries using submerged attached growth reactor. International Biodeterioration and Biodegradation, Volume 107, pp. 31–41.
75 Ranjan, N., Singh, P.K., and Maurya, N.S. (2022). Pharmaceuticals in water as emerging pollutants for river health: a critical review under Indian conditions. Ecotoxicology and Environmental Safety, Volume 247, p. 114220.
76 Aktar, W., Sengupta, D., and Chowdhury, A. (2009). Impact of pesticides use in agriculture: their benefits and hazards. Interdisciplinary Toxicology, Volume 2, pp. 1–12.
77 https://www.epa.gov/nps/nonpoint-source-agriculture (Last accessed 20.09.24).
78 Rahim, M.A., and Mostafa, M.G. (2021). Impact of sugar mills effluent on environment around mills area. AIMS Environmental Science, Volume 8, Issue 1, pp. 86–99. https://doi.org/10.3934/environsci.2021006.
79 http://swachhbharaturban.gov.in/writereaddata/SBM%20Plastic%20Waste%20Book.pdf (Last accessed 20.08.24).
80 Rizwan, K., Rasheed, T., Bilal, M., et al. (2022). Pulp and paper industry-based pollutants, and their adverse impacts. In: Bhat R, Nadda AK, Nguyen TA (eds) Micro and nanotechnologies, nanotechnology in paper and wood engineering, fundamentals, challenges and applications. Elsevier, pp. 143–160.
81 Santo, C.E., Vilar, V.J.P., Bhatnagar, A., et al. (2013). Biological treatment by activated sludge of petroleum refinery wastewaters. Desalination and water treatment, Volume 51, Issue 34–36, pp. 6641–6654.
82 https://books.rsc.org/books/edited-volume/937/chapter/741366/Industrial-wastewater-and-its-toxic-effects (Last accessed 20.09.24).

83 Ahmed, M., Musthafa, O.M., Giwa, A., et al. (2022). Recent developments in hazardous pollutants removal from wastewater and water reuse within a circular economy. Clean water, Volume 5, Article number 12.

84 Hua, M., Shujuan, Z., Bingcai, P., et al. (2012). Heavy metal removal from water/wastewater by nanosized metal oxides: a review. Journal of Hazardous Materials, Volume 211–212, pp. 317–331. https://doi.org/10.1016/j.jhazmat.2011.10.016.

85 Worch, E. (2012). Adsorption technology in water treatment: fundamentals, processes, and modeling, De Gruyter. https://doi.org/10.1515/9783110240238.

86 Adeleye, A.S., Jon, R.C., Kendra, G., et al. (2016). Engineered nanomaterials for water treatment and remediation: costs, benefits, and applicability. Chemical Engineering Journal, Volume 286, pp. 640–662. https://doi.org/10.1016/j.cej.2015.10.105.

87 Sahu, D., Pervez, S., Karbhal, I., et al. (2024). Applications of different adsorbent materials for the removal of organic and inorganic contaminants from water and wastewater – a review. Desalination and Water Treatment, Volume 317, p. 100253.

88 Datta, S., Gupta, B., Srivastava, S.K., et al. (2024). Recent advances on the removal of dyes from wastewater using various adsorbents: a critical review. Materials Advances, Volume 2, Issue 14, pp. 4497–4531. https://doi.org/10.1039/D1MA00354B.

89 Turksoy, R., Tergiozlu, G., Ertugrul, Y.I.E., et al. (2021). Removal of heavy metals from textile industry wastewater. Frontiers in Life Sciences and Related Technologies, Volume 2, Issue 2, pp. 44–50. https://doi.org/10.51753/flsrt.958165.

90 Hussain, Z., Fadhil, Z., Kareem, S., et al. (2020). Removal of organic contaminants from textile wastewater by adsorption on natural biosorbent. Materials Science Forum, 1662-9752, Volume 1002, pp. 489–497.

91 Vinayagam, V., Murugan, S., Kumaresan, R., et al. (2022). Sustainable adsorbents for the removal of pharmaceuticals from wastewater: a review. Chemosphere, Volume 300, p. 134597. https://doi.org/10.1016/j.chemosphere.2022.134597.

92 Naeini, A.L., and Moradi, S.A.H. (2023). Adsorption method for removal of pharmaceuticals from wastewater: review. Iranian Journal of Materials Science and Engineering, Volume 20, Issue 4, pp. 1–12. https://doi.org/10.22068/ijmse.3385.

93 Chauhan, S., Shafi, T., Dubey, B.K., et al. (2023). Biochar-mediated removal of pharmaceutical compounds from aqueous matrices via adsorption. Waste Disposal and Sustainable Energy, Volume 5, pp. 37–62. https://doi.org/10.1007/s42768-022-00118-y.

94 Rosli, F.A., Ahmad, H., Jumbri, K., et al. (2021). Efficient removal of pharmaceuticals from water using graphene nanoplatelets as adsorbent. Royal Society Open Science, Volume 8, Issue 1, p. 201076. https://doi.org/10.1098/rsos.201076.

95 Ahmed, I., Osman, A.A., Farghali, M., et al. (2024). Advanced adsorbents for ibuprofen removal from aquatic environments: a review. Environmental Chemistry Letter, Volume 22, pp. 373–418. https://doi.org/10.1007/s10311-023-01647-6.

96 Aprianti, T., Miskah, S., Komala, R., et al. (2018). Heavy metal ions adsorption from pulp and paper industry wastewater using zeolite/activated carbon-ceramic composite adsorbent. AIP Conference Proceedings, Volume 2014, p. 020127.

97 Kakkar, S., Malik, A., and Malik, A. (2018). Treatment of pulp and paper mill effluent using low-cost adsorbents: an overview. Journal of Applied and Natural Science, Volume 10, Issue 2, pp. 695–704. https://doi.org/10.31018/jans.v10i2.1769.

98 Upadhyay, K., Maru, S., and Yadav, R. (2012). Colour removal from paper mill effluent using activated carbon. Journal of Industrial Pollution Control, Volume 28, Issue 1, pp. 45–50.

99 Oladimeji, T., Fawole, G.M., Emetere, M.E, et al. (2023). Adsorption of heavy metals from paint industry wastewater using activated African cherry seed shell and brewers spent grain. https://doi.org/10.20944/preprints202312.1286.v1.

100 Vahabisani, A., and An, C. (2021). Use of biomass-derived adsorbents for the removal of petroleum pollutants from water: a mini-review. Environmental Systems Research, Volume 10, Issue 1, Article number 25.

101 Ziati, M., Khemili, S., and Metali, A. (2023). Organic pollutants removal from petroleum refinery effluents through adsorption using anthracite coal: kinetics, isotherms, and thermodynamic modeling, Scientia Iranica. https://doi.org/10.24200/SCI.2023.62631.7943.

102 Tolkou, A.K., Tsoutsa, E.K., Kyzas, G.Z., et al. (2024). Sustainable use of low-cost adsorbents prepared from waste fruit peels for the removal of selected reactive and basic dyes found in wastewaters. Environmental Science and Pollution Research International, Volume 31, pp. 14662–14689. https://doi.org/10.1007/s11356-024-31868-3.

103 Ahmed, H. (2022). Nanoparticles made from industrial waste remove water pollutants. AZoNano (04 August). https://www.azonano.com/news.aspx?newsID=39595) (Last accessed 24.11.2023).

104 Teng, L., Chuan, Y., and Ganwei, Z. (2023). Epoxied SiO_2 nanoparticles and polyethyleneimine (PEI) coated polyvinylidene fluoride (PVDF) membrane for improved oil water separation, anti-fouling, dye and heavy metal ions removal capabilities. Journal of Colloid and Interface Science. https://doi.org/10.1016/j.jcis.2022.09.148.

105 Lu, Q., Xiaoming, Z., and Ngie, H.W. (2022). Anti-biofouling polyvinylidene fluoride/quaternized polyvinyl alcohol ultrafiltration membrane selectively separates aromatic contaminants from wastewater by host–guest interactions. Separation and Purification Technology, Volume 296, p. 121387. https://doi.org/10.1016/j.seppur.2022.121387.

106 Ladewig, B.P., and Al-Shaeli, M. (2017). Fouling in membrane bioreactors, In: Fundamentals of membrane bioreactors, Springer, Singapore, pp. 39–85. https://doi.org/10.1007/978-981-10-2014-8_3.

107 Zainith, S., and Bharagava, R.N. (2021). Membrane-based hybrid processes in industrial waste effluent treatment. In: Shah MP, Rodriguez-Couto S (eds) Membrane-based hybrid processes for wastewater treatment. Elsevier, pp. 205–226.

108 Chang, M.H., Liang, Y., Meng, Y., et al. (2024). Research progress and trends on state-of-the-art membrane technologies in textile wastewater treatment. Separation and Purification Technology, Volume 333, p. 125853. https://doi.org/10.1016/j.seppur.2023.125853.

109 Yang, J., Monnot, M., Eljaddi, T., et al. (2021). Ultrafiltration as tertiary treatment for municipal wastewater reuse. Separation and Purification Technology, Volume 272, p. 118921. https://doi.org/10.1016/j.seppur.2021.118921.

110 Mohammad, A.W., Teow, Y.H., and Ang, W.L., et al. (2015). Nanofiltration membranes review: recent advances and future prospects. Desalination, Volume 356, pp. 226–254. https://doi.org/10.1016/j.desal.2014.10.043.

111 Oatley-Radcliffe, D.L. Matthew, W., Thomas, J.A., et al. (2017). Nanofiltration membranes and processes: a review of research trends over the past decade. Journal of Water Process Engineering, Volume 19, pp. 164–171. https://doi.org/10.1016/j.jwpe.2017.07.026.

112 Wilf, M., and Steven, A. (2000). Application of low fouling RO membrane elements for reclamation of municipal wastewater. Desalination, Volume 132, Issues 1–3, pp. 11–19. https://doi.org/10.1016/S0011-9164(00)00130-2.

113 Malaeb, L., and George, M. A. (2011). Reverse osmosis technology for water treatment: state of the art review. Desalination, Volume 267, Issue 1, pp. 1–8. https://doi.org/10.1016/j.desal.2010.09.001.

114 Ezugbe, E.O., and Rathilal, S. (2020). Membrane technologies in wastewater treatment: a review. Membranes (Basel), Volume 10, Issue 5, p. 89.

115 Giakoumis, T., and Voulvoulis, N. (2020). Wastewater treatment and reuse – present and future perspectives in technological developments and management issues. In: Verlicchi P (ed) Advances in chemical pollution, environmental management and protection. Academic Press, pp. 1–252.

116 Benito, A., Gerard, G., and Rafael, G. (2017). Fouling reduction by UV-based pretreatment in hollow fiber ultrafiltration membranes for urban wastewater reuse. Journal of Membrane Science, Volume 536, pp. 141–147. https://doi.org/10.1016/j.memsci.2017.04.070.

117 https://www.forbes.com/home-improvement/home/reverse-osmosis-water-pros-cons/ (Last accessed 14.05.24).

118 Jafarinejad, S., and Esfahani, M.R. (2021). A review on the nanofiltration process for treating wastewaters from the petroleum industry. Separations, Volume 8, Issue 11, p. 206. https://doi.org/10.3390/separations8110206.

119 Guohua, C. (2004). Electrochemical technologies in wastewater treatment. Separation and purification technology, Volume 38, Issue 1, pp. 11–41. https://doi.org/10.1016/j.seppur.2003.10.006.

120 Martínez-Huitle, C.A., and Panizza, M. (2018). Electrochemical oxidation of organic pollutants for wastewater treatment. Current Opinion in Electrochemistry, Volume 11, pp. 62–71. https://doi.org/10.1016/j.coelec.2018.07.010.

121 Feng, Y., Yang, L., Liu, J., et al. (2016). Electrochemical technologies for wastewater treatment and resource reclamation. Environmental Science: Water Resource and Technology, Volume 2, pp. 800–831. https://doi.org/10.1039/C5EW00289C.

122 Manna, M., and Sen, S. (2023). Advanced oxidation process: a sustainable technology for treating refractory organic compounds present in industrial wastewater. Environmental Science and Pollution Research, Volume 30, Issue 10, pp. 25477–25505. https://doi.org/10.1007/s11356-022-19435-0.

123 Parulekar, P.J., Sahasrabudhe, I., Hegiste, M., et al. (2024). Review of advanced oxidation processes for waste water treatment. International Journal of Engineering Research and Technology, Volume 13, Issue 2. https://doi.org/10.17577/ijertv13IS020100.

124 Malik, S.N., Ghosh, P.C., Vaidya, A.N., et al. (2020). Hybrid ozonation process for industrial wastewater treatment: principles and applications: a review. Journal of Water Process Engineering, Volume 35, p. 101193. https://doi.org/10.1016/j.jwpe.2020.101193.

125 Bazargan, A. (ed). Photocatalytic water and wastewater treatment. University of Tehran, IWA Publishing.

126 Wang, B., Su, H., and Zhang, B. (2021). Hydrodynamic cavitation as a promising route for wastewater treatment – a review. Chemical Engineering Journal, Volume 412, p. 128685. https://doi.org/10.1016/j.cej.2021.128685.

127 Szaja, A., Montusiewicz, A., and Lebiocka, M. (2022). Challenges of hydrodynamic cavitation of organic wastes. Applied Sciences, Volume 12, Issue 15, p. 7936. https://doi.org/10.3390/app12157936.

128 Priyadarshini, M., Das, I., Ghangrekar, M.M., et al. (2022). Advanced oxidation processes: performance, advantages, and scale-up of emerging technologies. Journal of Environmental Management, Volume 316, p. 115295. https://doi.org/10.1016/j.jenvman.2022.115295.

129 Brosler, P., Girao, A.V., Silva R.F., et al. (2023). Electrochemical advanced oxidation processes using diamond technology: a critical review. Environments, Volume 10, Issue 2, p. 15. https://doi.org/10.3390/environments10020015.

130 Brosler, P., Girão, A.V., Silva, R.F., et al. (2018). Integrated oxidation process and biological treatment for highly concentrated petrochemical effluents: a review, Chemical Engineering and Processing - Process Intensification, Volume 125, pp. 183–196. https://doi.org/10.1016/j.cep.2018.02.002.

131 Nidheesh, P.V., Ravindran, V., Gopinath, A., et al. (2021). Emerging technologies for mixed industrial wastewater treatment in developing countries: an overview. Environment Quality Management, Volume 31, Issue 3, pp. 121–141. https://doi.org/10.1002/tqem.21762.

132 Bilal, M., and Ferreira, L.F.R. (2022). Recent trends in integrated wastewater treatment for sustainable development. Frontiers in Microbiology, Volume 13, p. 846503. https://doi.org/10.3389/fmicb.2022.846503.

133 Chang, H.L., Liu, Y.L., Keng, C.J., et al. (2024). Challenges of industrial wastewater treatment: utilizing membrane bioreactors (MBRs) in conjunction with artificial intelligence (AI) technology. Journal of Industrial and Production Engineering, Volume 41, Issue 5, pp. 422–427. https://doi.org/10.1080/21681015.2024.2330401.

134 Afanga, H., Zazou, H., Titchou, F.E., et al. (2020). Integrated electrochemical processes for textile industry wastewater treatment: system performances and sludge settling characteristics. Sustainable Environment Research, Volume 30, p. 2. https://doi.org/10.1186/s42834-019-0043-2.

135 Garcia-Garcia, A., Martinez-Miranda, V., Martinez-Cienfuegos, I.G., et al. (2015). Industrial wastewater treatment by electrocoagulation-electrooxidation processes powered by solar cells. Fuel, Volume 149, pp. 46–54. https://doi.org/10.1016/j.fuel.2014.09.080.

136 Zazou, H., Afanga, H., Akhouairi, S., et al. (2019). Treatment of textile industry wastewater by electrocoagulation coupled with electrochemical advanced oxidation process. Journal of Water Process Engineering, Volume 28, pp. 214–221. https://doi.org/10.1016/j.jwpe.2019.02.006.

137 Décio, A.A. Jr., Carolina, E.D.O., Maicon, S.N.d.S., et al. (2021). Integration of improved methods for the treatment of wastewater from a soft drink industry. Biointerface Research in Applied Chemistry, Volume 11, Issue 5, pp. 12946–12957.

138 Doltade, S.B., Yadav, Y., and Jadhav, N.L. (2022). Industrial wastewater treatment using oxidative integrated approach. South African Journal of Chemical Engineering, Volume 40, pp. 100–106.

139 Bhandari, V.M., Sorokhaibam, L.G., and Ranade, V.V. (2016). Industrial wastewater treatment for fertilizer industry-a case study. Desalination and Water Treatment, Volume 57, Issue 57, pp. 27934–27944. https://doi.org/10.1080/19443994.2016.1186399.

140 Vymazal, J. (2014). Constructed wetlands for treatment of industrial wastewaters: a review. Ecological Engineering, Volume 73, pp. 724–751. https://doi.org/10.1016/j.ecoleng.2014.09.034.

141 de la Varga, D., van Oirschot, D., Soto, M., et al. (2017). Constructed wetlands for industrial wastewater treatment and removal of nutrients. In: A. Val del Rio, J.L. Campos Gómez, and A. Mosquera Corral. Technologies for the treatment and recovery of nutrients from industrial wastewater, IGI Global, pp. 202–230.

142 Dan, N.P., Thanh, B.X., and Truong, B.D. (2006). Case studies of groundwater pollution in southeast Vietnam. International Review for Environmental Strategies, Volume 6, Issue 2, pp. 361–372.

143 https://uacvoice.org/2015/04/elk-river-blues/ (Last accessed 14.05.25).

144 Harada, M. (1995). Minamata disease: methylmercury poisoning in Japan caused by environmental pollution. Critical Reviews in Toxicology, Volume 25, Issue 1, pp. 1–24. https://doi.org/10.3109/10408449509089885.

145 https://cecr.vn/en/minamata-the-deadly-disease-caused-by-mercury-contaminated-water-in-japan/ (Last accessed 14.05.25).

146 Ahmad, I., Ahmad, Z., Kamran, M.A., et al. (2012). Ground water quality profile of Kala Shah Kaku. Pakistan Journal of Science, Volume 64, Issue 2, pp. 103–107.

147 Tariq, M., Anayat, A., Waseem, M., et al. (2020). Physicochemical and bacteriological characterization of industrial wastewater being discharged to surface water bodies: significant threat to environmental pollution and human health. Journal of Chemistry, Volume 2020, pp. 1–10.

148 Tong, T., and Elimelech, M. (2016). The global rise of zero liquid discharge for wastewater management: drivers, technologies, and future directions. Environmental Science and Technology, Volume 50, pp. 6846–6855. https://doi.org/10.1021/acs.est.6b01000.

149 Chavan, R.B. (2017). Towards zero liquid discharge. Advance Research in Textile Engineering, Volume 2, Issue 1, p. 1015.

150 Idoko, F.A., Ezeamii, G.C., and Ojochogwu, O.J. (2024). Green chemistry in manufacturing: innovations in reducing environmental impact. World Journal of Advanced Research and Reviews, Volume 23, Issue 3, pp. 2826–2841. https://doi.org/10.30574/wjarr.2024.23.3.2938.

151 https://www.eea.europa.eu/en/european-zero-pollution-dashboards/indicators/best-available-techniques-bat-to-cut-the-use-and-impact-of-hazardous-chemicals (Last accessed 12.05.25).

152 https://www.epa.gov/eg/industrial-effluent-guidelines?utm_source (Last accessed 12.05.25).

153 https://chemtech-us.com/circular-economy-tapping-into-wastewater-treatment/ (Last accessed 12.05.25).

154 https://circulareconomy.europa.eu/platform/en/good-practices/circular-economy-approach-industrial-wastewater (Last accessed 12.05.25).

155 Das, S., Singh, C., Sodhi, K., et al. (2023). Circular economy approaches for water reuse and emerging contaminant mitigation: innovations in water treatment. Environment, Development and Sustainability, Volume 27, pp. 753–5794. https://doi.org/10.1007/s10668-023-04183-z.

156 Soares, A. (2020). Wastewater treatment in 2050: challenges ahead and future vision in a European context. Environmental Science and Ecotechnology, Volume 28, Issue 2, p. 100030. https://doi.org/10.1016/j.ese.2020.100030.

5

Heavy Metals in Wastewater – Toxicity Assessment and Remediation

Heavy metals get introduced into water by various point and nonpoint sources like industries, agricultural activities, and domestic wastewater. They can enter the environment naturally or by anthropogenic actions. The natural sources of heavy metal pollution include natural phenomena like volcanic eruptions, rock weathering, biogenesis, and forest fires. Natural sources also include wet and dry deposition of atmospheric salt, water–rock interaction, and water–soil interaction.

Trace elements are formed by geographical processes like melting, sedimentation, crystallization, etc. The geological, hydrological, and geochemical nature of the water body governs the presence of heavy metals in the water. Rocks like granite, basalt, andesite, and ores like magnetite (Fe_3O_4), calamine (Fe_2O_4Zn), arsenopyrite (FeAsS), and calcite ($CaCO_3$), when they interact with water, are introduced into the water bodies. Among the anthropogenic sources are industrial processes such as the manufacturing of electronic devices, smelting, electroplating, mining operations, inputs from industries involved in fertilizer manufacturing, nuclear fuel-consuming plants, thermal power plants, and chemical processes. Industries are the prime sources of heavy metal introduction into the environment, either through direct discharge or through leakage from industrial sites. Many heavy metals are deposited due to atmospheric deposits, which are transported to water bodies through precipitation. Mercury (Hg) is introduced by agricultural activities. Nickel (Ni) enters due to corroded metal pipes and containers. Lead (Pb) finds its way into the water system through paints, petrol, and aerosol precipitation.

All these have serious implications on aquatic and human health. Changes in precipitation and human behavior have led to fluctuations in the level of heavy metal pollution in the water system. It is imperative to conduct a detailed investigation to understand the dynamics in the wastewater composition and sources of heavy metal pollution. It will be helpful to identify the main contributors of heavy metal pollutants, which can facilitate efficient and effective strategies for the mitigation of heavy metal pollution and its adverse effects on the aquatic environment and human health.

5.1 Sources of Heavy Metal Pollution

5.1.1 Point Sources

These sources of heavy metal ions cause a high amount of pollution in water. Industrial setups near water streams are the prime sources of metal pollution in water.

Point sources can be analyzed in two categories [1, 2]:

1. **Industrial Sources:** Several industrial units like paper and pulp, sugar, textile, leather, drug, metallurgical plants, and foodstuff industries discharge their wastewater into water bodies. As per the Central Pollution Control Board (CPCB), India, about 260 million liters of industrial wastewater are discharged into the river Ganga every day [3]. A study conducted in 2019 by the Ministry of Ecology and Environment, China, also reported 25 billion tons of industrial wastewater discharge [4]. Similarly, a published report by the World Bank estimated 1.5 billion liters of wastewater discharge in river streams daily [5]. This wastewater is loaded with a number of heavy metals toxic to the living system.
2. **Domestic Sources:** The domestic wastewater composition varies depending on the nature of waste and its dumping. It contains a large amount of metals, including other contaminants. Based on recent information, the Yamuna River in India is exceedingly contaminated by household wastewater. Studies suggest that about 85% of the surface water in Bangladesh is contaminated due to untreated discharge from domestic sources [6].

5.1.2 Nonpoint or Diffuse Sources

Nonpoint sources are diffused and are not located at one explicit location. The diffused sources are difficult to analyze in terms of their characteristics and attributes. These sources mainly include runoff from fields and agricultural practices. Fertilizers and insecticides employed in various agricultural activities have several types of metal composition. The monsoon period plays a momentous function in leaching, groundwater runoff, and drainage, and acts as a pathway of transporting the contaminants to the water sources, both surface and underground. Such a diffused source has a miscellaneous array of substances and materials.

Though the dumping and landfilling of hazardous materials is done as per the guidelines issued by the concerned authorities, due to a lack of adequate characteristics of waste characteristics, the landfills can lead to severe pollution in the immediate and surrounding regions. A specific concern in open dumping is the leachate movements, which are prominently loaded with toxic metals and can contaminate ground and surface water, including soil and vegetation. Heavy metals like lead (Pb), cadmium (Cd), chromium (Cr), copper (Cu),and zinc (Zn) have potential risks associated with their occurrence in wastewater.

Heavy metal pollution has raised public concern for environmental and human health safety. Understanding the sources, chemical transformation, and pathways of transportation, along with modes of deposition, is required to mitigate the risk of heavy metal pollution. In China, in Yuman region, mining operations for metal extraction and smelting are the main contributors of heavy metal contamination in the water bodies, while in Egypt, agricultural drainage water and sewage effluent are the main sources of pollution. In the African and Asian continents, fertilizers and insecticides, along with rock weathering, have majorly contributed to about 56.7% and 97.1% of metal pollution, respectively. The combined contribution of heavy metal pollution in North America includes mining and manufacturing, along with chemicals employed for agricultural activities. This contribution exceeds 90.4% for North America, while for South America, it is 93.5%. Also, sources of heavy metal contamination in worldwide water varied between decades, from rock weathering in the 1970s to mining in the 1980s. Similarly, mining activity and manufacturing processing dominated in the 1990s, while rock weathering dominated in 2000 and 2010 [7].

There has been an incessant rise in the number of industries consuming heavy metals, due to increased requirements of electroplating industries, cosmetics, iron and steel products, printing, automobiles, nuclear developments, etc.

Several factors affect the presence, conversion, and metabolism of heavy metals concerning their occurrence in industrial effluents, water sources, and different types of wastewaters. The chief sources of heavy metals can be natural and synthetic. Figure 5.1 shows the key sources responsible for the disposal of heavy metals in water bodies.

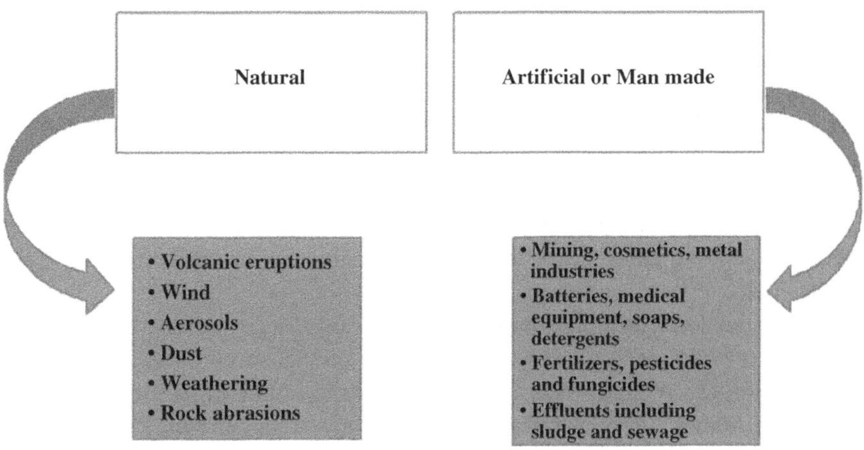

Figure 5.1 Causes of discharge of heavy metals in water resources.

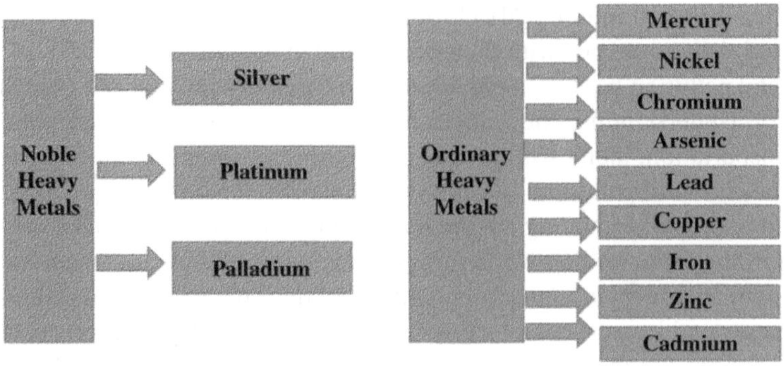

Figure 5.2 Heavy metals present in wastewater.

Heavy metals can be divided into noble and ordinary metals. Noble metals are precious, expensive, and can be used in making several costly items. The other set of heavy metals is ordinary metals, which include those metals that have routine applications in industries, nuclear processes, agricultural applications, etc. Heavy metal can be found in the form of sulfides, hydroxides, oxides, sulfates, phosphates, and silicates. These can enter different sections of the environment via various natural and anthropogenic activities. Figure 5.2 mentions different heavy metals present in wastewater.

5.2 Toxicity Assessment

As a result of increased consumption of water and improper treatment of wastewater before disposal, a plethora of health issues have arisen. The major hurdle is improper and partial treatment of wastewater before it is disposed into any potable water source. The effects caused on all forms of life due to heavy metals are substantial. The most prominent occurrence of heavy metals is due to human activities. These include several industrial manufacturing operations and agricultural activities. It has become mandatory to assess and curb the consumption of heavy metals as much as possible, or else, good treatment systems need to be developed so that these heavy metal contaminants can be remediated from the wastewater.

Heavy metal consumption induces toxicity and causes a disastrous effect on the lives of animals and human beings. It is crucial to eliminate these heavy metals from the treated wastewater completely to attain wholesome treatment of effluent water. Heavy metals are those that have a higher atomic number and a higher atomic weight. Heavy metals are carcinogenic, teratogenic, and lead to several health issues [8]. Some of the easily found heavy metals in abundance, which cause a great amount of toxicity to Earth's crust, include Cr, Ni, and Cd [9]. These heavy metals are profound carriers of toxicity and act as biomagnifiers [10]. Heavy metals like Cu, Ni, Hg, Cd, Pb, Cr, and Zn are highly soluble in water and similar aquatic environments. The nature of heavy metals is non-biodegradable and recalcitrant. This leads to bioaccumulation and biotoxicity in the consumers, including animals, plants, human beings, and other living beings.

5.2.1 Effect on Aquatic Organisms

Due to mounting anthropogenic activities elements/metal content in water has risen. Elements like Cd, Pb, Hg, Zn, and Cr cannot be converted into compounds with less or no toxicity, which alarms about the persistent presence of these in wastewater. On entering the water ecosystem, heavy metals get redistributed and accumulated in plants, sediments, and aquatic species. These can cause severe ecotoxicity, stress, and ill effects to humans and aquatic species. These metals inhibit the growth rate of fish larvae and juveniles. Low levels of metal contaminants can make fish lose weight, impede their growth, and reduce their ability to battle for food and territory. Metals can enter the fish body, either through the gills or through the digestive system of the fish. Heavy metals are instantly grasped by gills, while other modes may take time. Mostly, the accumulation of metal occurs in the liver, kidneys, gills, and muscles. This accumulation causes lesions in the organs of fish and disturbs their natural operation. These metals interfere with the growth of the embryo and its shape and other metabolic processes. The defects in the structure and functional aspects result in decreased larval hatching. The accumulation of metals like Zn may lead to a depressing influence on the respiratory tissues and leading to hypoxia and death. It also alters the structure of the lungs and heart. Cd is linked with cardiac illness in fish. Hg can lead to damage to blood vessels, hemorrhages, and reduced blood count. Other metals causing malfunctions in aquatic organisms include Ni, Cr, Pb, and As. Fin loss, underdevelopment of gills, and dysfunction of the liver, along with other physiological effects, are caused by the accumulation of metal in the body. It is also reported to cause immune system failure, infections, and decreased chances of survival on exposure. Heavy metal accumulation causes significant physiological, histological, and biochemical changes in fish and other species of water bodies. A drop in the hematological parameter indicates overexposure of the fish to Cr environment, which leads to anemia in *Labeo rohita* fish. Metals like Zn, Pb, Cd, Co, Hg, and Cd get accumulated in the reproductive organs and cause contamination of eggs, sperm, and reduced fertility and embryonic growth. Pb and Cu deposition causes the death of the newly hatched larvae. Neurotoxic impacts are observed due to mercury. Japan, with mercury-contaminated river water, has witnessed the impact of Minamata disease both in fish and human beings. Carcinogenic metals like Cd, As, and Cr can damage base pairs in DNA, leading to its mutation [11].

5.2.2 Effects on Plants

Metals like Cu, Co, iron (Fe), manganese (Mn), molybdenum (Mo), Ni, and Zn are required for the regular metabolism and plant development, but these could prove to be poisonous when present in excessive concentrations. Metal concentration in plant depends upon the type of plants and its metal uptake capacity. These metals are phytotoxic and result in diseases like chlorosis, yield reduction, and reduced nitrogen fixation. Pb may trim down the chances of seed germination. Cd and Ni are very hazardous to plants even at low concentrations. Metals affect the enzymatic makeup of the plant cell by swapping the metal ions and thus obstructing plant growth. Pb poisoning causes an imbalance in the water system of plants, affects membrane permeability, and inhibits enzymatic activity.

Plants exposed to arsenic (As) experience adverse effects, including process inhibition, reduced photosynthetic ability, biomass accumulation, and obstructed growth patterns. As affects seedling development. Unrestrained use of As in pesticides has increased its concentration in water bodies, and is bioaccumulated in the vital organs of the organisms, including flora. It causes a significant reduction in the algae composition and reduces the ecosystem health [12].

5.2.3 Effect on Human Health

Metals like magnesium (Mg), Mo, Mn, Cu, Fe, Ni, and Zn are vital nutrients required for an assortment of physiological activities in living organisms. Certain metals are needed by the living system, as they are enzyme co-factors, nutrients, and osmotic pressure regulators. Inadequate supply of these causes various deficiency diseases. However, these metals can be of potential toxicity if present in excess. For example, excessive Cu can cause damage at the cellular level leading to Wilson's disease. Cr and Cu both show a very low range of concentration between beneficial and toxic effects. Few other metals like antimony (Sb), aluminum (Al), barium (Ba), bismuth (Bi), titanium (Ti), lithium (Li), and vanadium (V) have no vital biological actions and are nonindispensable elements [13].

Heavy metals have an irreversible biotoxicity impact. They can enter the human body via absorption, skin contact, and inhalation, leading to mild to severe effects like anxiety, fatigue, stomach disorders, neurological issues, and cancerous growth. Heavy metal toxicity alters cellular functions, energy production, and metabolism. Chronic exposure to metals causes teratogenic impacts. Heavy metals are reported to affect the cell membrane, lysosomes, nuclei, and enzymes involved in metabolic activity. They interact with deoxyribonucleic acid (DNA) and nuclear proteins, causing conformational changes, which may trigger carcinogenesis.

As, Cd, Cr, Pb, and Hg show a high degree of toxicity and can provoke multiple organ damage. The level of harm caused depends on the bioavailability of the metals. These elements are called as, known or probable human carcinogens, as per the epidemiological experiments conducted by the International Agency for Research on Cancer (IARC) and the United States Environmental Protection Agency (USEPA) [13].

As is one of the prevalent elements present in drinking water and is toxic to human health. The World Health Organization (WHO) has reported adverse health conditions in areas with 10–50 μg/L of As concentration. Arsenic exposure can escalate the risk of skin malignancy, bladder and kidney cancer, and noncarcinogenic impacts like neurobehavioral effects and hormonal disturbance in human beings. Heavy metals can enter human bodies via the food chain (Figure 5.3). Absorption of heavy metals may lead to the production of sporadic fever, renal damage, hypertension, cramps, etc. Fish act as an important transfer medium of heavy metals to human beings. Heavy metals can be deleterious to human health due to their toxic and carcinogenic effects. These are linked with diseases like Alzheimer's disease and diabetes. Prolonged exposure may lead to muscular dystrophy and multiple sclerosis. Excess concentration can lead to reproductive complications and a threat to survival abilities. Humans are subjected to grave health issues, including mental disturbance.

A large concentration of metals can lead to acute and chronic health problems. Pb exposure can cause damage to the nervous system, inducing several development-related impairments. Cd exposure may cause bone damage and kidney malfunction. Hg causes neurological and behavioral disturbances. The health impacts of heavy metal exposure are exacerbated by socio-economic factors

Figure 5.3 Pathway of heavy metals entry into the living system.

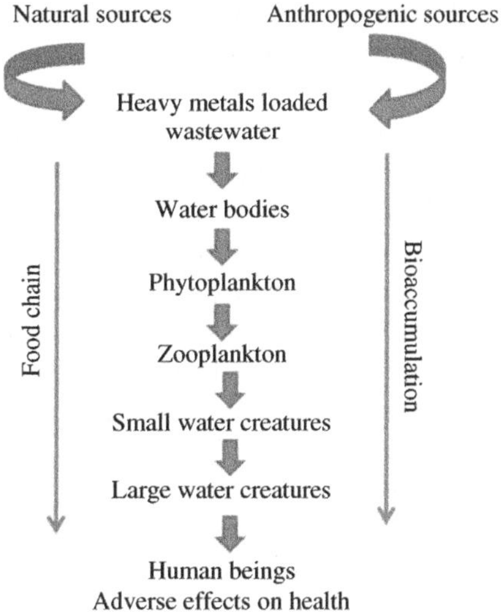

like inadequate healthcare facilities and a lack of awareness about the risks associated with metals in water bodies. The toxicity of heavy metals is exacerbated due to bioaccumulation in the food chain. Factors like pH, redox parameters, and the presence of organics govern the movement and availability of metals in the water environment. Cd can cause a hypoacetylation reaction leading to unacceptable gene expression and instability.

The outbreak of methylmercury poisoning (Minamata disease) was the outcome of metal pollution in water bodies. The poisoning caused sensory disturbances, hearing and speech-related issues, and loss of coordination and balance (cerebellar ataxia). Pregnant women exposed to toxic mercury delivered mentally disturbed and physically deformed babies. Similarly, in the Jinzu River region of Japan, an endemic illness, "Itai-itai," broke out due to high levels of cadmium, which mainly disturbed the skeletal and nephrological health, and was characterized by severe bone pain and fractures [2].

Lethal heavy metals such as As, which is found in several mining operations, agricultural wastes, etc., [14] can cause toxic effects, affecting the endocrine and hepatic systems. These are very perilous and can coagulate protein. Cr and Cu metals are commonly found in effluent from electroplating industries. Chromium is found in tanning industry effluents and other smelting activities [15]. The presence of Cr, along with other metals, can augment the glycogen level in different organs, displaying stress due to the metal contact.

Cu and Pb are found in plumbing appurtenances, and hence may be present in domestic wastewater due to leakage [16, 17]. Irreversible changes can be induced in the excretory and nervous systems due to higher concentrations of Pb. Hg is found as a result of urban runoff due to several activities, especially due to mining [18, 19]. The main target organ of the element is the central nervous system. Zn and Ni are present in several domestic appliances as well as in the effluent of several industrial processes [20–22]. Zn adds the risk of cardiovascular ailments. It can result in hypertension and digestive system damage. Zn is also accountable for neurotoxic impacts. Zinc can cause physical dysfunctions. A noticeable reduction in the weight is observed on exposure to Ni contamination. Cd is another heavy metal that is found in agricultural and urban runoff [23–25]. Elevated levels of Cd in water can induce prostate and lung cancer.

Different heavy metals can lead to several deteriorating effects. Table 5.1 shows the harmful effects of several types of heavy metals.

Understanding the biochemical interactions of heavy metals with the body is important for developing targeted solutions and preventive measures to combat their toxic effects. The recalcitrant property and toxic effects of heavy metals mandate research and development of methods and techniques for the treatment of wastewater containing heavy metals [40, 41]. Some of these methods have been depicted in Figure 5.4.

Table 5.1 Harmful effects and sources of heavy metals.

S. No.	Metal	Sources	Induced harmful effects	References
1.	Zinc	• Cosmetics • Soldering • Pigments • Galvanic industry • Alkaline batteries	• Immunity imbalance • Damages the nervous system • Gastrointestinal effects • Nausea • Hyperactivity • Psychological dysfunction	[26, 27]
2.	Lead	• Polyvinyl chloride pipes • Jewelry • Lead batteries • Lead paints	• Blood pressure issue • Carcinogenic effect • Reproductive problems • Memory loss • Anemic tendencies are induced • Damage is caused to the renal area	[28]
3.	Arsenic	• Preservatives • Pesticides • Oxidation of iron ores • Fertilizers • Metal smelting • Mining	• Heart problems • Cardiovascular issues • Inflammation of nerves • Neurotoxic influence • Damages protein structure • Cancerous	[29, 30]
4.	Nickel	• Battery manufacturing • Electroplating • Alloy production • Metal finishing • Coinage	• Nasal cancer • Hair problem • Dermal problems • Effects on the gastric system and gastrointestinal issues • Risk of cancer	[31, 32]
5.	Mercury	• Coal combustion • Waste incineration • Volcanic eruptions • Dental waste	• Brain damage • Neurological problems • Damage to lungs • Respiratory issues • Damage to the kidney • Abortion issue	[33, 34]

(Continued)

Table 5.1 (Continued)

S. No.	Metal	Sources	Induced harmful effects	References
6.	Cadmium	• Paint and pigment industries • Batteries • Photoconductors • Photovoltaic cells	• Troubles kidney • Hyperactivity problem • Bone related disorders • Joint pains • Respiratory issues • Carcinogenic	[35–37]
7.	Copper	• Fertilizers • Tanning industry • Photovoltaic cells	• Effects on the gastric system • Damage to the liver • Damages the nervous system	[38]
8.	Chromium	• Leather industry • Chrome plating activity	• Irritation to skin • Allergic reactions • Damage to the circulatory system • Lung cancer • Respiratory troubles	[39]

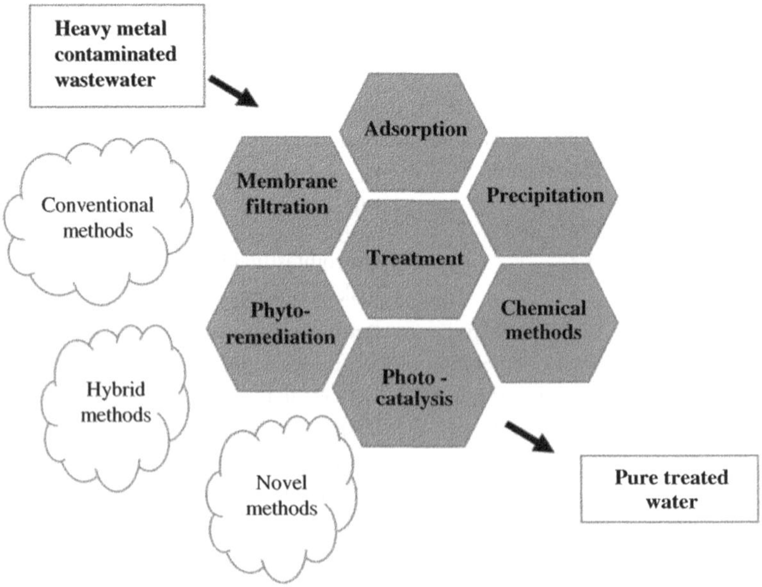

Figure 5.4 Different methods for remediation of heavy metals in wastewater.

5.3 Treatment Methods

The contamination of wastewater with heavy metals is a global problem and is found to be relevant worldwide in the industrial sector, agricultural areas, and at domestic levels. These need to be completely removed from the effluent wastewater. The removal of these heavy metals from wastewater becomes very crucial. Such methods of remediation ought to be cost-effective, green, and efficient. Figure 5.4 gives the different methods for the remediation of wastewater contaminated with heavy metals.

A major setback in the research of heavy metal contaminated wastewater remediation is the scarcity of studies on real wastewater, in comparison to synthetic wastewater. Many conventional approaches are available for the remediation of such contaminated wastewater. These include adsorption, ion exchange, coagulation, precipitation, membrane treatment, and several other chemical treatments. Several prominent issues are noticed while using conventional techniques, particularly in the post-treatment and the generation of a large sludge mass [42].

Adsorption is the most common technique; however, it has many shortcomings, like concurrent removal of ions, cycling stability, and a long retention period, which need to be tackled for targeting a successful remediation [43]. Membrane filtration is another conventional technique used at a large scale for the treatment of wastewater. The performance of the membrane can be improved by reducing scaling and fouling. This can increase the costs of operation and maintenance, due to the requirement of pretreatment and regular cleaning of membranes.

Many novel techniques have been explored, like flotation, electroflotation, ion exchange, bioremediation and photocatalysis. Many electrical and electrochemical methods have also been used, but these are difficult to explore practically, as they generate large number of wastes and are difficult to dispose off. More research and development thus is required to be done in this regard.

The need of the hour is to develop integrated hybrid methods [44] by merging conventional and advanced techniques, to compensate for the lacunas in solitary methods. These also promise the merits of both methodologies, making them self-sufficient approaches. Nanoscience and nanotechnology are the most recent fronts on effluent treatment [45].

The membrane filtration method is a conventionally applied method that is effective in the remediation of wastewater containing heavy metals. It requires less space, has low cost, and is highly feasible and environmentally friendly. Different types of membrane filtrations [46] used for removing heavy metals are reverse osmosis, nanofiltration, ultrafiltration, and electro dialysis [47]. The membrane separation processes offer a promising approach for pursuing heavy metal remediation from wastewater.

Several parameters were investigated by Shrestha et al. [48], in 2021, such as pH, initial concentration of heavy metals, interference of other ions, and temperature, which affect the rate of removal reaction. Chemical precipitation, adsorption, and ion exchange methods are other conventional methods that can be applied for the remediation of the above-mentioned heavy metals. Chemical methods like precipitation, flotation, coagulation, and flocculation consume a huge quantity of chemicals and are hence expensive.

One such research for removing several heavy metals like Fe, Cr, Cu, Pb, and Ni was performed by Abid et al. in 2011 [49]. Three different chemical precipitators were used, namely, magnesia, lime, and caustic soda. The mechanism responsible for the elimination of heavy metals was, single and multi-component hydroxide precipitation, along with adsorption. Magnesia was found to be the most efficient chemical precipitator, which removed nearly 100% of heavy metals. The most optimum doses of magnesia were 1 to −4 g/L, and the pH range was 9.5–10, which is the alkaline range.

Precipitation uses the concept of the formation of solid precipitates, which happens in cases where the solubility of the particular salt exceeds its saturation point. The chief mechanisms responsible for this are agglomeration, nucleation, and growth. Agglomeration occurs for a large number of particles in an environment that is typically supersaturated, while nucleation occurs when there is high supersaturation. Lower saturation levels lead to the growth of crystals [50].

Limitations of the traditional methods are that they are labor-intensive and costly. Table 5.2 portrays the advantages and limitations of some major conventional methods for the management of heavy metal-contaminated wastewater.

Table 5.2 Merits and demerits of conventional methods for heavy metal wastewater remediation.

S. No.	Technique	Merits	Demerits	References
1.	Membrane separation	Varied applications high efficiency	Issues of fouling, problems of scaling	[51, 43]
2.	Chemical precipitation	Large applications, useful over a wide scale of pH, can be used for several heavy metals	High operation and maintenance costs, more space, chemicals	[52]
3.	Membrane techniques, ion-exchange	Simple and efficient	Highly selective	[53]
4.	Adsorption	Many applications and a wide variety of heavy metal removal possible	Costly issues of desorption and recovery	[54]

5.4 Novel Methods for Aqueous Heavy Metal Remediation

Due to the increasing population, incessant urbanization, and industrialization, many novel methods are being explored and developed for efficient heavy metal treatment from industrial effluents and wastewater. These heavy metals are rampant in wastewater, and many new techniques like advanced oxidation processes, adsorption using waste materials, photocatalysis, ozonation, Fenton, electro-Fenton, and similar techniques are extensively being employed. Other arenas of novel methods consist of new and sophisticated bioremediation techniques like phytoremediation, membrane bioreactors, and other advanced forms of activated sludge.

5.4.1 Adsorption by Novel Adsorbents

Now-a-days, adsorbents made from waste materials need to be developed for remediating wastewater so as to gauge the sustainability potential of the adsorption system. Ahmed and Ahmaruzzaman, in 2016, carried out a review on the removal of heavy metals using waste materials from different industries, like slag, sludge, fly ash, red mud, lignin, black liquor lignin, flue dust, etc. [54]. A number of interesting factors were found to be important for the efficient removal, such as contact time, initial concentration of heavy metals, dosage, time of contact and interfering ions. Moreover, research has led to the development of hybrid adsorbents by a combination of new and old adsorbents. The only issue faced in adsorption, as noticed was the recovery and exclusion of heavy metal ions from the adsorbents.

Adsorption using a cost-effective waste adsorbents such as slag, has been studied for the removal of As [55]. It gave almost hundred percent efficiency for removal of As at an acidic pH of 2. The adsorbent was stable, and arsenic did not desorb easily from the surface of the adsorbent slag until relative reducing and complexing conditions were provided.

Another such waste material used for adsorption included carbon made from the husk of peanuts. It was useful for the confiscation of several heavy metals like Pb, Zn, Ni, and Cd, effectively from effluent wastewater. The initial concentration was around 0.15 mM, and the effect of the size of the adsorbent particle and the dosage of adsorbent was varied and studied. It was found that lead had the best affinity to the used adsorbent in comparison to other heavy metals like Cd, Ni, and Zn [56].

5.4.2 Phytoremediation

The phytoremediation method make use of plants for treating wastewater consisting of heavy metals. The process of phytoextraction is useful for uptake of pollutants in contaminated wastewater from roots of plants. This depends on a number of components like concentration of heavy metals, ability of plant for absorbing heavy metals, and accumulation tendencies of heavy metal. It enables the removal of contaminants for ensuring complete remediation and cleaning of toxic wastewater. The major merits of this emerging method are that it is both cost-friendly and a sustainable method, for the remediation of heavy metals from all types of wastewater effluents. The only concern is the ecotoxicity to the plant species, which can be taken care of by keeping concentration of the heavy metals within allowable, limitable, and nonharmful range.

In a research aiming at phytoremediation, aquatic plants *Azolla filiculoides Lam* and *Eichhornia crassipes* were employed for the removal of heavy metals Zn, Cu, Cd and Ni, which proved to be a successful trial. Similarly, applied terrestrial pants like *Prosopis* spp., *Salsola kali, Brassica juncea*, were found to be fruitful. There is also a possibility of removal and recovery of precious heavy metals like platinum (Pt), gold (Au), silver (Ag), and palladium (Pd) using phytoremediation, by using aquatic and terrestrial plants.

Significant research needs to be carried out in the field of phytoremediation for the efficient elimination of heavy metals from wastewater through laboratory and pilot-scale projects. Many heavy metal-resistant bacteria and beneficial microbes have also been useful for phytoextraction of heavy metals [57]. Another major area of heavy metal contamination and subsequent remediation is agricultural runoff from fields. It is necessary to understand the mechanism of removal, that is, the genetic behavior and biochemistry responsible in such cases.

The need of the hour is in the study of genetic engineering and classical breeding, for an understanding of the mechanism of phytoremediation [58]. Chandrasekhar et al., in 2005, studied through pot experiments the contamination of lead in soil containing wastewater, and its removal by *Hemidesmus indicus*. It was found that the lead accumulated in the roots as well as shoots, and for a long period, which was further removed by using suitable modifiers [59].

Phytoremediation involves the use of plants, which is an ecological remediation technology. It is an eco-friendly method that makes use of direct sunlight and does not require any special equipment. Also, this plant-based method is less expensive as compared to other technologies employed for removing heavy metals [60]. Various plants have been identified as highly effective in absorbing and accumulating toxic heavy metals, even if present in trace amounts in wastewater. Plants which are having the ability to accumulate large amounts of trace elements in the portion above ground are called hyperaccumulators. Macrophytes which are emergent, submergent, or floating plants growing in water have been used in wastewater treatment. Macrophytes like water hyacinth, water spinach, smart weed, water mimosa, *Mimosa pudica* are easy growing, quickly harvesting and good heavy metal accumulators.

Duckweeds (*L. minor*) have a large metal accumulation efficiency. Water hyacinth (*E. crassipes*) has shown a great potential for Hg, Cd, and As removal. Dry silver glass plant (*Miscanthus*) also showed good ability of heavy metals sequestration from wastewater, which was confirmed by Brunauer–Emmett–Teller method (BET) and other analytical techniques including elemental analysis, thermogravimetric analysis (TGA), X-ray diffraction (XRD), differential scanning calorimetry (DCS), inductively coupled plasma spectrometry (ICP), etc. [61].

In an investigation conducted by Balamoorthy et al., *Mimosa pudica* was selected for heavy metal elimination from wastewater. The roots of the plant created carbon disulfide, which inhibited the mycorrhizal growth in the rhizosphere of the plant. The plant root nodules have endosymbiotic diazotrophs. *Mimosa pudica* contains an alkaloid, Mimosine, which shows an antiproliferative effect, in addition to antioxidant and antibacterial features. Investigation reported that the plant *Mimosa pudica* accumulated up to 92% of Cd, 58% Pb, and 80% Cu during phytoremediation. The adsorption rate was controlled by the roots in the initial days of treatment, but later, translocation of a metal occurred by the shoots of the plant. The absorption of heavy metal was examined by using atomic absorption spectroscopy (AAS). Evidences revealed the plant to be an effective heavy metal accumulator [62].

Plants accumulating heavy metals survive through restriction mechanism. Literature indicates about 400 plants species as heavy metal accumulators, which includes plants of family *Asteraceae, Fabaceae, Euphorbiaceae*, and *Lamiaceae*. Efficient adsorption of heavy metals such as Cd, Zn, Co, Ni, and Se has been reported by *Thlaspi caerulescens* belonging to *Brassicaceae* family. Salix (hardwood), corn stalk, and *Brassica juncea* also exhibited excellent heavy metal tolerance and uptake efficiency. Plants like *Brassica oleracea* and *Lactuca sativa* accumulated metals like Cd in their leaves while *Brassica juncea* accumulated Cd, Cr, Pb, Ni, and Zn in its roots. *Helianthus annuus* belonging to Asteraceae family eliminated Pb, Cs, U, and Sr from hydroponic solution [63].

Phytostabilization reduces the mobilization of the heavy metals, and phytofiltration absorbs the metal ions, finally converting them into less toxic volatile compounds, releasing them into the environment. Volatilization occurs through leaves. Algae, like *Sargassum* sp., scalpelliformis sp., *Pterocladia* sp., and *Ulva* sp., have been evaluated for their bioremediation ability using virtual wastewater, as well as real wastewater, and it was concluded that these possess an appreciable water remediation capacity for heavy metal sequestration. Table 5.3 presents the potential of using algae for heavy metal removal from wastewater.

There are two types of protective mechanisms used by plants to detoxify heavy metals. They are avoidance and tolerance. The avoidance strategy for plants is their capacity to regulate the heavy metal adsorption and confinement of their translocation into the plant tissues through root cells. Root exudates have the potential

Table 5.3 Bioremediation using algal species.

Algae	Heavy metals	Operating parameters			Efficiency (%)
		Time	pH	Temperature (°C)	
P. capillacea	Cr (III)	45 min	1	25	85
Padina sp.	Cd (II), Zn (II), Ni (II)	1 hour	5.5	–	90
S. platensis	Cu (II)	1.5 hour	7	37	91
C. minutissima	Zn (II)	20 min	6	28	62
C. fastigiata	Pb (II)	1.5 hours	5	–	82
J. rubens	Hg (II)	1 hour	6	–	71
U. lactuca	Cr (VI)	3 hours	5	–	96
S. coronopifolius	Hg (II)	1 hour	6	–	90
A. filiculoides	Cr (VI)	100 min	2	–	83
S. crassifolium	Cd (II), Hg (II), Pb (II)	1 hour	2–9	–	99
O. pinnatifida	Cd (II)	1 hour	5	–	70
S. myriocystum	Pb (II)	1 hour	5	25	90
C. scalpelliformis	Zn (II)	–	5.7	30	90

Source: [63]/MDPI.

to combine with the heavy metals to form a stable complex in the rhizosphere. In this zone, the precipitation of heavy metals takes place, which reduces the heavy metal toxicity. Arbuscular mycorrhiza (symbiotic association between a plant and a fungus), absorbs, adsorbs, and chelates heavy metals in the rhizosphere and stops heavy metals from getting into the roots. The cell wall of the plants, composed of polygalacturonic acids having negative charge can effectively bind with the heavy metals. Tolerance strategy is employed to manage the harmful effects of heavy metal present in the cytosol (cytoplasmic medium). Inactivation, chelation, and compartmentalization help reduce the heavy metal toxicity. Amino acids present in the cells, and the xylem sap chelate with the heavy metal, facilitating their detoxification [64].

5.4.3 Photocatalysis

Photocatalysis is an effective method for a sustainable remediation of heavy metals. This can be performed by means of natural sunlight and can be applied in

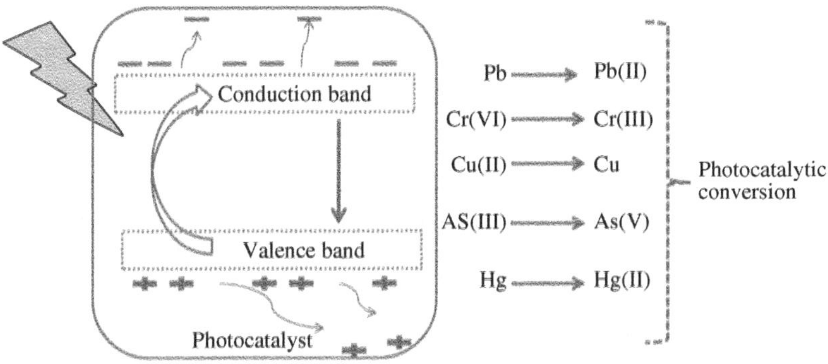

Figure 5.5 General mechanism of photocatalytic degradation.

areas where, sunlight is found in abundance. Many important parameters are to be taken into consideration in the process, which include the type of reactor, size, and configuration, efficiency of light, doping of photo catalysts, etc. [65]. There is an extensive application of photocatalysis for the removal of heavy metals. The most conventional photocatalysts include several semiconductors like titanium dioxide (TiO_2) and zinc oxide (ZnO). These can be applied with a good efficiency for ultraviolet light, and do not give appreciable results in visible light. Figure 5.5 illustrates the basic mechanism involved in the process.

Photocatalysis exhibits promising results for the removal of heavy metals and is thus popular among the advanced oxidation processes. Cr metal has applications in various industries like electroplating, textiles, wood preservation, and metallurgy. Long term exposure of Cr present in wastewater can damage nasal septum and can cause skin-related disorders. Successful photoreduction of Cr (VI) was done using immobilized titanium oxide with the use of UV radiation. Later, a high removal efficiency was reported for titanium oxide chelated with alizarin red dye. The process was independent of photon flux. The photocatalytic treatment of Cr (VI) was influenced by pH, presence of accelerators like organic compounds but was independent of the oxygen at acidic pH. In aqueous solution, As exists as As (III) and As (V). When pH is less than nine, As exists as H_3AsO_3, which is eliminated with difficulty by employing processes like coagulation, adsorption, and even precipitation, as reported. This is attributed to taking place, probably due to neutrality and complex ionization. A synergistic removal of Cr (VI) and As (III) was carried out by light-initiated ligand to metal charge transfer (LMCT) using UV radiation. Cr (VI) got reduced to Cr (III), and As (III) was oxidized to As (V), which got settled and removed from wastewater [66].

A removal efficiency of 80–86% was reported by impregnating the PET walls with titanium oxide. The UV radiation was provided for 6 hours at a pH of 7.8, which showed promising results. As (V) can also be reduced in the dark by UV irradiation of titanium oxide nanoparticles in alcoholic solutions. Heavy metal Hg (0) and Hg (II) have diverse photocatalytic removal tendencies. Hg (II) can be reduced by photocatalyst titanium oxide after 1 hour 5 min of irradiation, showing 70% of removal efficiency. The simple reduction method is not sufficient for the elimination of Hg ions. Due to lower water solubility and volatile nature, Hg (0) is complicated to remove. Bismuth-based photocatalysts are helpful in this regard. Au-blended titanium oxide also exhibited photocatalytic removal of Hg (II) in aqueous solution. It was inferred that it is more feasible to oxidize the Hg (0) to Hg (II) and remove it as HgO [66].

In addition, heavy metal ions like As, U, and Cd are difficult to biodegrade and are highly toxic even at very low concentrations. In one of the investigations, the removal of 100% Cd (I) in three hours at neutral pH was observed with Eosin Y sensitized titanium oxide photocatalyst. The solution pH is very important in heavy metal removal, as pH not only affects the surface charge of the catalyst but is also accountable for the destruction of metal ions, which helps in an interactive association between the catalyst and the contaminant. Alkaline pH effectively results in complete removal of Cd (II), Pb (II), and Cu (II) from wastewaters [67].

Pb (II) pollution has an anthropogenic origin. The photocatalytic reaction of Pb can be represented in Equation 5.1:

$$Pb^{2+} + H_2O \rightarrow Pb^0 + H^+ + O_2 \tag{5.1}$$

One promising method was designed for Pb removal from wastewaters using ZnO/ZnO_2 composites for photocatalytic degradation. The investigations revealed that a composite dosage of 100 mg/L at alkaline pH 10 reported 91% Pb removal from synthetic wastewater. ZnO is known for its good oxidizing strength and large energy gap, while ZnO_2 has better durability and resistance characteristics. The synergistic effect of the two photocatalysts resulted in a superior separation of the charges and large light energy absorption capacity, which led to a high removal efficiency. The process of photocatalytic Pb removal from wastewaters involved the generation of electrons and holes due to light radiation and absorption of Pb ions. This is followed by reduction to less soluble state which can be eliminated from the aqueous phase. Table 5.4 gives an overview of the photocatalytic technique employed for metal ion eradication from wastewater [68].

Nowadays, metal organic frameworks like g-C_3N_4 are used for photocatalysis to target the exclusion of several heavy metals from wastewater. The advantage of these photocatalysts is that they can be applied to visible light. However, the manufacturing of these heavy metals is complicated. The most significant task is to aim for a hybrid process consisting of several advanced and conventional processes, including photocatalysis, adsorption, phytoremediation, chemical and electrochemical methods, like electrocatalysis, photo electrocatalysis, and other thermodynamic processes.

Table 5.4 Photocatalytic technique for heavy metal removal.

Photocatalyst	Light source	Degradation of heavy metal	Efficiency
Titanium dioxide (TiO_2)/ porous boron nitride nanosheets	Simulated solar light and visible light	Cr (VI) → Cr (III), in 70 to 80 min	99% and 99%
TiO_2 (formic acid as electron donor)	Visible light	Cr (VI) → Cr (III), in 80 min	100%
TiO_2 nanoarrays/silver–silver bromide	Solar light	Cr (VI) → Cr (III), in 180 min	~59%
Zinc polytitanate $ZnTiO_3$/ $Zn_2Ti_3O_8$/ZnO	Full-spectrum light	Cr (VI) → Cr (III), in 150 min	47%
Zinc oxide/graphene	UV light	Cr (VI) → Cr (III), in 240 min	98%
Niobium pentoxide (Nb_2O_5)	Full-spectrum light	Cr (VI) → Cr (III), in 120 min	90%
Zirconium oxide (ZrO_2)	UV light	Cr (VI) → Cr (III), /Cr in 90 min	About 100%
TiO_2	UV light	As (III) → As (V) in 30 min	About 100%
TiO_2-ZrO_2	UV light	Cu (II) → Cu, and Cr (VI) → Cr (III)/Cr in 630 min	96.29 and 99.17%
α-Fe_2O_3/g-C_3N_4	Visible light	Hg (II) → Hg in 1 hour	90%
Bismuth oxyiodide (BiOI)/ Bismuth oxychloride (BiOCl)	Visible light	Hg → HgO/Hg (II) in 50 min	~72%
CeO_2/$BiOIO_3$	Visible light	Hg → HgO in 30 min	86.53%
Ag/TiO_2	UV light	Zn (II) → Zn, Mn (II) → Mn Ni (II) → Ni, Cu (II) → Cu Cd (II) → Cd, in 2 hour	65, 58, 96, 56%, and 100%, respectively
Tin oxide (SnO_2) nanoparticles	UV light	Co (II) → Co, in 1 hour	94%

Source: [68]/MDPI/CC BY 4.0.

5.4.4 Electrochemical Method

Electrochemical treatments for heavy metal elimination from wastewater are gaining increasing interest. These methods utilize electrostatic fields to dry metal ions and improve precipitation, adsorption, and separation. These electrochemical methods can be categorized into electrocoagulation, capacitive deionization, electroflotation, electrodialysis, and electrodeposition, depending on the mechanism involved [69]. Figure 5.6 illustrates the mechanism involved in the process.

1. **Electrocoagulation:** Electrocoagulation involves the in situ production of coagulants and respective hydroxides under direct electrical current. Electrocoagulation has low sludge production and maintenance costs, and is suitable for water quality variations. Also, it helps achieve heavy metal removal efficiency greater than 90%. But the dissolution of the sacrificial anodes requires frequent disposal, which limits their application. It is cost-intensive in terms of energy input.
2. **Electroflotation:** Electroflotation involves the separation of suspended contaminant particles by electrolytic gas bubble generation. The limitations of this process include the high cost of electrodes and the formation of a large volume of sludge. As the system involves a water splitting process, it is quite energy-consuming. Anode materials employed in electroflotation are stainless steel, nickel plates, and titanium. Dimensionally stable anodes are the most attractive options in electroflotation.
3. **Electrodialysis:** In this method, the electrode potential difference transports the metal ions between two solutions along the ion exchange membranes.

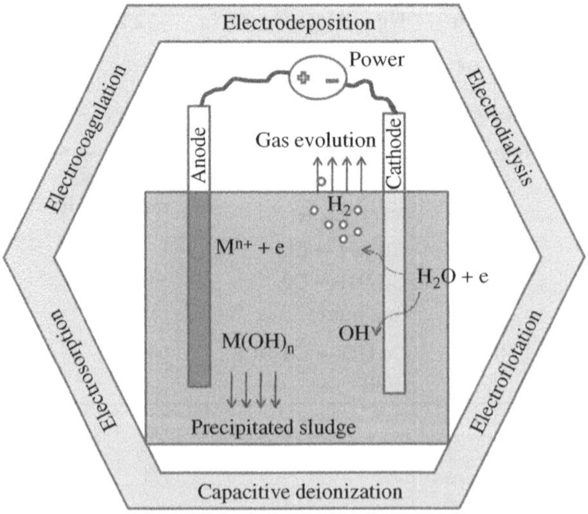

Figure 5.6 Electrochemical methods for heavy metal removal.

Electrodialysis is performed in a confinement known as an electrodialytic cell. The technique suffers from the limitation of scaling, membrane fouling and high energy input. Electrodialysis has been used to remove metals like Cr (VI), Zn (II), Ni (II), As (II), etc. It involves use of efficient ion exchange membranes with good mechanical durability, low electrical resistance, good permeability, and high selectivity.

4. **Electrodeposition:** Heavy metals are reduced and deposited on the electrodes in solid form. As (V), Cr (VI), Cu (II) have been effectively remove dusting this electrochemical method. However, the method has a drawback of side reactions, which are difficult to control.
5. **Capacitive Deionization:** It is an emerging technique for removing charged metal ions from wastewater under an electric potential and by using porous electrodes. It is based on the electrosorption of cations and anions on the electrode surface due to the electrical double layer. The specific conductance of the electrode and its surface area are the decisive factors in adsorption capacity or removal efficiency. Capacitive deionization has emerged as a promising technique, as it requires no chemicals and is energy efficient, but parasitic reactions can cause energy loss and water quality fluctuations, which hinder the practical application of the process. The technique is efficient, has a low cost involved, has a high recovery rate, and is cost cost-competitive than the other membrane filtration processes. Materials employed in capacitive deionization are carbon materials like graphenes, carbon aerogels, activated carbon, carbon nanofibers, carbon nanotubes, and carbon-based composites. Other suitable materials include heteroatom doped carbon, polymer-functionalized carbon, metal oxide supported carbon composites and polymer-based materials like polypyrrole–chitosan composite, polyaniline–attapulgite composites, etc.

5.4.5 Nanotechnology for Metal Removal

Nanotechnology is a promising future approach for achieving fast, efficient, and green pursuit of heavy metal remediation [45]. Nanofiltration membrane is another method that can be effectively utilized for heavy metal treatment. Moreover, this method of nanofiltration can be merged with nanoparticles to improve the performance of membrane filtration. It induces significant adsorption, improved permeability, selectivity, and enhanced rejection. This method is useful for the eradication of Pb, Zn, Cr, and Cu.

The fabrication of such nanofiltration membranes can be done by interfacial polymerization and phase inversion methods. In situ method of polymerization, assembling layer by layer, coating, blending, and embedding methods are efficient for manufacturing such nanomembranes in the removal of heavy metals like Cu, Pb, Cd, and Co [70, 71].

Addressing water pollution is one of the chief concerns in the twenty-first century. A number of anthropogenic activities are responsible for degrading the quality and condition of freshwater resources on a large scale. The issue is getting worse day by day due to rampant industrial development, commercial and domestic influence, and agricultural needs of the growing population. Heavy metal pollution needs to be curbed so as to limit its deteriorating effects on the health of human beings, animals, and plants.

Several novel techniques discussed in this chapter have great potential for becoming future techniques for aqueous heavy metal remediation. Advanced wastewater remediation techniques [42] should be suitably used to provide high-quality, contaminant free, safe, and potable water. Another interesting pursuit is the incorporation of nanotechnology and waste materials as adsorbents in the adsorption process. This could pave ways for bringing novelty in the conventional adsorption process. Several nanoadsorbents are used which include the oxides of Fe, Mn, Mg, Zn, and several metal organic frameworks. Nanomaterials have demonstrated a promising way to eliminate heavy metal elements from wastewater. Various nanomaterials have been explored and prepared in environmental protection field. In the nanoscale, these materials have a large surface area, small size, and macro quantum effect, which provides these nanomaterials extraordinary adsorption capacity [72, 73] (Figure 5.7).

1. **Carbon Nanotubes:** Carbon nanotubes, both single-walled and multi-walled, have been investigated for the mitigation of wastewater heavy metals. Such carbon-based nanomaterials have properties such as a high specific surface area, good adsorption ability, and rapid kinetics. To improve the efficiency of carbon nanotubes for heavy metals sequestration, chemical modification or heat treatment is performed. Removal of Cu (II) by functionalized carbon

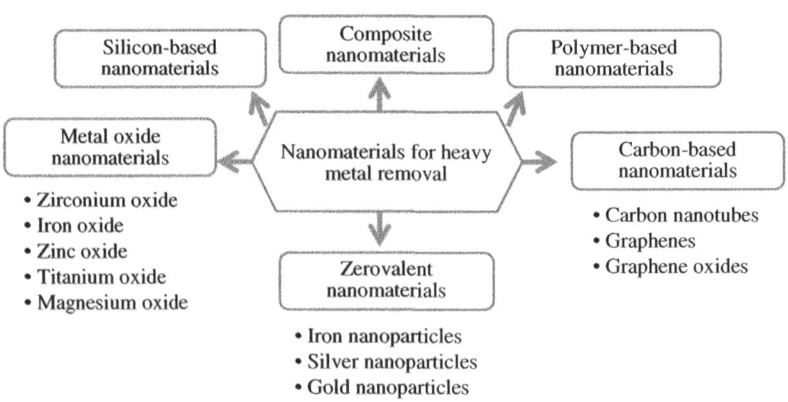

Figure 5.7 Nanomaterials for heavy metal removal.

nanotubes was carried out by researchers. The adsorption capacity of Hg (II) was 186.97 mg/g at a pH of 5.5, with 28 minutes of contact time, by employing eutectic solvent sonicated with carbon nanotubes [74]. Also, Cu (II) elimination was done using amino-functionalized carbon nanotubes. According to the Langmuir model, the highest adsorption capacity for Cu (II) was found as 30.49 mg/g, with an advantage of recovery of the adsorbent using a magnetic field [75]. The high cost of carbon nanotubes hinders their commercialization. Also, it suffers from the disadvantage of difficult recovery of the loaded carbon nanotubes from wastewaters.

2. **Graphene Nanomaterials:** These extraordinary materials are known for their mechanical toughness, firmness, elasticity, and electrical and thermal conductivity. Graphene oxide and reduced graphene oxide are nanomaterials employed to remove heavy metals. The presence of hydroxyl (—OH), carboxyl (—COOH), carbonyl (—CO), and epoxide groups makes graphenes apt for the removal of heavy metals. Graphene oxide is an oxidation product of graphene, while the reduction product of graphene oxide is reduced graphene oxide. pH, amount of the adsorbent, time of contact, temperature conditions, and presence of ionic species influence the performance of the graphene nanomaterials for heavy metal elimination. Graphene oxide is effective in the removal of Zn (II). Similarly, Co (II) and Cd (II) were effectively adsorbed by layered graphene oxide nanosheets, which was evident by a maximum adsorption capacity of 68.2 and 106.3 mg/g, respectively [76]. The results were further confirmed by thermodynamic parameters, suggesting endothermic and spontaneous removal of Cd (II) and Co (II). Graphene-based microbots could effectively remove Pb (II) and have the advantage of reusing the regenerated microbots [77].
3. **Silicon-based Nanomaterials:** Nontoxicity and surface characteristics of silicon-based nanomaterials makes them important materials for removing heavy metals. Silica nanospheres biofunctionalized by 3-aminopropyl and phenyl groups removed Cu (II) competently. Amino functionalized silica nano hollow spheres showed a maximum adsorption capacity for Pb, Cd, and Ni as 96.79 mg/g, 40.73 mg/g, and 31.29 mg/g, respectively [78].
4. **Iron Nanoparticles:** Iron zerovalent nanoparticles offer reactive sites and centers for electrostatic interaction with the heavy metals. Iron zerovalent nanoparticles have a good reducing capacity and specific surface area, which is responsible for their superior efficiency. For Pb (II) removal, the mechanism involved is reduction and sorption, while for Cr (VI), the removal is by reduction and precipitation. The mechanism is dependent upon the standard potential of the heavy metals. Iron zerovalent nanoparticles suffer from the disadvantage of getting oxidized by oxygen and water in an aqueous environment. These also form aggregates, which decrease the surface area and

mobility. Also, the separation of iron from wastewaters is difficult. Sodium dodecyl sulfate modified iron nanoparticles, Au doped iron nanoparticles, and bentonite-based iron nanoparticles showed better removal for heavy metals from wastewaters [79].

5. **Silver and Gold Nanoparticles:** Hg (II) removal was facilitated by using silver nanoparticles, and an enhanced removal efficiency was obtained for mercaptosuccinic acid modified Ag nanoparticles [80]. Au nanoparticles supported over aluminum and citrate-coated Au nanoparticles were also effective in the removal of Hg (II) [81].
6. **Polymer-based Nanomaterials:** Cellulose, dendrimer, and chitosan-based nanomaterials are eco-friendly materials used for heavy metal removal. Well-designed, porous, and fibrous structures, providing high surface area, enhance their ability to adsorb contaminants from wastewater. They have a biodegradable tendency, tunable functional groups, strong structural network, which makes it efficient to sequester a number of heavy metals from wastewaters. Such nanomaterials are fabricated using electrospinning, polymerization, and self-assembly approaches. Chitosan beads for Cu, and Cr, polyaniline fibers for As, metal organic frameworks for Pb and Cd, and pH-responsive hydrogels for Hg, have been investigated. Recently, biopolymers-based nanomaterials have been utilized for heavy metal removal. Tetramethylpiperidine oxide-oxidized 2,3,6-tricarboxylic cellulose nanofibers (T-CNFs) showed 82.12 and 97.34 mg/g of Pb and Cu ion removal efficiency, respectively [82]. Similarly, cellulose nanofibers and modified cellulose nanofibers confirmed their efficiency in eliminating Pb, Cd, Cu, and Cr from aqueous solution. Further research and innovation are essential to improve the effectiveness of these materials and advance their practical application on a large scale for wastewater treatment.
7. **Metal Oxide Nanoparticles:** Iron oxides like goethite, hematite, and magnetite, manganese dioxides, zinc oxide, aluminum oxides, zirconium oxides, etc., have been successfully employed for heavy metal remediation from wastewaters. Magnetic nanoparticles are known for their exceptional ability to trap heavy metals from wastewater and facilitate magnetic separation (Figure 5.8). A magnetic nanoparticle offers greater surface area, lower toxicity, easy dispersion, and more reliable water purification. The electrostatic interactions between the nanoparticles and the metals are influenced by operating conditions, which regulate the operating mechanism as adsorption, co-precipitation, magnetic separation, chelation, or surface complexation. Magnetic nanorods, nanowires, and nanotubes have demonstrated exceptional competence in removing heavy metals from wastewater. However, these suffer from the limitation of their practical implementation, due to the aggregation and oxidation features of bare magnetic nanoparticles.

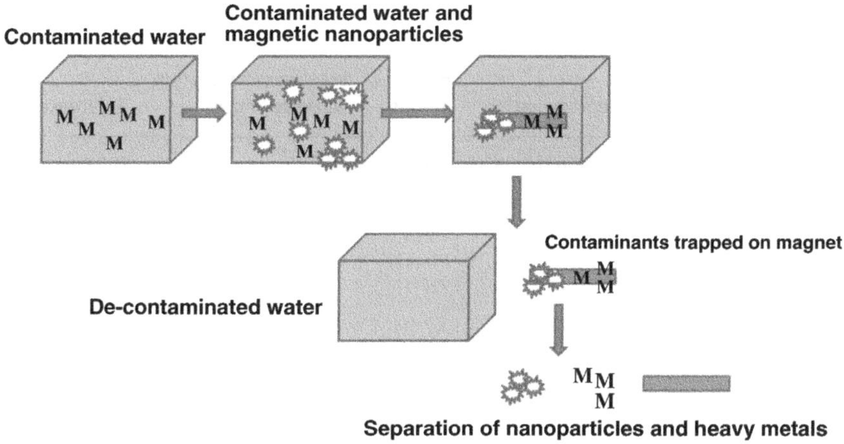

Figure 5.8 Magnetic nanoparticles and heavy metal separation.

The functionalization of the magnetic nanoparticles can help in the improvement of stability, adsorption capacity, and selectivity for the heavy metal removal. Table 5.5 summarizes a few metal oxides as heavy metal eliminators and their metal elimination efficacies.

Continuous research in this regard can ensure the development of nanoadsorbents having larger adsorption capacities [42]. Engineered nanomaterials have shown promising results in the elimination of heavy metals. Recently carbon carbon-layered silicate nanocomposites have offered enhanced metal removal properties. The membrane filtration technique integrated with nanoparticles has a substantial scope for the future. However, many challenges need to be addressed to explore the use of membrane filtration on a practical scale.

The removal and recovery of heavy metals from different types of wastewater samples is of paramount significance for a complete environmental cleanup. Plausible solutions need to be sought so as to curb the incessant growth of such recalcitrant heavy metals in water, as these are not only toxic but also have the potential of bio magnification.

5.5 Permissible Limits of Heavy Metals in Water Bodies

The guidelines for potable water standards and wastewater allowable limits are formulated after a comprehensive understanding and assessment of results and conclusions published by several researchers, along with an understanding of the presence of metals and their impact on the ecosystem.

Table 5.5 Metal oxides used for heavy metal elimination from wastewater.

Type of metal oxide nanoparticles	Representation	Heavy metals	Maximum adsorption capacity (mg/g)	References
Goethite	α-FeOOH	Cu (II), Cd (II)	149.25 167	[83, 84]
Hematite	α-Fe_2O_3	Cr (VI) Pb (II) Zn (II)	– – –	[85]
Maghemite	γ-Fe_2O_3	Ag (I) Hg (II)	260.55 237.60	[86]
Magnetite	Fe_3O_4	Cr (VI) Ni (II)	232.51 222.12	[87]
Hydrous iron oxide	HFO	As (V)	355	[88]
Hydrous manganese oxide	HMO	Pb (II) Cd (II)	– –	[89]
Nanocrystalline manganese oxide	MO	Pb (II) Cd (II)	318.7 105.1	[90]
Titanium oxide	TiO_2	Cr (VI)	12.6	[91]
Aluminum oxide	γ-Al_2O_3	Pb (II) Cd (II)	47.08 17.22	[92]
Magnesium oxide	MgO	Cd (II) Pb (II) Ni (II)	2294 2614 1684.25	[93,94]
Cerium oxide	CeO_2	Cr (VI)	121.95	[95]
Zinc oxide	ZnO	Cd (II) Pd (II) Co (II)	156.74 194.93 67.93	[96]
Zirconium oxide	ZrO_2	Co (II) Cu (II) Cd (II)	32.2 46.5 109.9	[97]

Several water quality criteria have been set to make water suitable for consumption, recreation, fishing, and other activities. These laws have been enforced to legalize and manage the maximum acceptable concentration of heavy metals in aqueous sources, sediment, or biotic species. The chief intention of these guidelines is to avert probable hazards associated with the presence of heavy metals in the water bodies. Thus, it is important to adhere to these standards to guarantee long-term well-being and sustainability. The establishment and enforcement of these standards safeguards and preserves the veracity and security of the water bodies. These laws help to manage the water resources both at the comprehensive and micro levels. The regulatory guidelines for a country vary depending on the economy and availability of resources. Every country generally sets up its national standards for water quality based on the WHO as a reference. Legislations formulated for drinking water require frequent and regular monitoring and assessment to control the presence of heavy metal pollutants. The Environmental Protection Agency (EPA), USA, and the Central Board of Pollution Control (CPCB), India, are the authoritative agencies that set the discharge standards and establish and enforce them. Organizations like the WHO and the US EPA have established maximum concentration limits (MCLs) for various heavy metals in drinking water. Also, the Environment and Health Organization in China and the Food and Agriculture Organization (FAO, 1985) have established acceptable limits of toxic metals in wastewater. Table 5.6 summarizes the allowable limits in the potable water source.

NTPA-002/2002 is a Romanian regulation that sets limit values for various pollutants in wastewater discharged into sewer systems and directly into wastewater treatment plants. The pollutants include heavy metals like Cu, Zn, Cd, Pb, and Ni, along with other important physicochemical characteristics like BOD, COD, TSS, and TDS. Table 5.7 depicts the maximum allowable concentration of heavy metals as directed by NTPA001/2002 [63].

Table 5.6 Allowable confines of toxic metals in potable water.

Metal	Concentration unit	EPA	European standard	WHO	India
Pb	mg/L	0.1	0.5	0.01	0.1
Ni	mg/L	0.1	0.1	0.07	0.1
Cr	mg/L	0.1	0.5	0.05	0.1
Cu	mg/L	1	3	–	0.01
Zn	mg/L	5	5	0.05	0.1
Cd	mg/L	0.005	0.2	0.003	0.001
Hg	mg/L	0.002	0.001	–	0.004
As	mg/L	10 μg/L	0.01	0.01	0.05
Ti	mg/L	–	–	0.05	–

Source: Adapted from [1, 98].

Table 5.7 Discharge limits for heavy metals in wastewater.

Metal ion	Pb	Mn	As	Cd	Cr	Fe	Cu	Co	Ni	Zn	Hg	Mo	Se
Limit (mg/L)	0.20	1	0.1	0.2	1.0	5.0	0.1	1	0.5	0.5	0.05	0.1	0.1

Source: [63]/MDPI.

5.6 Case Studies

5.6.1 Kutubdia Channel in Bangladesh: Contamination of Water and Sediments by Metals

An investigation for the evaluation of health risk due to nine different elements, viz. Zn, Cu, Fe, Mn, As, Pb, Co, Sr, and Cr were carried out for two different seasons. The main objective of the study was to evaluate the parameters of water quality of the site area and to explore the amount of metals in the water bodies and sediments, to comprehend the level of contamination. Energy Dispersive X-Ray Fluorescence (EDXRF) was employed to detect the elemental concentration.

Researchers calculated the pollution load index, along with probable ecological risk and hazard quotient. A measurable source of pollution was chosen in the Kutubdia channel, near Moheshkhali-Matarbari, Bangladesh, which was 4–8 km away from a thermal power firm. The samples were collected from a river stream, including coastal and offshore areas. The sampling period chosen was January to May and September to December. Multivariate statistical analysis, cluster analysis, and component analysis were carried out to determine the significance of the data obtained. The physicochemical parameters were assessed to evaluate the state of the natural ecosystem. Parameters like temperature, salinity, dissolved oxygen (DO), pH, and electrical conductivity were investigated. All nine elements were uniformly dispersed in the surface water. The concentration of Sr was the highest upto 0.511 mg/L, in the pre-monsoon season, which was reported as 0.413 mg/L after the monsoon season. As and Co were found to be present in low concentrations below their detection limit. The investigation site was free from As contamination, although Bangladesh is one of the most badly affected nations due to As contamination. The mean amount of Pb and Cr did not show variation in the two seasons investigated. Cr concentration was found to be less than the suggested value of 0.05 mg/L, as proposed by WHO. Cu concentration was slightly higher in pre-monsoon due to a decreased strength of the pollutant in surface water, because of rain and river influx. The mean concentration was below the permitted level of 5 mg/L. Zn is an essential metal needed for various physiological functions in the living system. The drinking water in the study area was not a good source of Zn. The level of Mn, which is a well-known mutagen, was found to be higher during the post-monsoon season. Overall, the level of heavy metals was under the customary levels for drinking water. This led to the conclusion that the available water of the channel was secure for consumption by humans. Also, the pH and dissolved oxygen values suggested it to be safe for drinking and fishing, all year. But salinity can be higher than the potable water guideline. The surface water of Kutubdia does not pose a threat for utilization in irrigation, or for fishing and human consumption, as revealed by potential ecological risk, which was found to be less than 4. The sediment pollution assessment revealed contamination due to Cu, Zn, and Pb. The human risk assessment revealed that the site sediment pose some noncarcinogenic risks in children and adults. Thus, continuous assessment and monitoring was believed to be essential, as proved by health risk index greater than 1 [99].

5.6.2 Heavy Metal Contamination of River Cauvery, Tamil Nadu, India

A study was conducted on the river Cauvery, which is one of the largest rivers in the state of Tamil Nadu, India, at three different regions. Regions of Erode, Salem, and Namakkal were chosen for determining the strength of metal ions present in the river water, in monsoon and pre-monsoon seasons. About 34 million liters of water is supplied to the state population from the river daily, which necessitated the investigation. Site one was Mettur in Salem district, where effluent from steel, cement, and leather mills was being discharged. The second site was the Namakkal district, where the dominance of the textile industry exists. Agraharam in Erode district was taken as the third sampling site, which had leather mills, garment factories, and oil refineries in its vicinity. The collection of the samples was done during post post-monsoon and monsoon seasons. The amount of heavy metals in the water was analyzed using Atomic Absorption Spectroscopy and Flame Spectrometry. Statistical package for the social sciences (SPSS) software was used to analyze the data statistically. The element Zn amount was found to higher than other metals, owing to steel production and coal fired stations in the nearby location of the sampling stations. The concentration was found to be within the acceptable limits.

Fe is naturally present in the water bodies. Fe content was found to be lower than the allowable limits, as given by WHO (0.3 mg/L). The concentration of Cu was found to be within the permissible limit of 1 mg/L, and ranged between 0.5 and 0.0005 mg/L in the post-monsoon and monsoon periods. During post post-monsoon season the amount of Cr was found to be higher than any other metal. The Cr level at site Namakkal, Agraharam, and Pallipalayam crossed the WHO limit during post post-monsoon period. The River Cauvery water was found to be heavily contaminated by Pb. Ni content was found to be within the permissible limit in the monsoon season, but during the post-monsoon, it crossed the limit of 0.02 mg/L, at Pallipalayam site. This reflected the contamination of the study area due to use of nickel acetate in textile processing activity. The concentration of Cd varied between 0.01 and 0.005 mg/L, and 0.007 and 0.002 mg/L in the post-monsoon and monsoon term, respectively. The site Mettur in Salem district and Agraharam in Erode district was found to exceed the limit. As crossed the limit in both the investigated seasons. River water was found to be toxic due to As discharge from industrial effluents. The Mettur site crossed the acceptable limit for Se and Hg during the monsoon season. The study revealed that the Cauvery River has become contaminated due to heavy concentrations of metals like Pb, Cr, As, Cr, Ni, and Se.

The study concluded that there was a need for regular monitoring of the river. This, however, was challenging as the river water was and is being used for various activities like drinking, fishing, etc. The presence of heavy metals in the river also created a need for the treatment of the industrial effluents before discharge into the river body [100].

References

1 Singh, V., Ahmed, G., Vedika, S., et al. (2024). Toxic heavy metal ions contamination in water and their sustainable reduction by eco-friendly methods: isotherms, thermodynamics and kinetics study. Scientific Reports, Volume 14, p. 7595. https://doi.org/10.1038/s41598-024-58061-3.

2 Aziz, K.H.H., Mustafa, F.S., Omer, K.M., et al. (2023). Heavy metal pollution in the aquatic environment: efficient and low-cost removal approaches to eliminate their toxicity: a review. RSC Advances, Volume 13, pp. 17595–17610. http://doi.org/10.1039/D3RA00723E.

3 Singh, R., Singh, A., Balomajumder, C., et al. (2025). Assessment of industrial effluent discharges contributing to Ganga water pollution through a multivariate statistical framework: investigating the context of Indian industries. Environmental Science and Pollution Research, https://doi.org/10.1007/s11356-024-35823-0.

4 Guo, X., Ho, M.S., You, L., et al. (2018). Industrial water pollution discharge taxes in China: a multi-sector dynamic analysis. Water, Volume 10, Issue 12, p. 1742. https://doi.org/10.3390/w10121742.

5 Hasan, Md. K., Shahriar, A., and Jim, K.U. (2019). Water pollution in Bangladesh and its impact on public health. Heliyon, Volume 5, Issue 8, p. e02145. https://doi.org/10.1016/j.heliyon.2019.e02145.

6 Anjum, S., and Rana, S. (2023). Impact of environmental pollutants on agriculture and food system. Applied and Environmental Microbiology Journal, Volume 2023, pp. 133–151. https://doi.org/10.1016/b978-0-323-91643-1.00005-3.

7 Zhou, Q., Yang, N., Li, Y., et al. (2020). Total concentrations and sources of heavy metal pollution in global river and lake water bodies from 1972 to 2017. Global Ecology and Conservation, Volume 22, p. e00925. https://doi.org/10.1016/j.gecco.2020.e00925.

8 Hama, A., Mustafa, K.H., Omer, F.S., et al. (2023). Heavy metal pollution in the aquatic environment: efficient and low-cost removal approaches to eliminate

their toxicity: a review. RSC Advances, Volume 13, Issue 26, pp. 17595–17610. https://doi.org/10.1039/D3RA00723E.

9 Shaheen, S.M., Eissa, F.I., Ghanem, K.M., et al. (2015). Metal ion removal from wastewaters by sorption on activated carbon, cement kiln dust, and sawdust. Water Environment Research, Volume 87, Issue 6, pp. 506–515. https://doi.org/10.2175/106143015x14212658614630.

10 Hegazi, H.A. (2013). Removal of heavy metals from wastewater using agricultural and industrial wastes as adsorbents. Housing and Building National Research Center Journal, Volume 9, Issue 3, pp. 276–282. https://doi.org/10.1016/j.hbrcj.2013.08.004.

11 Singh, A., Sharma, A., Verma, K.R., et al. (2022). Heavy metal contamination of water and their toxic effect on living organisms. In: Dorta DJ, de Oliveira DP (eds) The toxicity of environmental pollutants. IntechOpen, pp. 1–19. http://dx.doi.org/10.5772/intechopen.105075.

12 Ahmed, S.F., Senthil, P.K., Rozbu, M.R., et al. (2022). Heavy metal toxicity, sources, and remediation techniques for contaminated water and soil. Environmental Technology and Innovation, Volume 25, p. 102114. https://doi.org/10.1016/j.eti.2021.102114.

13 Tchounwou, P.B., Yedjou, C.G., Patlolla, A.K., et al. (2012). Heavy metal toxicity and the environment. Experientia Supplementum, Volume 101, pp. 133–164. https://doi.org/10.1007/978-3-7643-8340-4_6.

14 Sankhla, M.S., Kumar, R., and Agrawal, P. (2018). Arsenic in water contamination and toxic effect on human health: current scenario of India. Journal of Forensic Sciences and Criminal Investigation, Volume 10, Issue 2, pp. 001–005.

15 Vaiopoulou, E., and Gikas, P. (2012). Effects of chromium on activated sludge and on the performance of wastewater treatment plants: a review. Water Research, Volume 46, Issue 3, pp. 549–570. https://doi.org/10.1016/j.watres.2011.11.024.

16 Kanwal, A., Farhan, M., Sharif, F., et al. (2020). Effect of industrial wastewater on wheat germination, growth, yield, nutrients and bioaccumulation of lead. Scientific Reports, Volume 10, Issue 1, p. 11361. https://doi.org/10.1038/s41598-020-68208-7.

17 Liu, Y., Wang, H., Cui, Y., et al. (2023). Removal of copper ions from wastewater: a review. International Journal of Environmental Research and Public Health, Volume 20, Issue 5, p. 3885. https://doi.org/10.3390/ijerph20053885.

18 Tami-Pimiento, L.M., Joya-Herrera, L.M., Pérez-Chía, Y.I., et al. (2023). Selective electrochemical reduction of mercury (II) from a simulated traditional gold mining wastewater contaminated with cyanide and heavy metals. Journal of Solid-State Electrochemistry, Volume 28, pp. 2223–2232. https://doi.org/10.1007/s10008-023-05745-x.

19 Verma, R.K., Sakhla, M.S., and Kumar, R. (2018). Mercury contamination in water & its impact on public health. International Journal of Forensic Science, Volume 1, Issue 2, pp. 1–7.

20 Chan, P.C., de Toledo, R.A., Iu, H.I., et al. (2019). Effect of zinc supplementation on biogas production and short/long chain fatty acids accumulation during anaerobic co-digestion of food waste and domestic wastewater. Waste and Biomass Valorization, Volume 10, Issue 12, pp. 3885–3895. https://doi.org/10.1007/s12649-018-0323-9.

21 Hase, H., Nishiuchi, T., Sato, T., et al. (2017). A novel method for remediation of nickel containing wastewater at neutral conditions. Journal of Hazardous Materials, Volume 329, pp. 49–56. https://doi.org/10.1016/j.jhazmat.2017.01.019.

22 Rathore, G., Chopra, N., and Adhikari, T. (2014). Nickel as a pollutant and its management. International Research Journal of Environmental Sciences, Volume 3, Issue 10, pp. 94–98.

23 Hocaoglu-Ozyigit, A., and Genc, B.N. (2020). Cadmium in plants, humans and the environment. Frontiers in Life Sciences and Related Technologies, Volume 1, Issue 1, pp. 12–21.

24 Järup, L. (2002). Cadmium overload and toxicity. Nephrology Dialysis Transplantation, Volume 17, Issue (suppl_2), pp. 35–39. https://doi.org/10.1093/ndt/17.suppl_2.35.

25 Hassan, J.M., Raza, A.M., Ur Rehman, S., et al. (2020). Effect of cadmium toxicity on growth, oxidative damage, antioxidant defense system and cadmium accumulation in two sorghum cultivars. Plants, Volume 9, Issue 11, p. 1575. https://doi.org/10.3390/plants9111575.

26 Jain, G. (2013). Removal of copper and zinc from wastewater using chitosan (Publication No. 41628137), Doctoral dissertation, National Institute of Technology, Rourkela.

27 Trivunac, K., Sekulic, Z., and Stevanovic, S. (2012). Zinc removal from wastewater by complexation-microfiltration process. Journal of the Serbian Chemical Society, Volume 77, Issue 11, pp. 1661–1670. https://doi.org/10.2298/jsc120302037T.

28 Gunatilake, S.K. (2015). Methods of removing heavy metals from industrial wastewater. Journal of Multidisciplinary Engineering Science Studies, Volume 1, Issue 1, pp. 12–18.

29 Altowayti, W.A.H., Othman, N., Shahir, S., et al. (2022). Removal of arsenic from wastewater by using different technologies and adsorbents: a review. International Journal of Environmental Science and Technology, Volume 19, Issue 9, pp. 9243–9266. https://doi.org/10.1007/s13762-021-03660-0.

30 Mousavi, S.R., Balali-Mood, M., Riahi-Zanjani, B., et al. (2013). Concentrations of mercury, lead, chromium, cadmium, arsenic and aluminum in irrigation water wells and wastewaters used for agriculture in Mashhad, northeastern Iran. The International Journal of Occupational and Environmental Medicine, Volume 4, Issue 2, pp. 80–86.

31 Ahmed, A.M.M., Ali, A.E., and Ghazy, A.H. (2019). Adsorption separation of nickel from wastewater by using olive stones. Advanced Journal of Chemistry-Section A, Volume 2, Issue 1, pp. 79–93.

32 Białecka, B., Thomas, M., and Zdebik, D. (2023). Removal of copper, nickel and tin from model and real industrial wastewater using sodium trithiocarbonate. The negative impact of complexing compounds. Archives of Environmental Protection, Volume 44, Issue 1, pp. 33–47. https://doi.org/10.24425/118179.

33 Elahi, A., Arooj, I., Bukhari, D.A., et al. (2020). Successive use of microorganisms to remove chromium from wastewater. Applied Microbiology and Biotechnology, Volume 104, Issue 9, pp. 3729–3743. https://doi.org/10.1007/s00253-020-10533-y.

34 Rafati-Rahimzadeh, M., Rafati-Rahimzadeh, M., Kazemi, S., et al. (2014). Current approaches of the management of mercury poisoning: need of the hour. Journal of Pharmaceutical Sciences, Volume 22, Issue 1, p. 46. https://doi.org/10.1186/2008-2231-22-46.

35 Friberg, L. (2018). Cadmium in the environment. CRC Press. https://doi.org/10.1201/9781351070379.

36 Peana, M., Pelucelli, A., Chasapis, C.T., et al. (2022). Biological effects of human exposure to environmental cadmium. Biomolecules, Volume 13, Issue 1, p. 36. https://doi.org/10.3390/biom13010036.

37 Sharma, H., Rawal, N., and Mathew, B.B. (2015). The characteristics, toxicity and effects of cadmium. International Journal of Nanotechnology and Nanoscience, Volume 3, pp. 1–9.

38 Al-Saydeh, S.A., El-Naas, M.H., and Zaidi, S.J. (2017). Copper removal from industrial wastewater: a comprehensive review. Journal of Industrial and Engineering Chemistry, Volume 56, pp. 35–44. https://doi.org/10.1016/j.jiec.2017.07.026.

39 Kerur, S.S., Bandekar, S., Hanagadakar, M.S., et al. (2021). Removal of hexavalent chromium-industry treated water and wastewater: a review. Materials Today: Proceedings, Volume 42, pp. 1112–1121. https://doi.org/10.1016/j.matpr.2020.12.492.

40 Abu-El-Halawa, R., and Zabin, S.A. (2017). Removal efficiency of Pb, Cd, Cu and Zn from polluted water using dithiocarbamate ligands. Journal of Taibah University for Science, Volume 11, Issue 1, pp. 57–65. https://doi.org/10.1016/j.jtusci.2015.07.002.

41 Odumbe, E., Murunga, S., and Ndiiri, J. (2023). Heavy metals in wastewater effluent: causes, effects, and removal technologies. In: Joseph D (ed)

Trace metals in the environment, IntechOpen. https://doi.org/10.5772/intechopen.1001452.
42 Senthilkumar, S., Siva, V., Murugan, A., et al. (2024). Heavy metals in water: challenges and remediation. In: Srivastav AL, Grewal AS, Markandeya M, et al. (eds) Role of green chemistry in ecosystem restoration to achieve environmental sustainability. Elsevier, pp. 157–166. https://doi.org/10.1016/B978-0-443-15291-7.00014-6.
43 Qasem, N.A.A., Mohammed, R.H., and Lawal, D.U. (2021). Removal of heavy metal ions from wastewater: a comprehensive and critical review. npj Clean Water, Volume 4, Issue 1, p. 52. https://doi.org/10.1038/s41545-021-00144-z.
44 Fei, Y., and Hu, Y.H. (2023). Recent progress in removal of heavy metals from wastewater: a comprehensive review. Chemosphere, Volume 335, p. 139077. https://doi.org/10.1016/j.chemosphere.2023.139077.
45 Zamora-Ledezma, C., Negrete-Bolagay, D., Figueroa, F., et al. (2021). Heavy metal water pollution: a fresh look about hazards, novel and conventional remediation methods. Environmental Technology and Innovation, Volume 22, p. 101504. https://doi.org/10.1016/j.eti.2021.101504.
46 Zhao, M., Xu, Y., Zhang, C., et al. (2016). New trends in removing heavy metals from wastewater. Applied Microbiology and Biotechnology, Volume 100, Issue 15, pp. 6509–6518. https://doi.org/10.1007/s00253-016-7646-x.
47 Xiang, H., Min, X., Tang, C.J., et al. (2022). Recent advances in membrane filtration for heavy metal removal from wastewater: a mini review. Journal of Water Process Engineering, Volume 49, p. 103023. https://doi.org/10.1016/j.jwpe.2022.103023.
48 Shrestha, R., Ban, S., Devkota, S., et al. (2021). Technological trends in heavy metals removal from industrial wastewater: a review. Journal of Environmental Chemical Engineering, Volume 9, issue 4, p. 105688. https://doi.org/10.1016/j.jece.2021.105688.
49 Abid, A.B., Brboot, M.M., and Al-Shuwaik, M.N. (2011). Removal of heavy metals using chemicals precipitation. Engineering and Technology Journal, Volume 29, Issue 3, pp. 595–612. https://doi.org/10.30684/etj.29.3.15.
50 Lewis, A. (2017). Precipitation of heavy metals. In: Rene E, Sahinkaya E, Lewis A, et al. (eds) Sustainable heavy metal remediation. environmental chemistry for a sustainable world, Volume 8. Springer, pp. 101–120. https://doi.org/10.1007/978-3-319-58622-9_4.
51 Abdullah, N., Yusof, N., Lau, W.J., et al. (2019). Recent trends of heavy metal removal from water/wastewater by membrane technologies. Journal of Industrial and Engineering Chemistry, Volume 76, pp. 17–38. https://doi.org/10.1016/j.jiec.2019.03.029.
52 Elsheikh, M.A., and Al-Hemaidi, W.K. (2013). Approach in choosing suitable technology for industrial wastewater treatment. Journal of Civil

and Environmental Engineering, Volume 2, Issue 5, P. 1000123. https://doi.org/10.4172/2165-784X.1000123.

53 Dąbrowski, A., Hubicki, Z., Podkościelny, P., et al. (2004). Selective removal of the heavy metal ions from waters and industrial wastewaters by ion-exchange method. Chemosphere, Volume 56, Issue 2, pp. 91–106. https://doi.org/10.1016/j.chemosphere.2004.03.006.

54 Ahmed, Md. J.K., and Ahmaruzzaman, M. (2016). A review on potential usage of industrial waste materials for binding heavy metal ions from aqueous solutions. Journal of Water Process Engineering, Volume 10, pp. 39–47. https://doi.org/10.1016/j.jwpe.2016.01.014.

55 Oh, C., Rhee, S., Oh, M., et al. (2012). Removal characteristics of As (III) and As (V) from acidic aqueous solution by steel making slag. Journal of Hazardous Materials, Volume 213–214, pp. 147–155. https://doi.org/10.1016/j.jhazmat.2012.01.074.

56 Ricordel, S. (2001). Heavy metals removal by adsorption onto peanut husks carbon: characterization, kinetic study and modeling. Separation and Purification Technology, Volume 24, Issue 3, pp. 389–401. https://doi.org/10.1016/S1383-5866(01)00139-3.

57 Sharma, P., Bano, A., Singh, S.P., et al. (2023). Metal-tolerant microbial inoculants for improved phytoextraction. In: Sharma VK, Kumar A, Passarini MRZ, et al. (eds) Microbial inoculants. Elsevier, pp. 75–98. https://doi.org/10.1016/B978-0-323-99043-1.00005-0.

58 Bhargava, A., Carmona, F.F., Bhargava, M., et al. (2012). Approaches for enhanced phytoextraction of heavy metals. Journal of Environmental Management, Volume 105, pp. 103–120. https://doi.org/10.1016/j.jenvman.2012.04.002.

59 Chandrashekhar, K., Kamala, C.T., Chary, N.S., et al. (2005). Potential of *Hemidesmus indicus* for phytoextraction of lead from industrially contaminated soils. Chemosphere, Volume 58, Issue 4, pp. 507–514. https://doi.org/10.1016/j.chemosphere.2004.09.022.

60 Adiloğlu, S. (2018). Heavy metal removal with phytoremediation. In: Shiomi A (ed). Advances in bioremediation and phytoremediation. IntechOpen. http://dx.doi.org/10.5772/intechopen.70330.

61 Khan, A., Khan, A., Waris, A., et al. (2022). Phytoremediation of pollutants from wastewater: a concise review. Open Life Sciences, Volume 17, Issue 1, pp. 488–496. https://doi.org/10.1515/biol-2022-0056.

62 Balamoorthy, D., Velusamy, P., Rath, B., et al. (2022). Removal of heavy metals from wastewater by using phytoremediation technology. Journal of Civil Engineering Science and Technology, Volume 13, Issue 1, pp. 23–32. https://publisher.unimas.my/ojs/index.php/JCEST/article/view/4473.

63 Diaconu, L.I., Covaliu-Mierlă, C.I., Păunescu, O., et al. (2023). Phytoremediation of wastewater containing lead and manganese ions using algae. Biology, Volume 12, Issue 6, p. 773. https://doi.org/10.3390/biology12060773.

64 Islam, M.M., Saxena, N., and Sharma, D. (2024). Phytoremediation as a green and sustainable prospective method for heavy metal contamination: a review. RSC Sustainability, Volume 2, pp. 1269–1288. https://doi.org/10.1039/D3SU00440F.

65 Deng, Y., Feng, C., Tang, L., et al. (2019). Nanohybrid photocatalysts for heavy metal pollutant control. In: Tang L, Deng Y, Wang J, et al. (eds) Nanohybrid and nanoporous materials for aquatic pollution control. Elsevier, pp. 125–153. https://doi.org/10.1016/B978-0-12-814154-0.00005-0.

66 Gao, X., and Meng, X. (2021). Photocatalysis for heavy metal treatment: a review. Processes, Volume 9, Issue 10, p. 1729. https://doi.org/10.3390/pr9101729.

67 Ren, G., Han, H., Wang, Y., et al. (2021). Recent advances of photocatalytic application in water treatment: a review. Nanomaterials, Volume 11, Issue 7, p. 1804. https://doi.org/10.3390/nano110718.

68 Shakir, H.A., Alsaffar, M.A., Mageed, A.K., et al. (2024). Optimizing photocatalytic lead removal from wastewater using ZnO/ZrO_2: a response surface methodology approach. Chemical Engineering, Volume 8, Issue 4, p. 72. https://doi.org/10.3390/chemengineering8040072.

69 Xu, Y., Zhong, Z., Zeng, X., et al. (2023). Novel materials for heavy metal removal in capacitive deionization. Applied Sciences, Volume 13, Issue 9, p. 5635. https://doi.org/10.3390/app13095635.

70 Mahmoud, A.E.D., and Mostafa, E. (2023). Nanofiltration membranes for the removal of heavy metals from aqueous solutions: preparations and applications. Membranes, Volume 13, Issue 9, p. 789. https://doi.org/10.3390/membranes13090789.

71 Liu, L., Li, W., Song, W., et al. (2018). Remediation techniques for heavy metal-contaminated soils: principles and applicability. Science of Total Environment, Volume 633, pp. 206–219. https://doi.org/10.1016/j.scitotenv.2018.03.161.

72 Yang, J., Hou, B., Wang, J., et al. (2019). Nanomaterials for the removal of heavy metals from waste water. Nanomaterials, Volume 9, Issue 3, p. 424. https://doi.org/10.3390/nano9030424.

73 Alomar, M.K., Alsaadi, M.A., Hayyan, M., et al. (2017). Allyl triphenyl phosphonium bromide based DES-functionalized carbon nanotubes for the removal of mercury from water. Chemosphere, Volume 167, pp. 44–52.

74 Zhan, Y., Hai, H., Yi, H., et al. (2011). Novel amino-functionalized Fe_3O_4/carboxylic multi-walled carbon nanotubes: one-pot synthesis, characterization and removal for Cu (II). Russian Journal of Applied Chemistry, Volume 89, pp. 1894–1902.

75 Gao, W., Majumder, M., Alemany, L.B., et al. (2011). Engineered graphite oxide materials for application in water purification. ACS Applied Materials and Interfaces, Volume 3, pp. 1821–1826.

76 Zhao, G., Li, J., Ren, X., et al. (2011). Few-layered graphene oxide nanosheets as superior sorbents for heavy metal ion pollution management. Environmental Science and Technology, Volume 45, pp. 10454–10462.

77 Vilela, D., Parmar, J., Zeng, Y., et al. (2016). Graphene-based microbots for toxic heavy metal removal and recovery from water. Nano Letters, Volume 16, Issue 4, pp. 2860–2866. https://doi.org/10.1021/acs.nanolett.6b00768.

78 Najafi, M., Yousefi, Y., and Rafati, A.A. (2012). Synthesis, characterization and adsorption studies of several heavy metal ions on amino-functionalized silica nano hollow sphere and silica gel. Separation and Purification Technology, Volume 85, pp. 193–205.

79 Huang, P., Ye, Z., Xie, W., et al. (2013). Rapid magnetic removal of aqueous heavy metals and their relevant mechanisms using nanoscale zero valent iron (nZVI) particles. Water Research, Volume 47, pp. 4050–4058.

80 Sumesh, E., Bootharaju, M.S., Anshup, et al. (2011). A practical silver nanoparticle-based adsorbent for the removal of Hg^{2+} from water. Journal of Hazardous Materials, Volume 189, Issue 1–2, pp. 450–454. https://doi.org/10.1016/j.jhazmat.2011.02.061.

81 Ojea-Jiménez, I., López, X., Arbiol, J., et al. (2012). Citrate-coated gold nanoparticles as smart scavengers for mercury (II) removal from polluted waters. ACS Nano, Volume 6, p. 2253. https://doi.org/10.1021/nn204313a.

82 Karnwal, A., and Malik, T. (2024). Nano-revolution in heavy metal removal: engineered nanomaterials for cleaner water. Frontiers in Environmental Science, Volume 12, p. 1393694. https://doi.org/10.3389/fenvs.2024.1393694.

83 Chen, Y.H., and Li, F.A. (2010). Kinetic study on removal of copper (II) using goethite and hematite nano-photocatalysts. Journal of Colloid Interface Science, Volume 347, pp. 277–281. https://doi.org/10.1016/j.jcis.2010.03.050.

84 Khezami, L., M'Hamed, M.O., Lemine, O.M., et al. (2016). Milled goethite nanocrystalline for selective and fast uptake of cadmium ions from aqueous solution. Desalination and Water Treatment, Volume 57, pp. 1–9. https://doi.org/10.1080/19443994.2015.1010231.

85 Shipley, H.J., Engates, K.E., and Grover, V.A. (2013). Removal of Pb (II), Cd (II), Cu (II), and Zn (II) by hematite nanoparticles: effect of sorbent concentration, pH, temperature, and exhaustion. Environmental Science and Pollution Research, Volume 20, pp. 1727–1736. https://doi.org/10.1007/s11356-012-0984-z.

86 Madrakian, T., Afkhami, A., Zadpour, B., et al. (2015). New synthetic mercaptoethylamino homopolymer-modified maghemite nanoparticles for effective removal of some heavy metal ions from aqueous solution. Journal of

Industrial and Engineering Chemistry, Volume 21, pp. 1160–1166. https://doi.org/10.1016/j.jiec.2014.05.029.
87 Baghani, A.N., Mahvi, A.H., Gholami, M., et al. (2016). One-pot synthesis, characterization and adsorption studies of amine-functionalized magnetite nanoparticles for removal of Cr (VI) and Ni (II) ions from aqueous solution: kinetic, isotherm and thermodynamic studies. Journal of Environmental Health Science and Engineering, Volume 14, p. 11. https://doi.org/10.1186/s40201-016-0252-0.
88 Huo, L., Zeng, X., Su, S., et al. (2017). Enhanced removal of As (V) from aqueous solution using modified hydrous ferric oxide nanoparticles. Scientific Reports, Volume 7, p. 40765.
89 Wan, S., Wu, J., Zhou, S., et al. (2018). Enhanced lead and cadmium removal using biochar-supported hydrated manganese oxide (HMO) nanoparticles: behavior and mechanism. Science of the Total Environment, Volume 616–617, pp. 1298–1306. https://doi.org/10.1016/j.scitotenv.2017.10.188.
90 Wang, X., Huang, K., Chen, Y., et al. (2018). Preparation of dumbbell manganese dioxide/gelatin composites and their application in the removal of lead and cadmium ions. Journal of Hazardous Material, Volume 350, pp. 46–54. https://doi.org/10.1016/j.jhazmat.2018.02.020.
91 Seisenbaeva, G.A., Daniel, G., Nedelec, J.M., et al. (2012). High surface area ordered mesoporous nano-titania by a rapid surfactant-free approach. Journal of Materials Chemistry, Volume 22, pp. 20374–20380. https://doi.org/10.1039/C2JM33977C.
92 Tabesh, S., Davar, F., and Reza Loghman-Estarki, M. (2018). Preparation of γ-Al_2O_3 nanoparticles using modified sol-gel method and its use for the adsorption of lead and cadmium ions. Journal of Alloys and Compounds, Volume 730, pp. 441–449. https://doi.org/10.1016/j.jallcom.2017.09.246.
93 Xiong, C., Wang, W., Tan, F., et al. (2015). Investigation on the efficiency and mechanism of Cd (II) and Pb (II) removal from aqueous solutions using MgO nanoparticles. Journal of Hazardous Materials, Volume 299, pp. 664–674. https://doi.org/10.1016/j.jhazmat.2015.08.008.
94 Jing, F., Zou, L., Wang, Y., et al. (2015). Synthesis of high surface area, mesoporous MgO nanosheets with excellent adsorption capability for Ni (II) via a distillation treating. Journal of Colloid Interface Science, Volume 438, pp. 259–267. https://doi.org/10.1016/j.jcis.2014.10.004.
95 Recillas, S., Colón, J., Casals, E., et al. (2010). Chromium VI adsorption on cerium oxide nanoparticles and morphology changes during the process. Journal of Hazardous Materials, Volume 184, pp. 425–431. https://doi.org/10.1016/j.jhazmat.2010.08.052.
96 Somu, P., and Paul, S. (2018). Casein based biogenic-synthesized zinc oxide nanoparticles simultaneously decontaminate heavy metals, dyes, and

pathogenic microbes: a rational strategy for wastewater treatment. Journal of Chemical Technology and Biotechnology, Volume 93, Issue 10, pp. 2962–2976. https://doi.org/10.1002/jctb.5655.

97 Yalçınkaya, Ö., Kalfa, O.M., and Türker, A.R. (2011). Chelating agent free-solid phase extraction (CAF-SPE) of Co (II), Cu (II) and Cd (II) by new nano hybrid material (ZrO_2/B_2O_3). Journal of Hazardous Materials, Volume 195, pp. 332–339. https://doi.org/10.1016/j.jhazmat.2011.08.048.

98 Gautam, R.K., Sharma, S.K., Suresh Mahiya, S., et al. (2014) Contamination of heavy metals in aquatic media: transport, toxicity and technologies for remediation. In: Sharma S (ed). Removal of heavy metals present in water, RSC Publisher, pp. 1–24. https://doi.org/10.1039/9781782620174-00001.

99 Hossain, M.S., Ahmed, M.K., Liyana, E., et al. (2021). A case study on metal contamination in water and sediment near a coal thermal power plant on the eastern coast of Bangladesh. Environments, Volume 8, Issue 10, p. 108. https://doi.org/10.3390/environments8100108.

100 Myvizhi, P., and Aruna, D.P.S. (2020). Heavy metal contamination in water of the river Cauvery – a case study of erode, Salem and Namakkal districts, Tamil Nadu. Journal of Himalayan Ecology and Sustainable Development, Volume 15, pp. 190–202.

6

Dye Wastewater Characteristics and Treatment

Dyes have been in use for a long time in diverse industrial sectors like textile, cosmetics, leather, food, pharmaceutical, paper, and pulp. Almost 70 lakh tons of colorants are annually synthesized around the globe and are used in various applications [1].

Dyes are organic molecules chiefly employed as colorants for textiles, paper, leather, plastics, printing inks, paints, etc. They have been widely used to color fabrics, as they get absorbed into the pores of the textiles, not just adhere to the surface of the fiber. The dye molecules effectively get into the fibers due to the force of attraction operating between the dye and the fiber.

Until the 1850s, dyes were extracted from natural sources like vegetables, plants, trees, and lichens. But only some of these were widely used, due to the instability of the complicated mixtures and the unavailability of effective separation technology. These natural colorants provided a base for the synthesis of dyes, which dominated the textile sector by 1900. Natural dyes are eco-friendly, nontoxic, and biodegradable as they are obtained from animals, plants, and mineral sources. Alizarin is one such example, which is a natural dye obtained from the roots of the madder plant. Another well-known natural dye is Indigo, which is obtained from the leaves of the dyer's woad herb, *Isatis tinctoria*, or the indigo plant, *Indigofera tinctoria*. The other natural dyes are lutein from leaves of *Reseda luteola*, quercetin from bark of *Quercus tinctoria, the* oak tree, and crocetin (yellow dye) from stigmas of *Cocus sativus*. But today, heartwood extract from the bloodwood tree, *Haematoxylum campechianum,* is the only natural dye used for dyeing silk and leather.

The natural dying industry declined due to higher extraction expenses, application time, economic limitations, and unavailability of required resources. The advancement in organic chemistry and the development of new derivatives from coal tar revolutionized the colorant industry and provided a scientific foundation

Hazards and Treatment of Wastewater Contaminants, First Edition. Dipika Jaspal and Arti Malviya. Published 2026 by John Wiley & Sons, Inc.

for its expansion. The first commercially synthesized dye, mauve, discovered by Perkins, declined the dominance of natural dyes due to its associated advantages both in terms of production and application. Advances in understanding chemistry involved in the dye manufacturing also vigorously led to the development of synthetic dye industries.

The textile industry manufactures more than 3600 dyes to increase the aesthetic appeal of the fabric. Other than the colorants, several other chemical agents are also employed in the textile industry, in the entire dyeing and printing process. The research community has identified about 70 chemicals in the wastewater effluent that are being introduced into the water bodies from the textile dyeing industries. These chemicals are toxic to living beings, directly or indirectly, and many of them are recalcitrant. During the textile processing huge volume of water is also required. The daily water consumption of a moderate-sized textile plant, which produces about 8000 kg of fabric daily, is about 1.6 million liter. Specific water requirement ranges between 30 and 50 L/kg of the fabric, depending upon the chemical attributes of the dye. Dyeing units contribute to about 20% of the wastewater flow [2]. According to statistics obtained, about 21% of the industrial water pollution is caused by the dyeing facilities, paper and pulp industries contribute 10%, 8% is from the tannery and paint industry, and 7% from colorant manufacturers.

Dyes are widely used in the textile industry, along with other chemical substances. Among the various color contributors or sources, about 54% of the colored contaminants in the textile industry effluent are dyes. Due to improper dyeing procedures, the textile plants alone contribute 7×10^8 kg of these colorants to wastewater annually, during dyeing and finishing operations [1]. Recent statistics establish that Indian textiles produce 10^5 tons of dyes and pigments every year, which is projected to be 10000 cubic tons from all diverse forms of wastewater [2]. Dye molecules comprise of a dye agent that imparts color and the enhancer, which amplifies the solubility and surface attachment property of the dye. Several other chemicals being used in the textile industry are sizing, brightening, finishing agents, etc. Dyes find their way to the water bodies either due to nonbonding at the processing stage or discharge without prior treatment. This poses a serious ecotoxicological threat to living organisms. Majority of dyeing operations require a copious amount of water during dyeing, washing, and rinsing. A large amount of water being used in the textiles manufacturing units leads to the presence of a high amount of recalcitrant dyes in wastewater along with suspended solids, organics, metals, and salts. Insufficient and improper dyeing leads to a loss of 15–50% of the dyes that are not bound to the fibers and are released into the wastewater emerging from these industries. This leads to a severe threat to aquatic beings and humans, as these dyes are extremely hazardous. Small-scale textile industries dump the poisonous wastewater into the waterways and water bodies. About 100 out of 4000 dyes that have been tested for their toxicity are still available in the market, despite several prohibitions on their use in various applications.

Table 6.1 summarizes the percentage of dye wastewater discharged from various processing units operative in the textile manufacturing industry.

Table 6.1 Characteristics of dye wastewater from various processes.

Textile process	Dye wastewater contribution (%)	Components of wastewater	Characteristics of wastewater
Desizing (removal of sizing chemicals)	21	Starch, polyvinyl alcohol, enzymes, carboxyl methane cellulose, waxes, ammonia	High biological oxygen demand (BOD) and COD, presence of a large amount of dissolved solids
Scouring (removal of fatty oils and greases)	52	Waxes, sodium carbonate, surface active agents, anti-static agents	High COD, oily contaminants
Bleaching (discoloration/ whitening)	62	Sodium hypochlorite, surface active agents, sodium chlorite, hydrogen peroxide	Presence high percentage of alkalinity, causing ions, is highly alkaline
Mercerizing (swelling of fiber)	4	Na_2O_2, wax	Alkaline nature of water, Low BOD, high amount of dissolved solids
Dyeing	85	Colorants, pigments, oxidizing and reducing agents, leveling, and wetting agents	Presence of color in water, metals, high BOD, and lower suspended solids
Printing or localized dyeing	13	Softening, crosslinking agents, waterproofing agents, plasticizers	Presence of color in water, large BOD, and oily contaminants
Finishing (addition of special properties and aesthetic appeal)	58	Crosslinked resins, chlorine-containing compounds, waxes, softening agents, stiffeners, antistatic agents	The water has an alkaline pH, low BOD, toxic nature.

6.1 Classification of Dyes

Dyes are divided into various categories depending upon the chemical moieties present in the structure, and the presence of characteristic chromophores and auxochromes. They are classified as basic dyes, acid dyes, and direct dyes, or as derivatives having azo linkages, anthraquinone structure, sulfur, indigo, triaryl, and phthalocyanine dyes. Due to their easy application, wide variety, and high color fastness profiles, synthetic colorants have been used in industries worldwide in comparison to natural dyes. The multifaceted application of these materials in daily life is associated with their release from industrial effluents into water systems, which jeopardizes human health and the environment.

An array of synthetic dyes is classified considering different aspects. Synthetic dyes are commonly classified into various types based on their structural arrangement, dyeing process, and the target application (Figure 6.1). According to the chemical structures, dyes are azo dyes, phthalocyanine dyes, anthraquinone dyes, and indigo dyes. Based on dyeing operation procedure, they can disperse dyes, direct dyes, vat dyes, reactive dyes, basic dyes, or acid dyes. Based on their application, dyes can be the textile colorants, food dyes, cosmetic dyes, or special biological application dyes [1].

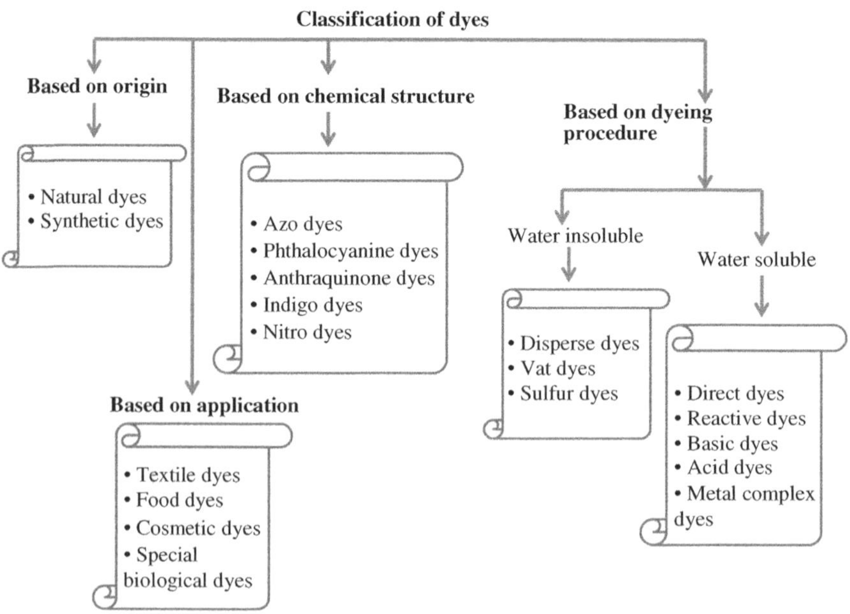

Figure 6.1 Classification of dyes.

1. **Acid Dyes:** Acid dyes are usually water-soluble anionic dyes applied to fibers such as silk, wool, nylon, etc. These possess good color fastness and have a strong attraction for the protein molecules in the fibers. They mainly include the sodium salts of sulfonic acids. If the concentration of acid dyes is 1 mg/L, they tend to increase the COD of the water body, causing esthetic disturbance to the aqueous environment. Anthraquinone dyes are an example of this category.
2. **Basic Dyes:** These dyes were the first category of dyes obtained from coal tar. These were used to dye silk, linen, hemp, etc. These dyes have high color intensity and possess brilliant shades. These dyes are cationic, which ionize in water and possess good light fastness, but poor wash stability. Congo red is one such basic dye characterized by intense color and brilliance, along with good tinctorial efficacy. Nowadays, these are mainly applied on acrylic fibers, although they can be used on other fabrics as well, along with mordants. These are not soluble in alkaline media. They have mainly imino or amino groups linked to triarylmethane or xanthene moieties.
3. **Direct Dyes:** These are named so, as they can be applied to fabric directly from their aqueous solutions and require no fixation phase. The mechanism involved is the establishment of the nonionic forces to attach the dyestuff to the fiber material. These are the basic dyes used on cotton, rayon, wool, silk, and nylon. They are usually water-soluble. They have simple application and have good wash fastness and low light fastness. To improve the color fixation property of the dyes, some dyes are complexed with the mordants. These dyes are associated with greater environmental risks and hence are limited in their use. These are also called substantive dyes.
4. **Azo Dyes:** Azo dyes are characterized by the presence of azo group linkages (—N=N—) as the chromophoric species. These are primarily used to provide bright color to the fabric. These are one of the most commonly manufactured and consumed dyestuffs, accounting for over half of the world's annual dye production. They are employed as colorants in the food industry, the pharmaceutical sector, textiles, cosmetics, and even the leather industry. When present in a water body, these prove to be carcinogenic, and once absorbed, are metabolized in the intestine by bacterial microflora of the liver to aromatic amines, leading to health issues. Numerous nations have legislative guidelines in place for the manufacturing and handling of azo dyes.
5. **Disperse Dyes:** These are sparingly water-soluble dyes and are developed by secondary cellulose acetate fibers. They require high pressure and temperature for their application. They are suitable for coloring hydrophobic fibers and are used for polyester, nylon, and acetate fibers. These being insoluble are applied after being ground into a fine powder, followed by their application

in the presence of a dispersing agent or carriers like alkyl sulfates, alkyl aryl sulfonates, etc. Being nonionic dyes, the mechanism of attachment involves weak van der Waals forces and dipole-dipole interactions. These portray fair light fastness and wash fastness. In comparison to all the dyestuffs, these possess the smallest molecular sizes. Nitro dye, anthraquinonoid dyes, and amino keto dyes belong to this category. Nitro dyes are polynitro derivatives of phenols employed for coloring wool. They constitute two or more aromatic rings.

6. **Vat Dyes:** Insoluble complex polycyclic molecules having quinone structures are known as vat dyes, and include carbazoles, benzanthrones, and acridones. These are usually prepared using indigo. They are applied to cotton, linen, rayon, wool, and silk. They are applied to the fiber in their reduced form and then re-oxidized once fixed to the fiber.
7. **Mordant Dyes:** These are not directly applied onto fabric; rather need a binding agent called mordant, which helps to bind fiber and color. Mordants can attach to both the fiber and the colorants and can be applied on wool as well as nylon. Potassium aluminum sulfate is one of the most commonly used mordants for dye fixation.
8. **Reactive Dyes:** Reactive dyes react with flax and cotton, which are cellulosic fiber materials to form a strong covalent linkage. Fabrics having a hydroxyl or amino group are usually dyed using such colorants. They are generally used for dyeing natural as well as synthetic fibers. These can be applied to linen, viscose, and silk. They undergo chemical reactions with the fibers they are applied to, and are water soluble. Effluents from the textile houses contain reactive dyes, which can be removed easily using an adsorbent with the presence of a hydroxyl group, due to the ability of the dye to chemically attach to the cellulose sites.
9. **Sulfur Dyes:** These dyes have complex molecular structures containing sulfur. Most sulfur dyes are insoluble in water. However, in the presence of a basic medium under the action of a reducing agent and elevated temperature, they become water soluble and can be easily applied to the fabric. They are extensively produced and employed in cotton-related industries due to their easy application, good color fastness, and inexpensiveness. Due to low color intensity and environmental considerations, these are limited in applications.
10. **Metal Complex Dyes:** Metal complex dyes or premetallized colorants are insoluble in water. They are mainly azo group-containing compounds with some attached hydroxyl, amino, and carboxyl groups. Due to the presence of these functional moieties, the coordination linkage with metals like cobalt, chromium, and nickel is possible. These are used to color wool, polyamides, silk, and nylon. But these dyes, when present in wastewaters, increase the toxicity level of the dye due to their associated metal species.

Based on the application of the dyes, they are classified as below:

1. **Textile Dyes:** These are the dyes used mainly to color the textile fibers. These include reactive dyes, direct dyes, disperse dyes, basic, and acid dyes.
2. **Food Dyes:** Several synthetic and natural dyes are employed in the food industry to provide color to the food products and beverages. This is done to enhance the esthetic appeal of the food article. These dyes have shown toxicological impact on living organisms.
3. **Cosmetics and Drug Industries:** These are employed in industries such as drug and cosmetics for coloration as well as identification. In the pharmaceutical, colorants are used in capsules, tablets, and oral dosage forms.
4. **Printing Inks:** This application involves the dispersion of the organic solvents. Both water and solvent-based inks are employed for printing on paper, textiles, and cardboards.
5. **Biological Stains:** Eosin and hematoxylin are dyes used as histological stains for biological examination. Others are crystal violet, Gram's iodine, fluorescent, and photochromic dyes.

Dyes are often categorized based on their chemical structures. Dyes are having azo (−N=N) groups, Nitro ($-NO_2$), carbonyl (−CO) and quinoid groups have light-absorbing properties. Fused aromatic ring systems containing dyes are the anthraquinone dyes. They can impart blue, violet, and red colors when used in textiles. Metal (like copper) incorporated dyes are the phthalocyanine dyes. They are often blue or green in color, and used in inks, pigments, etc.

Table 6.2 presents the classification of colorants based on their characteristics, along with associated toxicological impacts.

It is essential to understand the characteristics of the dyes to evaluate their behavior, application, and interaction with various substances to evaluate their elimination tendency from wastewaters. The properties of dyes include:

1. **Color Fastness:** This is the ability of the dyes to retain their color under environmental exposure to light, moisture, and chemicals.
2. **Solubility:** The structure of the dyes allows their dissolution in a medium to which they are applied, such as water or any other solvent. Water-soluble dyes are normally used in textile industries due to their compatibility with cotton and wool. Dyes are either water soluble or insoluble. Water-soluble dyes cause water pollution, and their elimination from water poses a big challenge for the research community.
3. **Affinity for the Substrate:** It is the degree to which the dye molecules interact with the material intended to be colored. There are various factors influencing this property, like the presence of certain groups, the size of the dye molecule, and the surface features of the fiber.

Table 6.2 Classification and toxicological impacts of dyes [1, 3–8].

S. No.	Dye	Examples	Source	Application	Toxicological impact
1.	Acidic dye	Acid yellow 36, Acid orange 7, Acid violet, Acid blue 7, Acid blue 83	Textiles, silk dyeing, leather industries, pharmaceutical industries, and polymer industry	Synthetic fiber dyes	Vomiting, stomach disorder, carcinogenic and mutagenic effects
2.	Basic dye	Basic yellow 2, Basic violet, Methylene blue, Basic red 1	Polymer industry, paper industry	Synthetic fiber dyes	Obstructs sunlight, affects floral growth, micro toxicity to biota
3.	Direct dye	Direct black 38, Direct red 28, Direct yellow 59, Congo red	Paper industry	Cellulose fiber dyes	Toxic to flora and fauna, carcinogenic effects, mutagenic effects, and dermal effects
4.	Vat dye	Vat blue 14, Vat blue 1, Vat acid blue 74	Dyeing of cellulosic fibers, paints and varnishes	Linen, rayon, cotton	Dermatitis, allergic effects
5.	Disperse dyes	Disperse orange dye, Disperse red 9 and 60, Disperse violet 1	Polyesters, polymer products	Synthetic fiber dyes	Mutagenic effect, asthma, carcinogenic, water pollution
6.	Nitro dye	Disperse yellow 26, Naphthol yellow (II)	Dyeing wool	Protein and cotton fibers	Impairs photosynthesis, decreases light infiltration, carcinogenic, and mutagenic.
7.	Mordant dye	Mordant blue, Mordant red 11, Mordant black 17	Textiles and the leather industry	Wool fibers	Allergic reactions
8.	Reactive dye	C.I. reactive red 147, Dichlorotriazine reactive dye, C.I. reactive blue 19, C.I. reactive red 120	Textiles	Cellulose fiber dye	Allergic reaction on skin and eyes, respiratory allergy

S. No.	Dye	Examples	Source	Application	Toxicological impact
9.	Sulfur dye	Sulfur black 1, Sulfur brilliant green, Sulfur blue, Leuco sulfur black 1, Phthalic anhydride	Dyeing industry	Cellulose fiber dye	Dermal irritation, carcinogenic
10.	Azo dyes	Methyl orange, Eriochrome black T, Direct black 22, Methyl red	Textile mills, food industry	Protein fiber dye	Carcinogenic to both humans and animals, it disturbs the digestive system

4. **Chemical Stability:** Dyes are resistant to chemicals or environmental factors. This is a desirable attribute as far as dyeing is concerned. The stability of the dyes, as designed, makes them recalcitrant, making their removal difficult from the textile effluent. Also, if the dyes are such that can be biodegraded, this results in the formation of compounds which directly or indirectly prove to be hazardous to health. Their stable nature and complex structure affect the ecosystem of the water bodies adversely.

6.2 Toxicity Assessment of Dyes

The untreated dye-loaded effluent, if disposed into the aqueous systems, has undesirable consequences on the photosynthetic performance, which hampers the growth of the flora in water bodies. Untreated discharge of textile dye can lead to eutrophication, a change in color, and foul odor, in addition to the turbidity issues in the water bodies. In the long term, persistence of dyes causes bioaccumulation and disruption of microbial ecosystem.

Dye contamination inhibits the algal blooms and disturbs the trophic energy transfer and nutrient cycle in the ecosystem. Various studies have confirmed the negative impacts of dyes on the macrophytes due to decreased electron transport in the chloroplast [2].

Dyes in effluents are considered to be highly toxic due to their recalcitrant nature, as well as their degradation products. Moreover, dyeing processes engage various supplementary chemicals like organic acids, soaps, softeners, stain removers, dyeing chemicals, and dye fixing agents, which have the potential to get converted into carcinogenic products. These have the ability to trigger allergic reactions. Studies have shown that dyes and related products get introduced into the food chain and can cause deleterious impact, biomagnification, and bioaccumulation. Protein sources like fish and other aquatic species with bioaccumulated contaminants cause hypertension and fever if consumed by human beings. It is reported that fish are prone to a number of diseases due to the hypoxic impacts on their immune system, due to dyes. The terrestrial habitats also get contaminated by dye deposition. The chemical composition of the dyes creates lethal effects on the aquatic organisms. Dyes have also been reported to have mutagenic, carcinogenic, nervous system related, skin related, and allergic effects on human as well. The disposed dyes contaminate the water bodies, decline its quality, and disturb aquatic life, and adversely affect the human health [9].

Consequently, scientists have focused their research in analyzing the harmful effects and carrying out remediation of dyes from the environment. The same is evident by a large number of publications based on "colorant removal" from 2016 to 2023 [2].

6.2.1 Toxicity in the Aqueous Environment

The textile industry is a major contributor to the deterioration of the water quality. The untreated effluent from the textile industries and other facilities contaminates the aqueous environment indefinitely. Dye loaded water is purged into the water resources changing the clean colorless water into contaminated colored water. The discharged water has negative impact on aquatic flora and fauna. Dyes obstruct the light from entering into the photic zones of water body. This alters the balanced ecosystem flourishing in the water body. Decreased penetration of the light drops the photosynthetic activity of the plants and impairs their performance. A high concentration of the dyes depletes the amount of oxygen in the aqueous systems and impairs the activity of the existing fauna. These dyes can get into the food chain and biomagnifies resulting in harmful impacts like cramps, fever and tension if consumed by humans. These usually increase the amount of suspended solids, which obstruct the flow of water through the gills of fishes and other aquatic organisms. This hampers the respiratory cycles and decreases the growth rate. The aquatic community due to a long-time exposure to these xenobiotics, develops a reduction in the food consumption, lowering the essential nutrients in the aquatic organisms, leading to a hampered growth. The stagnant aquatic bodies do not get sufficient oxygen which results in hypoxia in the aquatic species, leading to their large-scale death. Hypoxic condition arising due to dyes as contaminants affect the immune system, making the organisms susceptible to numerous diseases. These diseased fishes and other organisms on consumption by the humans, have a significant impact on the human health.

The ingested dyes re-metabolized in their bodies into toxic intermediates, which prove to be hazardous for the species and their predators [9].

The erroneous coloration of the water bodies by dyes destroys the aesthetic beauty of the water bodies. Dye contamination inhibits the growth of microalgae, which play an important role in aquatic health. The destruction of microalgae due to interference by these colorants disturbs the energy transfer in the aquatic environment and depletes the nutrient availability. The growth and pigment content of microalgae like *Spirulina platensis* and *Chlorella vulgaris* were reported to get diminished with an interaction with the dyes like methylene blue. Dyes significantly affect the rate of development of aquatic macrophytes. This happens due to the exposure of the plants to changed ecosystems created by dyes, which hampers the electron transfer cycle of the chloroplast. The phototoxicity of the dyes was confirmed by an investigation involving the effect of congo red and gentian violet on *Lemna minor* (*L. minor*), a common duckweed. The report suggests that the dye decreased the synthesis of biogenic amines, which inhibited the growth and development of *L. minor*.

Textile effluent alters the soil characteristics and disturbs the soil microfloral balance. The wastewater having a lower amount of air deprives the seedling of proper oxygen and disrupts the seedling's osmotic pressure balance. Dyes have the ability to destroy the chlorophyll content of the plants, which slows the plant's growth.

6.2.2 Toxicity Effects on the Living Organisms

Dyes are linked with a variety of human and animal diseases. They are responsible for problems like dermatological infections to severe central nervous system disturbances, etc. Workers coming in direct contact with the recalcitrant chemicals, suffer from contact dermatitis, conjunctivitis, and rhinitis. Azo dyes have been thoroughly investigated for their toxicity, and the dye and its derivatives are reported to be linked with human bladder cancer. The mutagenic potential has been reported for sudan 1, an azo dye, which got converted into carcinogenic aromatic amine by the intestinal microorganism. This had the potential to develop cancer. The same impact was also shown by disperse red dye which was mutagenic in human hepatoma cells. It caused DNA damage by base pair replacement and mutations. The cytotoxic effect of disperse dyes can cause apoptosis. The DNA structure could be intercalated by dyes like azure B. This dye on interaction with the lipid membrane of the animal cells can causes significant reversible inhibition of intracellular enzymes of central nervous system which are essential for human behavior. Other than this, triphenyl methane dyes are also carcinogenic to humans, as they produce aromatic amines during anaerobic dye degradation. They are also the causes of allergic dermatitis. Another triphenyl methane dye namely crystal violet, causes mitotic poisoning and chromosomal aberrations. It is also linked with the reticular cell damage of the organs including vagina and urinary bladder. The dye also has the potential to cause cystitis, digestive system problems, respiratory troubles, and renal failure in humans [3]. Table 6.3 portrays various toxicological impacts of colorants on living systems.

The treatment of dye-loaded wastewater is thus crucial to make it suitable for subsequent use. The treatment can be carried out by using physical, chemical, biological, or hybrid processes. The selection of the most efficient treatment processes or their hybridization is reliant on the nature of the dye and the methods adopted in dyestuff manufacture [10].

Table 6.3 Toxicological impact of colorants on living organisms [3].

Organ system	Toxicological impact	Responsible dyes
Nervous system	Inhibits the intracellular enzyme of the central nervous system Dyes incorporated with metal damage the tissues	Tartrazine food dye Allura red AC Brilliant blue FCF
Dermal system	Dermatitis Allergic conjunctivitis Rhinitis Occupational asthma	Reactive dye Basic red 9 Sunset yellow FCF
Enzymatic system	Inactivation of enzymes Aromatic amines are generated, which are carcinogenic Disturbs cellular redox balance	Azure B Indigo Carmine Tartrazine
Chromosomal effect	Genotoxic impact Interferes with DNA structure and RNA Mutagenic potential carcinogenic	Reactive green 19 Reactive blue 2 Disperse red 1
Reproductive system	Cytotoxicity of spermatozoa Decline in ovarian protein	Direct blue 14 Erythrosine Brilliant blue
Renal system	Sarcoma Urinary bladder cancer	Disperse yellow Crystal violet

6.3 Treatment of Dye Wastewater

Figure 6.2 shows the different types of techniques employed for dye removal.

6.3.1 Physical Treatment

The physical methods, like ion exchange, membrane technology, and adsorption, have been used for years to eradicate the dyes from wastewaters. The ion exchange process is popularly employed for the purification, separation, and eradication of the colorants from their solutions. This technique involves the exchange of ions from the external medium within a solid matrix through electrostatic interactions. It suffers from the limitation of high salt sludge, which requires careful disposal operations. Reverse osmosis, microfiltration, and nanofiltration are the methods successfully used for the treatment of colorants in water, but these suffer from the limitations of pore blockage and fouling issues.

Adsorption is also one of the most popular methods due to the large amount of adsorbent availability. Sawdust, a lignocellulosic material, was also examined for its ability to remove wastewater pollutants, and it was confirmed that the presence of amide linkage (—NH_2), carboxyl (—COOH), phenolic (—OH), and hydroxyl groups in it, favored the adsorption of a variety of dyes. Also, modification using acid and alkali further improved its adsorptive characteristics [11, 12]. In an investigation dyes named methyl orange and methylene blue were adsorbed onto modified carbon nanomaterials [13]. Another research described about the potential

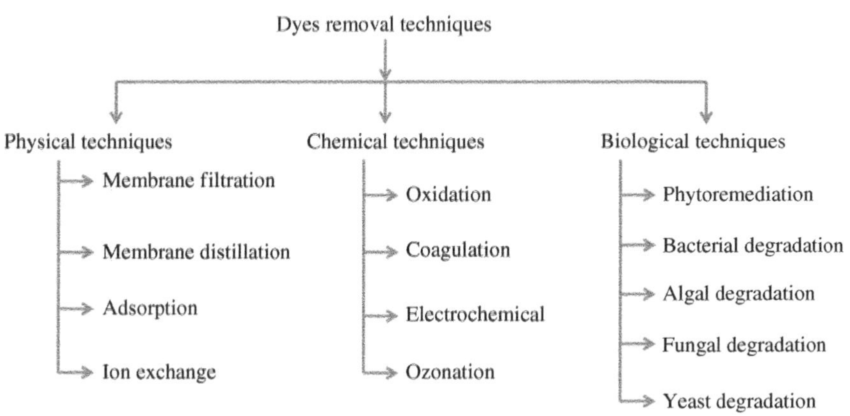

Figure 6.2 Dye removal techniques.

of multiwall carbon nanotubes for removing cationic and anionic dyes like scarlet 3R and crystal violet 2B. Adsorption was found to be governed by pH and time of contact. The studies on adsorption of the dyes onto various adsorbent materials are briefed in Table 6.4 [14].

The elimination of basic yellow 51 colorant by activated and chemically modified sugar beet pulp via ion exchange process was investigated. It was reported that activation increased the number of functional groups in pectin and cellulose, which enhanced the removal efficiency of the material. The process was found to be endothermic and spontaneous with a maximum adsorption capacity as 200 mg/g [18].

The ion exchange phenomenon for effective removal of the dyes was further supported by the synthesis of micro-sized ammonium phosphomolybdate resin. It was found that cationic dyes could be effectively removed from wastewater at an appropriate pH value. The ammonium ions and cationic dye moieties were involved in the exchange process. The removal efficiency of 94.6% was reported, withholding the efficiency till the 16th cycle [19].

A similar removal of the colorant was successfully investigated using various types of ion exchange resins. Resin D201 was found to show excellent performance in discoloration. The interactions involved during the ion exchange process also included electrostatic and π–π interactions along with hydrogen bonding [20].

A number of cationic and anionic dyes were removed from their prepared solutions using modified ion exchange monoliths. An adsorption potential of 606.3 mg/g was reported with a colorant exclusion efficiency of 72.3% suggesting the probability of removal of dyes from wastewater via cellulose resin [21].

6.3.2 Chemical Treatment

Chemical approaches for dye remediation include, coagulation, chlorination, advanced oxidation, etc. Chlorination of wastewater is carried out to destroy the dyes present in effluent. It is needed to neutralize the wastewater prior to its discharge. The chemical method requires high electrical input, chemical investment, and additional challenges of tackling the byproducts formed during the process. Natural coagulants like *Tamarindus indica* and *Azadirachta indica* have been demonstrated to have excellent potential for dye confiscation from wastewater effluents [22].

Advanced oxidation process involves the in situ production of hydroxyl (OH) radicals, as oxidizing agents, for dye-containing wastewater. The processes involved under this category include photo oxidation, Fenton process, ozonation etc. (Figure 6.3). These effectively remove dye even under harsh conditions with less or no sludge, but they have limitation of being expensive and pH dependent [22]. Ozonation leads to degradation of dyes to a biodegradable form.

Table 6.4 Adsorption characteristics of dye onto various adsorbents [15–17].

Target contaminant	Adsorbent	Operational conditions				Adsorption capacity (mg/g)
		Initial concentration (mg/L)	Adsorbent (mg)	Temperature (°C)	pH	
Rhodamine B	Graphite nanocomposite	10	10	–	–	–
Methylene blue	Graphite nanocomposite	5	10	–	–	–
Methylene blue	Magnetic carboxyl-carbon nanocomposite	100	20	25	–	254.83
Brilliant yellow	Magnetic polyurethane	1	100	30–65	1–12	328.00
Brilliant green		1	100	30–65	1–12	225.00
Malachite green	Magnetic cyclodextrin graphene oxide	5–800	5	45	7	990.10
Ponceau 4R	Polyamidoamine cyclodextrin	340	–	15	1–5	254.30
Reactive orange	Iron oxide nanosphere	20–80	5	–	7	32.50
Reactive yellow	Iron oxide nanosphere	20–80	5	–	7	25.00
Rhodamine 6G	Moroccan phosphate	100–400	–	–	5.2	6.84
Congo red	Moroccan phosphate	100–400	–	–	5.2	19.81
Methylene blue	Orange bagasse beads	10–500	20	25	–	2307.9
Disperse orange 25	Cenosphere zeolite	38	38	–	6.1	–
Disperse blue 79 : 1	Cenosphere zeolite	26.72	1.5	–	5.68	–
Methylene blue	Hierarchical zeolite	10	10	20	3–10	133.10
Reactive black 5		0.05	–	25	3	77.52
Congo red		0.05	–	25	3	34.36

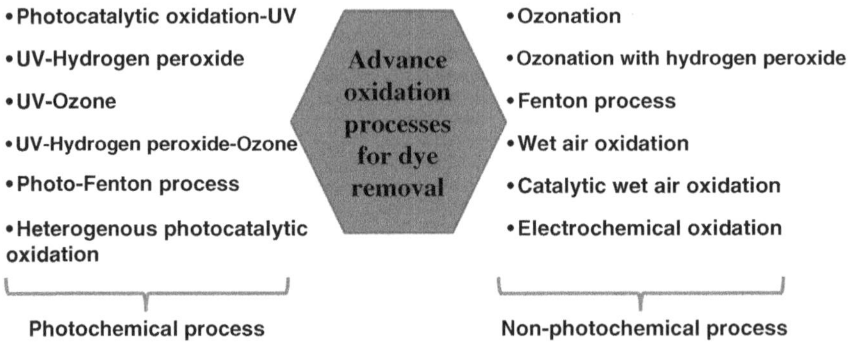

Figure 6.3 Advanced oxidation processes for dye elimination from wastewater.

6.3.2.1 Ozonation

Ozone is found to be an effective reagent for decolorization of the intensity of the wastewaters due to different chromophoric groups of the colorants. Lanzetta and team investigated the influence of ozonation process on the decolorization of wastewaters from a leather industry. They focused their study at different contact times and basic pH conditions. The removal efficiencies for color, and COD were observed to be 100% and 51%, respectively. In an uncontrolled pH condition, the removal efficiency of 92% was reported, which indicated that ozonation could be a viable option in post treatment of wastewater, especially for a biologically processed waste matter [23]. Ozone deocolorization is an irreversible process which indicates the degradation of the chromophoric part of the colorant molecules including any heterocyclic or carbocyclic moiety, whereby the dual purpose of color reduction and destruction of decay products of the dyes is achieved [24].

Ion exchanging capacity of zeolites due to its microporous structure is well documented. Zeolites can also act as excellent nanocatalysts. Ozonation supported by a catalyst can accelerate the degradation process due to creation of hydroxyl radicals. In the same regard, iron-based catalyst in ozonation have proven to show good catalytic action, because of their abundance, synthesis simplicity and non-toxicity. Electrophilic nature of ozone can initiate the reaction with negatively charged species like manganese dioxide. The low cost of abundantly present manganese oxide is of great importance in wastewater treatment [25].

In a similar investigation, the catalytic activity of ozonation-based bimetallic zeolite was tested to eradicate Disperse yellow dye. The zeolite employed was prepared by impregnation of iron and manganese on its surface. Dye removal was greatly affected by the contact time, catalyst amount, and dosage of ozone. It was reported that about 73% of the dye removal was recorded with a catalytic dose of 0.5 g and ozonation at the rate of 1.8 mg/min. The production of hydroxyl radical was governing the extent of dye removal efficiency. An increase in the COD removal percentage was also observed with catalytic ozonation [26].

Other than the catalytic ozonation, various other factors like nature of the pollutant, temperature, pH, and initiators also influence the efficiency of the process. A radical scavenger like inorganic salts decreases the effectiveness of the ozonation process. Properly managed contact time between the dye and ozone can help increase the performance of the ozonation for decolorization. The performance of physicochemical treatment can be enhanced by pretreating the wastewater by ozonation technique as it tends to lower down the organics of the wastewater along with turbidly [27].

Ozonation was performed in combination with other methods like anaerobic treatment which proved to be economical, and about 90% reduction in COD was observed, along with 84% of organics reduction. The removal of dye was found to increase with increase in the contact time and was found to be around 94% within 25 minutes. There was improvement in the BOD:COD as well. Oxalates, sulfates, and nitrates were obtained as ozonation byproducts during the combined treatment process [28].

6.3.2.2 Wet Air Oxidation

Wet air oxidation is used for removal of organic pollutants from wastewater. It oxidizes the compounds in air or oxygen at 0.5–20 MPa pressure and at a high temperature of 175–320 °C wherein the pollutants obtained are carbon dioxide ammonia and inorganic salts. The operational conditions can be moderated by using catalytic wet air oxidation. Varied types of heterogeneous and homogenous catalyst proved to offer better removal efficiency than the conventional air oxidation process [29].

In this regard, methyl orange dye was degraded by using mixed oxides of iron, cerium, and titanium catalyst and 98.09% of color removal was achieved. Moreover, the catalyst employed had high activity and good stability. Safranin T was eliminated employing polyoxometalate in catalytic wet air oxidation process. Ninety-eight percent of color removal and about 90% of COD was sequestered in less than 45 minutes under atmospheric pressure and normal temperature. Chlorides, bicarbonates, and nitrates were obtained as end products [30].

Similarly, congo red dye showed a 100% color removal from wastewater in presence of modified Y-zeolite. Optimum parameters of neutral pH and 1 g/L of catalyst were considered after carrying out a number of investigations for effect of pH, temperature, contact time catalyst dosage, etc. [31].

Few azo dyes like methyl orange and direct brown were completely removed using copper oxide-aluminum oxide mixed catalyst from wastewater. The contact time of 2 hours was sufficient for about 80% removal of both the dyes. The chromophoric group of the azo based dyes was oxidized by the hydroxyl radical generated. Nickel supported over magnesium–aluminum mixed oxides was used as a catalyst for the removal of crystal violet. After investigating different parameters like impact of pressure, concentration of colorant, temperature etc. the total organic carbon was found to vary from 60 to 20%, while the dye removal occurred from 63 to 18.4%, for an initial dye concentration taken between 10 and 50 mg/L. Similarly, methyl orange and methylene blue were eliminated using catalytic wet air oxidation method, using copper and molybdenum mixed oxide and phosphorus oxide. Out of the two colorants, methyl blue was found to show a better removal efficiency of 99.26%, at a neutral pH, with dye concentration of 0.3 g/L at normal temperature [32, 33].

6.3.2.3 Fenton Process

The Fenton process was employed for eliminating carmoisine dye from wastewater. The optimization was done using the Taguchi fractional method, and about 92.7% of the color removal was reported, with optimum conditions as 0.015 mmol of ferrous ions, an initial colorant concentration of 20 mg/L and pH 3.5.

Iron-supported zeolite was synthesized for decolorization of an azo dye. At pH 5.96, the decolorization obtained was 99.3% with 50 ppm initial dye concentration, 80 °C temperature, 15 g/L of catalyst and six minutes of contact time. Another azo dye named reactive black 5 was removed using iron supported over montmorillonite in the presence of hydrogen peroxide. The efficiency of removal was 99% within two and half hours [34].

Fenton process is extensively used in the degradation of organic compounds though it suffers from a limitation of acidic pH and high iron ion generation. It is essential that the iron must be eliminated from the reaction mixture using additional procedures. Other than iron-based Fenton process, copper and cobalt based Fenton processes are also used in wastewater treatment, but they have a lower rate of reaction as compared to the conventional method [35].

6.3.2.4 Photodegradation

Photocatalysts like titanium oxides, zinc oxides, and iron oxides can be employed for destruction of dye pollutants. In this regard titanium oxide nanoparticles have been widely used. The photodegradation of rhodamine B dye was done using titanium oxide hollow microspheres with promising results. A mixture of rhodamine B and a cationic dye methylene blue, was investigated for their removal using solar irradiated copper oxide nanozeolites. The decolorization was achieved with 0.1 g/L of catalyst at pH 5.3. Similarly, a COD reduction of 9.3% at neutral pH was obtained for methylene blue and bromocresol green using copper sulfide incorporated over clinoptilolite. The dye removal percentage of methylene blue and bromophenol was obtained as 61% and 32%, respectively, during two hours using similar nanoparticles [36, 37].

Photocatalytic ozonation was investigated for the removal of colorants from textile industries using titanium oxide as a catalyst. The percentage color and COD removal obtained were 98% and 64%, respectively, in 30 minutes. Copper ferrite nanoparticles at 25 °C were found to effectively remove reactive dyes at pH 3. Reactive red 198 and direct green 6 were eliminated to 100% in 20 minutes at pH 3, by employing nickel–zinc ferrite magnetic nanoparticles. Also, orange II dye when investigated for photocatalytic ozonation using gold supported bismuth oxide nanorods, where 75% color removal was observed [38].

Electrocoagulation, electro-Fenton, and anodic oxidation have evolved as new methods of eradicating the dyes, have less chemical investment and no sludge production. Doria and team carried out the degradation of indanthrene blue dye and reported encouraging results [39]. The only disadvantage associated was the high electricity cost. Similarly, the electrocoagulation approach was successfully demonstrated by researchers for a synthetic azo colorant [40]. The anodic direct oxidation process involved the removal of dyes by adsorption at the anodic areas, followed by degradation through electron transfer, while the indirect oxidation process involved strong oxidants for the purpose [1]. Efficiencies of various advanced oxidation processes involved in dye removal are presented in Table 6.5.

6.3.3 Biological Treatment

Biological treatment of dye-laden wastewater has gained importance in recent years. It has been found to be the superlative remediation method due to its prospective to destroy a diverse variety of colorants, cost effectiveness and an environmentally friendly approach. It can be carried out either by biological decolorization by microbes or by bacterial or fungal, or yeast degradation. Bacterial degradation can be brought about either by bacterial cells or by a bacterial consortium (Figure 6.4).

Azo dyes have been reported to be efficiently degraded by the enzymes like azoreductase, laccase and peroxidase. The end products of the reactions are nontoxic and can be obtained along with lesser volume of sludge. The method has gained momentum as enzymes increase the degradation reaction rate, though the limitation of low removal rate is still to be addressed. Similarly, fungal based degradation can be brought about by using ligninolytic fungi, *Aspergillus fumigatus, Penicillium janthinellum*, etc. Algal and yeast-based degradation approach is still to be explored, but algae can be seen as an alternatives to other techniques, since they can effectively degrade the dyes even under extreme environmental conditions. The potential of *Chlorella vulgaris* and *Saccharomyces* species can be explored for the removal of various dyes from wastewater [10].

Phytoremediation has evolved as an ecofriendly and cost-effective approach for dye cleanup from wastewater in comparison to other physicochemical and biological process. This is attributed to the extraordinary metabolic and removal

Table 6.5 Efficiencies of different advanced oxidation processes for dye removal.

		Parameters				
Dye	**Process**	**Agent**	**pH**	**Time**	**Initial dye concentration**	**Efficiencies**
Direct red 80	Ozonation	ozone	11	10 min	2000 mg/L	52–64% dye decolorization
Direct red 81	Ozonation	ozone	11	27 min	2000 mg/L	61% dye decolorization
Reactive red 239	Ozonation	ozone	7	20 min	50 mg/L	100% decolorization, 62% COD removal
Acid red 14	Ozonation	ozone	1.7	25 min	1500 mg/L	93% decolorization
Crystal violet	Ozonation	ozone	6.8	60	50 mg/L	78% decolorization
Reactive blue 19	Ozonation	ozone	7	–	–	60% COD removal
Reactive black 5	Ozonation	ozone	11	30 min	–	90% decolorization
Methyl orange	Catalytic ozonation	Ni-based hydroxide catalyst ozone	9	–	500 mg/L	96% dye elimination 72% COD reduction
Direct black 22	Catalytic ozonation	Zinc based slag, ozone	11	30 min	–	69% COD removal
Reactive black 5	Catalytic ozonation	Activated carbon	11	30 min	–	90% decolorization
Reactive blue 194	Catalytic ozonation	Activated carbon	5–12	40 min	–	100% decolorization, 90% COD removal
Alizarin red S	Catalytic ozonation	Iron oxide based activated carbon nano-composite, ozone	11	30 min	100–500 mg/L	95% dye removal
Methylene blue	Fenton process	Ferrous sulfate	3	30 min	10–50 mg/L	98.8% dye elimination, 85% COD reduction

Acid yellow 17	Fenton process	Ferrous sulfate	3	60 min	0.06–0.09 mM	89% degradation
Crystal violet	Fenton process	Iron nanoparticles dispersed on polyester fabrics	5	–	–	99% dye removal
Orange II	Fenton process	Calcium strontium copper loaded perovskite	–	90 min	10–100 mg/L	95% degradation, 40% carbon removal
Eriochrome black	Fenton process	Spinel framework of Ni-Co coated with iron phthalocyanine	6.7	–	20 mg/L	100% dye elimination
Rhodamine B	Fenton process	Spinel framework of Ni-Co coated with iron phthalocyanine	6.7	–	20 mg/L	90% TOC reduction
Congo red	Photo-Fenton process	$FeO.Cr_2O_3$	4	600 min	–	100% decolorization
Methylene blue	Photo-Fenton process	Fe_2O_3/TiO_2composite	7	40 min	–	98.86% decolorization, 65.5% TOC removal
Carmoisine	Photo-Fenton process	–	3.5	12 min	–	95.8% decolorization, COD 89%
Rhodamine B	Photo-Fenton process	Iron-modified rectorite	3	60 min	–	99% decolorization
Ponceau S	Chlorination	UV/Cl_2	4.5	5 min	–	99.2% decolorization
Azure C	Chlorination	UV/Cl_2	8	60 min	–	93.8% decolorization
CI Reactive red 2	Chlorination	UV/Cl_2	–	6 min	–	97.9% decolorization

Source: [41]/MDPI/CC BY 4.0.

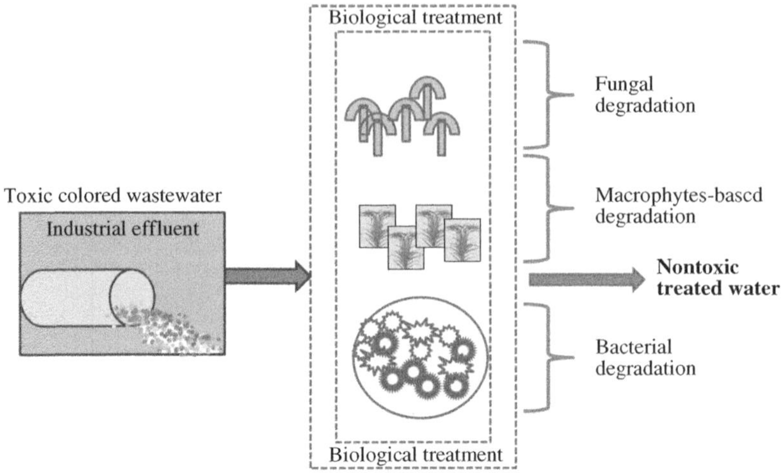

Figure 6.4 Biological treatment technologies for dye remediation.

capacity of plants. Commonly available plants like aster, petunia, ferns, and aquatic plants have been advocated for their colorant degradation ability [42].

Studies have reflected that aquatic plant species have a great potential of eradicating contaminants from wastewater due to their faster growth, pollutant uptake ability, and huge biomass production capacity. Species like *Phragmites australis, Typha angustifolia* can detoxify the wastewater, metabolize the colorant, and can curtail the effects of contaminants on the environment. *Eichhornia crassipes* was employed for the removal of an organic dye from wastewater and percentage efficiency of 90.8% and 62.8% was observed for cationic and the anionic dyes, respectively. Also, modified biological species proved to be good candidates for the removal of phenol red (67.4% removal) and gentian violet (71% removal). Similarly, *Lemna minor* also adsorbs contaminants and is considered to be a good bioadsorbent due to its easy availability, simple structure and morphology, quick growth, and has presence of carboxyl acid, polysaccharide, and cellulose facilitating the phytosorption of contaminant dye [43].

Ceratophyllum demersum plant was investigated for eradication of methylene blue dye and acid blue 92. The percentage removal after 15 days contact time with wastewater was obtained as 78.53%. A nanomagnetite supported over biochar from *Eichhornia crassipes* and *Phragmites australis* was also established to be efficient for dye elimination [44, 45].

Pistia species, Salvinia species, and *Hydrilla verticillata* macrophytes were examined for their phytoremediation capacity of removing pollutants from textile industries. The physicochemical characteristics like color, smell, pH, total dissolved matter, COD, fluoride, and so on were determined for the textile effluent. The total dissolved solid content, COD, and chloride concentration were found to be 6500, 2400, and 2050 ppm, respectively, which were beyond the discharge limits. Metals like chromium and lead were eliminated from wastewater using the above plants species. Among the investigated flora, *Pistia species* was found to be effective in eliminating 86% of the color and showed reduction of total dissolved solids (TDS), COD and chloride content as, 66%, 77%, and 61.33%, respectively [46].

Synthetic solution and industrial effluent containing methylene blue dye was treated with seaweed *Sargassum latifolium*. The dried biosorbent was characterized before and after the adsorption of the dye and it was found that an optimum temperature of 30 °C and adsorbent dose of 0.1 g/L removed 95% of the dye from the solution. The data were well confirmed by both Tempkin and Freundlich isotherm equilibrium models. Also, pseudo second order kinetics parameters showed high correlation coefficient values. The mechanism involved was endothermic in nature [47].

Macrophytes namely *Phragmites australis* and *Rheum rhabarbarum*, many ornamental plants, ferns, etc., have reported removal efficiency for acid orange 7, anthraquinone derivatives and other dyes. The role of plant cells of *Blumea malcolmii* and *Nopal cochenillifera*, furry roots of *Brassica juncea* and other species have been confirmed for dye degradation. It has been advocated that riboflavin reductase, lignin peroxidase enzyme, laccase enzyme, azo reductase, dichlorophenol indophenol reductase, tyrosinase, and an enzyme named alcohol oxidase are principle biotransforming enzymes. These are responsible for collapsing the intricate assembly of dyes, thus bringing about a considerable reduction in the COD, turbidity, and conductivity of the wastewater. Further, research based on improvement of bioengineered or modified plant species for effectual phytodegradation of fabric colorants must be focused [48].

The efficacy of different biological species for removal of various colorants is summarized in Table 6.6

6.3.4 Novel/Hybrid Treatment Methods

Recently, membrane distillation technology has evolved as one of the most effectual wastewater recovery techniques. This technique involves passing the volatiles from wastewater through microporous membranes and collection of distillate. The technology is effective not only in terms of water reclamation but also involves small sized equipment, low energy consumption, manageable operating conditions, and low investment. The main disadvantages associated are membrane fouling and scaling. Membrane distillation technology can be divided into four chief setups based on the procedure of induction of vapor pressure and collection of permeate [49]. Table 6.7 gives the different types of designs adopted in the process.

Table 6.6 Efficiencies of biological process for dye removal.

Dye	Biological species	Parameters	Initial dye concentration	Efficiencies
Brilliant crocein	*Providencia rettgeri*	Medium = low nutrient Inoculums = 2% volume per volume Time = 8 days	50 mg/L	100% decolorization
Reactive black 5	*Pseudomonas aeruginosa*	pH = 8 Inoculums = 5% *v*/*v* Temp. = 37 °C	50 mg/L	100% decolorization
Reactive blue 250	*Lysinibacillus* sp. KPB6	Temp. = 37 °C Over alginate beads	–	98.2% decolorization
Metanil yellow G	Bacterial consortium	pH = 10 Temp. = 40 °C Time = 16 hours	100 mg/L	93% decolorization
Congo red	*Bacillus* sp.	pH = 9 Carrier filling ratio = 60% volume per volume	10–100 mg/L	95.7%
Crystal violet	*Bjerkandera adusta*	Inoculums = 4 g Shaking = 150 rpm	50–250 mg/L	90% decolorization
Reactive yellow F3R	*Neurospora* sp.	Inoculums = 10% volume per volume 2% glucose = 2% pH = 8 Temp. = 30 °C Time = 5 hours	200 mg/L	48–89% colorant removal
Methylene blue	*Trametes hirsuta*	Speed of shaker = 150 rpm pH = neutral Time = 6 hours	10–50 mg/L	80% decolorization
Methyl orange	*Aspergillus iizukae*	Inoculum 5% volume per volume Agitation = 150 rpm Temp. = room temperature Time = 9 days	100–1200 mg/L	85% decolorization

Dye	Biological species	Parameters	Initial dye concentration	Efficiencies
Congo red	*Aspergillus flavus*	Oligotrophic medium Shaking speed = 120 rpm Temp. = 25 °C Time = 3 days	200 mg/L	97% color removal
Brilliant blue	*Lactarius deliciosus*	Shaking speed = 120 rpm pH = 5 Temp. = 28 °C Time = 12 hours	25–800 mg/L	99% decolorization
Reactive orange 16	*P. ostreatus* and *C. zeylanoides*	Shaking speed = 80 rpm Temp. = 28 °C Time = 11 days	25–200 mg/L	87.5% decolorization
Acid blue 29	*Trametes hirsuta*	Shaking speed = 100 rpm Temp. = 30 °C	100 mg/L	96% decolorization
Methyl red	*Charavulgaris*L	pH = 9.5 Time = 2 days	10–50 mg/L	70–100% decolorization
Malachite green	*Synechococcus elongatus*PCC 7942	Temp. = 30 °C Time = 12 hours Light intensity = 100 μmol m^{-2} s^{-1}	100 mg/L	99.5% dye removal
Direct red 31	Bionanocomposite	Light = constant Time = 150 min	40 mg/L	92% decolorization
Disperse blue 1	Bacterial-algal consortium	Anaerobic bioreactor and photobioreactor Time = 12 hours light/dark	–	78% TOC removal, 96–99% decolorization
Reactive orange 16	Activated sludge	Anaerobic condition carrier filling ratio = 40% volume per volume	5 mg/L	61% dye and 92% COD
Reactive yellow 2	Activated sludge	Conditions = anaerobic and aerobic	40 mg/L	100% elimination of color 81–100% COD reduction

(Continued)

Table 6.6 (Continued)

Dye	Biological species	Parameters	Initial dye concentration	Efficiencies
Reactive yellow 15	Activated sludge	Conditions = anaerobic and aerobic	40 mg/L	100% colorant and COD removal
Direct black 22	Anaerobic sludge	Microaeration	0.06 mM	69–79% elimination of color 59–78% COD reduction
Methylene blue	Anaerobic sludge	pH = moderate of range 6.8–7.2	15	80% elimination of color, 99.5% COD reduction
Reactive black	Anaerobic sludge	Temp. = 30 °C	–	58% decolorization, 90% COD

Source: [41]/MDPI/CC BY 4.0.

Though several treatment techniques are available and have been explored, the treatment process becomes difficult considering the matrix of dyes and chemicals emerging from the textile industries, posing challenges.

Various physical, chemical, and biological methods are combined to eliminate dyes from wastewater to achieve better removal performance. Hybrid methods eliminate a variety of dyes from wastewater, minimize environmental impacts and are practical, though these require careful operation and optimization. For example, activated sludge along with advance oxidation, electrochemical oxidation in conjunction with biological method can help better degradation of colorants from wastewater. Ozonation along with upflow anaerobic sludge blanket reduces colorants to about 94%. Membrane bioreactor and photocatalytic membrane reactor system shows good performance for dye matrix effluents. A hybrid treatment technology offers promising solution for degradation of colorants. Few other novel technologies developed for dye remediation are mentioned in Figure 6.5.

1. **Microbial Fuel Cells:** These are the cells which oxidize dyes at the anode and transfers electrons for reduction of oxygen at the cathode. In such systems, microbes are used to generate bioelectricity, where microbes act as oxidizing agents. The degradation is influenced by microbial activity and its type, the anode, the cathode, and substrates.

Table 6.7 Types of membrane distillation designs.

	Direct contact membrane distillation (DCMD)	Air gap membrane distillation (AGMD)	Sweeping gas membrane distillation (SWMD)	Vacuum membrane distillation (VMD)
Details	Thermally driven process. The feed and permeate solutions are in direct contact with the membrane	The hot feed solution is in direct contact with the porous hydrophobic membrane surface. An air gap demarcates permeate side and condensation surface	A cold sweep gas acts as driving force for evaporation of volatiles	Vacuum pressure is functional at the permeate side, and vapor condensation occurs external to the membrane module
Pros	Good rejection of nonvolatile, works at moderate temperature. Not influenced by concentration of influent. Low-cost input	Low conductive heat loss and membrane wetting	Reduce conductive heat loss	Volatile content can be effectively removed, high permeate flux, thermal and conductive heat loss is less, and low-cost input
Cons	Large heat loss	Low permeate flux	Large condenser requirement to accommodate sweep gas	Prone to wetting

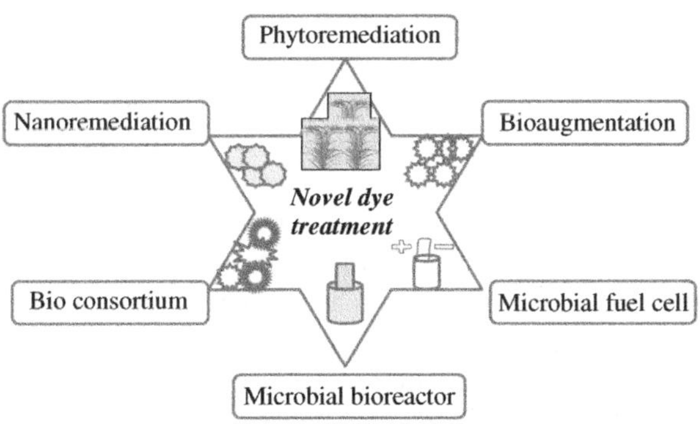

Figure 6.5 Novel/hybrid dye treatment methods.

2. **Nanoparticle-based Bioremediation:** Nanobioremediation is one of the viable alternatives for colorant elimination from the aqueous environment. The nanoparticles developed by microbes contribute to biodegradation of colorants.
3. **Membrane Bioreactors:** These show 99% removal of dyes using a modified anaerobic forward osmosis membrane bioreactor. These systems can be applied on a large scale for better abatement of dye pollution.
4. **Bioaugmentation:** Bioaugmentation refers to the use of genetically engineered microorganisms, which especially involves the use of bacteria, fungi, and yeast for remediation. A team of researchers reported a 10% increase in the degradation of wastewater containing azo dye with the use of engineered *E. coli* gene coding, for azo reductase enzymes, present in *P. luteola*. Similarly, 95% of the decolorization of dye acid violet was observed by transferring the gene code from *K. pneumonia* to *E. coli*.

6.4 Legislative Guidelines for Dye-loaded Wastewater

Recently, the increased awareness about the toxicological impacts of wastewater on the environment has directed the government agencies to formulate strict regulations related to the discharge standards of colored wastewater into the water bodies. Also, the arising scarcity of water has made industries dealing with the colorants opt for the treatment and reuse of the wastewater. Industries thus have to ensure that the discharged water from the industrial setups meets the required standards provided by the government regulatory bodies. In textile wastewater, the key contaminants of concern are the metal ions, dyes, and its colors. The discharge standards for the printing and dyeing industries vary for different countries.

According to Central Pollution Control Board (CPCB), the guidelines for dye effluent treatment plant includes a pH of 6.0–8.5, suspended solids as 100 mg/L, color in platinum cobalt units as 400 units, BOD for 3 days at 27 °C as 30 mg/L, and temperature of effluent discharged not exceeding 5 °C above the receiving water body.

The US Environmental Protection Agency (EPA) formulated textile mill effluent guidelines in 1974, which were amended in the years 1977 and 1982. The rules are framed for the direct discharge points. It is also a part of the National Pollutant Discharge Elimination System (NPDES). The standard of wastewater emission containing dyes is developed by the National Environmental Protection Department which is based on local conditions. This comprises of COD, BOD, pH, suspended solids, chrominance, total nitrogen, total phosphorus, sulfur, etc.

Table 6.8 Textile wastewater standards in various countries.

Country	Suspended solids (mg/L)	COD (mg/L)	BOD (mg/L)	pH	Temperature (°C)
Nigeria	25	60–90	30–50	6–9	40
Uganda	10	100	50	6–8	35
China	70	100	25	6–9	–
Malaysia	100	100	50	5.5–9	40
Germany	–	160	25	–	–
Thailand	120–400	20–60	20–60	5.5–9	40
Global limit	Less than 20	Less than 50	Less than 30	6–9	Less than 42

Source: [50]/Springer Nature.

As per Chinese legislation, the discharge limits are set as, COD = 100 mg/L, BOD = 25 mg/L, suspended solids = 70 mg/L, total nitrogen = 20 mg/L, total phosphorus = 1 mg/L, ammonical nitrogen 15 mg/L, sulfur = 1 mg/L, hexavalent chromium = 0.5 mg/L, aniline = 1 mg/L.

Discharge standards for water pollutants from dyeing and finishing of textile industry in Germany are COD = 160 mg/L, total phosphorus = 2 mg/L, ammonical nitrogen = 10, BOD = 25 mg/L, total nitrogen = 20 mg/L, nitrite = 1 mg/L.

The emission limits for discharge of textile wastewaters are shown in Table 6.8.

The Canadian legislation has provided the release guidelines for 25 azo disperse dyes and disperse yellow 3, with molar weights less than 360 g/mol, used in specific textiles. Some recommendations for best practices have also been provided. It suggests that in facilities the final point of discharge the concentration of the textile dye formulation should not be greater than 146.52 and 240 ppm for the dyeing processes. The daily use of various disperse dyes should not be greater than 90 kg/day for textile formulations and 10 kg/day for textile dyeing process.

Based on the Best Practicable Control Technology (BPT), various limits for different textile mill processing units have been given by the US Environmental Protection Agency (Table 6.9). These are the most effective and cost-efficient technologies for controlling the point source pollution discharges.

6.5 Case Studies

6.5.1 Textile Wastewater Treatment Plant – Arequipa, Peru

Real wastewater generated form a wool processing textile industry was investigated for its remediation using adsorption and membrane filtration technologies.

Table 6.9 Textile wastewater standards formulated by the US EPA, according to the best practical control technology [50].

	Printing and dyeing textile wastewater	Fabric printing and dyeing textile wastewater	Yarn printing and dyeing textile wastewater	Printing and dyeing textile wastewater	Fabric printing and dyeing textile wastewater	Yarn printing and dyeing textile wastewater
	Maximum			Average of 30 days		
Parameters	kg/tons (fabric)					
BOD	22.4	5	6.8	11.2	2.5	3.4
COD	16.31	60	84.6	81.5	30	42.3
TSS	35.2	21.8	17.4	17.6	10.9	8.7
Sulfur	0.28	0.2	0.24	0.14	0.10	0.12
Phenol	0.14	0.10	0.12	0.07	0.05	0.06
Chromium	0.14	0.10	0.12	0.07	0.05	0.06
pH	6–9	6–9	6–9	6–9	6–9	6–9

The adsorbent was sweet orange seeds obtained from a local market of Arequipa, Peru. The bioadsorbent was prepared by powdering the sweet orange seeds and subjecting to fat extraction in hexane. The dye effluent was collected and the analysis of wastewater for presence of color and other parameters like COD, total nitrogen, total phosphorus, surfactant, turbidity, and conductivity was conducted, using standard equipment and procedures.

The wastewater from wool processing industry showed purple tonality with a maximum absorbance wavelength at 530–536 nm. A COD value of 159 mg/L was observed. The optimum conditions obtained were pH of 2.6, adsorbent dosage of 0.58 g/L and temperature as 26 °C. The predicted dye removal calculated using by quadratic model was obtained as 100.00 ± 33%. The experimental values obtained were 95.9 ± 0.7%. The optimized conditions of pH, adsorbent dosage and temperature were used to scale up the textile wastewater to 10 L. Maximum dye adsorption occurred in first 30 minutes, where 92% of the colorant removal took place. Saturation of the adsorbent occurred after 28 minutes of contact. Water containing the bioadsorbent was sequentially subjected to microfiltration. By using ceramic membrane of 3 μm pore size, substantial results of removal were observed. The characteristics of textile wastewater after the sequential treatment revealed 97% dye removal by adsorption using bioadsorption. The anionic surfactant concentration by colorimetric method was reported to be less than 0.061 mg MBAS/L. The total dissolved solid concentration increased, probably owing to the partial dissociation of surface groups and was reported to be 2048 mg/L. The turbidity value was recorded to be 715 NTU after treatment, while overall 95.7% of removal of the colorant was carried out. Additional treatment setup like activated sludge and other biological processes were demanded to remove the total nitrogen from wastewaters. Investigations on a small scale for the real wastewater revealed that adsorption using orange seed as bioadsorbent combined with membrane filtration could remove dye effectively from real wastewaters [51].

6.5.2 Efficiency Analysis of Textile Wastewater Treatment Plant of Foshan, China

Wastewater sample was collected from a textile company at Foshan, China, with the production capacity of 75000 tons/day. The effluent from the textile industry was investigated for its pollutant contents, which revealed a pH in the range of 8.96–12.50, COD as 400–1000 mg/L, BOD as 200–450 mg/L, and color measured in terms of dilution times was 400–1500 times. Suspended solids were reported to be 50–250 mg/L.

The study involved monitoring and elimination analysis of different pollutants by using a combination sequence of a two-stage ozonation, ultrafiltration, and reverse osmosis, with an aim of reusing treated wastewater, thereby ensuring sustainability. The combined system comprised of a primary water reuse system with setup for ozonation, primary ultrafiltration and primary reverse osmosis, followed by secondary ozone aeration, secondary ultrafiltration, and secondary reverse osmosis. The characteristics of the wastewater before and after treatment were analyzed by using standard equipment and procedures. The primary and secondary filtering surfaces were around 450 and 700 m^2, respectively. The ozone dosage used was higher in comparison to the previous methods (80–100 g/m^3). The sand filtration tank was washed every 8 hrs. The water consumption in the primary and secondary filtration tank was around 1500 m^3 per day. Primary ultrafiltration removed turbidity and COD. The primary ultrafiltration system supplied 350 m^3/day of backwash water. The primary reverse osmosis system worked at 1–1.2 MPa pressure and was equipped with 5 μm size polypropene filter bag. The secondary ozone aeration tank was set at a retention time of 1.5 hour and ozone dosage as 100–120 g/m^3. The filtering surface in the next filtration tank was 280 m^2 and the wash water volume in the secondary reverse osmosis system was 350 m^3/day. This combined process showed good performance in the water reuse process.

The overall removal efficiency of secondary water reuse system observed was 99.62% for COD, 99.75% for color, with the decrease in the total nitrogen as

88.26%, total phosphorus removal of 99.12%, and 88.64% turbidity removal. The reverse osmosis system achieved the highest pollutant removal, while ozonation process decreased color in the wastewater enormously. The combined two stage water treatment was effective in elimination of colorants, facilitating the reuse of treated water for industrial processes. The effluent from the primary and secondary recycled water satisfied the reuse standards of recycled water and domestic drinking water.

The total operating cost involved reached 0.44 USD/m^3 for obtaining reusable water while the cost of color removal was merely 0.18 USD/m^3, which was much less than other processes used for the same purpose. A combination of the techniques used depicted the potential for application for large scale setups. In addition to this, the method proved to be an economically viable method. The case study endorsed the fact that recycling of treated wastewater and zero wastewater discharge was technically feasible and cost effective [52].

References

1 Dutta, S., Gupta, B., Srivastava, S.K., et al. (2021). Recent advances on the removal of dyes from wastewater using various adsorbents: a critical review. Material Advances, Volume 2, pp. 4497–4531. https://doi.org/10.1039/D1MA00354B.

2 Al-Tohamy, R., Ali, S.S., Li, F., et al. (2022). A critical review on the treatment of dye-containing wastewater: ecotoxicological and health concerns of textile dyes and possible remediation approaches for environmental safety. Ecotoxicology and Environmental Safety, Volume 231, p. 113160. https://doi.org/10.1016/j.ecoenv.2021.113160.

3 Sterenzon, E., Vadivel, V.K., Gerchman, Y., et al. (2022). Effective removal of acid dye in synthetic and silk dyeing effluent: isotherm and kinetic studies. ACS Omega, Volume 7, Issue 1, pp. 118–128. https://doi.org/10.1021/acsomega.1c04111.

4 Ho, S. (2020). Removal of dyes from wastewater by adsorption onto activated carbon: mini review. Journal of Geoscience and Environmental Protection, Volume 08, pp. 120–131. https://doi.org/10.4236/gep.2020.85008.

5 Elzahar, M.M.H., and Bassyouni, M. (2023). Removal of direct dyes from wastewater using chitosan and polyacrylamide blends. Scientific Reports, Volume 13, p. 15750. https://doi.org/10.1038/s41598-023-42960-y.

6 Alihosseini, F., Azimi, S.H., Mostajeran, M., et al. (2023). Removal of different vat dyes by *Aspergillus niger*: a comparative study focusing on the molecular structure. Journal of Microbiological Methods, Volume 208, p. 106720. https://doi.org/10.1016/j.mimet.2023.106720.

7 Markandeya, Shukla, S.P., and Mohan, D. (2017). Toxicity of disperse dyes and its removal from wastewater using various adsorbents: a review. Research Journal of Environmental Toxicology, Volume 11, pp. 72–89. https://doi.org/10.3923/rjet.2017.72.89.

8 Ali, N.S., Khader, E.H., Khudhur, R.H., et al. (2024). Removal of anionic azo dye from wastewater using Fe_3O_4 magnetic nanoparticles adsorbents in a batch system. Desalination and Water Treatment, Volume 317, p. 100033. https://doi.org/10.1016/j.dwt.2024.100033.

9 Bal, G., and Thakur, A. (2022). Distinct approaches of removal of dyes from wastewater: a review. Materials Today: Proceedings, Volume 50, Issue 5, pp. 1575–1579. https://doi.org/10.1016/j.matpr.2021.09.119.

10 Kathing, C., and Saini, G. (2022). A review of various treatment methods for the removal of dyes from textile effluent. Recent Progress in Materials, Volume 4, Issue 4, p.21. https://doi.org/10.21926/rpm.2204028.

11 Bożęcka, A.M., Orlof-Naturalna, M.M., and Kopeć, M. (2021). Methods of dyes removal from aqueous environment. Journal of Ecological Engineering, Volume 22, Issue 9, pp. 111–118. https://doi.org/10.12911/22998993/141368.

12 Chikri, R., Elhadiri, N., Benchanaa, M., et al. (2020). Efficiency of sawdust as low-cost adsorbent for dyes removal. Journal of Chemistry, Volume 2020, Issue 1, p. 8813420. https://doi.org/10.1155/2020/8813420.

13 Czubaszek, M., and Choma, J. (2018). Adsorption studies of selected dyes from aqueous solutions on nanoporous carbon materials under dynamic conditions. Ochrona Srodowiska, Volume 40, Issue 3, pp. 3–8.

14 Shabaan, O.A., Jahin, H.S., and Mohamed, G.G. (2020). Removal of anionic and cationic dyes from wastewater by adsorption using multiwall carbon nanotubes. Arabian Journal of Chemistry, Volume 13, Issue 3, pp. 4797–4810.

15 Piaskowski, K., Świderska-Dąbrowska, R., and Zarzycki, P.K. (2018). Dye removal from water and wastewater using various physical, chemical, and biological processes. Journal of AOAC International, Volume 101, Issue 5, pp. 1371–1384. https://doi.org/10.5740/jaoacint.18-0051.

16 Markandeya, Shukla, S.P., and Srivastav, A.L. (2021). Removal of disperse orange and disperse blue dyes present in textile mill effluent using zeolite synthesized from cenospheres. Water Science and Technology, Volume 84, Issue 2, pp. 445–457. https://doi.org/10.2166/wst.2021.216.

17 Ramezani, H., Azizi, S.N., and Cravotto, G. (2019). Improved removal of methylene blue on modified hierarchical zeolite Y: achieved by a "destructive-constructive" method. Green Processing and Synthesis, Volume 8, pp. 730–741. https://doi.org/10.1515/gps-2019-0043.

18 Turk, F.N., Ciftci, H., and Arslanoglu, H. (2023). Removal of basic yellow 51 dye by using ion exchange resin obtained by modification of byproduct sugar beet pulp. Sugar Technology, Volume 25, pp. 569–579. https://doi.org/10.1007/s12355-022-01207-2.

19 Joseph, J., Radhakrishnan, R.C., Johnson, J.K., et al. (2020). Ion-exchange mediated removal of cationic dye-stuffs from water using ammonium phosphomolybdate. Materials Chemistry and Physics, Volume 242, p. 122488. https://doi.org/10.1016/j.matchemphys.2019.122488.

20 Huang, R., Zhang, Q., Yao, H., et al. (2021). Ion-exchange resins for efficient removal of colorants in bis(hydroxyethyl) terephthalate. ACS Omega, Volume 6, Issue 18, pp. 12351–12360. https://doi.org/10.1021/acsomega.1c01477.

21 Yang, Z., Asoh, T., and Uyama, H. (2019).Removal of cationic or anionic dyes from water using ion exchange cellulose monoliths as adsorbents. Bulletin of the Chemical Society of Japan. Volume 92, Issue 9, pp. 1453–1461. https://doi.org/10.1246/bcsj.20190111.

22 Zhao, H., Cao, P., Lu, L., et al. (2022). Co-regulation of dispersion, exposure and defect sites on CeO_2(III) surface for catalytic oxidation of Hg^0. Journal of Hazardous Materials, Volume 424, Article 126566. https://doi.org/10.1016/j.jhazmat.2021.126566.

23 Lanzetta, A., Papirio, S., Oliva, A., et al. (2023). Ozonation processes for color removal from urban and leather tanning wastewater. Water, Volume 15, p. 2362. https://doi.org/10.3390/w15132362.

24 Belov, S., and Grigori Naumchik, G. (2020). The application of ozone to reduce the coloring Intensity of aqueous solutions of dyes used in the textile industry. E3S Web Conference, Volume 212, p. 01001. https://doi.org/10.1051/e3sconf/202021201001.

25 Ikhlaq, A., Zafar, M., Javed, F., et al. (2021). Catalytic ozonation for the removal of reactive black 5 (RB-5) dye using zeolites modified with $CuMn_2O_4/gC_3N_4$ in a synergic electro flocculation-catalytic ozonation process. Water Science and Technology, Volume 84, Issue 8, pp. 1943–1953. https://doi.org/10.2166/wst.2021.404.

26 Ali, Z., Ikhlaq, A., Qazi, U.Y., et al. (2023). Removal of disperse yellow-42 dye by catalytic ozonation using iron and manganese-loaded zeolites. Water, Volume 15, Issue 17, p. 3097. https://doi.org/10.3390/w15173097.

27 Shriram, B., and Kanmani, S. (2014). Ozonation of textile dyeing, wastewater – a review. Journal of Institute of Public Health Engineers, Volume 2014–15, Issue 3, pp. 46–50.

28 Venkatesh, S., Venkatesh, K., and Quaff, A.R. (2017). Dye decomposition by combined ozonation and anaerobic treatment: cost effective technology. Journal of Applied Research and Technology. Volume 15, Issue 4, pp. 340–345. https://doi.org/10.1016/j.jart.2017.02.006.

29 Fu, J., and Kyzas, G.Z. (2014). Wet air oxidation for the decolorization of dye wastewater: an overview of the last two decades. Chinese Journal of Catalysis, Volume 35, Issue 1, pp. 1–7. https://doi.org/10.1016/S1872-2067(12)60724-4.

30 Zhang, Y., Li, D.L., Chen, Y., et al. (2009). Catalytic wet air oxidation of dye pollutants by polyoxomolybdate nanotubes under room condition. Applied Catalysis B: Environmental, Volume 86, Issues 3–4, pp. 182–193. https://doi.org/10.1016/j.apcatb.2008.08.010.

31 Azimi, S.C., Shirini, F., and Ali, R.P. (2021). Advanced oxidation process as a green technology for dyes removal from wastewater: a review. Iranian Journal of Chemistry and Chemical Engineering, Volume 40, Issue 5, pp. 1467–1489. https://doi.org/10.30492/IJCCE.2020.43234.

32 Ovejero, G., Rodriguez, A., Vallet, A., et al. (2012). Ni Supported on Mg-Al oxides for continuous catalytic wet air oxidation of crystal violet. Journal of Applied Catalysis B. Volume 125, pp. 166–171. https://doi.org/10.1016/j.apcatb.2012.05.030.

33 Levec, J., and Pintar, A. (2007). Catalytic wet-air oxidation processes: a review. Catalysis Today, Volume 124, pp. 172–184.

34 Sohrabi, M.R., Khavaran, A., Shariati, S., et al. (2017). Removal of Carmoisine edible dye by Fenton and photo Fenton processes using Taguchi orthogonal array design. Arabian Journal of Chemistry, Volume 10, pp. S3523–S3531. https://doi.org/10.1016/j.arabjc.2014.02.019.

35 Carvalho, S.S.F., and Carvalho, N.M.F. (2017). Dye degradation by green heterogeneous Fenton catalysts prepared in presence of *Camellia sinensis*. Journal of Environmental Management, Volume 187, pp. 82–88. https://doi.org/10.1016/j.jenvman.2016.11.032.

36 Nezamzadeh-Ejhieh, A., and Zabihi-Mobarakeh, H.(2014). Heterogeneous photodecolorization of mixture of methylene blue and 3-bromophenol blue using CuO-nano-clinoptilolite. Journal of Industrial and Engineering Chemistry, Volume 20, Issue 4, pp. 1421–1431. https://doi.org/10.1016/j.jiec.2013.07.027.

37 Nezamzadeh-Ejhieh, A., and Moazzeni, N. (2013). Sunlight photodecolorization of a mixture of methyl orange and bromocresol green by CuS incorporated in a clinoptilolite zeolite as a heterogeneous catalyst. Journal of Industrial and Engineering Chemistry, Volume 19, Issue 5, pp. 1433–1442. https://doi.org/10.1016/j.jiec.2013.01.006.

38 Anandan, S., Lee, G.J., Chen, P.K., et al. (2010). Removal of Orange II dye in water by visible light assisted photocatalytic ozonation using Bi_2O_3 and Au/Bi_2O_3 nanorods. Industrial and Engineering Chemistry Research, Volume 49, Issue 20, pp. 9729–9737. https://doi.org/10.1021/ie101361c.

39 Doria, A.R., Pupo, M., Santos, G.D.O.S., et al. (2020).Electrochemical oxidation of indanthrene blue dye in a filter-press flow reactor and toxicity analyses with *Raphidocelis subcapitata* and *Lactuca sativa*. Ecotoxicology and Environmental Safety, Volume 198, Article 110659. https://doi.org/10.1016/j.ecoenv.2020.110659.

40 Hamad, H., Bassyouni, D., El-Ashtoukhy, E.S., et al. (2018). Electrocatalytic degradation and minimization of specific energy consumption of synthetic azo dye from wastewater by anodic oxidation process with an emphasis on enhancing economic efficiency and reaction mechanism. Ecotoxicology and Environmental Safety, Volume 148, pp. 501–512. https://doi.org/10.1016/j.ecoenv.2017.10.061.

41 Ledakowicz, S., and Paździor, K. (2021). Recent achievements in dyes removal focused on advanced oxidation processes integrated with biological methods. Molecules, Volume 26, Issue 4, pp. 870. https://doi.org/10.3390/molecules26040870.

42 Kaushal, J., Mahajan, P., and Kaur, N. (2021). A review on application of phytoremediation technique for eradication of synthetic dyes by using ornamental plants. Environmental Science and Pollution Research, Volume 28, pp. 67970–67989. https://doi.org/10.1007/s11356-021-16672-7.

43 El-Sadaawy, M.M., and Agib, N.S. (2024). Removal of textile dyes by ecofriendly aquatic plants from wastewater: a review on plants species, mechanisms, and perspectives. Blue Economy, Volume 2, Issue 2. https://doi.org/10.57241/2805-2994.1023.

44 Shaalan, M.H., Almansoory, A.F., and Al-Baldaw, I.A. (2023). Removing of methylene blue dye from wastewater by interaction between *Ceratophyllum demersum* and *Acinetobacter radioresistens*. IOP Conference Series, Earth and Environment Science, Volume 1215, p. 012042. https://doi.org/10.1088/1755-1315/1215/1/012042.

45 Doan, V.D., Tran, T.K.N., Nguyen, A., et al. (2021). Comparative study on adsorption of cationic and anionic dyes by nanomagnetite supported on biochar derived from *Eichhornia crassipes* and *Phragmites australis* stems. Environmental Nanotechnology, Monitoring and Management, Volume 16, p. 100569. https://doi.org/10.1016/j.enmm.2021.100569.

46 Ahila, K.G., Ravindran, B., Muthunarayanan, V., et al. (2021). Phytoremediation potential of freshwater macrophytes for treating dye-containing wastewater. Sustainability, Volume 13, p. 329. https://doi.org/10.3390/su13010329.

47 Mansour, A.T., Alprol, A.E., Abualnaja, K.M., et al. (2022). Dried brown seaweed's phytoremediation potential for methylene blue dye removal from aquatic environments. Polymers, Volume 14, p. 1375. https://doi.org/10.3390/polym14071375.

48 Khandare, R.V., and Govindwar, S.P. (2015). Phytoremediation of textile dyes and effluents: current scenario and future prospects. Biotechnology Advances, Volume 33, Issue 8, pp. 1697–1714. https://doi.org/10.1016/j.biotechadv.2015.09.003.

49 Yadav, A., Labhasetwar, P.K., and Shahi, V.K. (2022). Membrane distillation crystallization technology for zero liquid discharge and resource recovery:

opportunities, challenges and futuristic perspectives. Science of the Total Environment, Volume 806, p. 150692. https://doi.org/10.1016/j.scitotenv.2021.150692.

50 Khalidi-idrissi, A., Madinzi, A., Anouzla, A., et al. (2023). Recent advances in the biological treatment of wastewater rich in emerging pollutants produced by pharmaceutical industrial discharges. International Journal Environment Science Technology, Volume 20, pp. 11719–11740. https://doi.org/10.1007/s13762-023-04867-z.

51 Flores, A.M.A.D., Revilla, P.C., Garcia, B.K., et al. (2022). Efficient dye removal from real textile wastewater using orange seed powder as suitable bio-adsorbent and membrane technology. Water, Volume 14, Issue 24, p. 4104. https://doi.org/10.3390/w14244104.

52 Yin, H., Qiu, P., Qian, Y., et al. (2019). Textile wastewater treatment for water reuse: a case study. Processes, Volume 7, Issue 1, p. 34. https://doi.org/10.3390/pr7010034.

7

Contamination by Oil Spills

Over the years, several major and minor oil spills have occurred in the sea. Oil spills can happen due to unintentional or intentional causes such as natural seeping of oil, industrial discharges, urban runoff, marine accidents, activities related to oil extraction and refining, pipeline and storage tank leaks, etc. Spills are responsible for serious damage to life in the oceans. Oil pollutants can be in the form of crude oil, grease and lubricants, petroleum products, polycyclic aromatic hydrocarbons (PAHs), and oil–water emulsion. These lead to an oil slick, which is very hazardous as the emulsion of oil and water is very resistant. If an oil spill manages to reach the shores and coastline, it increases the effects of the catastrophe. Recovery from the oil spill naturally depends upon the size of the spill. The hazardous effects of oil spills have already been discussed in Chapter 2, Section 2.4.3.

7.1 Characteristics of Oils

Oil spills in the sea could be due to the introduction of refined petroleum products such as diesel, gasoline, crude oil, or waste oil, etc. The impact and severity of the oil spill depend on the type of oil, its chemical and physical properties, and the extent to which the spill has taken place [1]. After the spill, several physical and chemical changes take place, such as evaporation, weathering, etc., which lead to the production of chocolate mousse, along with several other products, which affect the recovery of oil from the water bodies.

Properties of oil, such as surface tension, color, and viscosity, which are physical, depend on the type of spilled oil. The spread of the oil in water depends on the viscosity and surface tension. Oils with less surface tension spread faster in water. In addition to this, the spread of oil is faster when the water is warm, in comparison to colder temperatures. Oil generally tends to float and spread horizontally on the surface of seawater due to its lower density. The black or brown color of oil gets converted to green, yellow, or red color. The lighter components of oil get evaporated with time, leading to an increase in the density and the settlement of heavy oils at the bottom of the sea [1].

Boiling point, melting point, flash point, solubility, and partition coefficient are some of the important chemical characteristics of oil. Oils are generally a complex mixture of hydrocarbons such as alkanes, alkenes, cycloalkanes, branched hydrocarbons, etc.

The Code of Federal Regulations and spill responders mention different categories of oils, along with their effects on the shoreline [2].

1. **Nonpersistent Light Oils:** These belong to Group 1 and comprise of condensate and gasoline. These are considered to be very volatile with a capacity to volatilize within one to two days. These contain a high concentration of toxic substances, leading to high impact on the water column, and do not leave any residue after evaporation. A very careful consideration of the cleanup efforts is necessary for this category, as these are highly flammable due to their volatile nature.
2. **Persistent Light Oils:** These are composed of light crudes, diesel, and fuel oil. In comparison to Group 1, these are moderately volatile and known to form Group 2. These leave residue after a few days, which is one-third of the spill amount, after evaporation. The concentration of toxic compounds is moderate. These can be effectively cleaned from the water bodies.
3. **Medium Oils:** These belong to Group 3 and comprise of most of the crude oils. These can cause severe contamination in the intertidal areas. Fur-bearing mammals and waterfowl could be severely affected. Effective cleanup is possible, but only if carried out quickly.
4. **Heavy Oils:** This Group 4 consists of Bunker C, Heavy Crude Oils, and No. 6 Fuel Oil. These do not dissolve or evaporate and can lead to heavy contamination of the tidal areas, with long-term impacts on the sediments. This category poses challenges in effective shoreline cleanup.
5. **Sinking Oils:** This is Group 5, comprising slurry oils and residual oils. As evident from the name given to this category of oils, these sink in water quickly with no shoreline contamination. As these are capable of sinking, therefore, they do not evaporate or dissolve when submerged. They affect the submerged life, such as animals, sediments, etc., with long-term impacts. Removal of these oils is possible by dredging.

7.2 Important Oil Properties

The main properties affecting the persistence of oil in water are shown in Figure 7.1.

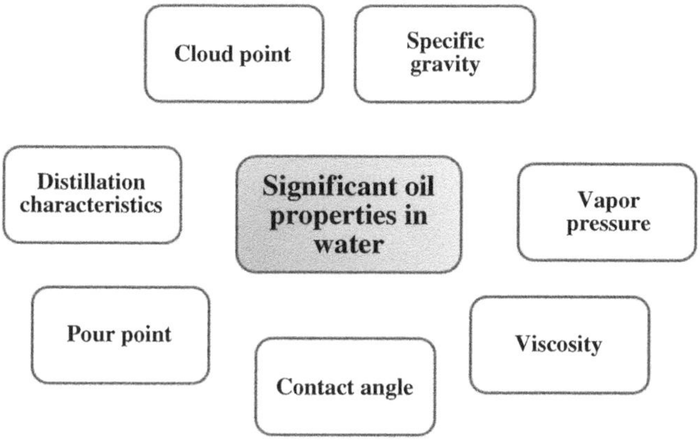

Figure 7.1 Properties affecting the persistence of oil in water.

Specific gravity refers to the relative gravity of oil in comparison to water. Degree API (The American Petroleum Institute gravity scale) (Equation 7.1) is a scale for the expression of specific gravity of a petroleum product.

$$^{\circ}\text{API} = \left(\frac{141.5}{\text{Specific Gravity}}\right) - 131.5 \tag{7.1}$$

Oils with a specific gravity around 1.025 are lighter than sea water and tend to float. The distillation characteristics and vapor pressure help in understanding the volatility of the oil samples, which is the resistance to flow. Another important trait is the viscosity. Less viscous oils tend to flow with ease and spread rapidly. The temperature at which an oil sample ceases to flow and behaves as a wax is known as its pour point, and the temperature where the wax forms crystalline structures is termed the cloud point [3]. Cloud point has been used for the removal of oil from the water streams [4]. Mechanical removal of oil is one of the important methods of oil removal in the various techniques adopted, and hence, the interfacial interactions play a vital role. The wetting and adhesion traits of the oil samples largely depend upon the contact angle [5]. On deposition of liquid on the solid surface, an angle (θ) is made by the liquid at the surface. With a decrease in the contact angle, the affinity of the solid for the liquid increases.

7.3 Oil Behavior in the Aquatic Environment

In order to understand the fate of oil pollutants, it is necessary to assess the magnitude, duration of exposure, and spatial extent.

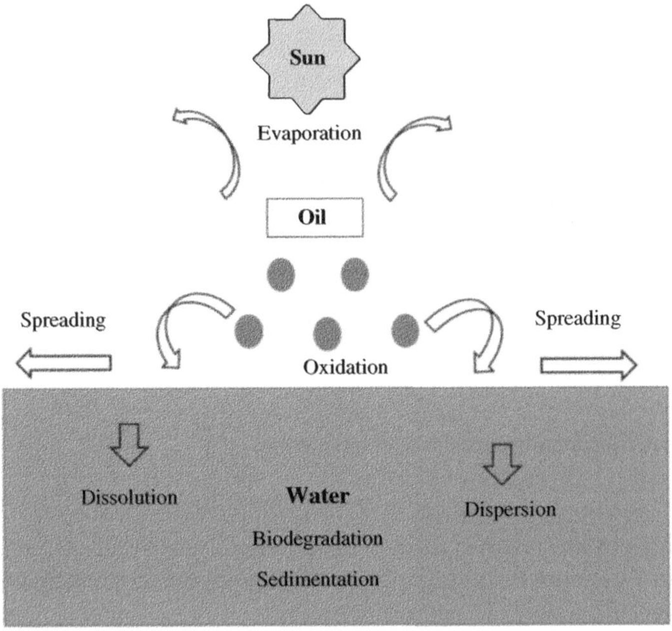

Figure 7.2 Changes taking place in oil after the spill.

As soon as the spill takes place, it starts spreading on the water surface immediately, with a series of chemical changes taking place (Figure 7.2). The extent of the spread depends on the viscosity of the spilled oil and the volume. Fluid oils tend to spread faster in comparison to viscous oils. These thin oils spread and in the form a coherent slick, which starts breaking up with time, with the change in appearance. Viscous oils, on the contrary, form patches which drift apart with the flow of water [6].

Crude oil, when introduced in the marine environment, involves weathering where processes such as emulsification, evaporation, oxidation, etc., take place. In addition to this, it spreads, disperses, undergoes entrainment, sinks, and tends to settle [6].

In weathering, there is a transformation of materials into different substances. Evaporation results in a reduction of volume from the original volume. The process of emulsification involves the formation of different combinations of water in oil, resulting in significant changes in the oil properties. The kind of emulsion formed is also an important factor in deciding the type of recovery method to be employed. Some of the oil components, such as the aromatic components, which are more toxic than aliphatic compounds, chemically stabilize by dissolution. By the process of oxidation, the formation of new substances takes place, depending upon the response of the organic compounds in the crude oil to oxidation. With time, and in the presence of plenty of oxygen, the ultimate components formed are carbon dioxide and water. Formation of ketones, alcohols, and organic acids takes place in the process of oxidation of crude oil, which are more soluble than the parent compounds. The energy for the oxidative reactions is derived from microbial oxidation and photo-oxidation. Both these processes mediate in the presence of sunlight and oxygen, whereas in the absence of sunlight and oxygen, anaerobic oxidation takes place. Horizontal and vertical dispersion, and transportation of the oil take place [6]. Some dispersed oil droplets become heavy enough after interaction with the organic and sediment particles to settle at the bottom of the seabed. Often, ingestion by the phytoplanktons results in the settlement of oil droplets at the bottom of the sea via their fecal pellets. In cases of rough seas, the oil droplets are usually washed off and remain in the subsurface, slowly resurfacing in a calm environment. Several microorganisms, such as yeast, bacteria, fungi, molds, protozoans, etc., are present in the sea that are capable of utilizing oil as a carbon and energy source, thereby leading to oil degradation [3].

The persistence of oil and weathering depends upon the conditions of the climate. Hence, the climate is one of the crucial factors governing the oil characteristics in water [7]. The degradation of oil in water is slowed down in cold weather conditions. Oil tends to spread on the upper shore with limited penetration into the soil [8]. Infiltration of oil can take place into the frozen soil through the voids not covered with ice. In the summers, the photodegradation of oil takes place at a faster rate, resulting in slick disappearance [9, 10].

7.4 Impacts of Oil Spills

There could be several categories of impacts of oil spills, ranging from the impacts on the environment to humans (Figure 7.3).

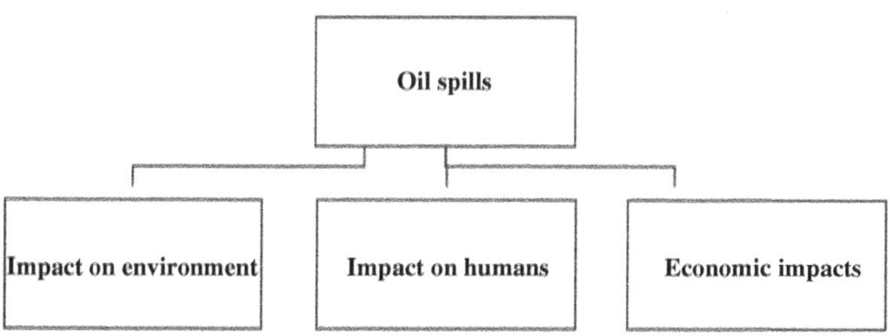

Figure 7.3 Categories of impacts of oil spills.

7.4.1 Impacts on the Environment

Oil spill incidents can have serious implications on the environment, organisms, ecosystems, and humans. The extent to which the oil spill can have an impact on the wildlife depends upon the oil exposure pathway, characteristics of an organism such as health, reproductive stage, and age, in addition to the chemicals employed for treating the oil spills [11]. There are three different ways in which the organisms exposed to oil spills can be affected. Oil can either be inhaled, ingested, or absorbed. The sections below discuss several impacts.

7.4.1.1 Marine Ecosystems

Oil spills affect the marine organisms as oils form a coat over the bodies of the living creatures. This adversely affects their normal functions such as insulation, respiration, repellence of water in feathers of the birds, etc. Some marine organisms, such as sea turtles, assume it for food and get trapped in oil [12]. Severe impacts on the reproductive capacity, immunity, lungs of whales and dolphins can result from inhaling spilled oil. Oil sometimes also tends to poison the animals and birds on ingestion.

Reduced growth patterns are witnessed in adult fishes, along with an alteration in respiration and heart rates. On ingestion of oil, the fishes also becomes unsafe for human consumption [12].

The marine food web also gets affected, leading to alterations in the nutrient cycle. Phytoplanktons, as primary producers, play a vital role in photosynthesis, and oxygen production is affected as a result of oils present, affecting their productivity. Similarly, zooplanktons, which are the sources of food for several marine organisms, such as larvae, etc., get adversely impacted, face mortality, or a reduction in their reproduction. This reduces their population, with an effect on the fish, due to a reduction in the food availability for higher trophic levels.

On rocky shores, oils can form a coating over the surface. This stranded oil hardens with time. This oil gradually gets eroded by waves but the oil in water remains in water for a long duration. Seeping of oil also takes place, impacting the muddy bottom. The marsh plants and the dependent species are also adversely affected.

7.4.1.2 Coastal Ecosystems

Oil spills result in adverse impacts on the food webs and lead to an alteration in the ecological balance and degradation of the natural habitat. Effects on the mangrove forests have been observed due to oil spills on Caribbean beaches in Panama over the years. The damages to the mangroves in this region were evaluated through aerial photographs, which were a consequence of two subsequent spills. The first spill took place due to the breaking up of the Witwater tanker in 1968, which affected mangroves in 49 ha, followed by another spill in 1986 due to the rupturing of the Refinería Panamá tank, affecting mangroves in a 69 ha area [13]. This degradation not only impacted the surrounding areas but also the species relying on such habitats.

7.4.1.3 Contamination of Water Bodies

In addition to the oil spill incidents taking place in the water bodies, oil spill incidents on land are also responsible for the introduction of oil into the water bodies. Oil, when it spills on land, becomes immobile after some time. Leaching of this oil takes place by rainfall or runoff, transferring oil into the water streams and eventually into the water bodies. The percolation of this oil also takes place in the water table, contaminating underground water. One such example has been the contamination of underground water by a crude oil pipeline burst 16 km northwest of Bemidji, Minnesota, where around 10700 barrels of crude oil spillage took place, covering a land area of about 7500 m^2. Around 400000 L of crude oil remained in the ground even after the cleanup efforts, which percolated into the water table [14].

7.4.2 Impacts on Humans

In the case of humans, the damage caused by the oil spills has been irreparable in some cases. These damages have been the damage to the liver, reproductive capacity, risk of cancer, etc. [15]. Mercury is considered to be a severe toxicant and contaminant. In a study, it has been found that the people involved in oil spill clean-up activities in Ecuadorian Amazon and Peruvian, showed high levels of mercury in their urine, which was twice as much as that of the normal population of the region. This also included the women population relying on the surface water for household use, rather than on other water sources [16]. Similarly, hazardous health effects have also been reported by many other researchers, such as reduced lung function and effects on mental health [17–20]. The impacts of oil spills on the workers involved in spill clean-up have been predominantly studied, in comparison to the nonworkers.

In case of nonworkers, oil spills have been responsible for immunotoxicity, carcinogenicity, mutagenic, and genotoxic effects [21]. An exposure to oil spills has led to eye and nose irritations, headaches, wheezing, an increase in miscarriages in pregnant women, an increase in mortality in newborn children, dizziness, and chest pains, etc. [22–25].

7.4.3 Economic Impacts

Even a single incident of an oil spills can affect the economy of the entire nation, which can be long-lasting. The effects could be direct or indirect. Direct effects include the costs involved in the clean-up efforts, a decrease in livelihoods, and obstruction in the businesses of the fishing industry, insurance premiums, and legal fines. Obstructions in offshore energy production, destruction of coral reefs, harm to shores, etc., which are heritage sites, are the indirect implications.

There have been losses of billions of dollars in various sectors such as property, tourism, fishing, etc., which were estimated in one such incident of Deepwater Horizon spill [26]. Also, in a study in 1996 [27], another such incident, the Exxon Valdez spill, affected other industries, including recreational fishing and tourism. A study by Sumaila et al. in 2012 [28] estimated the economic effects through input–output analysis, which indicated a loss of US$4.9 billion on commercial fishing and $3.5 billion on recreational fishing.

7.5 Detection of Oil Spills and Monitoring

Several methods have been used for the detection of oil spills in water. Some common technologies of oil spill detection [29] include visual imaging, which is of a relatively low cost, thermal imaging, laser spectroscopy, and radar [30]. Technologies such as machine learning (ML) and deep learning (DL) have also been explored [31]. Remote sensing, specifically optical remote sensing, laser fluorosensors, and microwave sensors [32] are important categories. Some important categories have been discussed below:

7.5.1 Remote Sensing

Conventional methods of oil spill detection, such as in situ sampling and visual imaging, are time-consuming with restricted coverage of the affected area. Techniques such as remote sensing help in the detection of oil spills without making any physical contact. In this method, satellites and aircraft mounted with sensors are used for the detection of oil spills by capturing data of the Earth's surface. Figure 7.4 gives the various steps involved in remote sensing.

There could be several categories of remote sensing data that can be used (Figure 7.5).

The light reflected by the Earth's surface is captured by the optical sensors. Oil spills (which appear darker in comparison to water can be identified with the help of the high-resolution images obtained. Microwave signals emitted by synthetic aperture radar (SAR) sensors due to the backscattered energy from the Earth's surface help in the detection of oil spills. Thermal infrared sensors make use of the emitted thermal radiation of the Earth's surface. The temperature of the spilled oil is usually warmer during the day and colder at night, which aids the detection. The method of using hyperspectral sensors captures the spectral bands, allowing the identification of oil, based on the spectral characteristics [33].

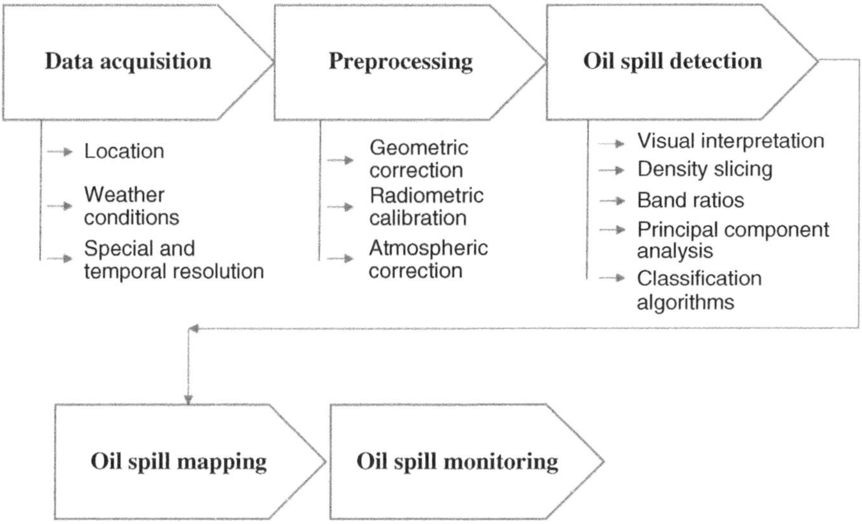

Figure 7.4 Steps involved in remote sensing.

7.5.2 Machine Learning and Deep Learning

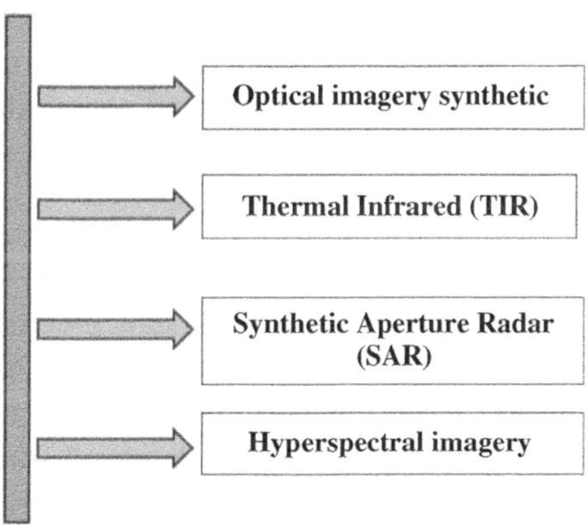

Figure 7.5 Types of remote sensing data.

ML helps in understanding the input–output relationship from representative samples, helping in the construction of different models. It helps in cases when the theoretical data is inadequate in spite of varied observations [34]. ML has recently been used for spill detection, overcoming the shortcomings of conventional methods. Several ML techniques have been used to detect oil spills from satellite images. ML models comprise of identification of dark spots, extraction of features, selection of features, model training, including validation, and classification of features. ML significantly improves the efficiency and pace of the detection of oil spills.

DL has shown its successful use in the classification of images. Various models of DL have been used, enabling efficient analysis of images [31]. DL also comes as a boon in automatically choosing features for the input data, with efficient handling of a large volume of data.

7.5.3 Laser Fluorosensors

Laser fluorosensors make the detection of oil possible in complex environments, with a clear discrimination between oiled and unoiled regions. Other sensors that have been used for the purpose generally depend on characteristics such as light reflection, emission of infrared energy, and microwave scattering, whereas this method focuses on the unique oil fluorescence spectral signature, which is a primary characteristic of an oil. These sensors could be active or passive depending upon whether the energy utilized is given by the sensors or the energy of the sunlight [35].

The aromatic compounds in petroleum become electronically excited by the absorption of ultra violet light. By fluorescence emission the release their absorbed energy which is then visible in the spectral visible region due to which the presence of oils is detected.

7.5.4 Ground Penetrating Radars (GPR)

In this method, an antenna is used for sending electromagnetic waves to a subsurface. The antenna is capable of movement on the medium surface under study. A diverging beam is sent to the sub surface which is collected by the receiver after reflection from material interfaces with differing electrical properties [36].

In regions where ice has a predominant occurrence, the technique of ground penetration radar is one of the most effective techniques [37]. This method is capable of detecting thick oil accumulations (approximately 2.5 cm) but has a limitation in the detection of thin slicks or oil trapped in thick ice, young ice etc.

7.5.5 Recent Techniques

Conventional methods, though precise but are generally time-consuming and involve tedious sample preparations. A study published in 2024 [38] deals with the use of direct analysis in real-time time-of-flight mass spectrometry (DART-ToF MS) for screening of petroleum oil spills and proving to be a rapid method. In general, samples of oils are collected in glass containers and analyzed in the laboratory. In the study carried out, a simple and rapid sampling method of using a hydrophobic filter paper was employed. This filter paper could be used directly by the immersion method, with the help of which, details of the spilled oil could be gathered, and the paper can be transported for analysis.

7.6 Technologies and Materials for Oil Spill Cleanup

Research is continuously being carried out to find more effective and efficient methods and materials for oil spill clean-up.

7.6.1 Some Important Studies

Purohit et al., in 2024, presented a review on the various technologies such as physical, chemical, biological, etc., being used for oil spill clean-up [39]. The US Environmental Protection Agency (EPA) is continuously working on the different methods of assessing and remediating oil spills [40]. Linfeng Piao et al. studied a rapid oil spill response system for the collection, recovery and storage of oil spills [41]. An implementation of the study was done first at lab scale followed by the actual field. Almost 80% oil recovery was achieved. A similar work has been done by University of Texas in the creation of technology more efficient than the currently existing ones [42]. A mesh roller two layers is combined with an induction heating. This separates oil and water with an efficiency of about 1400 kg of oil per hour, per square meter. Australian National Science Agency (CSIRO) is developing a technology with the use of sponges, which are super hydrophobic polymer coated with [43]. A gel-coated mesh filter was used by Yan et al. in 2023. In this, almost 99% of separation efficiency was achieved [44].

7.6.2 Innovative Techniques

In the Gulf of Mexico the sea grant program [45] several new ways are being adopted to use research tools such as unmanned surface vehicles (USVs). These do not require any crew on board. Sail drones and drifters are being employed as part of this technology, providing real-time information. Unmanned aerial vehicles (UAVs) are another category being widely used for environmental and disaster management [46, 47]. In this technology, drones have been used to monitor the spread of dyes in water, blimps, and balloons, which are slow and low-flying aircraft, are used, where monitoring is required for a long time. These technologies and methods have their advantages and disadvantages. The use of satellites is another method that has been used along with remote sensors to capture the changes in the shoreline and provide details about oil spills [48]. Techniques such as image processing have also been effectively used for the analysis and detection of oil spills [49].

Over 99% oil separation efficiency has been reported by a technology developed by University of Texas [50]. The technology is a combination of induction heating with a dual-layer mesh roller, showing tenfold efficiency.

Several innovative and quick methods for the removal of oil spills have been developed by several researchers, which include ultra-absorption sponges, magnetic soaps, and autonomous robots. Researchers believe that sponges, when coated with magnetic nanostructures, can be efficient materials as these are capable of retaining oil up to 30 times their weight, being able to withstand the force of the waves. Similarly, it is believed that the potential of magnetic soaps is magnificent [51]. The development of the Sea Swarm robot MIT Senseable City Lab has been a boon in this regard.

Seaswarm robots [52] are conveyor belts consisting of a mesh of thin nanowires for oil collection, which softly roll over water. These robots can absorb oil, 20 times their weight. These robots work in fleets, communicating their location using Wi-Fi and GPS without any human intervention. These can conveniently be used for locations hard to reach such as coast lines and estuaries. These are capable of digesting oil after absorption, reducing the time for trips to the shores for oil cleanup. Similar to this, Zahugi and coworkers employed a multi-robot system for cleaning oil spills [53]. The robot was designed to work on the water surface for fast cleanup and to prevent the oil from spreading.

An approach of nanotechnology was applied where a nanocomposite system, which was tea-bag shaped, was developed for oil spill cleanup [54]. The technology combined oleophilic graphite nanoplatelets and superhydrophobic nanomembranes. The tea-bag-shaped system revealed easy recyclability and high efficiency.

7.6.3 Integration of Methods

An integration of several technologies has also been an important development in this regard. In a cited research of 2024, an integration of containment barriers with sponges was done, in addition to the use of bioremediation, using microorganisms. This was done to accelerate the degradation process of oil in a recovery drill at Gran Tarajal Harbor [55]. There was a substantial reduction in the sedimented hydrocarbon concentration from 60000 mg/kg to less than 1600 mg/kg in a seven-month span. In addition to a significant reduction in the hydrocarbon concentration, there was also a contribution to the ecological recovery. In addition to this, there was a noticeable reduction in the concentration of heavy metals too.

A combination of conventional methods of oil spill treatment along with bioremediation can be helpful in maximizing the response to oil spills. An integration of bioremediation with electrochemical methods enables the efficient treatment of oil spills by forming bio-electrochemical systems, with the help of bacteria such as electricigens, exoelectrogens, or anode-respiring bacteria. These bacteria are capable of anaerobically oxidizing the organic pollutants by transferring electrons to the electrode. The efficiency of bioremediation processes can further be enhanced by the addition of substances such as polymers, inorganics, biosurfactants, etc. [56].

7.6.4 Materials for Oil Spill Cleanup

In addition to the various technologies being adopted, several materials are also being developed for oil spill clean-up. Natural and synthetic materials are being used for oil absorption [57]. Current advancements also involve nanomaterials, composites, or magnetic materials, which have also been investigated [58]. Natural materials such as sawdust, corn cob, coconut coir, cotton waste, wool waste, etc., have been successfully used. Materials such as foams, polypropylene, polyethylene, modified clay, etc., which are synthetic in nature, have also been explored [57].

Biomaterials are promising alternatives to synthetic materials as they are biodegradable, cost-effective, and environment friendly. Several biomass adsorbents that have been used are portrayed in Table 7.1 [59, 60].

Agricultural wastes such as corn stack, bagasse, corn silk, corn cane cob, rice husk, sawdust, sugarcane bagasse, wheat husk-based adsorbents, etc., have been used [70]. The potential of weeds such as fibers of *Asclepias syriaca* (milkweed) and leaves of *Salvinia minima* (water spangle) was investigated in salt and fresh water conditions [71]. Utilization of waste paper was done for the preparation of a cellulose-based adsorbent. The absorption efficiency of the material prepared was found to be efficient in comparison to other cellulose based adsorbents. The material could adsorb 16–28 times of its weight, with a potential for recycling of the oil collected. The adsorbent proved to be a promising option as it could be reused to over eight cycles [72]. Waste paper pellets and *Hibiscus cannabinus L.* (*Kenaf*) fibers were used for the removal of oil spills. Several parameters were analyzed for the prepared material and it was observed that a combination of 80% Kenaf pellets and 20% paper pellets optimally eliminated oil [73].

Table 7.1 Biomasses for oil spill cleanup.

Biomass	Oil	Removal	References
Banana skin	Crude oil	5–7 g/g	[61]
Potato peel	Lubricating oil	2.15 g/g	[62]
Rice husk	Crude oil	6.22%	[63]
Luffa	Diesel oil	>85%	[64]
Walnut shell	Mineral oil	0.56 g/g	[65]
Oil palm leaves	Crude oil	90%	[66]
Kapok	Gasoline	74.5%	[67]
	Diesel	66.8%	
	Soybean oil	64.4%	
	Paraffin oil	47.8%	
Bark	Oily compounds	95%	[68]
Wool	Oil	88.93%	[69]

In a green approach for the removal of oil spills, magnetic silica sorbent materials (MSSM) were developed using agricultural and industrial wastes. Rice husk or bagasse and red mud from steel factories were used for the preparation of the adsorbents [74].

Considering environmental stability, a unique material was developed utilizing industrial waste for cleaning oil spills. A combination of cooking oil and sulfur was used to synthesize a polymer that acted as a super sponge for the removal of diesel and crude oil from the sea, leading to the transformation into a gel. An organogelator–cellulose composite was also developed where oil was removed by processes such as absorption followed by gelation [75].

In addition to the several materials which were used, microbes such as fungi, bacteria and microalgae have also been actively used. Species of fungi such as *Lasiodiplodia, Fusarium, Fungi Penicillium, Rhizopus, Drechslera, Trichoderma, Curvularia, Aspergillus* etc., have been known to treat aromatic hydrocarbons. Bacteria such as *Alteromonas, Achromobacter, Arthrobacter, Burkholderia, Acinetobacter, Enterobacter, Bacillus, Pseudomonas,* and *Flavobacterium* have been investigated. Similarly microalgae of Chlorella, Selenastrum, and genera Scenedesmus have been used for the treatment of polycyclic aromatic hydrocarbons [76].

Moss based adsorbent, derived from sphagnum moss also has been a promising material due to its reusability and selective oil absorption. A sponge created by chemically modifying peat moss showed 90% oil removal capacity even after 10 cycles of reuse. The hydrophobicity of the sponge was increased by silane grafting. Significant removal of oil was noticed for organic compounds. Removal shown for motor oil was 22.756 g/g [77].

7.6.4.1 Nanomaterials and Nanotechnology

Nanomaterials have gained much attention in oil spill clean-up due to their small sizes with a high surface area, facilitating oil removal [78–80]. These are promising materials with great potential. Nanomaterials work via different mechanisms and break the oil particles by a combination of adsorption and absorption, by individual mechanism, or even by catalytic activity. These nanomaterials are also helpful in bacterial growth and nutrient supply [80]. These possess properties such as emulsion stabilization, reduction in the interfacial tension, change in surface wetting and enhancing oil breakdown.

Nanomaterials possess several adsorption sites and large surface area. Some bio-based nanomaterials are abundantly available, economical, and eco-friendly. Nanoparticles can interact with the oil particles in various ways, such as alteration of wettability, reduction in interfacial tension, and stabilization or destabilization of emulsion. The wettability of the solid substrate can be altered from oil wet to water wet, thereby facilitating the removal of locked oil in the substrate.

Figure 7.6 Categories of use of nanomaterials for oil spill removal.

Some nanomaterials, such as TiO_2 and ZnO, with photocatalytic properties when exposed to solar radiation, are capable of degrading petroleum hydrocarbons. The biodegradation of oil is also enhanced in the presence of some nanoparticles. Some of the examples being nanoclay, which behave as Pickering emulsifiers, thereby providing a conducive environment for the growth of oil consuming bacteria and the rate of biodegradation. These act as buffers, helping in maintaining the pH by the absorption of protons liberated as a result of the hydrocarbon breakdown [80].

Nanomaterials can be used in a variety of ways for oil spill removal. The different ways by which these can be used are shown in Figure 7.6.

Nanomaterials such as chitosan, nanocomposites, and modified nanoclay have been used as surface washing agents for oiled sand on the beach. These materials have proved to be nontoxic and efficient in oil removal. Magnetic nanomaterials have also proved to be very fruitful in this regard. Nanomaterials as dispersants are used for the reduction of interfacial interaction, but could be toxic in some cases. Nanodispersants can be silica, carbon black, naturally occurring clay etc. Nanomaterials as sorbents are used to remove oil by adsorption, absorption or a combination of both. Bio-based nanomaterials are generally biodegradable. Mineral clay, graphene, carbon nanotubes, etc., are efficient in removing the oil droplets, and at the same time are advantageous as they are recyclable. Strategies of bioremediation, with the involvement of nanomaterials, have proved to have an edge over the solitary processes of bioremediation. Clay nanomaterials such as nontronite, montmorillonite, and kaolinite have shown an enhancement in remediation with the use of microorganisms [80, 81].

7.7 Prevention

In order to avoid the high costs and challenges involved in the treatment of oil spills, it becomes important to prevent the occurrence of such incidents.

Certain technologies help to identify and detect oil spillage in less than few minutes, helping the prevention of further spread. The Hydrocarbon Early Automatic Detection System (HEADS system) by Repsol Technology Lab [82] is an example. Several agencies carry out dedicated work for preparedness in oil spill cases such as International Maritime Organization (IMO), an agency of the United Nations, Office of Response and Restoration (OR&R), in the United States of

America, National Oceanic and Atmospheric Administration (NOAA) European Maritime Safety Agency (EMSA) in Europe, International Petroleum Industry Environmental Conservation Association (IPIECA), in the United Kingdom, Occupational Safety and Health Administration (OSHA), in the United States of America, etc. The federal statutory framework for oil spills is given by the Oil Pollution Act (OPA) and the Clean Water Act (CWA).

Using leak-proof tankers and containers, equipment maintenance, employing leak detection systems, along with proper automated monitoring can prove to be beneficial. Rules for storage are to be properly abided by, and necessary training needs to be given to the workers working in the oil companies [83, 84].

7.8 Important Factors in the Cleanup Activity

Several factors affecting oil spill cleanup activities are shown in Figure 7.7.

7.8.1 Environmental Factors

The important factors affecting the oil spill cleanup activities are weather conditions, which play a vital role in the spread of the spilled oil, which can affect the cleanup priorities. Temperature is another important factor that makes the oil viscous or fluid. Rainfall and storms are also responsible for the dilution and spread of oil on the water surface.

7.8.2 Nature of the Oils

The nature and composition of oil, its density, viscosity, volatility, and solubility are of great importance in deciding the course of action for the cleanup efforts. The spread of light crude oil is quicker in comparison to heavy crude

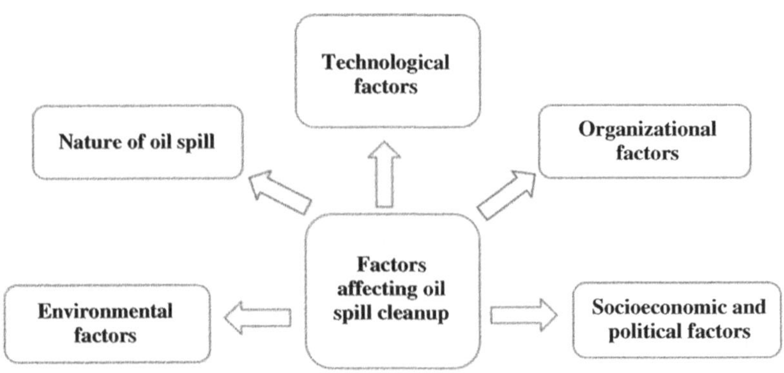

Figure 7.7 Factors affecting oil spill cleanup.

oil. Petroleum products such as kerosene are quickly evaporating components and need rapid elimination owing to their inflammability and high toxicity. The recovery methods for oils with less density could be simple, converting to complicated methods when the density is higher, and the oils sink to the bottom. Volatile oils may evaporate, could be inflammable and can emit harmful fumes. The methods of treatment adopted also depend upon the solubility.

7.8.3 Technological Factors

Depending upon the kind of oil spills various methods and technologies are adopted for cleanup. This depends upon the type of oil spilled, conditions of spill, volume of spill etc., before appropriate technologies are adopted. For application of any specific technology the availability of equipment, accessibility of the methods to the oil spill region, weather conditions, feasibility of oil recovery, scalability potential, costs involved in the application in different region become the aspects worth considering, for effective implementation.

7.8.4 Organizational Factors

The key factors in any oil spill clean-up activities are organizational planning, preparedness for contingencies, management, information sharing and communication, legal and regulatory compliances, and involvement of stakeholders. In addition to the mentioned, real-time monitoring by organizations and impact assessment postactivities are also crucial.

7.8.5 Socioeconomic and Political Factors

An important socioeconomic consideration for oil spill cleanup activities is the availability of funds, effect on the shore and port activities, availability of a skilled workforce, and health implications on the population. Important political factors such as regulatory frameworks, government regulations, cross-border limitations, the influence of oil companies, and their cooperation also play a vital role.

7.9 Treatment of Oil Spills

Different methods used to clean up oil spills can be categorized as below:

7.9.1 Use of Skimmers

Special devices known as skimmers can be employed for the removal of oil spills. These devices mechanically separate (skimming) the oil which floats on the surface. The oil floating on the surface is skimmed and transferred into the storage container of the device.

Some important aspects before making a choice of the skimmers are the type of oil targeted, viscosity, capacity, environmental conditions and conditions of the sea. The type of the skimmer which can be used could be oleophilic or nonoleophilic, depending upon the affinity or nonaffinity of the materials used in these devices to oil or to water respectively [85]. The oleophilic skimmers or oil attracting skimmers function by adhesion with the oil, which is then removed by wiping. Nonoleophilic gravity skimmers have been used in several industries to remove oils that are free-floating.

Skimmers can be successfully used for all kinds of oils and can work with efficiently. These work well in debris and rough ice, and are less influenced by waves (Oleophilic skimmers). The main limitations of the skimmers are that these are expensive, the oil collected needs to be further treated, are labor intensive, and require maintenance [86].

7.9.2 Oil-absorbent Pads

Oil-absorbing pads are another simple method for oil spill removal. These pads have a tendency to repel water in comparison to oil. These consist of mats or sheets of different materials capable of efficiently absorbing and desorbing different types of oils. The oil absorbing pads are quick in action and save the disposal expenses. These are reusable, low-cost cost and efficient environment environment-friendly alternatives for oil removal.

Oil-absorbing pads can be made of several categories of materials. Some materials which can be used are polypropylene, natural fibers such as cotton, Kenaf, wool, etc., cellulose-based pads and synthetic combinations of different materials.

Some of the advantages of oil absorption pads are selective absorption, efficient absorption, easy application, flexibility of use in various conditions and quick action.

7.9.3 Burning In Situ (Thermal Remediation)

It is a process of controlled burning of spilled oil. It is a rapid, efficient and simple means of oil removal, without the use of any sophisticated equipment. The method has no harmful impacts on the marine organisms. This is a very high removal rate technique which is very efficient by converting it into water and carbon dioxide without any need of pumps, collection devices and storage containers. Thick slicks can be removed with an efficiency of 90%. The rate of 2000 m^3/h is achievable with a fire area of 10000 m^2. This technique comes as a boon where the other techniques show some limitation and threat to the environment. The only necessary conditions for this method are fuel, oxygen and source of fire [87, 88]. The burning rate of the oils is reported as units of thickness per unit time (mm/min). The conditions having a direct impact on in situ burning are slick thickness, fire size (or diameter) and oil type. The environmental conditions also play a vital role in this regard. The efficiency of this method automatically reduces if the fire is not able to spread over the surface of the slick. The burning technique can be carried out in "batch mode" or "continuous mode" [89]. Several steps in batch mode consist of collection of oil using boom, transfer of the boom to safe distance for incineration, ignition of collected oil, flame extinguishing, residue collection and maneuvering the boom back for oil collection of the next batch. The continuous mode involves the positioning of the fire-resistant boom to a safe distance from the point of oil leakage where a continuous and an intermittent burning of the accumulated oil are carried out [90].

The thickness of the oil and availability of oxygen are the two important aspects for a successful and efficient use of the method. The process of combustion can be sustained with the use of additional agents such as light crude oil, gasoline and wicking agents such as wood, straw silica and glass beads [91]. Despite of the advantages of this method of thermal removal of oil spill, the method can lead to the formation of residues, which eventually settle at the bottom of the water body. These can be mechanically removed [91]. In addition to this limitation the hazard to the human life and environment, and the risk of secondary fire are also some of the shortcomings of the method.

7.9.4 Sorbents

Sorbents work in three different ways as follows [92]:

(1) Adsorption (2) Absorption (3) Both

Despite several techniques are available for the removal of oil spills, yet the use of sorbents such as saw dust are considered to be very convenient due to the ease of application, high removal capacity and reusability in some cases. Organic sorbents such as saw dust have proved to have an edge over other materials, owing to cost efficiency, oleophilic trait, and low density [93]. Several modifications of sawdust have been investigated for the enhancement of its efficiency to remove oil spills (g/g). Chemicals such as sodium hydroxide, ethanol, and hydrochloric acid have been tested as modifiers. Acid treated sawdust has revealed a high oil removal capacity in the order of 4.82 g/g, even after more than an hour of sorption, with significant reusability [61].

While choosing the different adsorbents, different factors that are important are as follows:

- **Rate of Absorption and Adsorption:** Different adsorbent materials possess different oil sorption capacities. The type of oils which are to be treated governs the selection of different adsorbents. The rates of adsorption and absorption are dependent upon the type of oil and the adsorbent.

- **Oil Retention:** The pores of the adsorbents release the trapped oil when the adsorbent is lifted from water causing secondary contamination. Oils which are lighter and less viscous have a greater tendency to get lost through the pores in comparison to the more viscous oils.
- **Ease of Application:** Certain adsorbents pose limitations to oil adsorption depending upon the type of oil and the environmental conditions. For example, some natural organic sorbents are too light to be used in windy regions.

The different categories of sorbent materials, which have been investigated for oil spill removal, are enlisted in Table 7.2 [61].

Table 7.2 Oil spill and sorbent materials.

S. No.	Types	Categories	Advantages	Disadvantages	References
1.	Natural inorganic: 4–20 times adsorption as per their weight	Vermiculites, volcanic ash and perlite	Non-inflammability, chemically inert, abundance	Reduced sorption capacity, difficult recovery	[94]
2.	Natural organic: 3 and 15 adsorption as per their weight	Fruit fibers, fruit peels, cotton, luffa, corn stalk, milkweed fiber, curcuma fibers, sugarcane bagasse and walnut shell, kapok, nettle fibers, palm fibers, pineapple leaves, rice husks, pistia leaves and roots, sawdust, wheat straw	Eco-friendly, cost effective, abundant, helpful in utilization of waste	Low removal capacity, low oleophilic nature	[95]
3.	Synthetic: 70 times adsorption as per their weight	Polypropylene, polyurethane	Moderate sorption, hydrophobic, reusability	Nonbiodegradable, complicated syntheses, expensive	[96, 97]

7.9.5 Hot Water and High Pressure Washing

This method is used in conditions where mechanical removal is not possible and where machinery is inaccessible. With the help of this method oil can be easily removed from seawalls, hard surfaces and big rocks.

Water is heated to approximately 170 °C and sprayed using high-pressure wands or nozzles. The oil gets flushed and is collected with the use of sorbents or skimmers. Careful use of this method becomes very essential as it can impact the aquatic flora and fauna, with short and long term effects. In order to avoid the harmful effects certain modifications in the process are done for example operating at a lower pressure, flushing the area with ambient temperature water and the use of a broader flow of water rather than a jet. While using this method, it also becomes essential to carry out proper collection of the mobilized oil and to have oil containment regions in the vicinity to avoid further spread.

7.9.6 Oil Spill Dispersants

Oil dispersants are a combination of emulsifiers and solvents, similar in action as soaps and detergents, which break the oil droplets into smaller droplets that are easy to disperse and biodegrade, for example, Corexit [97]. With the process of dispersion, there occurs an increase in the contact between water and oil, which leads to an enhanced interaction between the hydrocarbons and oil-degrading microbes. As in every environment, there are certain microbes which are capable of degrading oil. These dispersants degrade the oil into droplets of sizes <100 microns, making them easily degradable by increasing their surface area. The application of these dispersants can be surface or subsurface. Therefore, dispersants prevent the oil slicks from reaching the coast, preventing the shorelines, mammals, and sea birds from the harmful impacts of oil spills. These are generally used in combination with other techniques. The fastest application of dispersants is by spray through aircraft. The type of dispersant to be applied depends upon several factors such as volume of spill, type of oil, extent, and location of spill.

Dispersants can be conventional dispersants or concentrate dispersants. Conventional dispersants are made up of a mixture of nonaromatic hydrocarbon solvents. These are usually employed for shoreline cleaning. The concentrate category of dispersants comprises of a combination of oxygenates such as glycol along with nonaromatic hydrocarbons [98]. The use of dispersants can be toxic to some marine organisms [99], although for a short time, due to the highly inflammable traits of some dispersants. Also, with the possible harm to the coral reefs, the use of dispersants is restricted in the vicinity of the coral reefs. Though chemical dispersants can be used effectively still the solvents and surfactants used can be harmful to marine life [100]. Dispersants are also capable of contaminating the drinking water sources [101].

Considering the possible toxic effects of the surfactants, in the current scenario, efforts to prepare green surfactants have accelerated. These surfactants are made from renewable and naturally occurring materials with no harm to the aquatic life [102]. Surfactants prepared from marine waste, fish, fungal and bacterial strains, green solvents, vegetable oils, etc., have been explored. The efficiency of the dispersants depends upon the extent of hydrophilic or hydrophobic character. These dispersants are eco-friendly and potentially effective.

7.9.7 Oil Boom

Floating barriers are called Booms. The use of booms is a popular method for controlling the spread of oil. These equipment (known as containments) floats on the surface of water and behaves as temporary physical barrier to prevent the spread of oil. These are usually made up of metal, plastic, or other materials (Figure 7.8).

Oil booms perform varied functions. These help in oil containment by preventing the spread and helping in its accumulation, facilitating its collection and recovery. Booms also help in deflecting oil to the desired location for recovery by pumps and vacuum trucks. Protection of biologically sensitive areas is also a major aspect that is taken care of with the help of these barriers [103]. The deployment of booms is dependent upon the weather and sea conditions.

Oil booms can be of different shapes, sizes, and material, ranging from lightweight and small booms to expensive and large units, requiring heavy machinery

Figure 7.8 Oil boom. Credit: U.S. Fish and Wildlife Service/Wikimedia Commons/Public Domain.

for their movement. Booms can be categorized into Curtain and Fence booms. The curtain booms consist of a continuous sub-surface skirt or flexible screens. Fence booms generally have integral or external buoyancy. In addition to this, beach sealing or shore sealing booms also exist, where there are water-filled chambers instead of skirts.

Oil booms offer an advantage of their use for all categories of oils; however tend to be expensive, labor sensitive, have low stability in strong tides and winds, need to be towed, and need a combination of other techniques for oil removal [86].

7.9.8 Manual Labor

The method of mechanical recovery is the primary response method for oil recovery in incidents of oil spills. This method is generally used when there is difficulty in deploying heavy machinery along the shoreline. This method is generally used in heavily contaminated zones.

Manual separation of oil spills is a very common method where shovels, hoses, vacuum pumps, excavators, rakes, and bulldozers are used for the removal of oil. In this method, the oil slicks are concentrated using booms, skimmed, and collected in offshore vessels. Decantation of water is carried out, which is co-recovered, and finally, the recovered oil is transported for disposal [104]. Drums and buckets are used for the transfer of oil to the processing station.

The limitation of this method remains as this method separates oil only upto 30% of the spilled oil, whereas 70% remains in the water body. In addition to this, a large logistic requirement and manpower requirement are also some of the hurdles in the manual removal of oil spills.

7.9.9 Bioremediation

Bioremediation comes as an efficient, cost-effective, and eco-friendly alternative to other methods for oil spill removal. In this method, genetically modified or natural microorganisms, plants, or enzymes are used [105]. Various categories of organisms that have been used in this regard are shown in Table 7.3 [76]. In certain cases, microbes are added to aggravate the production of degrading microbe assemblies (bioaugmentation). Biostimulation is where nutrients are added to the area that is contaminated. This gives the microbes enough nutrients to grow in population to degrade oils [106–108].

Specific bacteria can be used for targeting specific contaminants depending on their specific affinities. Some of the examples are *Alcanivorax* sp. IO_7 for Alkane, *Alcaligenes aquatilis* BU33N, and *Cycloclasticus* sp. strain, for BG-2 Phenanthrene, *Cycloclasticus* sp. strain, for P1 naphthalene, phenanthrene and pyrene, *Halomonas* sp. strain MCTG39a, for hexadecane, *Halomonas pacifica*

Table 7.3 Different categories of biological species used for oil spill cleanup.

Organisms	Genus		
Bacteria	*Bacillus*	*Flavobacterium*	*Thalassolituus*
	Alcaligenes	*Enterobacter*	*Marinobacter*
	Pseudomonas	*Alcanivorax*	*Oleispira*
	Cycloclasticus	–	–
Microalgae/ Cyanobacteria	*Chlorella*	*Scenedesmus*	*Oscillatoria*
	Spirulina	*Synechocystis*	*Nannochloropsis*
	Spirogyra	*Chlorococcum*	*Selenastrum*
Fungi	*Fusarium*	*Aspergillus*	*Mucor*
	Curvularia	*Lasiodiplodia*	*Rhizopus*
	Drechslera	*Penicillium*	*Cryptococcus*
	Trichoderma		

strain Cnaph3, for naphthalene, *Oleispira antarctica* RB-8, for aliphatic alkanes, *Marinobacter hydrocarbonoclasticus* SdK644, for crude oil, *Pseudomonas pseudoalcaligenes* NP103, for phenanthrene and pyrene, *Pseudomonas* sp. sp48, for phenol, naphthalene, pentadecane, *Pseudomonas aeruginosa*N6P6, for phenanthrene and pyrene, *Pseudomonas aeruginosa* GOM1, for hexadecane and *Ralstonia pickettii*, for crude oil [76].

Though very cost-effective, the method of bioremediation can be associated with several disadvantages such as resistance of certain substances to biodegradation, such as asphaltenes, delayed degradation in some cases, biostimulation leading to eutrophication, unsustainability of bioaugmentation, and lack of substantial spilled oil quantities [109].

7.10 Some Major Oil Spill Incidents

Some important oil spill incidents that have taken place are discussed in this section.

7.10.1 Deepwater Horizon Oil Spill

In the Gulf of Mexico, on 20 April 2010, there was an explosion caused by the Deepwater Horizon oil rig [110], leading to a massive oil leakage. Approximately 4900000 barrels of oil leaked into the Gulf. The spread of oil took place for around 87 days.

The explosion also led to the loss of lives and caused injuries to people. The wildlife and habitats in the region were severely affected, including birds, fish, corals, etc., with a significant impact on the biodiversity. The oil leakage also harmed humans through fisheries, businesses, tourism, and recreational facilities [111].

Methods such as burning, the use of dispersants, and mechanical recovery were used to take control of the oil spread.

7.10.2 Exxon Valdez Oil Spill

The incident happened on 24 March 1989, in Prince William Sound, in the Gulf of Alaska, the United States, which was also considered to be one of the major disasters [112]. An oil tanker named Exxon Valdez struck a reef during its voyage. Eleven million gallons of crude oil spilled into the water [113].

Significant damage was caused to the aquatic wildlife, including seals, sea birds, bald eagles, and whales. Forty percent of the population of creatures such as sea otters were killed which never recovered for long. Severe harm to the business involving fisheries and shoreline activities resulted.

Methods such as oil skimming, use of dispersants, and bioremediation [114] were used.

7.10.3 Prestige Oil Spill

On 13 November 2002, Prestige, an oil tanker, broke and sank into the Atlantic Ocean, off the 260 km northwestern coast in Spain [115]. Around 77000 tons of oil spilled, spreading from Portugal to France [116].

The marine and coastal life was severely impacted. Fisheries and biological communities were heavily impacted. The regional economy was affected.

Booms and skimmers were used for clean-up. More than 300000 people volunteered for manual cleaning, heavy machinery such as trucks was used for shore cleaning activities, dispersants were used along with methods such as bioremediation, etc. It takes years to disperse the leftover oil by sunlight and waves.

7.10.4 Montara Oil Spill

This spill was considered to be the worst Australian disaster [117]. The incident occurred on 21 August 2009, when a well on the platform Montara blew out, spilling oil in the Timor Sea. The incident continued for about 74 days [118].

The spill led to heavy damage to the marine life, shoreline, coral reefs, dolphins, fish, and turtles, along with a substantial impact on the tourism and fishing industry [119].

Skimmers and dispersants were employed. The removal was challenging as the location was almost inaccessible. To report the recovery efforts, special rehabilitation funds are being given by the Australian government.

7.11 Disposal and Treatment of Wastes Generated

Substantial quantities and different categories (Figure 7.9) of wastes are generated after the treatment of oil spills, which need proper recovery or disposal and segregation. The wastes that are generated could be recovered oil, debris, used adsorbents, and contaminated equipment [120].

For oiled and hazardous waste, treatment methods such as open burning, portable incineration, landfilling, commercial incineration, recycling, and reprocessing can be applied [121]. The "Waste hierarchy" (Figure 7.10), an international framework for waste management, gives the level of priority for using several resources, considering the objectives of the Waste Avoidance and Resource Recovery Act 2001 [122].

For the management of wastes generated during oil spill treatment, below are the measures which can be adopted [123]:

1. **Reduce:** Selective technologies for oil spill cleanup can be used to reduce the quantity of waste generated. In addition to this, a careful consideration of the consumables and the adsorbents utilized will help manage of waste generated.
2. **Reuse:** The possibilities of reuse of the adsorbents and oiled equipment will help in reducing the amount of waste generated.

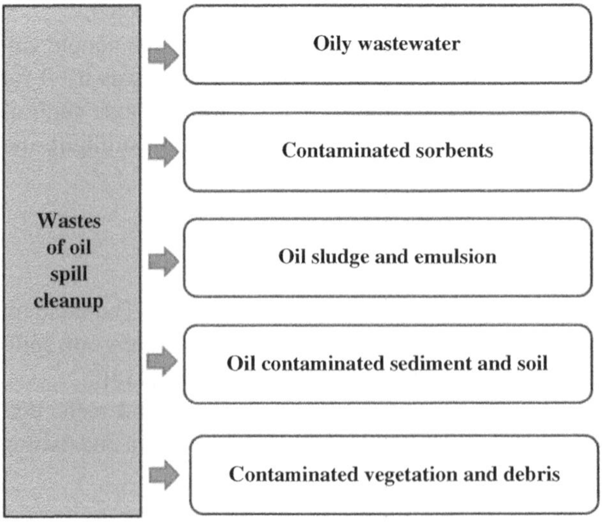

Figure 7.9 Different categories of generated waste after oil spill treatment.

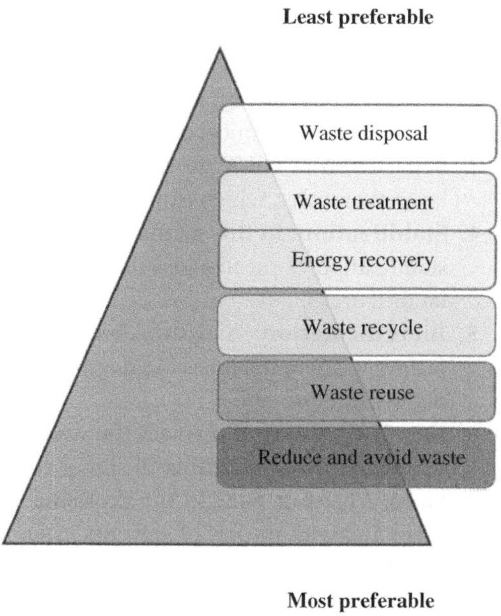

Figure 7.10 Waste hierarchy.

Table 7.4 Processes for oil spill cleanup and the wastes generated.

Process	Type of waste generated
In situ burning	Residues of the burnt oil Damaged booms Oiled PPE (personal protective equipment) and vessels
Shoreline cleanup	Oiled equipment, vessels, oily water, and vegetation, recovered oil sorbents
Use of dispersants	PPE, nonhydrocarbon waste, dispersant containers
Response operations in the sea	Oiled equipment, vessels, oily water, and vegetation, recovered oil sorbents

3. **Recycle:** The oil collected after recovery can be sent to the refineries and further utilized for different purposes.
4. **Recovery:** The wastes can be utilized for heat or power generation.
5. **Disposal:** Careful landfilling, composting, or incineration can be used for the waste that cannot be considered for any of the above-mentioned aspects.

Several categories of wastes are generated as a result of oil spill cleanup processes (Table 7.4) [124].

Methods that can be used for waste treatment are [124]:

1. **Reprocessing:** Oil is recovered by a recycling plant, which can be reused.
2. **Emulsion Breaking:** This is done to separate water from oil, either with the use of heat or chemicals. The oil is then reprocessed.
3. **Separation of Oil/Water:** This is usually done by gravity, where oil forms a layer over water, which can be separated with the help of skimmers.
4. **Stabilization:** In this method, several inorganic substances are used for the stabilization of oil to avoid its leaching and to facilitate its transport to the landfill areas.
5. **Bioremediation:** A natural breakdown of the oil is carried out by land farming, where the microbes degrade the oil, which is spread and mixed thoroughly into the soil.
6. **Beach Washing:** It involves the washing of pebbles and cobbles, which are washed and cleaned either in situ or away from the actual site.
7. **Sand Washing:** Sand in the shoreline is washed using sand cleaning equipment and with the use of solvents in some cases.
8. **Landfill:** Waste containing less than 5% oil can be disposed off along with the domestic waste and taken to the landfill sites. Incineration: The waste is destroyed by burning at a high temperature, converting it to safe ash. Thermal desorption and pyrolysis: This method involves heating the waste in the absence of oxygen, either at high temperature or low temperature.

Different separation and disposal options for liquid and solid wastes are mentioned in Table 7.5 [123].

Table 7.5 Separation and disposal methods for oil spill treatment generated waste.

	Liquid	Solid
Material	Nonemulsified oils Emulsified oils	Oil with sand Oil mixed with cobbles and pebbles Oil mixed with sorbents, plastics Oil with equipment
Separation	Settling, gravity separation Emulsion broken by heat treatment or by the use of chemicals	Liquid oil leaching Use of water or solvents for washing Sieving for the removal of solids
Disposal	Use of recovered oil as fuel, water is returned to the source Stabilization or incineration	Use of recovered oil as fuel Return water to a water source Use of composting or farming for degradation Landfill, incineration

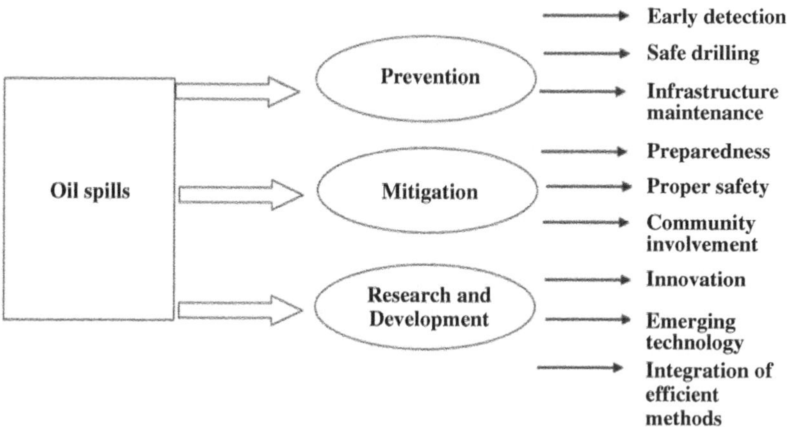

Figure 7.11 Measures for oil spill prevention and mitigation.

Along with the treatment of the waste generated it is equally important to rule out the possibility of secondary contamination where the oil spreads via equipment, transport people etc. Secondary contamination can be taken care of by checking leakages in pumps, storages, hose connections, decontamination of machinery and equipment and demarcation of "dirty" and clean zones.

7.12 Strategies for Prevention and Mitigation

Adoption of preventive measures and efficient mitigation strategies is the key to avoiding and recovering from such incidents. Figure 7.11 gives the key measures and techniques that can help in such incidents. Besides the mentioned, an advancement in the existing cleanup technologies and cooperation of the global bodies in instances of oil spill incidents is of utmost importance, to take control in these cases. A modification in the policies with the lessons learnt from the previous incidents also helps reduce the impacts.

It also becomes important to ensure the restoration of habitats, rescue of wildlife, and to consider economic compensations to the affected population.

7.13 Legislation and Role of Authorities in the Cleanup of Oil Spills

The management and control of oil spills is done with the collaboration between several government and non-government authorities. These authorities are responsible for the measures necessary to be taken for preparedness, prevention, and cleanup activities in incidents of oil spills.

Safety standards for offshore rigs and oil tankers are set by the authorities. In addition to this, they direct the regular necessary inspections. Government and regional organizations carry out surveillance activities, which could be aerial, satellite, or coastal visits for early detection of spills and their control measures. These are responsible for the initiation of a swift response to any oil spill incidents. In case of oil spill incidents, contingency plans could be activated by the organizations, in addition to international collaborations. Issuing of advisories, marking exclusion zones, along with providing updates on clean-up processes, are important aspects taken care of by the authorities. These studies address the impacts of oil spills with an assessment of environmental damage post-oil spill incidents. These government bodies also have the right to impose compensation and penalties on the parties responsible for the damage.

Some important agencies related to the management of oil spills are as follows:

1. International Maritime Organization (IMO)
2. UN Environment Programme (UNEP)
3. Environmental Protection Agency (EPA)
4. National Oceanic and Atmospheric Administration (NOAA)
5. Transport Canada
6. Maritime and Coastguard Agency (MCA)
7. Australian Maritime Safety Authority (AMSA)

Regulations related to oil spills in some major global regions are discussed in the following sections [86].

7.13.1 The US Regulations

The United States Environmental Protection Agency (EPA) has laid down the Facility Response Plan (FRP) rule and the Spill Prevention, Control, and Countermeasure (SPCC) rule. The FRP indicates the preparedness of the oil storage facilities for incidents of oil spills. Some key elements in FRP are the complete information of the facility, notifications for emergencies, equipment, and information for evacuation. Complete information on the spill hazards, along with any previous information, complete information on detection strategies, response training, exercises and drills, security preparedness and response plans, etc. [125].

The Spill Prevention, Control, and Countermeasure (SPCC) rule takes care of the prevention, preparedness, and response. This regulation deals with the containment of oil, preventing its spread [126] with the adoption of several procedures, use of equipment and methods by the facilities.

7.13.2 The UK Regulations

In the United Kingdom, there should be an emergency plan ready with the oil handling facilities, as per the International Convention on Oil Pollution

Preparedness, Response and Co-operation (OPRC). The plans by the different parties as part of the convention are expected to establish different procedures for handling oil spills and pollution incidents, along with collaboration with international and national or international cooperation. In case of any emergencies related to oil pollution, the parties are also expected to assist others [127].

References

1 Holakoo, L. (2001). On the capability of Rhamnolipids for oil spill control of surface water. Unpublished dissertation in partial fulfillment of the requirements for the degree of Master in applied Science, Concordia University, Montreal, Canada.

2 https://response.restoration.noaa.gov/oil-and-chemical-spills/oil-spills/oil-types.html (Last accessed 17.04.25)

3 https://www.itopf.org/fileadmin/uploads/itopf/data/Documents/TIPS_TAPS_new/TIP_2_Fate_of_Marine_Oil_Spills.pdf (Last accessed 17.04.25)

4 Zahugi, E.M.H., Shanta, M.M., and Prasad, T.V. (2012). Design of multi-robot system for cleaning up marine oil spill. International Journal of Advanced Information Technology, Volume 2, Issue 4, pp. 33–43.

5 Broje, V., and Keller, A.A. (2007). Interfacial interactions between hydrocarbon liquids and solid surfaces used in mechanical oil spill recovery. Journal of Colloid and Interface Science, Volume 305, Issue 2, pp. 286–292. https://doi.org/10.1016/j.jcis.2006.09.078.

6 National Research Council (US) Committee on Oil in the Sea: Inputs, Fates, and Effects (2003). Chapter 4: Behavior and fate of oil. In: Oil in the sea III: inputs, fates, and effects. National Academies Press (US), Washington (DC). https://www.ncbi.nlm.nih.gov/books/NBK220700/.

7 Zakaria, N.N., Convey, P., Gomez-Fuentes, C., et al. (2021). Oil bioremediation in the marine environment of Antarctica: a review and bibliometric keyword cluster analysis. Microorganisms, Volume 9, p. 419.

8 Zahri, K.N.M., Zulkharnain, A., Gomez-Fuentes, C., et al. (2021). The use of response surface methodology as a statistical tool for the optimisation of waste and pure canola oil biodegradation by Antarctic soil bacteria. Life, Volume 11, p. 456.

9 Iskander, L., Khalil, C.A., and Boufadel, M.C. (2020). Fate of crude oil in the environment and remediation of oil spills. STEM Fellowship Journal, Volume 6, pp. 69–75.

10 Asif, Z., Chen, Z., and An, C. (2022). Environmental impacts and challenges associated with oil spills on shorelines. Journal of Marine Science and Engineering, Volume 10, Issue 6, p. 762. https://doi.org/10.3390/jmse10060762.

11 https://edis.ifas.ufl.edu/publication/UW330 (Last accessed 11.04.25)

12 National Oceanic and Atmospheric Administration. https://oceanservice.noaa.gov/facts/oilimpacts.html (Last accessed 09.04.2025)

13 Duke, N.C., Pinzón M., Zulekia, S., and Prada, T., Martha, C. (2006). Large-scale damage to mangrove forests following two large oil spills in Panama. Biotropica, Volume 29, Issue 1, pp. 2–14. https://doi.org/10.1111/j.1744-7429.1997.tb00001.x.

14 Ground Water Contamination by Crude Oil near Bemidji, Minnesota. https://mn.water.usgs.gov/projects/bemidji/results/fact-sheet.pdf (Last accessed 11.04.2025)

15 https://amazonfrontlines.org/chronicles/health-oil/ (Last accessed 11.04.2025)

16 Webb, J., Coomes, O.T., Ross, N., et al. (2016). Mercury concentrations in urine of Amerindian populations near oil fields in the Peruvian and Ecuadorian Amazon. Environmental Research, Volume 151, pp. 344–350. https://doi.org/10.1016/j.envres.2016.07.040.

17 Gam, K.B., Engel, L.S., Kwok, R.K., et al. (2018). Association between Deepwater Horizon oil spill response and cleanup work experiences and lung function. Environment International, Volume 121, pp. 695–702. https://doi.org/10.1016/j.envint.2018.09.058.

18 Gam, K.B., Kwok, R.K., Engel, L.S., et al. (2018). Exposure to oil spill chemicals and lung function in Deepwater Horizon disaster response workers. Journal of Occupational and Environmental Medicine, Volume 60, Issue 6, pp. e312–e318. https://doi.org/10.1097/JOM.0000000000001292.

19 Gam, K.B., Kwok, R.K., Engel, L.S., et al. (2018). Lung function in oil spill response workers 1-3 years after the Deepwater Horizon disaster. Epidemiology, Volume 29, Issue 3, pp. 315–322. https://doi.org/10.1097/EDE.0000000000000808.

20 Kwok, R.R., McGrath, J.A., Lowe, S.R., et al. (2017). Mental health indicators associated with oil spill response and clean-up: cross-sectional analysis of the Gulf study cohort. Lancet Public Health, Volume 2, Issue 12, pp. e560–e567. https://doi.org/10.1016/S2468-2667(17)30194-9.

21 Sandifer, P.A., Ferguson, A., Finucane, M.L., et al. (2021). Human health and socioeconomic effects of the Deepwater Horizon oil spill in the Gulf of Mexico. Oceanography, Volume 34, Issue 1, pp. 174–191. https://doi.org/10.5670/oceanog.2021.125.

22 Peres, L.C., Trapido, E., Rung, A.L., et al. (2016). The Deepwater Horizon oil spill and physical health among adult women in Southern Louisiana: the women and their children's health (WaTCH) study. Environmental Health Perspectives, Volume 124, Issue 8, pp. 1208–1213. https://doi.org/10.1289/ehp.1510348.

23 Kim, B.M., Parks, E.K., Lee, S.Y., et al. (2009). BTEX exposure and its health effects in pregnant women following the Hebei Spirit oil spill. Journal of

Preventive Medicine and Public Health, Volume 42, pp. 96–103. https://doi.org/10.3961/jpmph.2009.42.2.96.

24 Bruederle, A., and Hodler, R. (2019). Effect of oil spills on infant mortality in Nigeria. Proceedings of the National Academies of Sciences of the United States of America, Volume 116, Issue 12, pp. 5467–5471. https://doi.org/10.1073/pnas.1818303116.

25 Langhinrichsen-Rohling, J., Selwyn, C.N., Jackson, S., et al. (2017). The prevalence and mental and physical health correlates of posttraumatic stress disorder symptoms in vulnerable and underresourced primary care patients referred to a behavioral health provider. Journal of Public Health Management and Practice, Volume 23, pp. S32–S39. https://doi.org/10.1097/PHH.0000000000000654.

26 https://www.epa.gov/enforcement/deepwater-horizon-bp-gulf-america-oil-spill (Last accessed 30.01.25)

27 Gill, D.A., Picou, J.S., and Ritchie, L.A. (2011). The Exxon Valdez and BP oil spills: a comparison of initial social and psychological Impacts. American Behavioral Scientist, Volume 56, pp. 3–23. https://doi.org/10.1177/0002764211408585.

28 Sumaila, U.R., Cisneros-Montemayor, A.M., Dyck, A., et al. (2012). Impact of the *Deepwater Horizon* well blowout on the economics of US Gulf fisheries. Canadian Journal of Fisheries and Aquatic Sciences, Volume 69, pp. 499–510. https://doi.org/10.1139/f2011-171.

29 https://www.engineering.com/a-review-of-oil-spill-detection-techniques-on-water/ (Last accessed 30.01.25)

30 Mushtaq, Z., Ali, I., Shah, R., et al., (2020). Detection, localization and analysis of oil spills in water through wireless thermal imaging and spectrometer based intelligent system. Wireless Personal Communications, Volume 111, pp. 679–698. https://doi.org/10.1007/s11277-019-06880-3.

31 Yekeen, S.T., and Balogun, A.L. (2020). Advances in remote sensing technology, machine learning and deep learning for marine oil spill detection, prediction and vulnerability assessment. Remote Sensing, Volume 12, Issue 20, p. 3416. https://doi.org/10.3390/rs12203416.

32 Pilžis, K., and Vaišis, V. (2016). Oil spill detection with remote sensors, Conference for Junior Researchers Science – Future of Lithuania. https://doi.org/10.3846/aainz.2016.20.

33 https://hogonext.com/how-to-detect-and-map-oil-spills-using-remote-sensing-data/ (Last accessed 13.04.25)

34 Al-Ruzouq, R., Gibril, M.B.A., Sanabelah, A., et al. (2020). Sensors, features, and machine learning for oil spill detection and monitoring: a review. Remote sensing, Volume 12, Issue 20, p. 3338.

35 Brown, C.E. (2011). Laser fluorosensors. In: Fingas M (ed) Oil spill science and technology, Gulf Professional Publishing, pp. 171–184. https://doi.org/10.1016/B978-1-85617-943-0.10007-3.

36 https://www.sandberg.co.uk/site/gpr/gpr-principles/ (Last accessed 13.04.25)

37 https://www.itopf.org/knowledge-resources/documents-guides/arctic-cold-climates/detecting-tracking-oil-in-ice/ (Last accessed 13.04.25)

38 Rabinovitch, L., Saturos, G., McCallum, P., et al. (2024). Rapid analysis of spilled petroleum oils by direct analysis in real time time-of-flight mass spectrometry with hydrophobic paper sample collection. Environmental Systems Research, Volume 13, p. 34. https://doi.org/10.1186/s40068-024-00361-8.

39 Purohit, B.K., Tewari, S., Prasad, K.S.N.V., et al. (2024). Marine oil spill clean-up: a review on technologies with recent trends and challenges. Regional Studies in Marine Science, Volume 80, p. 103876. https://doi.org/10.1016/j.rsma.2024.103876.

40 https://www.epa.gov/emergency-response-research/oil-spill-research (Last accessed 30.01.25)

41 Piao, L., Park, C.J., Kim, S., et al. (2023). Development of rapid and effective oil-spill response system integrated with oil collection, recovery and storage devices for small oil spills at initial stage: from lab-scale study to field-scale test. Journal of Environmental Management, Volume 345, p. 118833. https://doi.org/10.1016/j.rsma.2024.103876.

42 https://news.utexas.edu/2023/09/06/oil-capturing-technology-offers-10x-improvement-cleaning-up-hazardous-spills/ (Last accessed 30.01.25)

43 https://www.csiro.au/en/research/environmental-impacts/sustainability/new-oil-spill-response-technology (Last accessed 30.01.25)

44 Yan, K., Zhao, F., Pan, L., et al. (2023). High-throughput clean-up of viscous oil spills enabled by a gel-coated mesh filter. Nature Sustainability, Volume 6, Issue 12, pp. 1654–1662. https://doi.org/10.1038/s41893-023-01217-2.

45 https://masgc.org/oilscience/oil-science-tech-air-water.pdf (Last accessed 30.01.25)

46 Allen, J., and Walsh, B. (2008). Enhanced oil spill surveillance, detection and monitoring through the applied technology of unmanned air systems. International Oil Spill Conference Proceedings, Volume 2008, Issue 1, pp. 113–120. https://doi.org/10.7901/2169-3358-2008-1-113.

47 Klemas, V.V. (2015). Coastal and environmental remote sensing from unmanned aerial vehicles: an overview. Journal of Coastal Research, Volume 31, Issue 5, pp. 1260–1267. https://doi.org/10.2112/JCOASTRES-D-15-00005.1.

48 Streett, D.D. (2011). NOAA's satellite monitoring of marine oil. In: Liu Y, Macfadyen A, Ji Z-G, et al. (eds) Monitoring and modeling the deepwater horizon oil spill: a record-breaking enterprise. American Geophysical Union, pp. 9–18. https://doi.org/10.1029/2011GM001104.

49 Battbootti, M.J.H.A., Goga, N., and Marin, I. (2022). Detection and analysis of oil spill using image processing. International Journal of Advanced Computer Science and Applications, Volume 13, Issue 4. https://doi.org/10.14569/IJACSA.2022.0130445.

50 https://www.austinchronicle.com/daily/news/2023-09-06/ut-designed-tech-could-revolutionize-oil-spill-cleanup/ (Last accessed 17.04.25)

51 May, K.T. (2012). 6 Very promising oil spill cleanup innovations. *TED Blog* (July 5). https://blog.ted.com/6-very-promising-oil-spill-clean-up-innovations/

52 https://senseable.mit.edu/seaswarm/ (Last accessed 17.04.25)

53 Ghouas, H., Haddou, B., Bouabdesselam, H., et al. (2010). Elimination of fuel spills from effluent using cloud point extraction methods. Journal of Hazardous Materials, Volume 180, Issues 1–3, pp. 188–196. https://doi.org/10.1016/j.jhazmat.2010.04.012.

54 Avila, A.F., Munhoz, V.C., de Oliveira, A.M., et al. (2014). Nano-based systems for oil spills control and cleanup. Journal of Hazardous Materials, Volume 272, pp. 20–27. https://doi.org/10.1016/j.jhazmat.2014.02.038.

55 Cisneros-Aguirre, J., and Afonso-Correa, M. (2024). Integrated application of innovative technologies for oil spill remediation in Gran Tarajal Harbor: a scientific approach. Waste, Volume 2, Issue 4, pp. 414–450. https://doi.org/10.3390/waste2040023.

56 Sharma, K., Shah, G., Singh, H., et al. (2024). Advancements in natural remediation management techniques for oil spills: challenges, innovations, and future directions. Environmental Pollution and Management, Volume 1, pp. 128–146. https://doi.org/10.1016/j.epm.2024.08.003.

57 https://www.icecleaning.co.uk/blog/what-type-of-absorbents-are-used-for-oil-spills (Last accessed 30.01.25)

58 Singh, H., Bhardwaj, N., Arya, S.K., et al. (2020). Environmental impacts of oil spills and their remediation by magnetic nanomaterials. Environmental Nanotechnology, Monitoring &Management, Volume 14, p. 100305. https://doi.org/10.1016/j.enmm.2020.10030.

59 Doshi, B., Sillanpää, M., and Kalliola, S. (2018). A review of bio-based materials for oil spill treatment. Water Research, Volume 135, pp. 262–277. https://doi.org/10.1016/j.watres.2018.02.034.

60 Odunlami, O.A., Odiakaose, E.O., Owoicho, I.A., et al. (2022). Treatment of oil spills with natural sorbents: a review. International Journal of Recent Research in Physics and Chemical Sciences (IJRRPCS), Volume 9, Issue 1, pp. 16–25. https://doi.org/10.5281/zenodo.6586925.

61 El-Din, G.A., Amer, A.A., Malsh, G., et al. (2018). Study on the use of banana peels for oil spill removal. Alexandria Engineering Journal, Volume 57, Issue 3, pp. 2061–2068.

62 Tontiwachwuthikul, P., AlZubaidi, I., Rennie, E., et al. (2016). Remediation of water from waste lubricating oil spill using potato peels, Proceedings of the 3rd International Conference on Fluid Flow, Heat and Mass Transfer (FFHMT'16), Ottawa, Canada, May 2–3, Paper No. 163.

63 Vlaev, L., Petkov, P., Dimitrov, A., et al. (2011). Cleanup of water polluted with crude oil or diesel fuel using rice husks ash. Journal of the Taiwan Institute of Chemical Engineers, Volume 42, Issue 6, pp. 957–964. https://doi.org/10.1016/j.jtice.2011.04.004.

64 Abdelwahab, O. (2014). Assessment of raw luffa as a natural hollow oleophilic fibrous sorbent for oil spill cleanup. Alexandria Engineering Journal, Volume 53, Issue 1, pp. 213–218. https://doi.org/10.1016/j.aej.2013.11.001.

65 Srinivasan, A., and Viraraghavan, T. (2008). Removal of oil by walnut shell media. Bioresource Technology, Volume 99, pp. 8214–8220. https://doi.org/10.1016/j.biortech.2008.03.072.

66 Sidik, S., Jalil, A., Triwahyono, S., et al. (2012). Modified oil palm leaves adsorbent with enhanced hydrophobicity for crude oil removal. Chemical Engineering Journal, Volume 203, pp. 9–18. https://doi.org/10.1016/j.cej.2012.06.132.

67 Wang, J., Zheng, Y., and Wang, A. (2013). Coated kapok fiber for removal of spilled oil. Marine Pollution Bulletin, Volume 69, Issue 1–2, pp. 91–96. https://doi.org/10.1016/j.marpolbul.2013.01.007.

68 Haussard, M., Gaballah, I., Kanari, N., et al. (2003). Separation of hydrocarbons and lipid from water using treated bark. Water Research, Volume 37, Issue 2, pp. 362–374. https://doi.org/10.1016/s00431354(02)00269-5.

69 Radetić, M.M., Jocić, D.M., Jovančić, P.M., et al. (2003). Recycled wool-based nonwoven material as an oil sorbent. Environmental Science and Technology, Volume 37, Issue 5, pp. 1008–1012. https://doi.org/10.1021/es0201303.

70 Wagih, N., Mahmoud, M.M., Elbaz, A.A., et al. (2023). Oil removal from water using agricultural wastes-based adsorbents for the application in reverse osmosis desalination systems. Desalination and Water Treatment, Volume 316, pp. 335–355.

71 Kathir, K., and Carroll, M. (2023). From trash to treasure: a sustainable approach to oil spill clean-up. Journal of Emerging Investigators, Volume 6, pp. 1–5.

72 Liu, J., and Wang, X. (2019). A new method to prepare oil adsorbent utilizing waste paper and its application for oil spill clean-ups. Bioresources, Volume 14, Issue 2, pp. 3886–3898. https://doi.org/10.15376/biores.14.2.3886-3898.

73 Kolajo, T.E., and Quadri, Q.L. (2023). Remediation of oil spills using pellets from waste papers and kenaf (*Hibiscus cannabinus* L.) fibres. FUOYE Journal of Engineering and Technology, Volume 8, Issue 4, pp. 483–488. https://doi.org/10.46792/fuoyejet.v8i4.1143.

74 Kumara, N., Amritphale, S.S., Matthews, J.C., et al. (2021). Oil spill cleanup using industrial and agricultural waste-based magnetic silica sorbent material: a green approach. Green Chemistry Letters and Reviews, Volume 14, Issue 4, pp. 634–641. https://doi.org/10.1080/17518253.2021.1993349.

75 https://falveyinsurancegroup.com/blog/trends/two-revolutionary-new-methods-of-cleaning-up-oil-spills/ (Last accessed 17.04.25)

76 Dell'Anno, F., Rastelli, E., Sansoneet, C., et al. (2021). Bacteria, fungi and microalgae for the bioremediation of marine sediments contaminated by petroleum hydrocarbons in the omics era. Microorganisms, Volume 9, Issue 8, pp. 1695. https://doi.org/10.3390/microorganisms9081695.

77 Ren, J., Yang, X., Wang, Y., et al. (2025). Sustainable hydrophobic bio-based adsorbent from modified sphagnum moss for efficient oil-water separation. Scientific Reports, Volume 15, p. 11792. https://doi.org/10.1038/s41598-025-96059-7.

78 Gao, Y., Zhou, L., Ouyang, S., et al. (2024). Environmental applications and risks of engineered nanomaterials in removing petroleum oil in soil. Science of the Total Environment, Volume 946, p. 174165. https://doi.org/10.1016/j.scitotenv.2024.174165.

79 Bi, H., An, C., Mulligan, C.N., et al. (2022). Treatment of oiled beach sand using a green and responsive washing fluid with nonionic surfactant-modified nanoclay. Journal of Cleaner Production, Volume 333, p. 130122.

80 Bi, H., Mulligan, C.N., Lee, K., et al. (2025). Nanotechnology for oil spill response and cleanup in coastal regions. Environmental Science, nano, Volume 12, Issue 1, pp. 41–47. https://doi.org/10.1039/D4EN00954A.

81 Chaerun, S.K., Tazaki, K., Asada, R., et al. (2005). Interaction between clay minerals and hydrocarbon-utilizing indigenous microorganisms in high concentrations of heavy oil: implications for bioremediation. Clay Minerals, Volume 40, pp. 105–114. https://doi.org/10.1180/0009855054010159.

82 https://www.repsol.com/en/sustainability/sustainability-pillars/environment/risk-prevention/index.cshtml (Last accessed 30.01.25)

83 https://www.pca.state.mn.us/sites/default/files/c-er2-01.pdf (Last accessed 30.01.25)

84 https://acmeboom.com/articles/oil-spill-prevention-and-preparedness-for-companies/ (Last accessed 30.01.25)

85 https://www.itopf.org/fileadmin/uploads/itopf/data/Documents/TIPS_TAPS_new/TIP_5_Use_of_skimmers_in_oil_pollution_response.pdf (Last accessed 12.02.24)

86 Dave, D., and Ghaly, A.E. (2011). Remediation technologies for marine oil spills: acritical review and comparative analysis. American Journal of Environmental Sciences, Volume 7, Issue 5, pp. 423–440.

87 Mullin, J.V., and Champ, M.A. (2003). Introduction/overview to *in situ* burning of oil spills. Spill Science and Technology Bulletin, Volume 8, Issue 4, pp. 323–330.

88 Buist, I. (1998). Window of opportunity for in situ burning, *Proceedings of the In Situ* Burning of Oil Spills Workshop. MMS, NIST SP.

89 Buist, I., McCourt, J., Potter, S., et al. (1999). In situ burning. Pure Applied Chemistry, Volume 71, Issue 1, pp. 43–65.

90 Walton, W.D., and Jason, N.H. (eds) (1998). In situ burning of oil spills workshop. Proceedings. National Institute of Standards and Technology and Minerals Management Service, November 2–4, 1998, New Orleans, LA. Washington, DC 20401-0003: Government Printing Office, 1999.

91 Davidson, W., Lee, K., and Cogswell, A. (2008). Oil spill response: a global perspective, NATO Science for Peace and Securities Series-C: Environmental Security, Springer, Netherlands, pp. 24.

92 https://www.absorbentsonline.com/spill-containment-blog/what-are-sorbents-for-oil-spills/ (Last accessed 12.02.24)

93 Che, S.N., Hussein, M., Othman, N.H., et al. (2019). Study of acid treated mixed sawdust as natural oil sorbent for oil spill. Materials Today: Proceedings, Volume 19, Part 4, pp. 1382–1389. https://doi.org/10.1016/j.matpr.2019.11.156.

94 Ouyang, D., Lei, X., Zheng, H., et al. (2023). Recent advances in biomass-based materials for oil spill cleanup. Nanomaterials (Basel), Volume 13, Issue 3, p. 620.

95 Zamparas, M., Tzivras, D., Dracopoulos, V., et al. (2020). Application of sorbents for oil spill cleanup focusing on natural-based modified materials: a review. Molecules, Volume 25, Issue 19, p. 4522.

96 https://archive.epa.gov/emergencies/content/learning/web/html/sorbents.html (Last accessed 12.02.24)

97 https://www.oilspillprevention.org/oil-spill-cleanup/oil-spill-cleanup-toolkit/dispersants (Last accessed 12.02.24)

98 https://www.marineinsight.com/environment/different-types-of-dispersants-used-in-an-oil-spill/ (Last accessed 17.04.25)

99 https://response.restoration.noaa.gov/oil-and-chemical-spills/oil-spills/resources/10-what-are-some-environmental-impacts-dispersants.html (Last accessed 22.04.25)

100 Jung, S.W., Kwan, D.Y., Joo, C.K., et al. (2012). Stronger impact of dispersant plus crude oil on natural plankton assemblages in short-term marine mesocosms. Journal of Hazardous Materials, Volumes 217–218, pp. 338–349. https://doi.org/10.1016/j.jhazmat.2012.03.034.

101 National Research Council(1989). Using oil spill dispersants on the sea. The National Academies Press, Washington, DC. https://doi.org/10.17226/736.

102 Giwa, A., Chalermthai, B., Shaikh, B., et al. (2023). Green dispersants for oil spill response: a comprehensive review of recent advances. Marine Pollution

Bulletin, Volume 193, p. 115118. https://doi.org/10.1016/j.marpolbul.2023.115118.

103 Use of booms in oil pollution response, Technical Information Paper, ITOPF.

104 Etkin, D.S., and Nedwed, T.J. (2021). Effectiveness of mechanical recovery for large offshore oil spills. Marine Pollution Bulletin, Volume 163, p. 111848. https://doi.org/10.1016/j.marpolbul.2020.111848.

105 Arora, S., Saxena, S., Sutaria, D., et al. (2022). Chapter 20: Bioremediation: an ecofriendly approach for the treatment of oil spills. In: Das P, Manna S, Pandey JK (eds) Advances in oil-water separation, A complete guide for physical, chemical, and biochemical processes, Elsevier, pp. 353–373.

106 Beolchini, F., Dell'Anno, A., De Propris, L., et al. (2009). Auto- and heterotrophic acidophilic bacteria enhance the bioremediation efficiency of sediments contaminated by heavy metals. Chemosphere, Volume 74, pp. 1321–1326. https://doi.org/10.1016/j.chemosphere.2008.11.057.

107 Adams, G.O., Fufeyin, P.T., Okoro, S.E., et al. (2015). Bioremediation, biostimulation and bioaugmention: a review. International Journal of Environmental Bioremediation and Biodegradation, Volume 3, pp. 28–39. https://doi.org/10.12691/ijebb-3-1-5.

108 Daccò, C., Girometta, C., Asemoloye, M.D., et al. (2020). Key fungal degradation patterns, enzymes and their applications for the removal of aliphatic hydrocarbons in polluted soils: a review. International Biodeterioration and Biodegradation, Volume 147, p. 104866. https://doi.org/10.1016/j.ibiod.2019.104866.

109 Macaulay, B.M., and Rees, M.D. (2014). Bioremediation of oil spills: a review of challenges for research advancement. Annals of Environmental Science, Volume 8, pp. 9–37.

110 https://www.britannica.com/event/Deepwater-Horizon-oil-spill (Last accessed 30.01.25)

111 https://oceanservice.noaa.gov/education/tutorial-coastal/oil-spills/os04-sub-01.html (Last accessed 29.01.25)

112 https://www.britannica.com/event/Exxon-Valdez-oil-spill (Last accessed 29.01.25)

113 https://www.history.com/topics/1980s/exxon-valdez-oil-spill (Last accessed 29.01.25)

114 Atlas, R.M., and Hazen, T.C. (2011). Oil biodegradation and bioremediation: a tale of the two worst spills in U.S. history. Environmental Science and Technology, Volume 45, Issue 16, pp. 6709–6715. https://doi.org/10.1021/es2013227.

115 Arenaz, M.P., Bellas, J., and Vázquez, E. (2009). Effects of the prestige oil spill on the biota of NW Spain: 5 years of learning. Advances in Marine Biology, Volume 56, pp. 365–396. https://doi.org/10.1016/S0065-2881(09)56005-1.

116 https://uisjournal.com/prestige-oil-tanker-spains-biggest-environmental-disaster-what-happened-22-years-ago/ (Last accessed 29.01.25)

117 https://www.greenpeace.org.au/article/the-massive-australian-oil-spill-youve-probably-never-heard-of/ (Last accessed 29.01.25)

118 Spies, R.B., Mukhtasor, M., and Burns, K.A. (2017). The Montara oil spill: a 2009 well blowout in the Timor Sea. Archives of Environmental Contamination and Toxicology, Volume 73, pp. 55–62. https://doi.org/10.1007/s00244-016-0356-7.

119 https://terminautical.com/the-montara-oil-spill-a-disaster-in-the-timor-sea/ (Last accessed 29.01.25)

120 https://www.itopf.org/knowledge-resources/documents-guides/disposal/ (Last accessed 25.04.25)

121 https://www.fema.gov/cbrn-tools/key-planning-factors-chemical-incident/kpf7/4/4-3 (Last accessed 25.04.25)

122 https://www.epa.nsw.gov.au/Your-environment/Recycling-and-reuse/warr-strategy/The-waste-hierarchy (Last accessed 25.04.25)

123 Disposal of oil and debris, Technical information paper, ITOPF. https://www.itopf.org/fileadmin/uploads/itopf/data/Documents/TIPS_TAPS_new/TIP_9_Disposal_of_Oil_and_Debris.pdf (Last accessed 25.04.25)

124 Guidelines for Oil Spill Waste Minimization and Management, IPIECA Report Series, Volume 12.

125 https://www.epa.gov/oil-spills-prevention-and-preparedness-regulations/facility-response-plan-frp-overview (Last accessed 25.04.25)

126 https://www.epa.gov/oil-spills-prevention-and-preparedness-regulations/overview-spill-prevention-control-and#OPP%20Overview (Last accessed 25.04.25)

127 https://www.imo.org/en/About/Conventions/Pages/International-Convention-on-Oil-Pollution-Preparedness,-Response-and-Co-operation-(OPRC).aspx (Last accessed 25.04.25)

8

Pesticide Elimination from Hydro Environment

With an increase in the global population in the last few years, high demands have been set for the agricultural sector to meet the increased necessity of food and nutrition. Modern agricultural techniques employ pesticides to meet the increasing demand of the food. Pesticides have helped in achieving the target to an economically profitable level, but at the same time, an overdose and mishandling of these have degraded the quality of the environment. Rivers and other water bodies are getting polluted by the increased levels of pesticides from agricultural field and landscapes. Use of pesticides in contemporary agriculture has infested water bodies with high amount of toxic chemicals. Also, the tools used for preparation and usage of the chemicals including washing setup and spraying equipment enrich the pesticides in the wastewaters. Trace amount of pesticides can lead to accumulation in the wastewater from domestic sectors. Domestic wastewater includes the washings of pesticides contaminated fruits and vegetables during cleaning process. The factories involved in production of these substances produces wastewater called as pesticide production wastewater. The amount of these chemicals from different sources varies between 0.1 and 107 mg/L [1].

Pesticides are chemical or biological substances that prevent, destroy or inactivate pests, and unwanted species or disease vectors. These are biologically dynamic chemical substances as they are involved in control of biological species. These include insecticides, herbicides, rodenticides, nematicides, molluscicides, bactericides, avicides, insect repellents, animal repellents, antimicrobial substances, fungicides, etc. [1]. These substances provide the means of combating against those organisms that interfere with the fulfillment of needs of man not only for food and fiber but are also responsible for toxicity to man, livestock, and vegetation. They are vital for human and ecological progress, but the consumed pesticides drastically influence unintentional species. Thus, pesticides are recognized as one of the emerging pollutants in the natural environment and are posing alarm in many regions in the globe.

8.1 Classification of Pesticides

Understanding the nature and potential risks associated with the pesticides it is important to adopt sustainable agricultural practices. Pesticides are classified based on their origin, primary targets, modes of action and characteristics [1].

The following are the approaches to categorize different pesticides:

1. Source or origin of the pesticide
2. Chemical structure of the pesticide
3. Degradation ability of the pesticide
4. Entry pathway
5. Pesticide activity and the organisms they destroy

On the basis of chemical composition pesticides are classified into following types:

1. **Organochlorines:** These are one of the earliest pesticides formulated. Many of these are banned due to their bioaccumulation tendency and unrelenting nature. These are very stable and persistent in the environment. They accumulate in the adipose tissue. In human beings, these disturb the electrophysiological properties. For example, dichlorodiphenyltrichloroethane (DDT) was employed as a prominent pesticide to increase crop production but later researchers discovered the mass death of several species of water birds, eagles, premature egg shell rupturing, and lethal effects on fishes due to the bioaccumulation and biomagnifications of the pesticides in living tissue. Lindane and endosulfan are among the most common biodegradable organochlorine pesticides still used.
2. **Carbamates:** These are characterized by inhibition of acetylcholinesterase enzyme which disturbs the central nervous system of pests leading to their death. It is capable of effecting nerve signal transmission in non-targeted species. These are organic esters derived from dimethyl N-methyl carbamic acid. Carbamates are short lived pesticides.
3. **Organophosphates:** These function in a similar manner as the carbamates and over stimulate the nerve action, paralyzing the target pest. These are esters derived from phosphoric acids. The intoxication effects of the pesticides include nausea, dizziness, convulsions, and even death.
4. **Pyrethroids:** These are natural insecticides derived from the extracts of chrysanthemum flowers. These cause fluctuations in the ion exchanging dynamics of the nerve cell which results in neuronal hyper excitation. These are active substances employed in a number of commercial products along with use as spray for homes and farms. These insecticides are widely used in agricultural

fields and pose risks to human beings on prolonged exposure. They also target the central nervous system of the insect.

5. **Neonicotinoids:** These are newly evolved pesticides and are related to nicotine chemically. They bring about the collapse of target organism by binding with nicotinic acetylcholine receptor. However, they are associated with low risk to nontarget organisms.

On the basis of target species, pesticides are known by the following terms:

1. **Fungicides:** These inhibit the growth and reproduction of fungi. Mancozeb and difenoconazole are fungicides which damage the cell membranes, inactivate cellular enzymes, and obstruct the normal biological processes in fungi.
2. **Herbicides:** These are the weedicides which kill weeds in the agricultural fields. Metolachlor, atrazine, and 2,4-dichlorophenoxyacetic acid are certain selective herbicides which eliminate specific weeds and leave the desired plantation unharmed. They can be further categorized based on their time of use and method of application.
3. **Nematicides:** Nematodes are parasitic worms which attack the crop root system and reduce the growth of the plants. Nematicides kill or repel the target species. Nowadays, nonfumigant (nonvapor phase) nematicides are applied against nematodes which are water soluble and nematistatic.
4. **Insecticides:** These chemicals are used to kill insects like aphids, caterpillars, beetles and vectors like mosquitoes and ticks. These are neurotoxic and can affect both insects and nontarget organisms, including human beings, birds, and wildlife. A number of health issues are related with the exposure of the living system to insecticides. For instance, acephate can cause respiratory problems, aldicarb leads to dizziness, and diarrhea, endosulfan may trigger itching, confusion and respiratory illness.
5. **Rodenticides:** Mice, rats, squirrels and bats can destroy an appreciable amount of yield in the agricultural fields. Rodenticides function by causing internal bleeding in the organisms. Warfarin and zinc phosphide are few commonly used rodenticides which interfere with the central nervous system of the rodents. These cause equal negative impacts on the nontarget organisms in the water bodies, wildlife, and human beings.
6. **Biopesticides:** These are recognized as good alternatives to synthetic pesticides due to their specific, safer, and sustainable nature. These are extracted from natural sources like microbial species and plant extracts. These include plant-incorporated biopesticides, microbial biopesticides, and biochemical pesticides. They do not cause any negative impacts to the nontargeted species.

Spinosad, neem oil, azadirachtin, and ladybugs are a few biopesticides used in pest management.

Pesticides can be further divided into various types depending upon the mode of their influence on the target system. The major categories are as follows:

1. **Systemic Pesticides:** These are absorbed by living cells and transferred to the untreated tissues. Pesticides like 2,4-dichlorophenoxyacetic acid and glyphosate are rapidly absorbed by living cells and further transported to the target organ or tissue.
2. **Contact Pesticides:** These show a response on coming in contact with the pest, for example, paraquat and diquat.
3. **Stomach Poisons:** These pesticides enter the pest body by ingestion through mouth and disturb the digestive system.
4. **Fumigants:** These are vapor producing pesticides which enter the pest body through the tracheal system. These include phosphines.
5. **Repellents:** These pesticides keep away the target organism from the protected site or plant but do not kill them. An example of such repellent is methiocarb.

On the basis of mode of action, pesticides can be categorized into the following types:

1. **Physical Impact:** They kill the target pest by putting physical pressure or effect. Activated clay is usually employed as pesticide chemical which acts by a physical impact.
2. **Protoplasmic Poison:** These precipitate the protein of the cell. Arsenicals are capable of acting on the protoplasmic protein and destroying the functionality of the cell.
3. **Respiratory Poison:** Respiratory poisons like hydrogen cyanides inactivate the respiratory enzymes when in contact with the target pest.
4. **Nerve Poison:** These affect the central nervous system and inhibit the nerve impulse conduction. Malathion is the pesticide primarily impacting the central nervous system of the organisms.
5. **Chitin Poison:** It inhibits the strength of the natural polysaccharides found in the exoskeleton of the pest insect or crustacean. This results in the destruction of the outer protective body of the organisms. Pesticides like diflubenzuron can readily kill insects through this mode of action.

Pesticides are available in solid, liquid, or gaseous forms. Pesticide formulations are available as baits, dry or water dispersible granules, dust, vapor strips, pellets, soluble powder, aerosols, gels, and micro-encapsulated materials, fumigants (phosphine and phostoxin).

Table 8.1 gives an outline of the classification of pesticides [1, 2].

Table 8.1 Classification of pesticides.

Basis of classification	Types of pesticides		Examples
Degradability	Biodegradable		Pyrethrum, Azadirachtin, Dalapon, Chloroneb
	Nonbiodegradable		DDT, BHC, Aldrin, Parathion, Endrin
Sources	Synthetic	Carbamate	Propoxur, Bendiocarb, Carbaryl
		Organophosphate	Malathion, Temephos, Fenthion, Dichlorvos, Fenitrothion, Primiphos methyl
		Organochlorine	DDT, BHC
		Pyrethroid	Deltamethrin, Cyfluthrin, Bifenthrin, Lambda-cyhalothrin, Permethrin
	Biopesticides	Biochemical pesticides	Azadirachtin, Capsaicin
		Microbial pesticides	*Trichoderma* fungi, *Pseudomonas fluorescens* *Bacillus thuringiensis*
		Plant-incorporated protectants	*Bacillus thuringiensis*
Chemical structure	Carbamate, Thiocarbamates, Aryl *N*-methyl carbamate, Hetero cyclic mono carbamates, Hetero cyclic dimethyl carbamates, carbamoylated oximes, Thiocarbamates		Bendiocarb Propoxur Carbaryl
	Organophosphate, Phosphates Thiophosphates, Dithiophosphates, Phosphorodithioates		Malathion Temephos Fenthion Dichlorvos Fenitrothion Pirimiphos methyl
	Organochlorine		Dichlorodiphenyltrichloroethane (DDT), BHC
	Pyrethroid		Deltamethrin, Cyfluthrin, Bifenthrin, Lambdacyhalothrin, Permethrin

(Continued)

Table 8.1 (Continued)

Basis of classification	Types of pesticides		Examples
Mode of entry	Gastric toxins		Malathion
	Contact exposure		Diquat dibromide, Paraquat
	Expectorant	Fumigant	Aluminum phosphide (wheat pills), Soil fumigants
	Evaporators	Repellants	Methiocarb
Target pest	Avicides	Kill birds	Aminopyridine
	Algaecides	Regulates the growth of algae	Copper sulfate
	Acaricides	Kills mites or ticks	Dicofol
	Fungicide	Removes or kills fungi, blights, mildews, and molds	Azoxystrobin, Chlorothalonil
	Herbicides	Destroys weeds	Atrazine, Glyphosate, 2,4-dichlorophenoxyacetic acid
	Insecticides	Kill insects and other arthropods	Aldicarb, Carbaryl, Imidacloprid
	Larvicide	Kills larvae	Methoprene
	Termiticides	Destroy termites	Fipronil
	Rodenticide	Repels or kills mice and other rodents	Warfarin
	Nematicides	Kill plant parasites like nematodes	Aldicarb, Ethoprop
	Chemosterilant	Obstructs the reproductive system of the pest by making it infertile	Diflubenzuron
	Desiccants	Has the ability to dry the tissue of the pest organism	Boric acid
	Lampricides	Kills larvae of lamprey	Nitrophenol
	Repellants	Repel vertebrates and invertebrate pests	Methiocarb

8.2 Scenario of Global Pesticide Consumption

As per the report of the Food and Agriculture Organization of the United Nations (FAO, 2021), 4.21 Mt of pesticides were globally used in the fields to increase productivity in the year 2018, which later declined to 2.7 Mt in 2020. The depression was due to revised Chinese legislation policy, which influenced other nations as well. Globally, 25% of the cultivated areas are consuming pesticides. According to the World Trade Organization (WTO), a 30% hype in disinfectant trading occurred during the COVID-19 pandemic. In comparison to the 1990s, the average global pesticide consumption has increased by 50%. However, awareness about the increasing risks linked with the application of chemical pesticides led to a decline in the consumption of harmful organochlorines and organophosphates around the world. As per reports, Asia is the largest consumer of pesticides, accounting for 50% of the global consumption. Asia was also seen as the largest exporter of pesticides (3.7 Mt) during the COVID-19 pandemic in 2020. The unit area cropland consumption of these chemicals is less than 1 kg/ha for most of the Asian countries except Japan, Taiwan, South Korea, and China. As per FAO, the American agricultural industry consumes the highest pesticide concentration of 1 Mt average annually and the highest unit area of cropland consumption (2.83 kg/ha), including unit capita consumption (1.17 kg/person). Pesticide consumption for others follows the order Europe, Oceania regions, and Africa. The Directorate of Plant Protection, Quarantine and Storage, India reported 52466 Mt of pesticides production in the country for the year 2022 [3]. Figure 8.1 illustrates the pesticide consumption status of various countries in the year 2020.

The consumption depends upon the demand and regulatory policies operating in various nations.

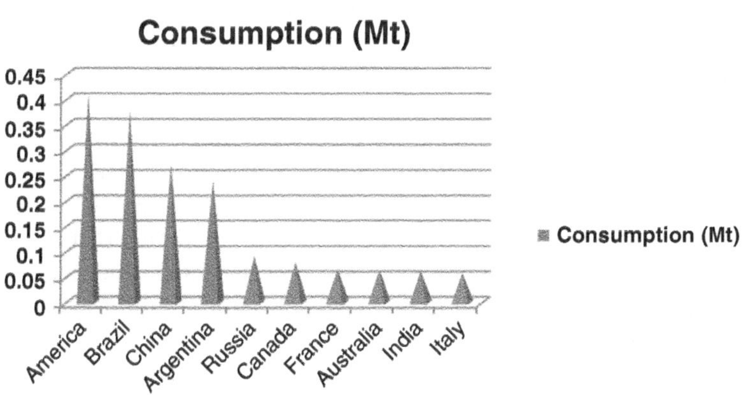

Figure 8.1 Country-wise pesticide consumption.

8.3 Toxicity Assessment of Pesticides

The factors influencing the impact of pesticides include the amount of pesticides and duration of exposure which helps to understand whether the toxicity of pesticide is acute or chronic. Acute toxicity refers to the exposure of the living system for a short time which can be fatal. Chronic toxicity is observed when individuals are exposed to a pesticide over an extended period of time. The assessment of chronic toxicity is challenging as it involves lengthy analytical procedures. A measure scale called LD_{50} is used by World Health Organization (WHO) to categorize pesticides. It has classified the toxic effects of pesticides from class I to class III. Class I grade pesticides are banned in developed nations due to their observed detrimental impacts.

According to FAO (1990), various factors regulate the impact of pesticides on water quality. These factors include the nature of the substance present inside the active ingredient of the pesticides. The active substance present in the pesticide formulations decides the strength of the action on target species. The water quality is damaged due to the intermediate compounds or end products generated after the degradation of the pesticide active ingredient. The water parameters also get disturbed due to the inclusion of supportive additives like preservatives, adhesives, and buffers present in the pesticide formulations [1, 2].

The erratic use of pesticides has emerged as an alarming concern for public health. The stability of pesticides in ecosystem has affected avian population, threatened wild life and other life forms. The components in pesticides can be easily transported via air and water from the point of application to elsewhere, thus aggravating the condition of spread of toxicants.

Pesticide wastes include empty containers, application equipment, leftover slurry, rinsate from containers and equipment, ready to use formulations, and expired or unwanted pesticides. Improperly disposed materials have the potential to cause ground and surface water contamination [4]. The fate of pesticides on entering the environment is illustrated in Figure 8.2.

Extensive use of pesticides can lead to leaching of the toxic chemicals from the soil to water bodies. In addition, domestic wastewater and industrial wastewater can also disperse these contaminants into the environment. The confiscation of pesticides from aqueous bodies is crucial due to the mobile, toxic, carcinogenic, and mutagenic nature of the pesticides. It is reported that more than 20% of the pesticides reach far beyond their target species via air, water or soil routes. The pesticides can transfer to the ecosystem either by diffusion, dispersion, or by permeation [5].

The principal sources of pesticides in the natural environment are agriculture, forests, and woodlands. It is reported that agricultural practices have a key

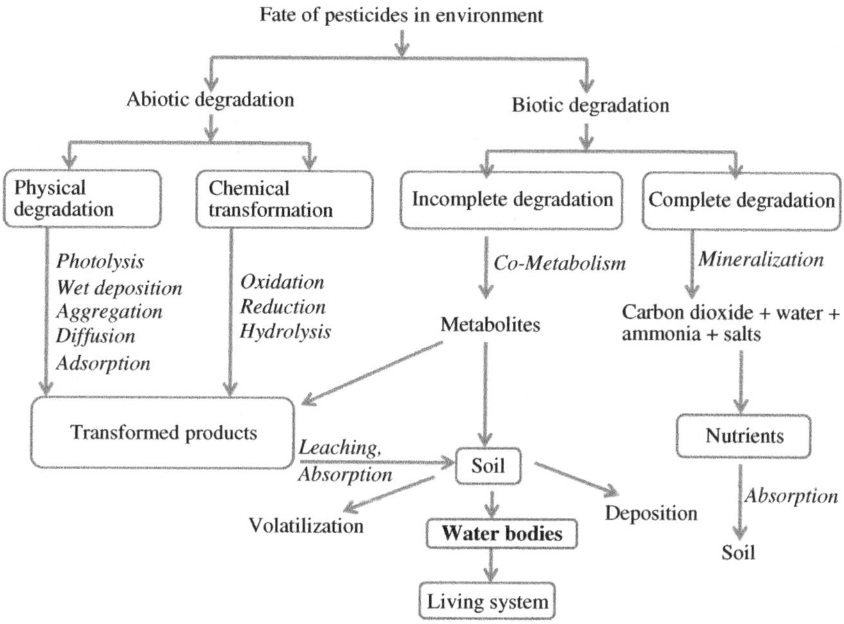

Figure 8.2 Pathways of pesticide degradation.

implication on the quality of water bodies and are responsible for critical environmental concerns. Pesticides reach the water system through sprinkling glide, sprinkling done aerially, wearing down of soil and casual dumping of rinsates [6]. Other than agricultural origin pesticides get introduced into the environment due to accidental spillage, domestic cleaning activities, and household pest controls.

Pesticides usually have complex molecular structures and can undergo various reactions to completely degrade or may result in formation of some intermediates whose degradation depends on the nature, physical and chemical properties of the pesticides. The main processes implicated in the degradation mechanism of the pesticides are oxidation, reduction and hydrolysis. Thermal energy, phytodegradation by sunlight and even microbial activity can disintegrate the chemicals.

Pesticides-loaded wastewater requires appropriate treatment before discharge into water bodies. Therefore, pesticide treatment is a critical investigatory domain due to recalcitrant, bioaccumulative and toxic nature of pesticides. The chemical oxygen demand (COD) of the wastewater is higher than the biological oxygen demand (BOD) and thus the natural species are not strong enough to clarify the water bodies naturally. The major challenges encountered in remediation of pesticides are high total suspended solids, and highly acidic nature of the pesticide contaminated water. Even the intermediates formed from the pesticides are stable and recalcitrant. It is of great concern to the public as well as authorities due to their negative impacts even at very low concentrations. The remediation of pesticides from water bodies and wastewater needs to be seriously acted upon before the condition is uncontrollable [7].

The impact of pesticides can either be acute or chronic depending upon the time of contact with the contaminants. Chronic diseases that affect the nervous system, renal, cardiovascular, and respiratory systems in humans are reported which establishes the link between exposure to the contaminants and their impacts [8, 9]. Pesticides can exert noteworthy effects on human health. Risk related to these chemicals exaggerates due to inappropriate protection strategies and unskilled supervision during their application and disposal. The examples include the indiscriminate handling of superfluous organochlorine pesticides in the Republic of Tajikistan in central Asia, which resulted in loading the soil morphology with pesticides concentration greater than 10 ppm, and around 24% of the used pesticides getting deposited on fruits and vegetables in Thailand [10, 11]. As per the World Health Organization (WHO), 10^6 people suffer from acute poisoning due to pesticide and each year 0.4–2.0% death rate is recorded, with the responsible factors being tumors and nervous disorders in the long term as side effects of pesticides [12, 13]. Mounting incidences of cancers and various other life threatening diseases in the last few years have been ascribed to the disproportionate use and exposure to the pesticides.

One of the herbicides named paraquat has marked toxicity and leads to multiple organ failure, deadly blood cancer and respiratory system disorder resulting in breathing malfunction and deaths [14]. Similarly, chronic exposure to pesticides like atrazine, and oxyfluorfen is reported to inhibit the photosynthetic activity in plants and is associated with cardiovascular issues, renal problems and muscular degeneration in human beings [15]. The pesticides can lead to failure of the reproductive system in some fish species, and thinning of egg shells of birds like falcons, kites and eagle owls. Carbamate pesticides reportedly interact with the immune systems of vertebrates and hamper thermoregulation and other metabolic actions. Paraquat causes growth-related issues in bird populations and is also responsible for eggshell thinning, along with dichlorodiphenyltrichloroethane (DDT). Pesticides destroy the friends of farmers "earthworms", which reduces the fertility of the soil due to the decreased population of the beneficial worms. A pesticide, pentachlorophenol, is responsible for killing of symbiotic rhizobium bacteria, which reduces the tendency of the plant for nitrogen fixation, thereby decreasing the productivity of the plants.

The challenges encountered in the treatment of pesticide-loaded wastewater include wastewater composition, variability of these chemicals in their structure and pH of the wastewater. COD of the pesticide constituting wastewater varies between 150 and 33750 mg/L, while the BOD is found to be between 30 and 11590 mg/L [3].

The toxicity effects of the pesticides emphasize the need to explore effective wastewater remediation methods for not only removing these pollutants from the wastewater but also to reuse and recycle water to meet the objective of zero liquid discharge. It becomes necessary to explore the extraction and detection methods employed for estimating pesticides in the water bodies.

The detection of multiclass pesticides can be done using methods such as gas chromatography, flame ionization detector, electron capture detector, and high pressure liquid chromatography (HPLC) technique. The extraction process employed includes single drop microextraction, solid phase extraction and liquid–liquid extraction.

8.4 Treatment of Pesticide Wastewater

The treatment of pesticides from wastewater involves various physical, chemical, and biological approaches (Figure 8.3). The disposal method involves the destruction of pesticide pollutants extraction, followed by stabilization to stop the movement of contaminants into the niche. Pesticides are disposed by land cultivation, dumping in soil or concrete pits, using evaporation beds and land filling, which are nonecofriendly and lead to contamination of ground water. Other methods include incineration, chemical treatments, physical methods, and biological treatments. The wastewater which has low level of pesticide is usually disposed off by land cultivation, disposal pits, evaporation ponds, and land filling.

Land cultivation involves placing waste liquid in low zone of soil for weathering, while disposal pits involve weathering in air contact. Liquid wastes are also

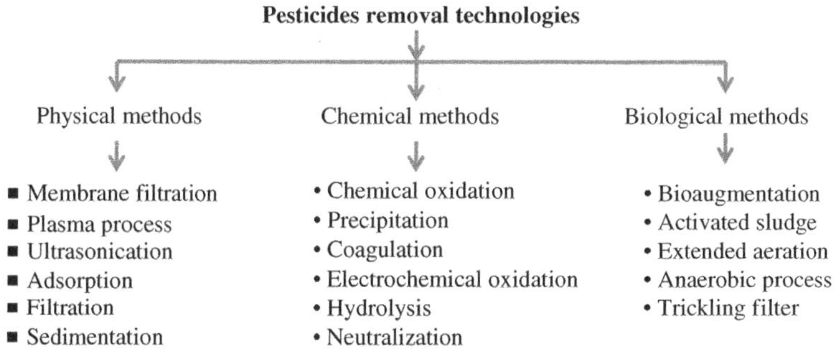

Figure 8.3 Technologies for pesticide removal from wastewater.

subjected to weathering in air by placing them in lined ponds. Though all the above methods involve onsite use of the reclaimed wastewater and are simple to execute, they suffer from the disadvantages of runoff, leaching and unsteady slow decomposition. Weathering is influenced by climatic variations. Reports have been published wherein pesticides like bentazon, mecoprop, and atrazine have been found in leaches in Japan from landfills, as they are recalcitrant to the landfill conditions. The disadvantage of leaching in soil pits is minimized by opting plastic lined pits. Reports have shown that no accumulation occurs in pits and pesticides are completely degraded by microbes. However, the systems like lined pits or concrete pits were not found suitable for various pesticides, including atrazine, and required long time for degradation. The degradation area demand was also high and complicated. Land filling is also one of the methods of dissipation of pesticides, but not effective in making wastewater completely free from chemicals. Care needs to be practiced on application, disposal and remediation of pesticides. Other methods are thus essential to ensure careful fast and effective pesticide remediation. Researchers have alarmed about the significance of understanding and examining the pathways of pesticides for contamination of water bodies and related health issues. The acute ill effects of pesticides on living system demand the elimination of pesticides from wastewater.

Except for the hydrophobic compounds like pyrethroids, conventional treatment processes are ineffective in the elimination of pesticides. There are diverse pesticide treatment and management techniques for the removal of pesticides from wastewater and including a blend of physical, chemical, and biological methods. The appropriateness of method is influenced by the amount of specific pesticide and the type of waste generated. Thermal treatment of pesticides containing wastewater is rapid but involves high capital costs and cannot be efficiently applied to all chemicals. The effectiveness as well as specificity can be enhanced by employing chemical methods through oxidation, reduction, hydrolysis, or use of catalytic reagents.

Each treatment technique is associated with related pros and cons not only in terms of economy, but also in terms of efficacy, ease of operation, consistency, ecological influence, pretreatment necessities, and the generation of manageable byproducts. Techniques known as advanced oxidation processes are known for their less sludge production, fast rate of reaction, and less detention time, but suffer from the limitations of high capital involvement [16]. Fenton process, though simple, works only at low pH and is related to a large amount of sludge formation [17]. Chlorination is cost effective but results in the formation of more toxic byproducts. Similarly, bioreactor technology is eco-friendly but demands high energy and cost input [18]. Figure 8.3 portrays different physical, chemical, and biological methods for the removal of pesticides from wastewater.

8.4.1 Physical Treatment Methods

8.4.1.1 Membrane Filtration

This method has the benefit of its implementation at any stage during the wastewater management process, depending upon the type of membrane and the targeted contaminants. Nanofiltration was evaluated for the filtration of organophosphate insecticides, dimethoate and atrazine, and the herbicide glyphosate. A thin film nanocomposite membrane was investigated for the confiscation of diazinon after adding triethylamine as an accelerator. Two flat sheets of nanofiltration membrane having a negative charge were used for the removal of atrazine, simazine, and diuron, and due to lower molecular cut off, these retained pesticides at a high level. The removal was carried out in the presence of complex naturally occurring humic acid, whose effect was governed by the structural arrangement of the pesticides and molecular membrane cutoff. A linear structure pesticide (diuron) having high molecular weight and high dipole moment was least retained in presence of humic acid. A thin film composite made up of polyamide showed rejection of pesticides more than 95%. The utilization of membrane technology geared up when the chlorine containing pesticides were recovered using acetylated cellulose membrane during reverse osmosis process. In most of following pesticides elimination investigations, noncellulosic and thin film composites were found to be more efficient and reliable in comparison to acetylated cellulose membranes. A number of novel membranes evolved and showed better chemical steadiness and amplification in the salt rejection capacity. Some treatment plants in France and Paris are employing nanofiltration technique on large scale for resurgence of pesticides from water bodies since 1999. Nowadays, nanofiltration and reverse osmosis based membranes are available commercially and are used for removing fungicide, insecticide, and herbicide from wastewater [15].

8.4.1.2 Plasma Process

This process offers a quick and competent degradation of pesticide residue in water, as it is non-selective and requires no chemical input. However, the performance is governed by system as the degradation of the pesticides is done by reactive oxygen and nitrogen species. The practical exploitation is still under exploration.

8.4.1.3 Ultrasonication

This technique develops less noxious and more biodegradable byproducts. The method can handle a number of variable wastewater composition and flow rate. It is also having the advantage of no sludge production. The disadvantage associated is prerequirement of the addition of oxidant and managing the heat generated during the process.

8.4.1.4 Adsorption

Adsorption technique is a promising and a most widely used process for wastewater treatment due to its rapid execution as well as the possibility of the method being adopted onsite. The area of research still to be explored for this technique involves the minimization of the byproducts or concentrated adsorbents reclaimed during the process. Supercritical carbon dioxide extraction has been employed to remove organophosphate pesticides like fenitrothion, chlorpyrifos, fenthion, etc. from agricultural wastewater. It was observed that a combination of extraction and Fenton degradation could be a better alternative for the removal process [19]. Nanomaterials have proved to be favorable materials for the elimination of pesticides from water bodies, for there use and recovery of water. They exhibit excellent thermal and chemical stability and thus can be competently used for an extensive diversity of pesticides. Carbon and polymer based nanomaterials, silica, iron and chitosan-based nanoparticles, cellulose-based nanocrystals and hybrid nanocomposites are much explored options in the field and are having good potential of removing the pesticides. A parasympathomimetic insecticide malathion was removed using functionalized multi-walled carbon nanotube [20]. Graphene oxide-based silica-coated magnetic nanoparticles functionalized with 2-phenylethylamine were investigated for the adsorption of organophosphorus pesticides namely, parathion, chlorpyrifos, and malathion. The optimum time, adsorbate concentration and dosage were reported as 15 minutes, 1 μg/mL, and 15 mg, respectively. Greater than 86.9% recovery using the adsorbent was reported, with real wastewater [21]. Pesticides with organophosphates as chemical composition like, fenthion, malathion, and diazinon were investigated for their adsorption onto waste cellulose. The factors influencing the adsorption characteristics included pH, time of contact, adsorbent dose, initial concentration and operating temperature [22]. A metal organic framework based adsorbent was fabricated and experimented successfully for elimination of glyphosate. The adsorption capacity of 256.54 mg/g was reported [23]. Rapid removal of DDT (50%) in two hours was reported onto adsorbents based on hollow mesoporous silica. The pore size and connectivity in the adsorbents influenced the adsorption characteristics. The research team also reported synthesis of hollow mesoporous silica with iron oxide magnetic functionalization. Better results in less time were observed in this case [24]. The toxic and highly mutagenic and carcinogenic nature of pentachlorophenol demanded its eradication form the industrial wastewaters. Bioaugmentation by *Pseudomonas putida* proved to be effective in the process [25]. Activated waste fibers of hemp showed 2192 m^2/g surface area, which was large, prepared by carbonization and activation with the ratio of 2 : 1 for KOH/carbonized waste. Good adsorption tendency was observed at 900 °C for pesticides like nicosulfuron, acetamiprid, carbofuran, dimethoate, and atrazine [26]. Agricultural solid waste utilization for pesticide elimination involves process of ion exchange, hydrogen bonding and complexation. The potential of chestnut shells were examined for removal of pirimicarb, thiamethoxam, imidacloprid, and acetamiprid pesticides, and promising results were reported [27]. A natural montmorillonite was employed for ametryn extraction from wastewater revealing the highest monolayer adsorption capacity as 188.81 mg/g [28]. Zeolites when incorporated with polymers have proven to be important scavengers of pesticides from wastewaters. This is attributed to their high specific surface area, greater porosity, and negatively charge-balanced exchangeable cations. Though use of zeolites suffers from the limitations of appalling adsorption performance of anions and organics, modification using acid or base treatment and surfactant impregnation improves adsorption capacity. In the same direction, zeolite and tungstophosphoric acid composites made by impregnation method followed by ultrasonication and calcination portrayed a removal capacity as 12.1–25.8 mg for per gram of nicosulfuron [29]. Other than the above mentioned technologies, biosorption has also proven to be effective in eliminating pesticides wherein algae, fungi and bacteria biomass are utilized. The degradation of nine different pesticides was carried out by immobilization of *Aspergillus laccase* (25 g/L) with wheat straw and peanut shell which reported 65.9% and 54.5% of removal, respectively, with maximum degradation rates as 92.0% and 92.9%, respectively. Thus, biosorption coupled with laccase degradation presented a successful pesticide exclusion from wastewater [30]. Figure 8.4 illustrates various categories of adsorbents being used for the removal of pesticides from wastewater.

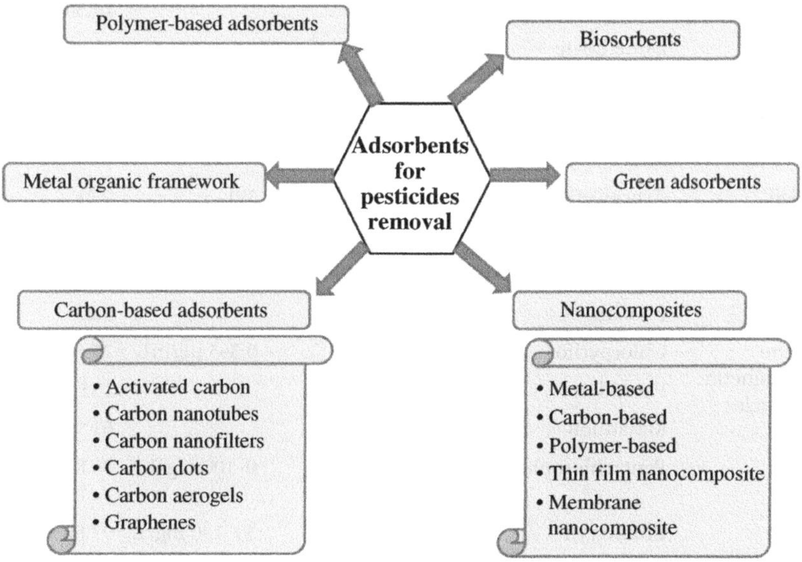

Figure 8.4 Categories of adsorbents employed for the removal of pesticides.

The competence of diverse adsorbent differs as it is governed by characteristics of the adsorbents and the composition and nature of the pesticides. Different adsorbents have revealed different adsorption efficiencies as shown in Table 8.2.

Table 8.2 Adsorption efficiencies of various adsorbents for pesticide elimination.

Adsorbents	Pesticide	Temperature (K)	Initial concentration	Maximum adsorption (mg/g)
Activated carbon-cloth	Diuron	298	6.5×10^{-5} mol/L	213.06
	Aldicarb			421.58
	Dinoseb			301.84
	Ametryn			354.61
Granular activated carbon	Carbofuran	303	50–225 mg/L	96.15
Date seed			25–250 mg/L	137.04
Waste rubber tire	Methoxychlor	298–318	2–12 mg/L	112.00
Waste hemp	Acetamiprid	298	10–50 mg/L	12.20
	Atrazine			15.50
	Carbofuran			15.40
	Nicosulfuron			19.50
	Dimethoate			11.80
Coconut shell	Malathion	303–353	7 μg/L	909.10
Graphitic carbon	2,4-dichlorophenoxyacetic acid		9–300 mg/L	77.13
Modified mesoporous silica	Avermectin	298–308	391–48430 μg/L	9.21
	Imidacloprid			6.40
	Acetamiprid			5.11
	Phoxim			8.23
Graphene based magnetic nanoparticles	Chlorpyrifos	298	0.3–5 μg/mL	25.60
	Parathion			135
	Malathion			61.90
Modified chitosan	Pentachlorophenol	293	0–100 mg/L	36.85
Modified microcrystalline cellulose	Prometryn	303–323	30–150 mg/L	97.80

Adsorbents	Pesticide	Temperature (K)	Initial concentration	Maximum adsorption (mg/g)
Bioadsorbent Fungus *Pleurotus mutilus*	Metribuzin	298	100–400 mg/L	3.30
Natural montmorillonite	Ametryn	303–323	25–150 mg/L	188.81
NH_4Cl-induced activated carbon	Diazinon	298–313	2–10 mg/L	250
Chestnut shells	Imidacloprid	298	5 mg/L	8.51
	Acetamiprid			4.69
	Thiamethoxam			14.31

Source: Adapted from [31].

8.4.1.5 Evaporation Beds

Evaporation beds works by photochemical and biological degradation. This method was found to be effective for some pesticides, but not all. This process offers the advantage of being economical, demands low maintenance and monitoring only annually. The noneffectiveness of this technique for various pesticides is its limitation.

8.4.2 Chemical Treatment Method

Chemical methods help is hydrolyzing the pollutants into nontoxic products by various chemical transformations. Some of the pesticides elimination techniques are mentioned below:

8.4.2.1 Iron-enhanced Sand Filters

A number of contaminants can be recovered by subjecting the contaminated water onto iron filing loaded sand filters. The removal of pesticides which are hydrophobic in nature can be achieved up to 10–30%. Though this method was not significantly reducing the levels of contaminants and is not considered a promising method for removal of pesticides [32].

8.4.2.2 Chlorination

This is mainly a tertiary water treatment method with the chief objective of disinfection of water for its use, especially for drinking purposes. The wastewater having pesticides, when treated with chlorine has proved to show negative impacts.

In the prechlorination of pesticide belonging to organophosphorus category, oxidation to more toxic oxons results, which are difficult to get rid off from water. Similar toxic product formation occurs for the organophosphate pesticides like Methyl parathion.

Few studies have reported the removal of the pesticides. Specifically, 100% removal of isoproturon, diuron, and parathion was observed while atrazine was eliminated to only 20%. Thus, a conclusion can be drawn that this method can be included as secondary treatment method, following the method of oxidation, which aids in the oxidation of the pesticide. Still, further investigations are needed to reduce the disinfection byproducts to acceptable levels.

8.4.2.3 Advanced Oxidation Process

The advanced oxidation process utilizes oxidizing agents to oxidize contaminants (Figure 8.5).

Radicals involved in the pesticide removal are hydroxyl radical (OH) and sulfate radical (SO_4^{2-}). This process does not involve the formation of any sludge or the transformation of the contaminant from one phase to another. It offers less retention time and greater rate of reaction in comparison to other conventional treatment processes. However, the technique is associated with high operational and maintenance cost. The system designing involves complex chemical methods. These include ozone treatment, ultraviolet (UV) treatment, electrochemical process, catalytic and photo advanced oxidation process.

Sodium persulfate was used to oxidize various pesticides like pymetrozine, cymoxanil, spinosad, fluopyram, spiromesifen, acrinathrin, etc., with initial concentration in prepared water as 0.02–1.17 mg/L. The formation of sulfate

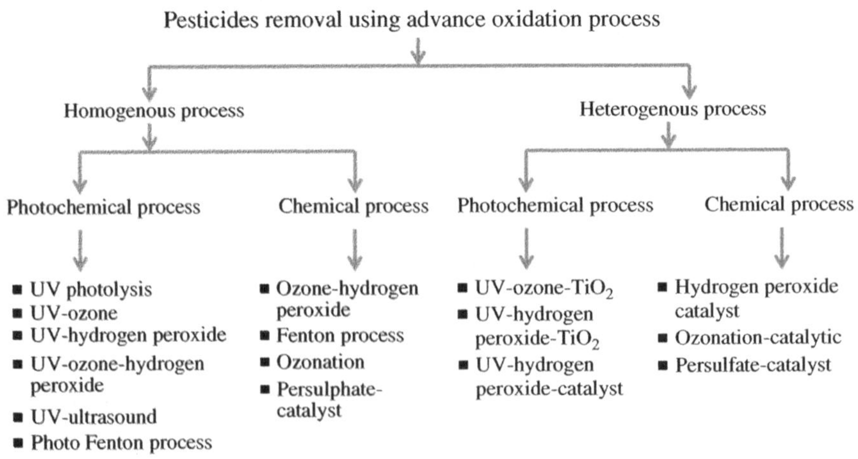

Figure 8.5 Techniques involved in the advanced oxidation process.

radicals from persulfate led to greater contact time between the contaminant and the radical and better degradation of the pesticides. Ultimate mineralization occurs from parent pesticides to carbon dioxide and water. The reduction of dissolved organic carbon content occurred by 87%. The above result can be achieved by activating the persulfate chemical using sunlight. It is reported that iron catalyzed photoactivation of the oxidant could remove 90% of the pesticides like atrazine from wastewater, having 4 mg/L of atrazine. Other oxidants include magnesium dioxide by bisulfate activation, peroxymonosulfate, activation of persulfate by zerovalent iron etc. Photocatalytic activation of peroxymonosulfate in presence of titanium oxide attached with ferrite showed good efficiency of elimination 2,4-dichlorophenoxyacetic acid from wastewater, within half an hour with a removal rate of 97.2% [33].

8.4.2.4 Photochemical Degradation

UV light is employed along with other techniques for the elimination of pesticides from wastewater. An organic thiophosphate like triazophos can be photodegraded in presence of titanium oxide catalyst. In the method employed various degradation byproducts are generated during the treatment process. The photocatalyst in presence of UV radiation results into excitement of photogenerated electrons from the valence band to the conduction band. The photogenerated holes oxidize the pesticide directly to simple products. Similar results were obtained from carbofuran present in water with concentration varying between 50 and 250 mg/L. The complete mineralization process was confirmed by the HPLC technique.

In a study zinc oxide and tin oxide nanocomposite, as photocatalyst, could eliminate triclopyr herbicide. The heterojunction formation between these oxides resulted in a decreased recombination of charge carriers, which increased the photodegradation activity. Investigations have successfully demonstrated complete degradation for acephate pesticide using nanocomposite derived from rice husk based silicon oxide gel and a nanocomposite, from an extract of peach leaves, as reducing agent. The complete degradation involved the formation of nitrate (NO_3^-), sulfate (SO_4^{2-}) and phosphate (PO_4^{3-}), as intermediate species at an irradiation of 70 minutes. Dicofol was degraded to about 85% using recyclable Molybdenum disulfide/ZnS/N–S doped carbon nanocomposite catalyst via hydrothermal and post calcinations processes, under sunlight. Photocatalytic degradation is not a suitable approach for all types of pesticide contaminants and demands extensive analysis before its application on a large scale. Also, the process may lead to agglomeration of nanoparticles leading to a difficult separation. High band gap energy also comes as a disadvantage of the process.

8.4.2.5 Ozonation

Ozonation has shown promising results for removal of pesticides from wastewater. A number of pesticides like dimethoate, diuron, vinclozolin, parathion methyl, atrazine, etc., were removed using this method. The results improved when ozonation was combined with activated carbon. A pilot scale reactor was designed to eliminate five pesticides in wastewater, which included atrazine, alachlor, diuron, isoproturon, and chlorfenvinphos. All the pesticides could be completely removed in less than four and half hour, while total organic carbon showed only 26% reduction after 17 hours of exposure to ozone. Ozone produces free radicals in wastewater which are highly reactive, parallel to having disadvantage of onsite production and increased cost of treatment. For the above-mentioned pesticides, ozonation could effectively serves as a primary treatment. In addition to this photodegradation or oxidation processes could increase the overall performance. Some researchers have successfully used ozonation solitarily for complete water clearing up. However, ozonation in combination with other methods could increase the removal of pesticides like fluroxypyr to 92% [34].

8.4.2.6 Fenton Process

Fenton reagent is a combination of hydrogen peroxide and a catalyst, usually ferrous sulfate. Redox reactions occur during the treatment, resulting in the elimination of complex pesticides from wastewater. The technology has shown efficiency in the complete removal of organophosphate pesticide like fenitrothion from wastewater, including the removal of organochlorine pesticides. Photo-Fenton method has been used to successfully eliminate pesticide lindane and reduce the COD from wastewater. The reaction involves liberation of chlorine during the photooxidation process. A novel method employing heterogeneous electro-Fenton catalyst showed high removal rates then other Fenton methods. Photo-assisted Fenton process offers an economic advantage, due to lower energy demand, lesser byproduct formation and low cost input. The optimum pH for substantial efficiency lies between 2 and 4. It leads to formation of iron containing sludge which requires further treatment. In order to make electro-Fenton process more economical, the electrode forming cathode is made up of carbonaceous substances which also enhances the efficiency, durability, and surface area for interaction [15].

8.4.2.7 Electrocoagulation

Electrocoagulation or advanced oxidative electrochemical method is employed to eradicate various pesticides present in the agricultural runoff. However, the method is selective for removal of few pesticides. It is a hazard free method of eliminating the pesticide contaminants. Anodic oxidation and electro-Fenton processes are such technologies. In anodic oxidation method water undergoes electrochemical oxidation to form hydroxyl radicals which are absorbed on the anodic surface and become susceptible to oxidation. The pesticide molecules on reaching the surface get oxidized by the radical species and are eliminated.

Table 8.3 summarizes the pros and cons of the various methods adopted as chemical methods for treatment of pesticide wastewater.

Table 8.3 Pros and cons of various wastewater treatment technologies.

S. No.	Treatment method	Pros	Cons
1.	Advanced oxidation process	• Effective for industrial real water • Easy degradation possible • Mineralization of pesticides occurs • Methodology and principles established	• Leads to the formation of sludge as waste • pH dependent • Favorable only in acidic conditions • Prior neutralization is demanded before oxidation
2.	Chlorination	• Cost effective • Simple to operate • Easy to maintain • Easy availability of reagents	• Associated with pungent odor • Taste of clarified water alters • Leads to formation of toxic byproducts • Involves post-treatment methods also
3.	Photochemical degradation	• Involves no sludge production • Rapid method • One of the eco-friendly methods for degradation • Mechanism and technological soundness	• By-product formation occurs • High energy requirement • Cost-intensive process • Choice and utilization of the correct wavelength is a prior requirement
4.	Adsorption	• Simple technology • Easy to execute • Cost-effective • Good efficiency • Wide availability • Regeneration and reusability of adsorbents are possible	• Destruction of harmful chemicals does not occur • Contaminants accumulate on the surface of adsorbents • Regeneration is not very efficient
5.	Membrane filtration	• Compact setup • Highly efficient	• High investment and operational cost • Membrane fouling problems

(Continued)

Table 8.3 (Continued)

S. No.	Treatment method	Pros	Cons
6.	Biological treatment	• Cost-effective method • Efficient elimination • Eco-friendly • Complete mineralization of pesticides to CO_2 and H_2O • No intermediates, toxic, or sludge formation	• The biomaterials are subjected to decomposition and degradation • Depends on microbial activity • Factors influencing microbial activity, like temperature, substrate, and time of contact, influence the process

Source: Adapted from [35].

8.4.3 Biological Treatment Methods

Many researchers have worked on applying microbial populations for the eradication of pesticides and have shown remarkable removal of these chemicals, but the effectiveness has been variable, as the removal is dependent upon the microbial activity.

A biological purification method has been designed to eliminate pesticides through microbial activity. It is usually carried out by using a humic acid-rich substance capable of providing good retention to the pesticide, soil, and microbial species. The microorganisms existing in the biological mixture can be either endogenous bacteria or exogenous bacteria. The treatment method can be either aerobic or anaerobic treatment process. Aerobic treatment is usually found effective for dechlorinated pesticides. The biological method usually involves an oxidation process, the rupture of the ether bonds, and the hydroxylation of the chlorine-containing pesticides. In most cases, a conversion from chlorophenol to chlorocatechol occurs [36]. The ring rupture facilitates easy metabolism by the bacteria, with water and carbon dioxide as end products. A pretreatment method is always favored before exposing the wastewater to the biological process. Aerobic biological treatment can serve as an efficient secondary treatment process if it is conducted after primary photocatalytic degradation of pesticides.

8.4.3.1 Pesticide Degradation Using Ligninolytic Fungi

As pesticides are integrated into lignin-based formulations, a biomixture targeting pesticides can be feasible using ligninolytic fungi. For example, *Trametes versicolor,* a polypore mushroom, can be used for bioaugmentation and treatment of several fungicides [37].

8.4.3.2 Pesticide Degradation Using Bioaugmented Activated Sludge

The bioaugmentation of sludge can improve the removal efficiency of pesticides to an appreciable amount. For instance, 2,4-dichlorophenol degrading bacteria augmented activated sludge and showed 88–89% sequestration efficiency, at an initial concentration of 10–100 mg/L. The removal percentage of oxyfluorfen herbicide obtained was up to 80%, after more than 2 days of exposure. Similarly, glyphosate was completely removed in 4, 13, and 18 hours with concentrations of 0.1, 0.5, and 1 g/L, respectively, from wastewater. However, when the concentration of herbicide increased, the removal effectiveness decreased due to inhibition of bacterial cell growth in the sludge [38].

8.4.3.3 Pesticide Degradation Using Microbial Consortium

An active consortium from *Pseudomonas, Flavobacterium*, and *Vibrio strains* was capable of digesting hexachlorocyclohexane, an insecticide, in liquid culture. Similarly, 97.9% of atrazine was removed using a bacterial consortium in a submerged biological aerated filter [39]. Others include mineralization of phenylurea herbicide using *Diaphorobacter* species and *Achromobacter* species. The consortium also degraded the pesticide byproducts. The two-member consortium of *Arthrobacter globiformis* and *Variovorax* remediated diuron and linuron from water.

8.4.3.4 Pressurized Activated Sludge for Pesticide Elimination

This biological method is highly recommended and is a recognized treatment technique. The installation of the system does not demand a large land area, is less cost-intensive, and more eco-friendly than other conventional treatment methods. However, the requirement of skilled supervision and proper sludge disposal is a barrier to its application on a large scale. The disadvantages of the activated sludge method can be taken care of by pressurized activated sludge, as it involves a greater amount of oxygen due to an enhancement in the total operative pressure. This results in a greater COD removal than the activated sludge. About 92.5% of COD removal can be achieved by subjecting the wastewater to 0.2 MPa pressure for 360 minutes of exposure to air [40].

8.4.3.5 Combined Aerobic and Anaerobic Biological Treatment

The efficacy of bioremediation was investigated for the purification of wastewater by removal of triadimenol fungicide. When the contaminant containing wastewater was subjected to aerobic conditions at 22 °C for 172 days, 96% of triadimenol removal took place. On the other hand, the same pesticide-loaded water required a longer retention time when operated under anaerobic conditions. Thus, alternation of the two processes can increase the efficiency of the treatment by decreasing the wastewater holding time. Simulation statistics inferred that 12 hours of anaerobic treatment followed by one day aerobic handling gives a better removal percentage for triadimenol in wastewater [41].

8.4.3.6 Membrane Bioreactor for Pesticide Removal

This is a combination of membrane filtration technique (micro or ultrafiltration membranes) and biological treatment. The membrane bioreactor showed a 99% removal rate for one-week retention time for abroad spectrum herbicide named ametryn. In this, the mixed liquor was filtered through a membrane with a pore size less than 0.01 μm. The method demonstrated efficient removal and successfully maintained the quality of permeate with varied COD and biomass content.

An investigation was carried out to evaluate the potential of the membrane bioreactor for elimination of sixteen pesticides including triazines, organochlorines and linuron from a synthetic urban residual water. The operational conditions maintained were an average flow of 5.44 L/m^2/hour and a retention time of 1–2 months. The results reflected a relatively lower degree of removal for triazines and linuron (73% and 82%), while organochlorines were eliminated to a greater extent (99%). Membrane bioreactors are quite expensive and are not capable of completely eliminating all classes of pesticides.

Biological treatment of pesticides is often difficult if the wastewater is associated with antibiotics. Such a situation is prevalent if agricultural fields are sprayed with antibiotics to avoid bacterial spoilage of crops. In such conditions, antibiotics can limit the activity of the microbes in the biomixture. Thus, the choice of biological method for pesticide remediation from wastewater is influenced by the presence of antimicrobial agents. Also, it is required to couple the method with other treatment process for better removal efficiencies.

A summary of explored methods along with the results obtained is portrayed in Table 8.4.

The bacterial activity of metabolizing pesticides like isoxathion, alachlor, and monocrotophos has been well documented, but has been successful only for low concentrations.

Though several treatments have been successfully utilized for the elimination of pesticides, simple, cost-effective, reliable, and efficient wastewater treatment methods still need to be explored. Adsorption, being easy to handle and operate, is often seen as the choice of the research community in addition to being an economically affordable technology. But the disposal of pesticide-loaded waste needs urgent attention. Intense investigations featuring the reusability of the adsorbents are recommended for the technique to actually become economically viable and suitable for a cleaner environment.

The utilization of composites and several combined technological applications is also to be seen as a future wastewater treatment technology for the treatment

Table 8.4 Biological techniques for the treatment of pesticide-contaminated wastewater.

Technique	Type of pesticide	Pesticide	Initial concentration	Removal (%)	Other factors
Bioaugmentation	Fungicide	Tebuconazole	0.1–1 g/L	No significant result	Ligninolytic fungi as an agent
		Carbendazim			
		Metalaxyl			
		Edifenphos			
		Epoxiconazole			
	Herbicide	2,4-dichlorophenol	10–100 mg/L	88–89	Activated sludge bacteria
		Fluoxil	85–500 mg/L	80	
		Glyphosate	0.1–1 g/L	100	
	Insecticide	Hexachlorocyclohexane	0.01–10 mg/L	97.9	*Pseudomonas*, *Flavobacterium*, *vibrio* strains, microbial consortium
		Atrazine			
	Polyurea herbicide	Linuron	4–56.4 μM	88–99.9	*Diaphorobacter*, *Achromobacter*, a microbial consortium
		Diuron			
		Chlortoluron			
		Siduron			
		Fluometuron			
Pressurized biochemical process	Insecticide	Omethoate	2500–5000 mg/L	85–92	–
		Fenvalerate		92.5	6 hours aeration

(Continued)

Table 8.4 (Continued)

Technique	Type of pesticide	Pesticide	Initial concentration	Removal (%)	Other factors
Aerobic treatment	Fungicide	Triadimenol	–	96	22 °C, 172 days
Anaerobic treatment				30 °C	30 °C, 230 days
Combined treatment				100	12 h aerobic and 2 h anaerobic
Anoxic hybrid membrane technology and biological treatment	Herbicide	Ametryn	1–4 mg/L	83–99	1.5–7.5 days, 15 h
Anaerobic hybrid membrane technology and biological treatment		Atrazine	1–4 mg/L	6.8	–
		Linuron	1–4 mg/L	10.5	–
Composting	Insecticide	Carbaryl	–	100	–
		Parathion		100	–
		Endosulfan		96.3	–
		Aldrin		–	–
Phytoremediation	Herbicide	Atrazine	–	65	*Lemna minor*
		Dimethomorph		2.5–50	
		Carbofuran		–	Rhizophore fungi
		Trichloroethylene		95	Hybrid polar tree
		Simazine		77.1	Constructed wetlands
		Metolachlor		82.4	

Source: Adapted from [42].

of pesticide wastewater, which will prove to be more efficient and feasible, with easy application.

8.4.4 Hybrid Technologies

Due to the varied chemical characteristics of pesticides, it is difficult to eliminate them using conventional wastewater treatment methods. If the generation of pesticides in wastewater is controlled or is treated as soon as generated, the volume of water to be directed for treatment decreases. Also, instead of mixing wastewaters from various industrial and agricultural sources, the installation of specialized treatment units in every sector discharge point will allow convenient selection of the appropriate treatment method. Some of the successful attempts in combination with the treatment technologies for pesticide removal are summarized in Table 8.5.

Table 8.5 Performance of combined technologies for the removal of pesticides from wastewater.

Combination of technologies	Key observations
Activated carbon + ultrafiltration	84% removal of pesticides
Dissolved air flotation and powdered activated carbon + coagulation	A high level of decrease in the pesticide metaldehyde concentration
Magnetic activated carbon + powdered activated carbon + ultrafiltration	Efficient removal of paraquat and linuron
Granular activated carbon + ultrafiltration	100% retention of 2,4-dichlorophenoxyacetic acid (herbicide) was achieved in a granular activated carbon and ultrafiltration hybrid system. The results showed a high retention percentage but less flux.
Magnetic activated carbon + ultrafiltration	Magnetic activated carbon was considered a suitable pretreatment process in the ultrafiltration systems because of its good adsorption performance, inexpensiveness, and effortless elimination from the liquid medium.
Photo-Fenton + aerobic biodegradation	100% pesticide removal under examination except for pyrimethanil. Also, 98% COD and BOD removal
Chemical precipitation + activated sludge	Effective removal of pesticides, namely chloridazon, tebufenozide, and irgarol

(Continued)

Table 8.5 (Continued)

Combination of technologies	Key observations
Consortia of bacterial species + protozoa	Additional UV radiation in the presence of titanium oxide catalyst enhanced the efficiency of biodegradation of 2,4-dichlorophenoxyacetic acid
Activated sludge + filtration	93% elimination of triadimefon while less than 50% removal of cyproconazole, penconazole, tebuconazole, and pyrimethanil.
Mechanical treatment + activated sludge treatment sand filtration	Removal of fungicides like, tebuconazole, penconazole, triadimefon, cyproconazole, and pyrimethanil from municipal wastewater was investigated. The removal rate of 93% was obtained for triadimefon in comparison to other fungicides tested.
Fenton reagent + UV method	The treatment was optimized at pH 5, ferrous ion (Fe^{2+}) concentration as 0.8 mmol/L, and hydrogen peroxide concentration 100 mmol/L. A decline in the COD value was reported as 79.8%. dimethoate, triazophos, and malathion were completely eliminated.
Membrane bioreactor + granular activated carbon + UV oxidation	The solitary method led to 30% and 70% removal of fenoprop herbicide by the membrane bioreactor technique and granular activated carbon, respectively. In the case of pentachlorophenol, 65% was eliminated by the membrane bioreactor, and the remaining was removed by granular activated carbon. A combination of membrane bioreactor technique and UV oxidation showed an elevated removal rate for fenoprop (90%) and pentachlorophenol (100%).

Source: [15]/with permission of Elsevier.

Thus, it is clear that the key to fabricating and implementing the best wastewater treatment process is to compare the physical and chemical properties of the pesticides with their environmental impact, solubility, and reactivity with the free radicals or specific chemicals or radiation used for treatment. Also, the nature of intermediate degradation products is very important to be recognized in order to decide the appropriate technique or hybrid methods to be used for remediation. Full-scale treatment of wastewater demands an understanding of the correlation between the contaminant characteristics, such as structure, reactivity, solubility, molecular weight, and degradability. This will be crucial in deciding the best technique for optimum pesticide elimination, with cost-effectiveness and efficiency.

Currently, optimization of the technique for 100% pesticide remediation demands extensive scientific investigations and analysis. The selection of an appropriate technique for a full-scale wastewater system is yet to be explored and analyzed for sustainable wastewater treatment. Recent recommendations suggest that an onsite treatment will not lead to an increase in the concentration of contaminants and thus, handling, transport, and treatment costs will be lowered.

The emphasis on the concept of zero liquid discharge, aiming at the reusability and recyclability onsite, of actual wastewater, along with pesticides remediation, demands public awareness, government initiatives, along acceptance by industrialists, agriculturists, and environmentalists. A decentralization of the wastewater system would ensure better pesticide remediation.

8.5 Legislative Framework of Pesticides

To minimize the contamination of water resources national government authorities introduced pesticide residue limit guidelines. Regulatory limits comprise of particulars of the type of wastewater, residue information, analytical method for detection of residue, and limits derived from applying a safety factor [43].

The guidelines aim at setting up standards of water quality for human consumption. In the United States, the maximum contaminant level (MCL) is taken as the maximum permissible limit of the pesticide in water. Fact sheets issued by the Canadian government explicitly define the residue limits to which the concentrations are acceptable. As per the World Health Organization (WHO), nine pesticides were evaluated for their toxicity, and guidelines were set for drinking water. The assumption was the consumption of 2 L/day water by an average adult of 70 kg weight and 10% of the acceptable daily intake in drinking water. The document suggests that a 0.3 μg/L concentration in potable water may induce the risk of malignancy and thus should not be present in ground or surface water, as this may lead to contamination of the drinking water. Other pesticides examined with their permissible limits are bentazone (25 μg/L), 2-methyl-4-chlorophenoxyacetic acid (MCPA) (0.5 μg/L), metolachlor (5 μg/L), pendimethalin (17 μg/L), propanil (175 μg/L), pyridat (60 μg/L), simazine (17 μg/L), and trifluralin (170 μg/L).

The guidelines for the set standards differ for different regions, due to the influence of local geographical features, dietary practices, social influences, and economic and industrial conditions. The guidelines consider tolerable daily intake (TDI) and acceptable daily intake (ADI) in concurrence with pesticides showing threshold toxicity effects. In this, a 60 kg adult weight with a daily consumption of 2 L was considered to calculate the guideline value for pesticides with bioaccumulation ability. One percent total daily intake was allocated to drinking water, while for others default of 10% total daily intake was taken. Table 8.6 mentions the guideline values of a few pesticides in drinking water, as per the World Health Organization.

In Australia, the National Health and Medical Research Council sets the Australian drinking water guidelines (Table 8.7). The guidelines are set under two types, viz. Guideline values and Health values. The principles are formulated to comply with the needs of consumers and application at the consumption site. The guidelines do not apply to packaged water.

Table 8.6 World Health Organization guidelines for pesticides in drinking water.

Pesticides	Guideline value µg/L	Pesticides	Guideline value µg/L
Aldicarb	10	Lindane	2
Atrazine	2	MCPA	2
Carbofuran	7	Mecoprop	10
Chlordane	0.2	Methoxychlor	20
DDT	2	Molinate	6
Chlortoluron	30	Permethrin	20
Diquat	10	Simazine	2
Fenoprop	9	Terbuthylazine	7

Table 8.7 Australian drinking water guidelines.

Pesticides	Health value µg/L	Guideline value µg/L	Pesticides	Health value µg/L	Guideline value µg/L
Aldicarb	1	1	Lindane	20	0.05
Atrazine	20	0.5	MCPA	–	–
Carbofuran	10	5	Mecoprop	–	–
Chlordane	1	0.01	Methoxychlor	300	0.2
DDT	20	0.06	Molinate	5	0.5
Chlortoluron	–	–	Permethrin	100	1
Diquat	5	0.5	Simazine	20	0.5
Fenoprop	10	–	Terbuthylazine	–	–

The Environmental Protection Agency (EPA) ambient water standard criteria are set to safeguard aquatic organisms in rivers, streams reservoirs. Criterion maximum concentration (CMC) for fresh water is the highest concentration of the pollutant that the fresh water organisms can be exposed to, for one hour, with no harmful effects on growth or survival. The EPA guideline criteria for aquatic organisms for freshwater sources are given in Table 8.8.

The New Zealand Ministry of Health considers maximum acceptable values (MAV) for pesticides in drinking water (Table 8.9). Most of these values are aligned with the standards laid by the World Health Organization.

Table 8.8 Environmental Protection Agency (EPA) ambient water standard for freshwater.

Pesticides	Acute value μg/L	Chronic value πg/L
Chlordane	2.4	0.0043
Endosulfan	0.22	0.0056
Endrin	0.19	0.061
Lindane	2	0.08
Malathion	–	0.1
Mirex	–	0.001
Parathion	0.065	0.013
Camphechlor	0.73	0.0002

Table 8.9 New Zealand drinking water guidelines.

Pesticides	Maximum acceptable values μg/L	Pesticides	Maximum acceptable values μg/L
Aldicarb	10	Lindane	2
Atrazine	2 (provisional)	MCPA	2
Carbofuran	8	Mecoprop	10
Chlordane	0.2	Methoxychlor	20
DDT	2	Molinate	7
Chlortoluron	40	Permethrin	20
Diquat	10	Simazine	2 (provisional)
Fenoprop	10	Terbuthylazine	8

The environmental quality standard and effluent standards are set by the Japanese government for target levels and permissible limits, respectively, for the protection of human health and the environment. Effluent permissible limits for phosphorus pesticides are set as 1000 μg/L. The drinking water standards have been set by considering a 2 L/day consumption of water by a 50 kg person. The drinking water standard for 1,3-dichloropropene is 20 μg/L, for simazine is 30 μg/L, thiobencarb is 200 μg/L, and thiram is 6 μg/L. The agency has set limits for rice field discharge water with a 10-fold dilution factor and a drinking water limit.

Taiwan effluent standards from industries and municipal treatment facilities have set maximum effluent limits as 30, 1000, 500, and 500 μg/L for endosulfan, herbicides such as butachlor, paraquat, total aminomethylcarbamate, and total organophosphorus compounds, respectively.

Table 8.10 portrays Canadian and interim guidelines for pesticide residues in irrigation runoff water, livestock wastewater, and fresh water. The Canadian government has circulated three conventions for the derivation of water quality guidelines [43].

A comprehensive legislation framework was established in the European Union as Regulation (EC) No. 396/2005. Later, it approved the water reuse regulation (EU) 2020/741 to establish monitoring requirements for water reuse. According to the Water Framework Directives for groundwater water a quality standard of 0.1 μg/L is set for pesticides in groundwater.

Table 8.10 Canadian wastewater guidelines.

Pesticides	Irrigation wastewater μg/L	Livestock wastewater μg/L	Fresh water μg/L
Aldicarb	54.9	11	1 (interim)
Atrazine	10 (interim)	5 (interim)	108
Carbofuran	–	45	108
Chlorpyrifos	–	24 (interim)	0.0035
Lindane	–	4	0.01
MCPA	0.025	25 (interim)	2.6 (interim)
Metochlor	28 (interim)	50 (interim)	2.6 (interim)
Simazine	0.5 (interim)	10 (interim)	10
Triallate	–	230	0.24
Dicamba	0.006	122	10

Table 8.11 EPA guidelines for contaminants.

Pesticides	MCL µg/L	MCLG µg/L	Pesticides	MCL µg/L	MCLG µg/L
Aldicarb	7	7	Lindane	–	–
Atrazine	3	3	MCPA	40	40
Carbofuran	40	40	Mecoprop	–	–
Chlordane	2	0	Methoxychlor	–	–
DDT	20	20	Molinate	4	4
Chlortoluron	50	50	Permethrin	–	–
Diquat	0.2	0.2	Simazine	–	–
Fenoprop	–	–	Terbuthylazine	–	–

The United States Environmental Protection Agency (EPA) disseminated initial effluent guidelines (Table 8.11) and standards for pesticides and chemicals in 1978 and revised them in 2007. The Code of Federal Regulations (CFR), Title 40, Chapter 1, part 455, sets effluent limits for pesticides. These guidelines are designed to protect living systems and their environment. Under the National Primary Drinking Water Regulation (NPDWR), a non-enforceable health goal is set by the EPA, known as the Maximum Contaminant Level Goal (MCLG), for the reduction is the level of certain chemicals to zero. The maximum contaminant level (MCL) is set as 4.0 (ppt) for certain harmful chemicals such as perfluorooctane sulfonic acid (PFOS) and perfluorooctanoic acid (PFOA). Such conditions and contaminant levels are being set for other chemicals, too, considering their hazardous traits.

In India, the first noticeable law was enforced in 1970 for the protection of environmental resources by setting up of National Committee on Environmental Planning and Coordination (NCEPC). The Water (Prevention and Control of Pollution) Cess Act, 1977, and the Environmental Protection Act of 1986 were implemented. The Environmental Protection Act is a broad legislation providing insights for water protection (Table 8.12).

8.6 Case Studies

8.6.1 Removal of Pentachlorophenol from Wastewater

A case study on the pesticide pentachlorophenol was conducted for its removal from wastewater, using aquatic plants. According to the EPA, pentachlorophenol,

Table 8.12 Indian water standard guidelines (BIS standard).

Pesticides	Maximum residue limit as per BIS 10500 standards μg/L
Atrazine	2
alpha-HCH	0.01
beta-HCH	0.04
delta-HCH	0.04
Chlorpyrifos	30
DDT	1
Endosulfan	0.4
Malathion	190
Lindane	2
Methyl parathion	0.3
Aldrin/Dieldrin	0.03

which is employed as a fungicide, insecticide, and bactericide, is considered one of the priority pollutants. The pesticide pentachlorophenol is chemically stable and cannot be degraded due to its stable aromatic ring structure. Pentachlorophenol can come in contact with humans through inhalation, dermal absorption, and consumption of water and food infected with the pesticide. These can cause toxic impacts. Due to the stretched half-life period of the pesticide, it accumulates in the environment as it is lipophilic. Pentachlorophenol toxicity includes a fatal impact on the liver, immune system, and reproductive system. It is also associated with carcinogenic effects. Pentachlorophenol is also a degradation product of a popularly used pesticide called lindane and polychlorinated phenolic substances. The International Agency for research has labeled this as a possible class 2B human carcinogen.

Pentachlorophenol is usually treated using biological and physicochemical approaches. Due to the expenses associated with physical and chemical processes, the biological approach can be practiced due to its cost-effectiveness and eco-friendly nature. The biological treatment usually carried out for pentachlorophenol removal includes phytoremediation, along with biostimulation and bioaugmentation. Phytoremediation uses plants to decrease the concentration of contaminants and to degrade the pesticides. Under this, Typhus species, which are monocotyledonous macrophytes, have been used successfully under various climatic conditions for the protection of water bodies. Another plant called duckweed is suitable for shallow water contaminants. Duckweed is reported as an excellent accumulator of xenobiotic pesticides. Microalgae-based systems are also extensively adopted to treat the pentachlorophenol-loaded wastewaters. Photobioreactor-based microalgae could eliminate 95% of doxycycline from wastewater. This system uses organic matter and nutrients from water bodies for its growth and development. They are additionally having the benefit of acting as biofuel, and feed for farm animals as they all have good degrees of protein, starches, and cellulose.

Phytoremediation is an eco-friendly, sustainable method. It is based on natural energy input, thus, low equipment cost and clean operation occur. Also, the process can remove a variety of contaminants from wastewater indiscriminately. The growth of the plants is additionally linked to the advantage of taking up nutrients from wastewater, which can turn up into a number of high-value nutrients to achieve the resource recovery objective. The biomass employed is fast-growing and has a higher potential than any other method. The microbial population surviving in the leaves, roots and stems of the plants used can heighten the pollutant extraction from the wastewater. The removal of pentachlorophenol using phytoremediation suffers from the limitations of environmental pollution due to chelation, slow, and seasonally variable phytoremediation. Moreover, the handling and disposal of the contaminated biomass is the crucial footprint of this technique. It is required to address the issues related to the technique as it could act as a potential way of removing Pentachlorophenol using aquatic macrophytes and weeds with low economic burden and environmental disturbances [44].

8.6.2 Pesticide Removal from Secondary Wastewater of Waste Treatment Plant, Tunisia

Researchers investigated the elimination of pesticide pentachlorophenol from secondary wastewater using the competence of bacterial strain *Pseudomonas putida*. The findings highlighted the potential of *Pseudomonas* bacteria in removing harmful pesticides. The wastewater was sampled from the El-Hamma-Gabes wastewater treatment plant, Tunisia, and was investigated for its physical and chemical properties. The bacterial strain was isolated from an industrial wastewater and examined by the gene sequencing method.

The selection of *Pseudomonas putida* was done by checking its resistance and the ability to remove the pesticides. The secondary treated wastewater had a pH of 8.7, a COD value of 632 mg/L, and a BOD as 380 mg/L. The pesticide removal percentage was 24.23% during the exponential phase, which was seen during 48 hours of contact. This supported the potential activity of the microorganisms. The pentachlorophenol removal increased after 48 hours and reached 75.33% and finally 92.36% after 168 hours of incubation. The results confirmed the potential of the bacterial strain to remove the pesticides from secondary wastewater.

Iron sulfate was added, which showed the formation of neutral iron-bacterial strain, *Pseudomonas putida* complex, contributing to the growth of bacteria and decreasing the pollution load. Also, the investigations confirmed the formation of bacterial biofilm, which enhanced their degradation ability. Further, research is to be focused on the bacterial weathering-based remediation process. The monitoring of the bioaugmentation process was executed using assessment of chloride content via HPLC, biofilm production and pyoverdine (a major siderophore). An appreciable removal rate was observed at a concentration of 100 mg/L of the pesticides through bioaugmentation using the strain. Pentachlorophenol was efficiently removed by this method, with an increasing concentration of iron from 5 to 20 mg/L. The chlorine level was found to be proportional to the iron concentration. At the end of the experiments chloride level decreased from 3.8 to 3 g/L in the secondary treated wastewater. The results were supported by greater degradability of the strain. The presence of iron enhanced chelation and ion transfer, which facilitated bacterial growth. pH of the media was also important in pentachlorophenol mineralization, and pH ranging from 6.5 to 8, increased biodegradation by elevating its solubility. The study revealed the production of pyoverdine by the *Pseudomonas putida* strain, as confirmed by the spectrophotometric technique. Biofilm systems are operated by various electrostatic and sorption interactions. Pentachlorophenol pesticide could be chemically de-chlorinated by iron, generated due to the reduction process of iron-reducing organisms. Siderophores were capable of binding with iron (Fe II) produced, which helped in the enhancement of the degradation process. Thus, iron added to pentachlorophenol loaded wastewater initiated iron–pentachlorophenol interactions, which produced pyoverdine siderophore [25]. Biological methods thus could be helpful in eradicating harmful contaminants from wastewater, if properly controlled and managed.

References

1 Hassaan, M.A., and El Nemr, A. (2020). Pesticides pollution: classifications, human health impact, extraction and treatment techniques. Egyptian Journal of Aquatic Research, Volume 46, Issue 3, pp. 207–220. https://doi.org/10.1016/j.ejar.2020.08.007.

2 Yadav, I.C., Devi, N.L., Syed, J.H., et al. (2015). Current status of persistent organic pesticides residues in air, water, and soil, and their possible effect on neighboring countries: a comprehensive review of India. Science of the Total Environment, Volume 511, pp. 123–137. https://doi.org/10.1016/j.scitotenv.2014.12.041.

3 Rajan, S., Parween, M., and Raju, N.J. (2023). Pesticides in the hydrogeo-environment: a review of contaminant prevalence, source and mobilization in

India. Environmental Geochemistry and Health, Volume 45, pp. 5481–5513. https://doi.org/10.1007/s10653-023-01608-6.

4 Lofrano, G., Libralato, G., Meric, S., et al. (2020). Occurrence and potential risks of emerging contaminants in water. In: Sacco O, Vaiano V (eds) Visible light active structured photocatalysts for the removal of emerging contaminants, Elsevier, pp. 1–25.

5 Ioannides, C. (1991). Public-health impact of pesticides used in agriculture-WHO. Journal of Royal Society of Health, Volume 111, Issue 5, p. 206. https://doi.org/10.1177/146642409111100516.

6 Warra, A.A., and Prasad, M.N.V. (2020). African perspective of chemical usage in agriculture and horticulture—their impact on human health and environment. In: Prasad MNV (ed) Agrochemicals detection, treatment and remediation, pesticides and chemical fertilizers, Butterworth-Heineman, pp. 401–436. https://doi.org/10.1016/B978-0-08-103017-2.00016-7.

7 Manasa, R.L., and Mehta, A. (2020). Wastewater: sources of pollutants and its remediation. In: Gothandam K, Ranjan S, Dasgupta N, et al. (eds) Environmental biotechnology, Volume 2, Springer International Publishing, Cham, pp. 197–219.

8 Ionel, I.L., Mara, G., Stefania, R., et al. (2019). A hazard to human health–pesticide residues in some vegetal and animal food stuff. Journal of Biotechnology, Volume 305, pp. S22–S23. https://doi.org/10.1016/j.jbiotec.2019.05.089.

9 Andrunik, M., and Bajda, T. (2021). Removal of pesticides from waters by adsorption: comparison between synthetic zeolites and mesoporous silica materials. a review. Materials (Basel), Volume 14, Issue 13, p. 3532. https://doi.org/10.3390/ma14133532.

10 Barron, M.G., Ashurova, Z.J., Kukaniev, M.A., et al. (2017). Residues of organochlorine pesticides in surface soil and raw foods from rural areas of the Republic of Tajikistan. Environmental Pollution, Volume 224, pp. 494–502. https://doi.org/10.1016/j.envpol.2017.02.031.

11 Thapinta, A., and Hudak, P.F. (2000). Pesticide use and residual occurrence in Thailand. Environmental Monitoring Assessment, Volume 60, pp. 103–114. https://doi.org/10.1023/A:1006156313253.

12 Eddleston, M. (2020). Poisoning by pesticides. Medicine, Volume 48, Issue 3, pp. 214–217. https://doi.org/10.1016/j.mpmed.2019.12.019.

13 Bertero, A., Chiari, M., Vitale, N., et al. (2020). Types of pesticides involved in domestic and wild animal poisoning in Italy. Science of the Total Environment, Volume 707, p. 136129. https://doi.org/10.1016/j.scitotenv.2019.136129.

14 Rajmohan, K.S., Chandrasekaran, R., and Varjani, S. (2020). A review on occurrence of pesticides in environment and current technologies for their remediation and management. Indian Journal of Microbiology, Volume 60, Issue 2, pp. 125–138. https://doi.org/10.1007/s12088-019-00841-x.

15 Saleh, I.A., Zouari, N., and Al-Ghouti, M.A. (2020). Removal of pesticides from water and wastewater: chemical, physical and biological treatment approaches. Environmental Technology and Innovation, Volume 19, p. 101026. https://doi.org/10.1016/j.eti.2020.101026.

16 Marican, A., and Durán-Lara, E.F. (2018). A review on pesticide removal through different processes. Environmental Science and Pollution Research, Volume 25, Issue 3, pp. 2051–2064. https://doi.org/10.1007/s11356-017-0796-2.

17 Goodwin, L., Carra, I., Campo-Moreno, P., et al. (2017). Treatment options for reclaiming wastewater produced by the pesticide industry. International Journal of Water and Wastewater Treatment, Volume 3. https://doi.org/10.16966/2381-5299.149.

18 Ahmed, M., Zhou, J., Ngo, H., et al. (2017). Progress in the biological and chemical treatment technologies for emerging contaminant removal from wastewater: a critical review. Journal of Hazardous Materials, Volume 323, pp. 274–298. https://doi.org/10.1016/j.jhazmat.2016.04.045.

19 Yu, J.J. (2002). Removal of organophosphate pesticides from wastewater by supercritical carbon dioxide extraction. Water Research, Volume 36, Issue 4, pp. 1095–101. https://doi.org/10.1016/s0043-1354(01)00293-7.

20 Massad, Y., Hanbali, G., Jodeh, S., et al. (2021). The efficiency of removal of organophosphorus malathion pesticide using functionalized multi-walled carbon nanotube: impact of dissolved organic matter (DOM). Separation Science and Technology, Volume 57, Issue 1, pp. 1–12. https://doi.org/10.1080/01496395.2021.1881118.

21 Wanjeri, V.W.O., Sheppard, C.J., Prinsloo, A.R.E., et al. (2018). Isotherm and kinetic investigations on the adsorption of organophosphorus pesticides on graphene oxide-based silica coated magnetic nanoparticles functionalized with 2-phenylethylamine. Journal of Environmental Chemical Engineering, Volume 6, Issue 1, pp. 1333–1346. https://doi.org/10.1016/j.jece.2018.01.064.

22 Aksas, H. (2024). Removal of organophosphorus pesticides from wastewater on waste cellulosic matrice; kinetic models approach. *Authorea*. https://doi.org/10.22541/au.170670486.60792737/v1.

23 Yang, Q., Wang, J., Chen, X., et al. (2008). The simultaneous detection and removal of organophosphorus pesticides by a novel Zr-MOF based smart adsorbent. Journal of Materials Chemistry A, Volume 6, pp. 2184–2192. https://doi.org/doi.org/10.1039/C7TA08399H.

24 Tian, H., Li, J.J., Zou, L.D., et al. (2009). Removal of DDT from aqueous solutions using mesoporous silica materials. Journal of Chemical Technology and Biotechnology, Volume 84, pp. 490–496. https://doi.org/10.1002/jctb.2067.

25 Ammeri, R.W., Kloula, S., Rauso Simeone, G.D., et al. (2023). Pesticide removal by the bioaugmentation process in secondary treated wastewater. Water Quality

Research Journal, Volume 58, Issue 2, pp. 111–127. https://doi.org/10.2166/wqrj.2023.001.

26 Vukcevic, M.M., Kalijadis, A.M., Vasiljevic, T.M., et al. (2015). Production of activated carbon derived from waste hemp (*Cannabis sativa*) fibers and its performance in pesticide adsorption. Microporous and Mesoporous Materials, Volume 214, pp. 156–165. https://doi.org/10.1016/j.micromeso.2015.05.012.

27 Cobas, M., Meijide, J., Sanromán, M.A., et al. (2016). Chestnut shells to mitigate pesticide contamination. Journal of the Taiwan Institute of Chemical Engineers, Volume 61, pp. 166–173. https://doi.org/10.1016/j.jtice.2015.11.026.

28 Shattar, S.F.A., Zakaria, N.A., and Foo, K.Y. (2017). Utilization of montmorillonite as a refining solution for the treatment of ametryn, a second generation of pesticide. Journal of Environmental Chemical Engineering, Volume 5, Issue 4, pp. 3235–3242. https://doi.org/10.1016/j.jece.2017.06.031.

29 Bajuk-Bogdanović, D., Jović, A., Vasiljević, B.N., et al. (2017). 12-Tungstophosphoric acid/BEA zeolite composites – characterization and application for pesticide removal. Materials Science and Engineering B, Volume 225, pp. 60–67. https://doi.org/10.1016/j.mseb.2017.08.011.

30 Chen, X., Zhou, Q., Liu, F., et al. (2019). Removal of nine pesticide residues from water and soil by biosorption coupled with degradation on biosorbent immobilized laccase. Chemosphere, Volume 233, pp. 49–56. https://doi.org/10.1016/j.chemosphere.2019.05.144.

31 Garba, Z.N., Abdullahi, A.K., Haruna, A., et al. (2021).Risk assessment and the adsorptive removal of some pesticides from synthetic wastewater: a review. Beni-Suef university Journal of Basic and Applied Sciences, Volume 10, p. 19. https://doi.org/10.1186/s43088-021-00109-8.

32 Fairbairn, D.J., Elliott, S.M., Kiesling, R.L., et al. (2018).Contaminants of emerging concern in urban stormwater: spatiotemporal patterns and removal by iron-enhanced sand filters (IESFs). Water Research, Volume 145, pp. 332–345.

33 Vela, N., Fenoll, J., Garrido, I., et al. (2019). Reclamation of agro-wastewater polluted with pesticide residues using sunlight activated persulfate for agricultural reuse. Science of the Total Environment, Volume 660, pp. 923–930.

34 Pérez-Lucas, G., Aliste, M., Vela, N., et al. (2020). Decline of fluroxypyr and triclopyr residues from pure, drinking and leaching water by photo-assisted peroxidation. Process Safety and Environmental Protection, Volume 137, pp. 358–365.

35 Zieliński, B., Miądlicki, P., and Przepiórski, J. (2022). Development of activated carbon for removal of pesticides from water: case study. Scientific Reports, Volume 12, p. 20869. https://doi.org/10.1038/s41598-022-25247-6.

36 Javaid, M.K., Ashiq, M., and Tahir, M. (2016). Biodegradation and environmental impacts of pesticides. Scientifica, Volume 2016, Article ID 1598325, p. 9. http://dx.doi.org/10.1155/2016/1598325.

37 Murillo-Zamora, S., Castro-Gutierrez, V., Masis-Mora, M., et al. (2017). Elimination of fungicides in biopurification systems: effect of fungal bioaugmentation on removal performance and microbial community structure. Chemosphere, Volume 186, pp. 625–634. https://doi.org/10.1016/j.chemosphere.2017.07.162.

38 Tazdait, D., Salah, R., Grib, H., et al. (2018). Kinetic study on biodegradation of glyphosate with unacclimated activated sludge. International Journal of Environmental Health Research, Volume 28, Issue 4, pp. 448–459. https://doi.org/10.1080/09603123.2018.1487043.

39 Baghapour, M.A., Nasseri, S., and Derakhshan, Z. (2013). Atrazine removal from aqueous solutions using submerged biological aerated filter. Journal Environmental Health Science Engineering, Volume 11, Issue 1, p. 6. https://doi.org/10.1186/2052-336X-11-6.

40 Pan, Z., Yu, S., Huang, H., et al. (2004). Treatment of high concentration pesticide wastewater by pressurized biochemical process. Asian Pacific Journal Chemical Engineering, p. 624.

41 Shawaqfeh, A.T. (2010). Removal of pesticides from water using anaerobic-aerobic biological treatment. Chinese Journal Chemical Engineering, Volume 18, Issue 4, pp. 672–680. https://doi.org/10.1016/S1004-9541(10)60274-1.

42 Al Hattab, M.T., and Ghaly, A.E. (2012). Disposal and treatment methods for pesticide containing wastewaters: critical review and comparative analysis. Journal of Environmental Protection, Volume 3, pp. 431–453. https://doi.org/10.4236/jep.2012.35054.

43 Hamilton, D., Ambrus, Á., Dieterle, R., et al. (2003). Regulatory limits for pesticide residues in water (IUPAC Technical Report). Pure and Applied Chemistry, Volume 75, Issue 8, pp. 1123–1155. https://doi.org/10.1351/pac200375081123.

44 Werheni, A.R., Souid, F., Hajjeji, F., et al. (2023). Potential of aquatic plants for pesticide removal in wastewater: a case study on pentachlorophenol. In: Shukla G, Bhat JA, Chakravarty S, et al. (eds) Floristic Diversity, Intech Open. https://doi.org/10.5772/intechopen.107064.

9

Radioactive Contaminants

Considering the potential risk that radioactive contaminants and radioactive wastes pose, the treatment of wastewater containing radioactive contaminants differs from the conventional treatment for industrial or domestic wastewater. Radioactive wastewater is generated from nuclear facilities, mainly where the radioisotopes are employed. The radioactivity level of the waste present in the water depends on the chemical composition and the operational aspects of the isotope. Radionuclides, when present in water, are highly mobile and can enter groundwater if not adequately taken care of. Figure 9.1 shows some important categories of radioactive nuclides.

9.1 Detection of Radioactive Contaminants

It becomes very essential to detect and monitor the amounts of radionuclides in the water samples, considering the hazardous effects associated with them. Several analytical methods are used for the detection of radioactive contaminants in wastewater. Figure 9.2 portrays the different detection techniques used for various categories of nuclides.

9.1.1 Gamma Ray Spectrometry

This method is a powerful tool for the analysis of emitters of gamma radiation [1, 2]. It is a nondestructive and quick method for the detection of radionuclides. This method is used for quantitative and qualitative measurement of radionuclides emitting gamma radiation, such as Cobalt $\left(^{60}Co\right)$, Iodine $\left(^{131}I\right)$, Cesium $\left(^{137}Cs\right)$, etc. A gamma ray spectrometer (GRS) (Figure 9.3) measures the intensity distribution of gamma rays, which are ultra-high frequency light radiations emitted by radioactive substances.

Hazards and Treatment of Wastewater Contaminants, First Edition. Dipika Jaspal and Arti Malviya. Published 2026 by John Wiley & Sons, Inc.

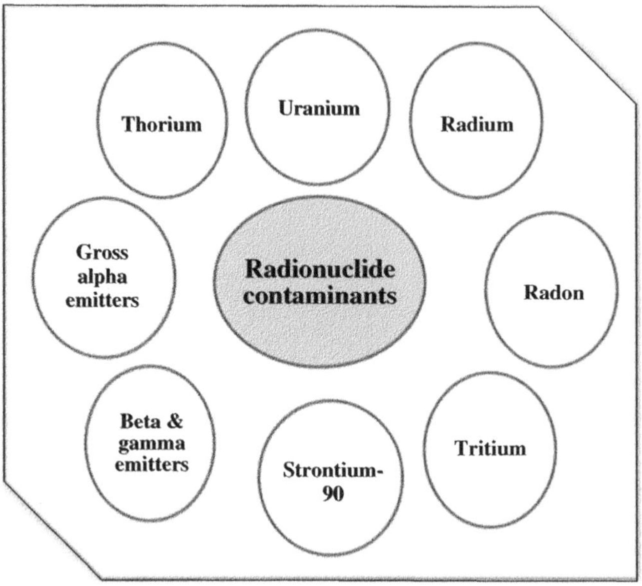

Figure 9.1 Important radionuclides as contaminants.

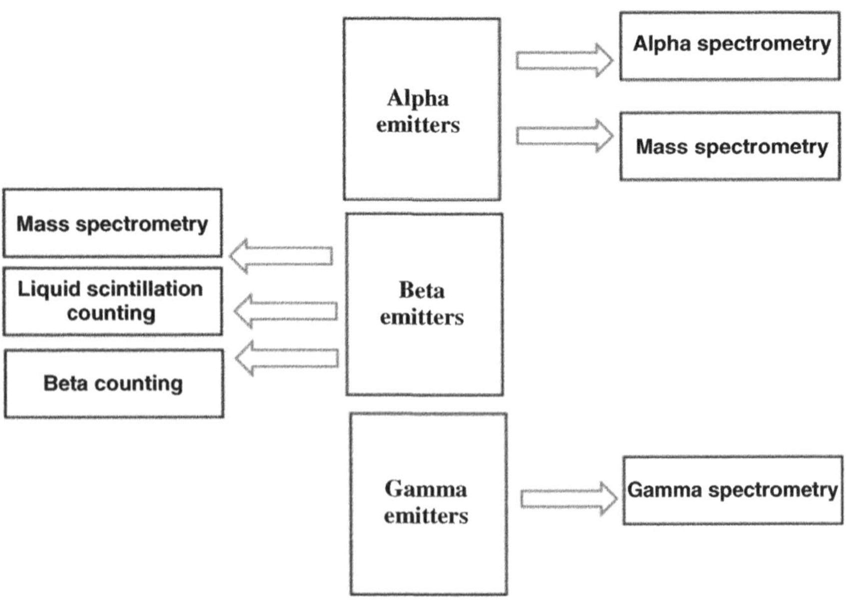

Figure 9.2 Detection methods for nuclides.

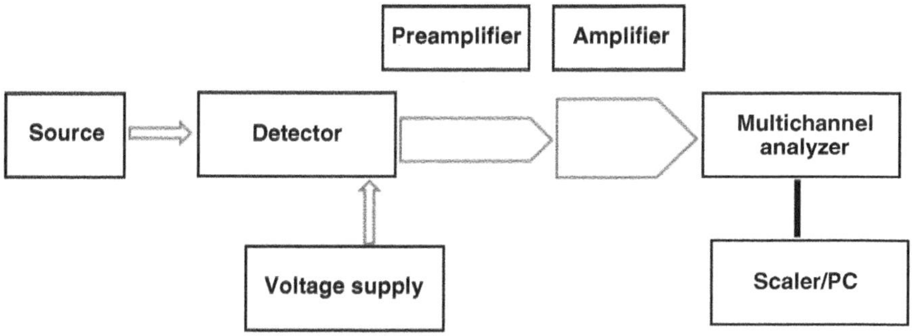

Figure 9.3 Block diagram of a gamma spectrometer.

In this method, the energy of incident gamma rays is measured with the use of a detector. By comparing the energy measured and the known energy of radioisotopes emitting radiation, identification of the emitting radionuclide can be done, after analysis of the spectrum.

In a study, such radionuclides were detected in sewer sludge [3]. New tools have been developed to provide an efficient means to detect the radionuclides [4], with spectral unmixing. Spectral unmixing covers the entire range of gamma radiations in comparison to the conventional peak-based methods.

9.1.2 Liquid Scintillation Counting (LSC)

This method quantifies the low-energy, generally beta or alpha-emitting isotopes. Specific cocktails are required that absorb energy into detectable light pulses. These should comprise of aromatic organic solvents (scintillators or fluors). The aromatic solvent molecules undergo excitation by the released energy of the radioactive decay. These solvent molecules can absorb the nuclear radiation and convert it into photons of light, which are converted into electrical pulses by photomultiplier tubes (Figure 9.4).

The method of liquid scintillation counting has been used for the successful detection of radionuclides such as tritium (^{3}H) [5]. This method has also been successfully employed for the identification of radiopharmaceutical waste [6] and in the case of radionuclides in biochemical, pharmaceutical, and environmental samples [7, 8].

9.1.3 Mass Spectrometry

Mass spectrometry (MS) comes as an efficient method for the detection of trace amounts of radionuclide samples in environmental samples with efficiency [9].

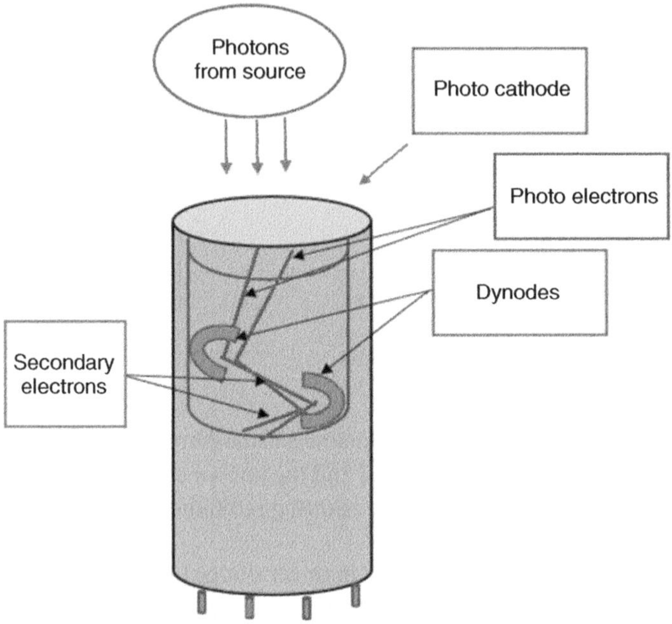

Figure 9.4 Photomultiplier tube.

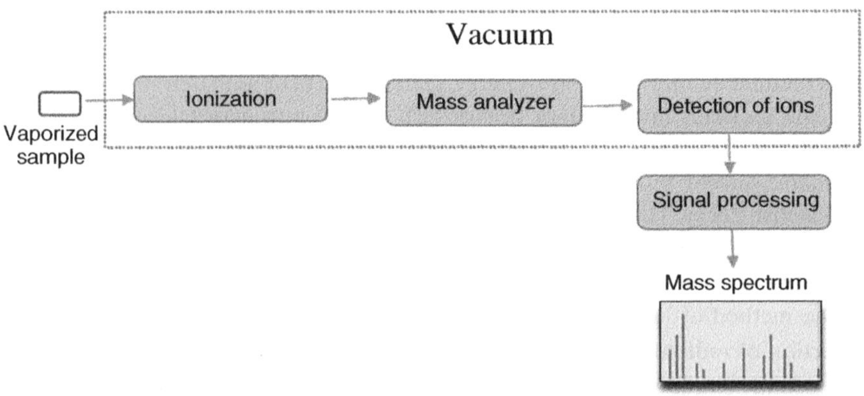

Figure 9.5 Schematic representation of a mass spectrometer.

Different ions moving in a magnetic field can be segregated based on their mass-to-charge ratio. Figure 9.5 gives a schematic representation of a mass spectrometer.

A coupling of mass spectrometry to liquid chromatography (LC) and gas chromatography (GC) can be done for the detection of organic contaminants in wastewater. This coupling enables the mass detection of many wastewater contaminants in very low concentrations, even below parts per trillion. LC-MS has further been enhanced in recent years with the incorporation of ultra-high-performance liquid chromatography (UPLC or UHPLC). This technique utilizes the same methodology as the conventional technology of high-pressure LC (HPLC).

The sample to be analyzed is exposed to high vapor pressure, leading to ionization and fragmentation of the sample. An acceleration of these samples takes place in the ionizer, due to which they undergo movement in the mass analyzer according to their mass. These ions then enter the detector, where the ions with similar velocity follow either a circular or perpendicular path. The ions possessing similar electric and magnetic charge pass undeflected and are recorded by the system [10].

Several categories of mass spectrometric techniques are being used, such as inductively coupled plasma mass spectrometry (ICP-MS), gas chromatography-mass spectrometry (GC-MS), and liquid chromatography-mass spectrometry (LC-MS).

9.1.4 Alpha Spectrometry

It consists of low-level semi-conducting detectors with a surface of 100–600 mm^2 connected to a vacuum chamber and a multichannel analyzer. The alpha spectrometers (Figure 9.6) are capable of measuring energy in the range of 3–8 MeV [11].

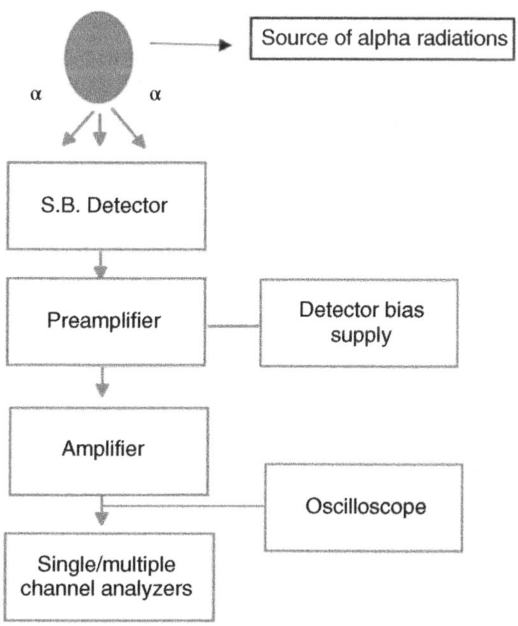

Figure 9.6 Schematic representation of an alpha spectrometer.

This method has been used to determine the presence of uranium isotopes [12] from wastewater. Isotopes such as ^{235}U, ^{234}U, and ^{238}U have been determined in wastewater using this technique. The detection of radionuclides from sewage wastewater and wastewater from nuclear industries has been possible using this technique. The extraction of uranium has been carried out using nitric acid, and exposure to this technique is used for its determination. Thorium isotopes such as ^{230}Th, ^{232}Th, and ^{228}Th [13] have also been detected with the use of this method. The samples to be investigated are first acidified and then concentrated. Pre-concentration of the thorium isotopes was carried out by co-precipitation using yttrium fluoride. The fluoride precipitate obtained was dissolved using nitric acid and converted into sulfates. Extraction of the thorium isotopes was then carried out using tri-octyl methyl ammonium nitrate (TOA)/xylene, and re-extraction was carried out using hydrochloric acid. Separation of iron and uranium has been carried out by anion exchange using hydrochlorated solution. Isotopes of thorium were extracted from the solution by electrodeposition. Activities of thorium isotopes are determined by alpha spectrometry.

Alpha spectrometry proves to be a powerful tool for the analysis of radioactive nuclides. This method provides excellent peak resolutions, identification, along quantification of radionuclides [14].

9.1.5 Beta Counting

In this type of decay, the radionuclide emits beta particles, thereby converting into isobars. A more stable ratio of protons and neutrons is obtained by this method.

The radionuclides in water, which are those naturally occurring in underground, surface, and seawater, can be determined by Gross beta counting. "Gross beta counting" encompasses all the radionuclides present in water. The World Health Organization (WHO) limit of 1 Bq/L is set in drinking water. The samples on a suitable filter are counted using a beta counter. The method is cost-effective and simple, but it is incapable of giving specific concentrations. Therefore, if the concentration of radioactivity exceeds the permissible limits, specific analysis is required. In radiochemistry, beta counting, the element under study is segregated from the other elements using methods such as precipitation, ion exchange, etc., and then counted with a beta counter. This gives the count of the beta activity of the segregated element [15].

9.2 Methods of Treatment

Before the radioactive wastewater is discharged, it is exposed to several treatment methods such as reduction of volume, radioactivity, and hazardous traits. The methods discussed below are the methods that are usually employed in the predischarge or predisposal phase. After the removal of hazardous contaminants from the waste, it is stored for some time before final discharge. This storage can be for short-term or long-term, up to years.

The removal methods for nuclides not only aim at the removal of radionuclides for their safe exclusion from the environment, but also aim at reducing the physiological burden, radioactivity. Many chemical, physical, or biological processes can be introduced for the treatment of nuclides in wastewater [16]. One of the methods involves the elimination of the pollutants without altering their chemical nature. In another method, chemical modification was carried out, and the pollutant was converted into certain harmless products. The dilution method has also been used in radioactive wastewater treatment, where only the concentration of the pollutant is reduced to bring contaminants to harmless levels. The basis of all the treatment processes is the conversion of radionuclide into smaller volumes of concentrates, followed by further treatment methods, to fulfill the wastewater discharge standards. Figure 9.7 shows the various methods that can be adopted for the treatment.

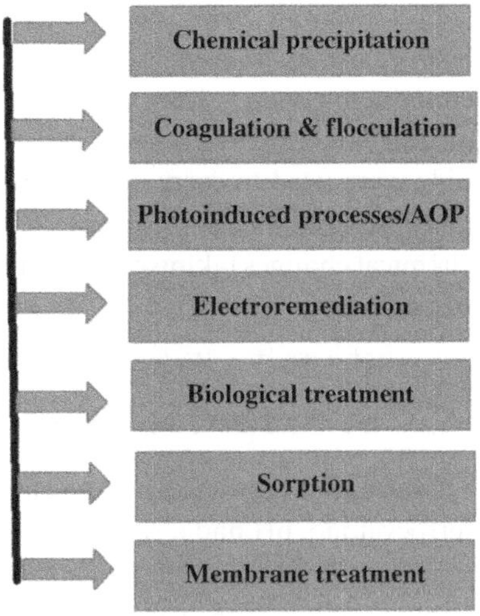

Figure 9.7 Methods to treat radioactive contaminants.

Several categories of treatments that are carried out for radioactive wastewater are discussed in the following sections.

9.2.1 Chemical Precipitation and Coagulation

In this method, the radionuclides are co-precipitated along with the precipitant [17, 18]. Chemicals such as soda, phosphates, iron salts, etc., are used. Factors such as the quantity of the precipitating agent, time, pH, and stirring speed affect the efficiency of the process. The enhancement of removal can be done with the use of additional substances such as coagulants and electrolytes [19]. Some nuclides, such as cesium (Cs), ruthenium (Ru), and iodine (I), etc., are not efficiently removed by this method and require integration of treatment technologies [20].

Coagulation and flocculation have shown promising results using polyacrylamide flocculent and ferric hydroxide for radionuclides such as antimony (Sb), cobalt (Co), ruthenium (Ru), and manganese (Mn) [21]. The possible drawback of this method has been the development of a large amount of sludge and the risk of secondary pollution.

In a study in 2023, nuclides such as ^{60}Co, Strontium (^{90}Sr), and ^{137}Cs have been removed from wastewater generated by chemically treating radionuclide-containing concrete [22]. Chemical precipitation was carried out for the removal of nuclides. Chemicals such as barium sulfate ($BaSO_4$) and potassium ferrocyanide were used to carry out precipitation. Iron ferrocyanide, existing in soluble and insoluble forms, effectively removes ^{137}Cs. Equations 9.1 and 9.2 show the chemical changes taking place during the removal by soluble and insoluble forms, respectively.

$$K^+ + Fe^{3+} + [Fe(CN)_6]^{4-} \rightarrow KFe[Fe(CN)_6] \quad (9.1)$$

$$4Fe^{3+} + 3[Fe(CN)_6]^{4-} + xH_2O \rightarrow Fe_4[Fe(CN)_6]^3 \cdot xH_2O \quad (9.2)$$

Substantial removal of ^{137}Cs can be obtained by direct injection of potassium ferrocyanide. pH had a significant impact on the removal tendency. At pH 3, the injection of potassium led to a substantial elimination of ^{137}Cs, along elimination of ^{60}Co, which was suboptimal. At pH 8, opposite results were observed. Enhancement of the removal could be achieved for ^{137}Cs and ^{60}Co by carrying out the injection of potassium ferrocyanide first, at pH 3, and finally adjusting the pH to 8, for precipitating. In order to meet the wastewater discharge criteria, the insoluble form of iron ferrocyanide (0.4 g/L) was needed after the pretreatment of the complex solution for impurity removal. An optimal direct aqueous method was employed for the removal of ^{137}Cs and ^{60}Co to be in agreement with the discharge standards. This was mainly because the separation was easier in comparison to the use of the adsorbent, due to the larger particle size formed, with the use of the injection of potassium ferrocyanide.

The removal of radionuclides has also been studied with the technique of coagulation and precipitation in a waterworks in Scania province [23]. Coagulation, flocculation, and sand filters for filtration were used. The coagulants being used are currently aluminum sulfate [$Al_2(SO_4)_3$] and ferric chloride [$FeCl_3$]. Once the precipitates are formed filtration is carried out with the use of sand filters using the method of slow and rapid filtration. The removal efficiency of thorium, polonium, uranium, and plutonium, as analyzed were >90%, >90%, about 85%, and >95%, respectively. Cesium (Cs), radium (Ra), and strontium (Sr) did not show any change in the concentrations.

In a study by Kim et al. [21], polyacrylamide and ferric hydroxide were used for the treatment of seawater with radioactive waste by the coagulation–flocculation method. The study was carried out on the seawater containing radioactive waste, after the accident of the nuclear power plant in Fukushima Daiichi. In this study, the optimum conditions were determined for the removal of radionuclides such as Mn, Sb, Sr, Ru, Co, which were left in the solution after the adsorptive removal of Cs, from the radioactive sea wastewater, generated after the incident. Almost 99% removal for the targeted radionuclide was achieved at a pH of 8 for the radionuclides such as Mn, Sb, Ru, and Co.

In addition to the several precipitants being examined by several researchers, cuprous chloride (CuCl) has been investigated to precipitate radioactive iodide from aqueous solutions [24]. Factors such as dosage, initial concentration of iodide and bicarbonate, and reaction time were studied. The removal efficiency achieved was 95.8%, with the dosage of 150 mg/L, a concentration of 5–40 mg/L, and a reaction time of 15 minutes, which decreased with the addition of bicarbonate.

9.2.2 Membrane Technique

Membrane techniques prove to be advantageous as they can be applied to large volumes, are simple to apply, and have low energy consumption. The membrane process is selective to the kind of contaminants to be removed and work based on the pore sizes [25]. Membrane processes that can be used are microfiltration, nanofiltration, reverse osmosis, and ultrafiltration.

Membrane processes, which are pressure driven [26], such as nanofiltration, reverse osmosis, microfiltration, and ultrafiltration are preferred over other membrane processes, especially by the nuclear industry. This is mainly due to their efficiency in treating industrial wastewaters, the possibility of optimizing performance, and process design being flexible. Further, their integration with the conventional treatments is possible, with a wide range of membrane materials available. These pressure-driven processes are suitable for radioactive waste with varied characteristics.

A combination of membrane techniques with other methods can also be more effective [27, 28]. Reports have suggested that radioactive liquid can be treated using a membrane process, which has the advantage of a high removal coefficient, large operational flexibility, simplicity, and low energy input [29]. A hybrid technique employing flocculation, sedimentation, and microfiltration was investigated by Zhao and coworkers, who reported a 99.9% removal rate for plutonium by careful control of pH [27].

In 2023, Ahmad et al. [30] explored a thin film composite (TFC) RO membrane for the removal of cesium ions as a radionuclide model. This membrane was developed by using a polyamide (PA) surface and coating it with titania nanosheets (TNS), which were oppositely charged. A thin film nanocomposite layer (TFN) with 15 ppm TNS bilayer revealed a 45.8% permeability. The TFN membrane showed CsCl rejection of almost 99.32%, and all the TFN membranes gave almost complete rejection.

9.2.3 Sorption

As in the case of non-radioactive contaminants, sorption has revealed to have an edge over the other methods, due to cost effectiveness and ease of application. Radioactive elements such as cesium and strontium have been successfully removed with the use of different categories of adsorbents [31, 32]. Adsorption of the contaminants however is dependent on several factors such as medium, contaminant type, porosity, etc.

Jun and coworkers in 2021 tested the method of adsorption for uranium removal [33]. Nanomaterials [34] and several inorganic materials [35] as adsorbents also have proven their worth in the elimination of radioactive contaminants. Polymer composites have been found to have good ability and selectivity to take up radioactive pollutants. The ability can be further enhanced by incorporating conducting polymers and magnetic nanoparticles. Zeolites have been found to have a decontamination coefficient in the range 62–68 and a high reproducibility. Strontium removal efficiency of 142.9 mg/g was achieved using melamine-styrene-based polymeric adsorbent. Similarly, metal functionalized magnetic adsorbent was utilized for the removal of radioactive cesium (166.67 mg/g) [32].

A material such as anionic clay hydrotalcite has been explored as an adsorbent for the elimination of radionuclides from liquid waste solutions [36]. Sorption of selenium (^{75}Se) and ^{131}I radionuclides was studied using magnesium iron hydrotalcite (Mg/Fe HTlc). The material was prepared by the method of co-precipitation. Characterization of the material was done using techniques such as XRD, FT-IR, SEM, BET, DTA, TGA, and XRF. The process of adsorption proceeded at a very fast pace, with the highest sorption capacities of 9.25 and 21.45 mg/g for ^{75}Se and ^{131}I, respectively. Prepared sorbent was found to be suitable for consideration as a potential material for effluent polluted with ^{75}Se and ^{131}I.

The removal of $^{85}Sr^{2+}$, $^{137}Cs^{+}$, and americium ion $\left(^{241}Am^{3+}\right)$ has been carried out using alginates of barium (Ba), calcium (Ca), and strontium [37]. Highest adsorption was found in the pH range of 5–6, with an optimum time of 2 hours using calcium alginate and 4 hours for the remaining adsorbent. Maussa et al. investigated [38] the use of a nanocomposite as an adsorbent for the sorption of radionuclides such as $^{152+154}Eu$ (III) and 60Co (II) from aqueous solution prepared by polyacrylic acid/maleic acid hybridization with nano copper oxide (PAACMA/CuO). Fungus *Aspergillus terreus* was used for the production of Nano-CuO. A monolayer spontaneous and endothermic adsorption of 21.54 mg/g took place for Eu (III) and 11.04 mg/g for Co (II). For efficient sorption, the optimum conditions identified were an optimum time duration of 24 hours, temperature 20 °C, and pH 3.53, for $^{152+154}Eu$ and pH4.5 for ^{60}Co.

Chitin, its derivatives, and a fungal biomass have been used as sorbents of radionuclides from wastewater and other contaminated sources [39]. The properties of chitin, such as flexible structure, presence of many functional groups, and high hydrophobicity, are the properties that enable it to be a good adsorbent.

Xu et al. in 2022 [40] discussed the use of nanomaterials for selective sorption followed by catalytic reduction of radionuclides. Nanomaterials such as metal–organic frameworks (MOFs), carbon-based nanomaterials, and covalent organic frameworks (COFs) have been explored for batch sorption experiments, along with the mechanism of sorption. MOFs are crystalline materials with extensive porosity and large internal surface areas of more than 6000 m^2/g. This makes them excellent materials for varied applications. COFs are also highly porous with ordered structures. These produce strong organic networks on interaction with organic precursors. Carbonaceous nanomaterials are capable of intervening in terrestrial and aquatic environments via effluents, surface runoff, etc. Graphene nanosheets and carbon nanotubes (CNTs) are the two main categories of nanomaterials.

Several techniques have been used for the synthesis of nanomaterials, such as the bottom-up method, blending, and postsurface modification. Elimination of radionuclides such as U (VI), $^{99}TcO^{4-}$, ^{90}Sr, Eu (III), etc., has been carried out with the use of these nanomaterials. Carbonaceous nanomaterials [41] are stable, strongly acidic, and resilient on exposure to radiation and heat.

One of the most important radionuclides used in a part of nuclear industry is uranium. Uranium is hence present in the wastewater of nuclear industries, salt lakes, oceans, etc., making its removal and recovery necessary. Table 9.1 shows the key results and the category of such nanomaterials as sorbents for the elimination of uranium (VI), which have been used.

9.2.4 Photo-induced/AOP Processes

Photo-induced technology is another method where the reduction or degradation of the radioisotopes can be carried out. Ultraviolet (UV) light or the abundant availability of sunlight has been a boon in this regard. Advanced oxidation processes with combinations such as the use of ozone (O_3) + hydrogen peroxide, UV + O_3, O_3+TiO_2, etc., can prove to be efficient processes for the elimination and treatment of radioactive nuclides. The results obtained from some photo-induced processes for U (VI) have been portrayed in Table 9.2 [16].

Techniques such as photo catalysis have been successfully investigated for the elimination of uranium [59]. U (VI) was extracted from seawater using TiO_2 as a photocatalyst [57]. The maximum extraction capacity achieved for uranium without saturation was 3960 mg/g. Factors such as pH were found to have an impact on the process of extraction. A pH greater than 6 facilitated the process. High salinity, high dose of TiO_2, and high initial concentration of U (VI) hindered the process. Substances such as methanol and humic acid positively impacted the

Table 9.1 Removal of uranium (VI) using MOF sorbents.

MOF		
Material	**Key results**	**References**
Zn-based MOFs	Removal efficiency 0.53 mg/g in 1 L artificial seawater, the carboxyl group is responsible for material efficiency	[42]
Amidoxime-functionalized MOF	Sorption of 2.68 mg/g U (VI) from real seawater, binding due to the coordination of U (VI) with the amidoxime ligands	[43]
Hierarchical-porous Cu-MOF	Sorption efficiency 563 mg/g at pH 5.5, binding due to micro- and macroporous interaction of U (VI)	[44]
UiO-66-derived MOFs	High sorption efficiency (6.85 mg/g) from real seawater, binding due to carboxyl and amino functional groups, and MOF skeleton	[45]
COF		
Material	**Key results**	**References**
Polyarylether-based COFs	Sorption efficiency 5.12 mg/g, highest adsorption energy of −19.38 kcal/mol	[46]
COFs/GO/CH aerogel nanocomposite	Maximum sorption 1191 mg/g at pH 6, pseudo-second-order kinetic model followed along with Langmuir isotherm, regeneration and reuse possible for more than five successive cycles.	[47]
Carbon-based nanomaterial sorbents		
GO-CNTs	The process followed the Langmuir isotherm with a maximum adsorption of 86 mgU/g. The second-order rate equation was a good fit for the process.	[41, 48]
GO nanosheet	Sorption increased with temperature increase, maximum sorption was 97.5 mg/g at pH 5.0 ± 0.1 and temperature 20 °C	[41, 49]
Oxidized MWCNTs	pH and ionic strength governed sorption, outer-sphere surface complexation/cation exchange was responsible for the process	[41, 50]

MOF		
Material	**Key results**	**References**
Pristine CNTs	Acidic treatment enhanced the sorption of uranium (VI). In aqueous suspension, surface treatment affected the process substantially	[41, 51]
LDH@GO	Maximum sorption capacity was 277.80 mg/g, the process was spontaneous and endothermic, pseudo-second-order rate equation described the process.	[41, 52]
Grafted MWCNT (MWCNT-g-CMC)	Showed a higher sorption capacity than raw MWCNT	[41, 53]

Table 9.2 Photo-induced methods used for U (VI).

Method adopted	Compound used for photo-induced process	Key findings	References
Photo reduction	Graphene oxide – Potassium titanates ($K_2Ti_6O_{13}$), (GO-KTO)	99.99% at pH 10, after 210-minute, U (VI) was reduced to U (IV) which was less toxic	[54]
Adsorption-Photocatalytic reduction	Niobate/titanate nanoflakes (Nb/TiNFs), 0.2 g/L	Nb/TiNFs revealed rapid adsorption kinetics, 89% of U (VI) reduced by Nb/TiNFs under solar light.	[55, 56]
Photocatalysis/ Photoconversion	TiO_2, 0.6 g/L	Extraction capacity was more than >3960 mg/g by the photocatalytic method, the efficiency increased with the increase in the dose of TiO_2.	[57]
Heterogeneous photocatalytic reduction	TiO_2, 1 g/L	Efficiency 94% pH 3.0 after 120 min, formation of U (III) was also apparent in the absence of the photo catalyst	[58]

process under specific conditions. The extraction was enhanced with an increase in the irradiation. Photocatalysis proved to be a sustainable, eco-friendly, facile, and economical method for the extraction of uranium from sea water. TiO_2 [60], iron oxide [61], and g-C_3N_4 [62, 63] have been actively used for the treatment of radioactive contaminants such as uranium. Salomone and team reported a quartz photo reactor to reduce uranyl nitrate by 98% within half an hour in comparison to the uranyl perchlorate-perchlorate system and uranyl acetate-acetic acid system [59].

In another research, titanium oxide photo catalyst increased the adsorption characteristic of uranium (VI) and achieved a maximum adsorption capacity of 44 mg/g. Reduction of soluble uranium (VI) to insoluble uranium (IV), based on a photo catalyst, is a simple and efficient method for treating radioactive wastewater. Nickel-incorporated porphyrin-based organic framework removed 86.74% of uranium (VI) from real radioactive wastewater during two hours of photoinduction [64].

9.2.5 Bioremediation

The chemical and physical methods prove to be expensive at times for the removal of radioactive contaminants. Bioremediation is the most cost-effective and efficient method for the control over radionuclides in wastewater. This method is not only beneficial for the removal of radioactive substances but also for the removal of inorganic and organic pollutants. It usually involves the use of plants or microorganisms. The use of bioremediation can prove to be eco-friendly, where microbes are used for radioactive wastewater [65]. The biological agents can easily transfer the radionuclides into stable isotopes. In processes such as bioreduction, the radionuclides can be immobilized due to the enzymes of the biological species [66]. This involves removal via biotransformation and bioaccumulation using plant species (phytoremediation) or microbial cells. It helps in the reduction of radioactive waste to a great extent.

In the bioremediation processes, such as biosorption, bioaccumulation, biotransformation, biomineralization, and phytoremediation, are commonly used. Table 9.3 below mentions various species which has been used for the elimination of different radioisotopes from wastewater [16].

In an investigation conducted by Ferreira, bacterial colonies cultured in the soil of radioactive mining areas could adsorb 90–100% of radioactive elements [80]. This suggests the suitability of biotechnology in the processing of radioactive wastewater. Reduction in the organic load was also ensured through the process. The process was quite complicated. *Bacillus subtilis* for Strontium, *Rhizopus* for thorium and uranium, and moss for strontium elimination have been reported by researchers [81, 82].

Table 9.3 Some microbial and plant species used for bioremediation.

Microbial and plant species	Method used	Target radionuclide	Key results	Type of sample	References
Mucor circinelloides (sporangiospores), 0.3 g/lL	Bioaccumulation/ Bio-sorption/ Biotransformation	U (VI) (30 mg/L)	pH_{zpc} 3.0, removal efficiency 90% after 48-hour, pH 7, temperature 25 °C	Wastewater	[67]
Saccharomyces cerevisiae	Bio-mineralization/ Bio-sorption	U (VI) (10 mg/L)	Removal efficiency 92% after 28.5 hours, temperature 30 °C, pH 5.5	Water	[68]
Polyphosphate accumulating microorganisms *(PAO) Leifsonia sp. Biochar (BC)* from sawdust, 1.5 g/L	Bio-mineralization/ Bio-sorption	U (VI) (42 μmol/L)	99.5% removal at 3.5 pH after 12 hours	Water	[69]
Water primrose (Ludwigia stolonifera), 4 g	Bioaccumulation/ Phytoremediation	^{60}Co (100 Bq) ^{137}Cs (250 Bq)	98% efficiency at pH 3.0, after 100 hours 63% removal after 225 hours	Tap water	[70]
Saccharomyces cerevisiae, 0.5 g/L	Bio-sorption/ Bio-mineralization	U (VI) (10 mg/L)	Removal efficiency 92% after 28.5 hours 5.5 pH, 30 °C temperature	Water	[71]
Chlorella vulgaris Chlorella sorokiniana	Bio-mineralization by microalgae	^{137}Cs (2100 Bq/mL) ^{90}Sr (2000 Bq/mL) ^{137}Cs (2100 Bq/mL)	70% removal, 30 °C under 1800 lux illumination 95% efficiency after six days and 35% after six days	Water	[72]

(Continued)

Table 9.3 (Continued)

Microbial and plant species	Method used	Target radionuclide	Key results	Type of sample	References
Clostridium types	Biosorption/ Biotransformation	U (VI) (22 mg/L)U (VI) (22 mg/L) + SO_4^{2-} (1800 mg/L)	Removal efficiency 18.18% after 20-days pH 7.0–7.2, abiotic conditions 13.63% removal efficiency after 20 days U (VI) (22 mg/L) + SO_4^{2-}(1800 mg/L) Anaerobic environment	Water	[73]
Arthrobacter G975 strain (4×10^{10} cells)	Bio-sorption/ Bioaccumulation/ Biomineralization	U (VI) 61.76 (μM/L)	Removal efficiency 85% at pH 7.3 after 24 hours at temperature 25 °C	Groundwater	[74]
Napier grass (Pennisetum purpureum Schum.)	Phytoremediation/ Bioaccumulation	^{137}Cs/ ^{133}Cs (50, 150, 300, 1000, 3000 μmol/L)	97% removal at pH 4.9	Soil and water	[75]
Water hyacinth (Eichhornia crassipes), 5 g	Phytoremediation/ Bioaccumulation	^{137}Cs (119 Bq/mL) $^{60}C0$ (135 Bq/mL)	65% removal efficiency at pH 4.9	Drinking water	[76]
Bacteriogenic iron oxide (BIOS) assisted by *Gallionella ferruginea* and *Leptothrix ochracea*	Bio-sorption/ Bio-transformation	U (VI) (2.1×10^{-7}mol/L)	9% removal after 30-min by 1 g/L 100% (Groundwater), pH 7.090% after 30-min by 1 g/L Groundwater with 90% distilled water	Groundwater	[77]
Reed *(Phragmites australis L.)*	Phytoremediation/ Bioaccumulation	^{137}Cs (14 Bq/L)	15% removal after 32 days	Water	[78]
Magnetotactic bacteria (MTB)	Bioaccumulation	U (VI) (100 μmol/L) Plutonium (Pu)	Removal efficiency 98% after 2 h. 52% after 148 min	Wastewater	[79]

In addition to the species mentioned above, microbes such as *Haematococcus pluvialis,* Red cyst, *Haematococcus pluvialis,* Green cells, *Chlorella vulgaris, Anabaena* sp., *Chengiopanax, Sciadophylloides, Quercus serrata, Acer crataegifolium, Ilex macropoda, Fraxinus sieboldiana, Desulfovibrio desulfuricans (SRB), Amaranthus chlorostachys, Helianthus annuus L., etc.* have also been used for the same purpose.

9.2.6 Electrokinetic Remediation (EKR) and Ion Exchange

In this method, an electric current is used for the removal of contaminants. The methods have shown substantial efficiency and cost effectiveness, especially for less permeable substrates [83].

An electric current is applied to create the movement of ions in the subsurface. Initially, a direct removal was carried out for the elimination of radionuclides, in addition to several other contaminants such as heavy metals, organic contaminants, sediments, etc. Electrokinetics has been applied recently to carry out in situ remediation with the help of methods such as dichlorination (ERD) and in situ chemical oxidation (ISCO) [84]. These methods require the use of several inert electrodes installed in aquifers. The electric field and voltage gradient are created with the application of low-voltage direct current (DC). The positively charged electrodes are anodes, whereas the negatively charged electrodes are cathodes. The application of current leads to two processes, which are electro-migration and electro-osmosis. Electromigration involves the migration of charged ions, which are dissolved, toward the opposite charged electrode, through the aqueous medium, and electroosmosis osmosis which involves the movement of the ions dissolved within a porous medium from anode to cathode.

This technique has been successfully tried for contaminants such as uranium and cesium. The method involves the implantation and use of EKR electrodes to which the contaminated source is exposed. In the presence of a direct current movement of water, particles and ions are placed, thereby accumulating the contaminants near the electrode.

By the process of ion exchange, almost 90% of the radionuclide removal is possible. The water samples are allowed to pass through resins that contain charged particles. The contaminants are retained in the resin after exchange reactions. Figure 9.8 gives a diagrammatic representation of the ion exchange process.

The ion exchange process is usually suitable for liquids with less salt content [85]. Exchangers, which are organic or inorganic, natural or synthetic, or membrane exchangers, can be used. Composite ion exchangers have also recently drawn attention. Inorganic ion exchangers are more suitable for radioactive wastewater treatment due to their greater chemical inactivity and radiation resistance. They are usually very efficient for many monovalent and divalent metal cations.

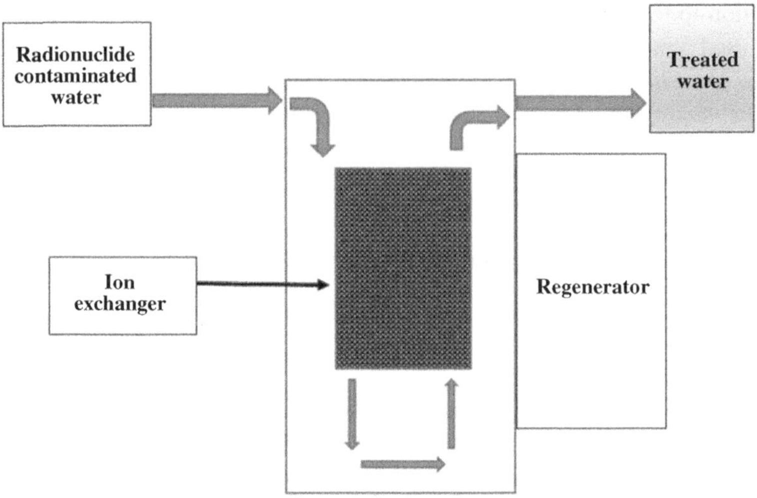

Figure 9.8 Ion exchange process.

Some examples of ion exchangers are bentonite, illite, and kaolinite, which are natural inorganic; ellagic acid, cellulose, straw, and peat, which are organic; titanates and silico-titanates, zeolites, transition metal hexacyanoferrates, which are synthetic inorganics, and polyacrylonitrile gel, which is a prepared composite ion exchanger.

The main requisite and the important aspect to be considered for ion exchangers, for aqueous liquids [86], for efficient removal, is that the concentration of suspended solids should be low; there is a need for prefiltration. In addition to this, the total dissolved salt content should be low (1–2 g/L) for the proper exchange of radionuclides by the ion exchange method. Organic contaminants (oil and grease) should be in very small amounts, as these can lead to fouling. Also, certain oxidizing compounds, such as nitrates, can lead to degradation in the presence of metallic ions and need to be regulated.

The application of ion exchangers can be done in batch, column, and membrane application modes. Batch operations involve simple construction and prove to be good for small-scale treatment without the need for granular materials. Column operations, on the other hand, offer the use of a variety of media applications and enable their use at high temperatures and pressure. With the use of membrane processes also there is no need for granular materials and a high level of decontamination is possible [86].

Inorganic resins have been successfully tested for the removal of thorium and uranium. Bhattacharya and coworkers reported the adsorption characteristics of thorium and uranium on a cationic exchange resin and inferred a better removal of thorium [87]. The ion exchange capacity of strontium over resorcinol-formaldehyde resin was found to be 2.28 meq/g at alkaline pH (7.5–8.5). Fang et al. in 2016 [88] used a zeolite 4A for the treatment of laundry wastewater containing organics, suspended impurities, and contaminants such as Cs^+, Sr^{2+}, and Co^{2+}. Almost 90% of the removal was achieved in a neutral environment and at room temperature, which was greater than 98.7%, when the environment was alkaline. The adsorption order of the zeolite 4A for contaminants was of the order solids > organics > radionuclides, as per orthogonal experimentation. Removal of tritium radionuclide from liquid waste was investigated using a sequential ion exchange resin [89]. A strong acidic resin (KU-2) was used to study the removal of tritium from aqueous solutions containing radioactive substances. Studies on the effect of contact time, resin weight, and pH were carried out. The pH for maximum removal was 7.8 with an equilibrium at 1.5 hours. Desorption studies were also carried out using eluents such as 2M NaOH and 3M HNO_3.

Ion exchangers such as titanosilicates, zeolites, hydrous metal oxides, bentonite/clays, and hexacyanoferrate metal oxides have been studied for the removal of radioactive cesium, including the use of several composites using titanosilicates/zeolites, ammonium phosphomolybdate (AMPs), and hexacyanoferrates [90]. The removal of Cs has been studied using a zeolite made from fly ash by El-Naggar et al. [91]. A calculation of parameters such as entropy ($\Delta S°$), enthalpy ($\Delta H°$), and Gibbs free energy ($\Delta G°$) was done. The experimental data fitted well with homogeneous particle diffusion model and pseudo-second-order models. Montmorillonite and bentonite were used by Galambo et al. [92] for Cs containing wastewater.

9.2.7 Evaporative Method

This technology is generally used to concentrate the radioactive waste. The radioactive waste is sent to the evaporator, which is heated. Subsequent heating and cooling are carried out, resulting in condensed water. The non-volatile radionuclides are concentrated, discharged, and exposed to subsequent solidification treatment. In this method, no additives are used; hence, this method can be operated without using any additional additives, reducing the chances of secondary pollution [93]. The method has great potential for combination with other technologies [94]. Evaporation separates the solvent from the solute. This results in the removal of radioactive gases, which are collected and directed to the waste gas system. The leftover solution, called concentrate, is further processed and is transferred to radioactive waste system for solidification. Wastewater with high concentration of radioactive substances is significantly removed using evaporative concentration. Researchers have made use of and explored various types of evaporators for use in the process.

Kettle-type evaporators, rising film evaporators, and natural circulation type evaporators are commonly used. Solar-based evaporators are also being used in new versions [95]. Direct heating using evaporators consumes a lot of electrical energy. MVR (mechanical vapor recompression) technology works on the principle of pumping of heat. Hou et al. [96] studied the use of MVR technology. The decontamination effect and efficiency in terms of energy utilization were evaluated. For optimization, the heat transfer temperature difference and evaporation temperature were selected. The results obtained demonstrated the feasibility of the use of mechanical vapor recompression for radioactive wastewater treatment. Ions such as cobalt, cesium, and strontium were evaporated at a temperature of 90 °C to maximum ionic decontamination factors of 1.45×10^4, 1.74×10^4, and 5.90×10^3, respectively. The study proved to be a guide toward the treatment of the nuclear wastewater. An evaporation plant for radioactive wastewater with a production capacity of more than 100000 gallons/month was constructed [97]. A forced evaporation system was developed for the treatment of radioactive wastewater [98], consisting of units such as a loading module for pumping wastewater into the heat exchanger, an injection module for steam, and a module for condensation. This method can be used for the purification of radioactive waste with high concentration, as the operation cost is high and the equipment is susceptible to corrosion.

9.2.8 Integrated Methods

A greater interest has recently been drawn to the integration of different methods for greater efficiency. Figure 9.9 shows some examples of integration of methods. For example, coupling of membrane with precipitation, adsorption, flocculation, and microfiltration, combining precipitation with membrane process and adsorption, ultrafiltration, and reverse osmosis, and reverse osmosis and evaporation, etc.

In a study by the Institute for Nuclear Research and Nuclear Energy, a combination of sorption and precipitation was done for radioactive wastewater treatment. Radionuclides that were targeted were ^{90}Sr, ^{137}Cs, and ^{60}Co [99]. For sorption, several zeolites were tested for their sorption capacity. The materials chosen had different sorption capacities. Out of the zeolites chosen for the study, higher and faster sorption was shown on clinoptilolite, which was hence selected as a sorbent for the treatment of radioactive waste containing ^{90}Sr, ^{60}Co, ^{137}Cs, ^{226}Ra, ^{110m}Ag, and ^{133}Ba. Depending on the radionuclide in the radioactive waste, the distribution coefficients ranged from 10^2 to 10^4 mL/g. The sorption capacity could be improved by conversion of clinoptilolite into the sodium-containing form. Same as zeolites, a comparison between several precipitation methods was drawn and mixed hydroxide–calcium phosphate precipitation was the main method

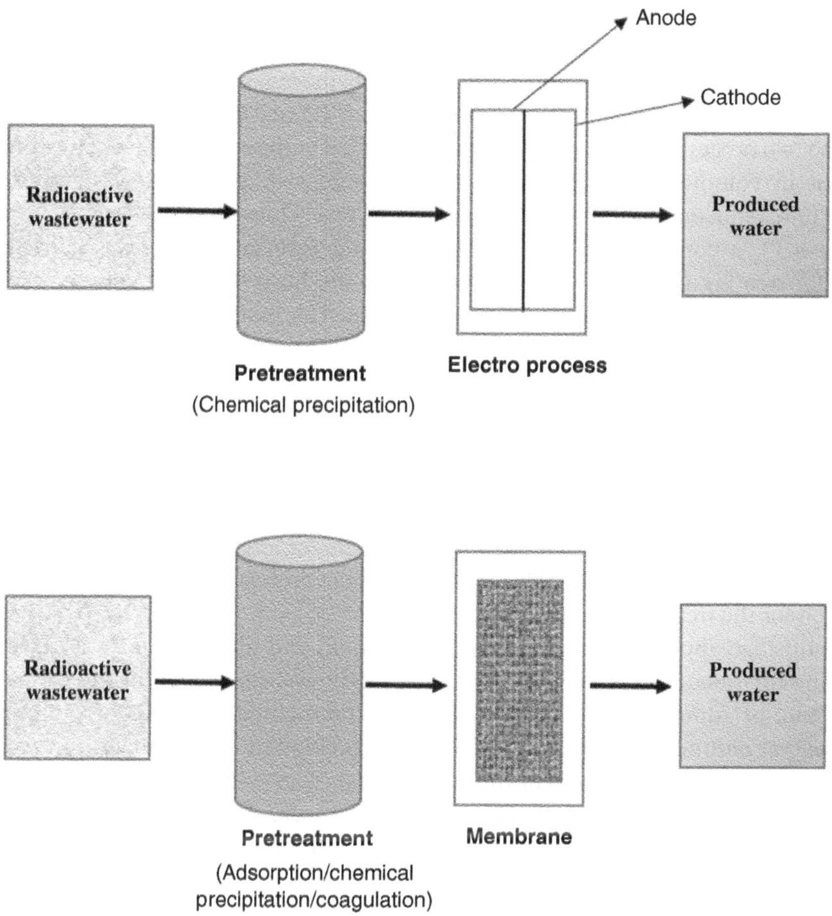

Figure 9.9 Examples of integration of methods.

of precipitation, which was selected. The decontamination factor was high for ^{65}Zn, ^{60}Co, ^{144}Ce, and ^{54}Mn.

A combination of co-precipitation and filtration was also investigated for the removal of strontium (5 mg/L) from aqueous solution by Cao et al. in 2010 [100]. The precipitating agent used in this case was sodium carbonate. For enhancing the removal, sodium ferric chloride was used (20 mg/L), which, when added to the membrane separator, reduced membrane fouling with a dosage of 10 mg/L. Two intermittent tests were carried out with sodium carbonate doses of 1000 and 2000 mg/L. For each of the concentrations, the mean decontamination factors (DFs) and the mean concentration factors (CFs) were 158 and 462, and 237 and 288, respectively.

Arnal et al. [101] studied the combination of reverse osmosis and evaporation for the treatment of radioactive water from nuclear facility dismantling using a pool-type reactor. The work aimed at declassifying the maximum volume of water, thereby decreasing the liquid radioactive waste. A simulation of the combination of the two processes was first carried out, which led to the results that a reduction factor in volume higher than 600 could be achieved. With real wastewater, a volume reduction factor of approximately 170 was achieved in 194 days.

In 2013, a combination of ultra-filtration and reverse osmosis was carried out for treating radioactive liquid wastes from the hospital RIA (radioimmune assay), which were potentially infectious wastes. The main purpose behind the treatment was radioactive nuclide removal along with pathogen elimination [102]. Adsorption and microfiltration together (AMF) were tested in the laboratory for the removal of cesium from aqueous solution in lab-scale and jar tests [103]. The adsorbent used for the process was K_2Zn_3 [Fe $(CN)_6]_2$. The process followed the Freundlich isotherm well. The concentration of cesium ions in effluent and raw water was 0.59 and 106.87 μg/L, respectively. About 99.44% cesium removal was obtained in the lab scale tests. The quality of effluent after treatment was good, with turbidity lower than 0.1 NTU, proving that the process (AFM) was a feasible option for the treatment of radioactive waste.

Multistage application of sorption, flocculation, and ultrafiltration was used for the treatment of radioactive waste. The study was carried out at the China Institute of Atomic Energy (CIAE), Beijing. The method consisted of the use of several natural or synthetic sorbents using precipitation, flocculation in the first stage, and ultrafiltration in the second stage. Sorption on natural sorbents was the third stage of treatment. Investigations were done for both real and simulated liquid waste. Chemical precipitation in combination with sorption was also studied at the Malaysian Institute for Nuclear Technology Research (MINT). Radium, uranium, and thorium were removed from the liquid waste with decontamination factors of 391, 80, and 234, respectively. At a pH of 9.9, which was optimum for the process, the effluent after treatment was as per the permissible discharge limits. Thorium, as its hydroxide, dominated during precipitation [99].

Different treatment processes for radioactive contaminants and their advantages and disadvantages are portrayed in Table 9.4 [33, 104]. Some limitations in the use of organic resins may be chemical resistance and heat resistance [105], where inorganic resins such as zeolites, metal oxides, and clay are more advantageous [106].

Considering the advantages and disadvantages of different treatment methods, attention is now drawn to an integration of different techniques to provide the most efficient radioactive wastewater treatment. Integration can provide high treatment efficiency and cost effectiveness.

Table 9.4 Treatment methods for radioactive contaminants [104].

S. No.	Treatment method	Advantages	Disadvantages
1.	Chemical precipitation	Low cost, easy, and can be used for large volumes	Unsuitable for anionic radionuclides
2.	Membrane separation	Easy processing	Selectivity, not cost-effective, membrane contamination
3.	Sorption	Easy cost effective	Selectivity
4.	Ion exchange method	Easy and simple operation	Difficult to regenerate, affected by salinity, problem of secondary waste
5.	Photocatalytic method	Efficient, no secondary pollution and cost effective	Affected by sunlight utilization, efficiency of charge separation is low
6.	Bioremediation	Environmentally friendly, no secondary pollution	Less toleration of radiation by microbes
7.	Evaporation	Higher volume reduction more flexibility, wide range of application, can be combined with different technologies	High consumption of heat, not cost effective, not suitable for liquids that foam

Source: Adapted from [33].

9.3 Case Studies on Radioactive Contaminants

9.3.1 Hanford Site, Washington, USA

Hanford site was a large site in the United States which was established for the production of plutonium during World War II, and also in the first atomic bomb. The original operation of this site, as a unit of Manhattan Project, was done by the US Army Corps of Engineers, with administration done by civil government agencies at later stages. After ending of World War II, Cold War between the United States and the Soviet Union began. Several plutonium production reactors were built. During these years more than 400 billion gallons of contaminated liquids which were dumped into the Columbian river. Since 1940 to present, the Hanford nuclear site, in the eastern region of Washington has been a contributor to significant amount of nuclear waste, owing to the activities of the department of energy and defense, especially during World War II.

The site contained 177 massive tanks, underground, holding hot, corrosive and extremely hazardous radioactive wastes which approximately around 56 million gallons, the largest being of capacity 3785000 L [107]. The tanks which were installed first developed leaks with time. Some of the hazardous waste was dumped into the ground. The spent nuclear fuel was stored in containers which were corrosion prone kept within water filled basins near Columbian River. Other contaminated wastes included clothes, rail cars etc. which were buried in pits. Though Environmental Laws came in place in 1970 but a self-regulating site like Hanford did not abide by these laws [108].

There has been leakage of radioactive materials into the Columbia River in addition to the contamination to soil [109]. There has been groundwater contamination of billions of gallons, for which pumps have been deployed for treatment [110]. The release of the radioactive substances has affected the air, the population close to the water streams and the aquatic life.

The Hanford site has now been taken over by the US Department of Energy (DOE) since 1977. In 1989, the formal cleanup started by the Environmental Protection Agency, DOE and Washington state. The remediation efforts are still ongoing on after beginning in 1989, ensuring the proper treatment of soil and water.

9.3.2 Fukushima Daiichi Nuclear Disaster, Japan

This incident in Fukushima Daiichi plant of Japan one of the second-largest nuclear incident which took place. This facility consisted of six reactors with boiling water. Out of which only three were functional in 1971–1979. The other four worked as a temporary storage for used fuel rods.

The incident occurred on 11 March 2011, after an earthquake of magnitude 9.0 occurred in Japan. This lead to Tsunami waves, [111] which lead to a damage in the backup generators of the Fukushima Daiichi nuclear plant, as reported. The successfully operating reactors were shut down successfully, but within the first few days there was a failure of the cooling system in each of them. This led to the melting of the fuel rods in the three reactors with release of radiation. Explosions resulted on 12 March and 14 March. No fly zones were declared by the government officials, with the evacuation of 232 square miles [112]. After the third explosion on 15 March, the government officials decided wider zones asking the residents to remain indoors. As a result of subsequent explosions higher release of the radiations took place from the plant leading to the evacuation of around 47000 people. Several attempts were being made by the workers of the plant to cool the reactors, including the spraying of water from helicopters. These efforts only reduced some amount of radiations temporarily which again mounted. At the end of March the ocean water was declared to be contaminated with substantial levels of ^{131}I, resulting from radioactive water leaking from the plant to the ocean. Owing to this incident the government officials also mentioned some regions might become inhabitable.

The incident impacted human health, economic consequences, and environmental effects. Learning lessons from the incidence, reinforcement to the importance of earthquake and tsunami resistant structures was laid.

9.3.3 Mayak Facility and the Techa River, Russia

On 29 September 1957, there occurred an explosion in a liquid radioactive waste tank in one of the leading military enterprises of the Soviet Union, Mayak, owing to poor technology being adopted for storage [113]. Before the Chernobyl disaster, this was considered as the biggest catastrophe related to radiation in the world. Around 20 million curies of radioactive waste was released into the environment, out of which 2 million was outside the production territory. This event was in continuation to the release of radioactive waste into the Techa River, which was a major dumping of radio waste. This release from Mayak was in addition to the regular release of radioactive wastes into the Techa River. As a consequence of the catastrophe, known as the "Kyshtym catastrophe," thousands of people (272000 from 217 settlements) of various towns and villages were exposed to radiations. A radioactive trace with a width of 30–50 km, stretched for 300 km. Almost 1000 km^2 area omitted from economic production.

As per the officials, the exposure of the population was within "regulatory norms", however the health status of these people showed a significant difference in comparison to the rest of the population of Russia. As a result of the explosion, thousands of people from various villages and towns were exposed to harmful radiations [114], an example being Muslyumovo, a settlement, where an exposure to internal and external radiations occurred on the population. The internal exposure to radionuclides, mainly Sr^{90}, was due to the river water and the consumption of local products. This contaminant accumulated in the tissues and bones, as known from the autopsy reports of the inhabitants. About 153 case of chronic illness were reported in 50–60-year-old. The coefficient of cancer motility rates in the inhabitants increased from years 1950–1982 to 129.0 persons per 100000, the control figures being 114.9 persons per 100000 [115, 116].

As control and remedial measures evacuation and resettlements of population were done, excavation of the contaminated soil was done and was stockpiled in enclosures with fencing which were known as "graveyards of the earth" [117]. The affected areas were prohibited to access by the Soviet government in 1968 by creation of East Ural Nature Reserve. A study in 2002 mentions that the population by the Techa River and the workers of Mayank are still affected [118].

9.3.4 Chernobyl Disaster, Soviet Union

The Chernobyl disaster is considered to be one of the worst incidences which occurred in the nuclear power generation history [119]. The power station was located at the Pryp‘yat’ settlement, almost 65 miles from north of Kyiv, Ukraine. There were four reactors in the station. Each of the reactors had the capacity of production of electric power of 1000 megawatts. This disaster occurred at the Chernobyl nuclear power station (Soviet Union) on 25–26 April 1986.

There was a poorly designed experiment in one of the reactors. The power regulation of the reactor was shut down and the control rods were withdrawn, while the reactor continued running at 7% power. Chain reaction occurred in the core going uncontrollable, leading to several explosions, with the release of the radioactive material (radionuclides in the range of 50–185 million curies) into the atmosphere. This was said to be even more than that released when atomic bombs were dropped on Hiroshima and Nagasaki, in Japan.

About 30000 inhabitants from Pryp‘yat’ were evacuated. Radioactivity was reported to be wind transmitted, as reported by Swedish monitoring station on 28 April 1986. The containment activities began on 4 May, where the burying of the radioactive debris was done in several hundreds of sites and the reactor core was enclosed in steel and concrete.

A result of the exposure to radiation several people died, some experienced severe radiation sickness, millions of acres of farmlands, forests and water bodies were contaminated, cancer deaths, deformities in live stocks occurred over the period of time. In addition to this substantial ground water contamination also occurred [120]. Moreover the spread of the radioactive contaminants took place with wind covering Russia, Ukraine, Belarus, Italy and France.

A circular exclusion zone was created by the Soviet Union, with the power plant in the center, covering around 18.6 miles. This was later expanded. The nuclear power station was decommissioned in the year 2000. The site is open to scientists, researchers, etc., with necessary permissions for a limited time. Parts of the exclusion zone were opened by the Ukrainian government in 2011 for group tours with the destinations of Pryp‘yat’ and Chernobyl known as “dark tourists.”

9.4 Technologies and Materials Studied for Radioactive Contaminants

Several categories of wastewater contaminants are present, which pose a direct or indirect impact on the living organisms [121]. Research has been carried out using several materials and technologies to treat the radioactive contaminants in wastewater.

9.4.1 Novel Technologies

The Korea Atomic Energy Research Institute (KAERI) has developed a new method to separate the nuclides with a device using robots and sensors [122]. The analysis is carried out as a three-step process comprising preprocessing, which is followed by separation, and finally, the measurement. According to KAERI, efficient separation of species such as strontium-90, niobium-94, technetium-99, nickel-59 and nickel-63, and iron-55 is possible from a single sample. The separation of niobium, strontium, nickel, and technetium was three folds rapid than the conventional methods, giving a nuclide rate of recovery of 83–97%.

Technologies such as forward osmosis (FO) are being tested for the elimination of radionuclides from wastewater. This method has an edge over the conventional method of reverse osmosis. A semi-permeable membrane is used for the separation of two solutions with different concentrations. Processes like this can reduce the consumption of energy as they are driven by osmosis. Researchers are also focusing on the integration of technologies with FO [123]. A recent study has been carried out on capacitive deionization (CDI), an emerging technology [124], along with membrane CDI [125]. This technology has been tested for the elimination of radioactive ions of cesium, cobalt, iodine, uranium, and strontium. These techniques require less energy and are based on electron spin. In a research study in 2022, copper hexacyanoferrate was used to improve the cesium ion removal from a cationic mixture. An electro-sorption of 397 mg/g was noticed at a voltage of 1.4 V.

A study cited in 2020 describes the use of shock electrodialysis for the separation of radionuclides from wastewater [126]. The method was used for the removal of cesium and cobalt from a mixed feed of boric acid, cobalt, cesium, and lithium, as a continuous process. The system contained a weakly charged porous medium. This method was a membrane-less separation with no physical barriers.

There are many studies being carried out on the integration of several techniques for better management of radioactive species in wastewater. Chaalal and Islam in 2001 worked on the integration of techniques such as treatment with biomass, membrane method, and the use of a treatment fluidized bed reactor for the treatment of strontium ions [127].

There have been some biotechnological technologies and innovations also, which have been explored for the treatment of radioactive waste [128]. Genetically modified microbes are being created for the utilization of their potential from metabolic activities for immobilization of radionuclides or their conversion into less harmful forms, proving to be a major advancement in the field.

Another novel method is the continuous electrode ionization (CEDI) technology [129]. Low-level radioactive wastewater (LLRW) with radionuclides as contaminants is generated by nuclear power plants. These radioactive contaminants need to be removed safely before the discharge. CEDI has proved its worth for the treatment of LLRW and the treatment of radioactive waste. This technology proves to be fruitful for the treatment of water with low salinity, with a high decontamination factor. The CEDI setup consists of an alternate arrangement of dilute and concentrated compartments with ion exchange membranes in between. In a study [129], this method has been used for the removal of specifically for the elimination of Sr (II), Cs (I), and Co (II). In this study, migration of Sr (II) and Cs (I) took place into the particles of the resin by the electric field applied. With the increase in the current, the removal capacity is enhanced. The percentage removal for Sr (II), Cs (I), and Co (II) was observed to be 98.1%, 99.7%, and 77.1%, respectively, with the application of 0.5 A current. A deposition of $Co(OH)_2$, formed during the process, took place over the surface of the cation exchange resin. Adsorption tendency of the radionuclides was of the order Cs (I) >Sr (II) >Co (II).

A hydrate-based water purification process has also been employed for the elimination of multiple nuclides such as Cs^+, Co^{2+}, I^-, and Sr^{2+} from the aquatic environment [130]. The total radionuclide reduction achieved was 85% in 70 minutes, without the impact of the presence of other ions. Gas hydrates are crystalline inclusion compounds. The intermolecular interactions between water and gas molecules stabilize these hydrates. In this method, ice-like crystals are formed by water and gas molecules at a temperature beyond the freezing point of water. This effectively separates pure crystals from impure or contaminated liquids. This method has an edge over the other methods in terms of reduction in secondary waste and efficiency at high concentrations.

Shock treatment is another method that has been tested for the removal of radionuclides from nuclear reactor wastewater [126]. Deionization shockwaves are used in a water tube for pushing the electrically charged ions in the tube linings of a porous material. This method can help in the selective, continuous, and efficient removal of cesium and cobalt from a feed. The feed consisted of cesium, cobalt, lithium, and boric acid, which were a formulation of the water from nuclear processes and light water reactors. In the process, 99.5% removal of cobalt, with 43% water recovery, could be achieved.

9.4.2 Materials for Radionuclide Wastewater Remediation

Several materials have proved their worth as sorbent materials for the elimination of radionuclides. In 2022, Chakraborty et al., in a review, discussed the use of activated carbon as an adsorbent for the elimination of radionuclides [131]. Factors such as pressure, temperature, pH, and concentration were the factors of utmost importance in the process of adsorption. Materials such as bamboo charcoal have been used for the elimination of cesium [132]. The material showed

98% removal efficiency. Other such material, rice stem, was also investigated for uranium removal, which gave 97.6% removal efficiency [133]. In another study removal of radionuclides from acidic solutions was carried out using impregnation of methyl and carboxy-benzotriazoles over activated carbon. Activated carbon has been successfully investigated as an adsorbent for different radionuclides such as iodine, cobalt, uranium, strontium, etc. [134]. Results reveal the method to be a cost-effective, efficient means for the eradication of radionuclides such as uranium [135].

In addition to activated carbon produced from different materials, zeolites have also been tested for efficient removal of cesium, strontium, uranium, etc. [136]. Modifications in the zeolites or their activation using chemical agents or synthetic zeolites have also been successfully tested for the removal [137].

Graphene oxide [138, 139] has been used to treat wastewater with radioactive contaminants. In recent studies, metal carbides, which are two-dimensional, nitrides, along with carbon nitrides, known as MXenes, have gained focus for the treatment of radionuclides [140]. These possess excellent adsorption capacity and a high surface area for radionuclide adsorption. The use of MXenes is cost-effective and efficient, promising reuse potential. These are emerging materials also used for making composites for wastewater treatment [141]. In a study [142], a composite of MXene/silver nanowire (AgNW) was prepared and investigated for the removal of cesium (Cs^+) and iodide (I^-). Substantial adsorption capacities were portrayed by the material. Many other types of composites of MXenes have been prepared and investigated for the removal of radionuclides [143, 144].

Nanomaterials have also been excellent in the elimination of radioactive species from wastewater. Materials such as metal nanoparticles, nanocomposite membranes, nanotubes, and nanofibers have also been used [145]. MOFs, COFs, and amorphous porous organic polymers (POPs) have also been tested for radionuclide removal [146]. Polymeric materials have also shown their potential to remove radionuclides. One such synthetic polymer has been a melamine styrene-based polymer [32] for the removal of strontium.

Yao et al. in 2022 synthesized [147] novel Prussian blue tetragonal nanorods by solvothermal methods for the removal of Sr and Cs radionuclides. Analyses such as Fourier transform infrared spectroscopy (FTIR), thermogravimetric analysis (TGA), scanning electron microscopy (SEM), transmission electron microscopy (TEM), surface analysis, and zeta potential were done. These nanorods were found to be highly crystalline, stable under alkaline conditions, with a few crystal defects. The process of adsorption was endothermic and spontaneous, following the pseudo-second-order rate equation. The nanorod exhibited a maximum adsorption efficiency of 256.62 and 194.26 mg/g for Sr^{2+} and Cs^+, respectively. The adsorption process was varied out at a temperature of 298 K and pH 7.0 at an adsorbate concentration of 500 mg/L.Cs^+ entered the nanorod crystals with the generation of a new phase, and Sr^{2+} ions were trapped in the internal crystal.

Radio cesium was removed using a nanozeolite in 2016 [148]. Adsorption equilibrium was attained in 1 minute. The nanozeolite exhibited a higher reaction rate in comparison to that of larger zeolites, as supported by $k_{2,\text{bulk}}/k_{2,\text{nano}}$ ratio, and initial concentration of Cs, which gave a correlation coefficient of 0.99. Nanozeolite precipitation could easily be carried out using ferric flocculant.

9.5 Role of Regulatory Authorities

The regulatory authorities set protocols in cases of any accidental release and coordinate the efforts toward response for such incidents. The regulatory authorities also collaborate with several international organizations to implement best practices for the protection from radioactive contamination. The role of these authorities and policymakers becomes crucial for the successful implementation of discharge standards and appropriate risk management. Several standards are imposed by the regulatory authorities regarding the discharge limits of radioactive contaminants.

Several authorities such as US Environmental Protection Agency (EPA) and Department of Energy (DOE) in the United States, Office of Nuclear Regulation (ONR) and Scottish Environment protection Agency (SEPA) in the United Kingdom, Nuclear Regulation Authority (NRA) in Japan, and International Atomic Energy Agency (IAEA), work with international collaboration. These provide technical expertise and standards related to the use of radioactive materials. These agencies also help to cater to several radiological incidents. The World Health Organization (WHO) also guides safety in incidents of nuclear and radiological emergencies. The International Atomic Energy Agency has issued a safety guideline to protect the environment and the people. In addition to this, to ensure that the safety laws are abided by, licenses are issued to the facilities that deal with radioactive materials. In addition, these agencies carry out inspections for verification of compliance with the safety protocols. For example, the Nuclear Energy Agency (NEA), a special agency in the Organization for Economic Co-operation and Development (OECD), which takes care of contamination by radioactive substances by looking after the management of radioactive wastes, emphasizes the compliance of measures for radioactive waste management [149]. It disseminates information regarding radioactive waste to stakeholders. The management of radioactive nuclides in wastewater in Europe is governed by an integrated system of European Union (EU) directives, policies, and initiatives for research and strategies for engagement of public engagement. As per the EU's Radioactive Waste and Spent Fuel Management Directive (2011/70/Euratom) program [150], all the

EU countries should have a policy for the management of the radioactive waste from the point where they are generated to disposal. The European Commission encourages and supports research in the field for efficient implementation. One such initiative is the European Joint Programme on Radioactive Waste Management (EURAD), 2019 [151]. The European Commission and authorities pay due emphasis on the involvement of local communities in nature for radioactive waste management. The EU guides National regulatory bodies to oversee the management of radioactive nuclides in wastewater.

In India, for radiation protection and nuclear safety, the Atomic Energy Regulatory Board (AERB) is the central authority [152]. Strict guidelines are enforced by AERB to ensure proper management of wastewater containing radionuclides with different strategies. The key responsibilities of AERB are the development of guidelines and standards for safety, providing authorizations and licensing for agencies handling radioactive waste, developing facilities for radioactive waste management, and encouraging and supporting research in the field.

9.6 Challenges and Future Needs

With the rise in nuclear technologies, radioactive wastewater is becoming a concern. Also, the treatment of radioactive wastewater is more challenging than municipal and industrial wastewater, owing to the radioactivity of the contaminants and their potential risk to humans.

Decontamination of radioactive waste is about the solidification or stabilization of radioactive compounds for disposal. Although several physical, chemical, and biological methods have been explored, the challenges associated with radioactive contaminants and radioactive wastewaters are the removal of contaminants, followed by their stabilization. The removal is being emphasized rather than a complete treatment by stabilization of the removed radionuclides, which also requires attention. Further, the ability to remove multiple nuclides, treatment of high concentrations, and secondary waste production remain a challenge for the majority of the treatment processes.

Due to the complexity of wastewater containing organics, multiple radionuclides, heavy metals, etc., the complete and efficient treatment becomes challenging with the use of certain technologies, limiting the use of certain materials. Also, considering the limitations of different techniques for the detection of radionuclides, such as high cost, low sensitivity, selective removal, production of sludge, etc., it becomes essential to develop more cost-effective and sensitive solutions for the removal of radionuclides. A need for advanced remediation techniques and materials also arises to ensure safe disposal and treatment of radionuclides.

Further, to minimize the risks associated with the storage and disposal of radioactive contaminants, their volume reduction also plays a vital role.

Therefore, appropriate and efficient methods for the treatment of removed/solidified waste also need to be developed. In addition to this, the development of advanced materials for the removal of radioactive contaminants would play a pivotal role. An integration of technologies will aid effective and complete removal of radioactive contaminants to minimize the risks associated with them. The use of innovative technologies, which can be applied on a larger scale, will be a must. Robust systems for monitoring will help to keep in check the safety associated with different methods and technologies.

References

1 Touranlou, A.F., Moghimani, M., Marhamati, M., et al. (2024). Detection and measurement of radioactive substances in water and food: a narrative review. Italian Journal of Food Safety, Volume 13, Issue 1, p. 11651. https://doi.org/10.4081/ijfs.2024.116e51.

2 Saad, H., and Aref, N. (2021). Measuring of radioactive contamination by gamma-ray spectroscopy for many water projects in Baghdad City. Journal of Physics: Conference Series, Volume 2114, p. 012013. https://doi.org/10.1088/1742-6596/2114/1/012013.

3 Rumpel, N., Kammerer, F., Wagmann, M., et al. (2015). Gamma ray spectrometry of sewer sludge a useful tool for the identification of emission, sources in a city. Chimia, Volume 69, Issue 5, p. 301. https://doi.org/10.2533/chimia.2015.301.

4 Xu, J., Bobin, J., Ott, A.V., et al. (2022). Analysis of gamma-ray spectra with spectral unmixing, part II: recalibration for the quantitative analysis of HPGe measurements. Applied Radiation and Isotopes. Volume 182, p. 110082. https://doi.org/10.1016/j.apradiso.2021.110082.

5 Mao, J., Chen, L., Xia, W., et al. (2024). Measurement techniques for low-concentration tritium radiation in water: review and prospects. Sensors, Volume 24, Issue 17, p. 5722. https://doi.org/10.3390/s24175722.

6 Khan, M., and Um, W. (2015). Liquid scintillation counting methodology for ^{99}Tc analysis: a remedy for radiopharmaceutical waste. Analytical Chemistry, Volume 87, Issue 17, pp. 9054–9060. https://doi.org/10.1021/acs.analchem.5b02279.

7 https://openmedscience.com/illuminating-the-unseen-liquid-scintillation-counting-for-radioactive-measurement/ (Last accessed 28.01.25)

8 https://www.osti.gov/biblio/1243172 (Last accessed 28.01.25)

9 Bu, W., Ni, Y., Steinhauser, G., et al. (2018). The role of mass spectrometry in radioactive contamination assessment after the Fukushima nuclear accident. Journal of Analytical Atomic Spectrometry, Volume 33, pp. 519–546. https://doi.org/10.1039/C7JA00401J.

10 Strathmann, F.G., and Hoofnagle, A.N. (2011). Current and future applications of mass spectrometry to the clinical laboratory. American Journal of Clinical Pathology, Volume 136, pp. 609–616. https://doi.org/10.1309/AJCPW0TA8OBBNGCK.

11 Skwarzec, B. (2009). Determination of radionuclides in aquatic environment. In: Namiesnik J, Szefer P (eds) Analytical measurement in aquatic environment. CRC Press, Taylor and Francis Group, pp. 241–258. https://doi.org/10.1201/9781420082692-C11.

12 https://www.bmuv.de/fileadmin/Daten_BMU/Download_PDF/Strahlenschutz/leitstelle_h_alpha_spekt_awass_01_v1992_09_en_01.pdf (Last accessed 28.01.25)

13 https://www.bmuv.de/fileadmin/Daten_BMU/Download_PDF/Strahlenschutz/leitstelle_h_alpha_spekt_awass_05_v1993_12_en.pdf (Last accessed 28.01.25)

14 RCA Regional Office (RCARO) Korea Institute of Radiological & Medical Sciences (KIRAMS) (2021). Guideline for radioactivity measurements in the environment and individual dose assessment following a nuclear or radiological emergency.

15 METHOD 114 - Test Methods for Measuring Radionuclide Emission From Stationary Sources (EPA). (2017). https://www.epa.gov/emc/method-114-radionuclides-stationary-sources (Last accessed 21.03.25).

16 Hossain, F. (2020). Natural and anthropogenic radionuclides in water and wastewater: sources, treatments and recoveries. Journal of Environmental Radioactivity, Volume 225, p. 106423. https://doi.org/10.1016/j.jenvrad.2020.106423.

17 Ikeda-Ohno, A., Harrison, J.J., Thiruvoth, S., et al. (2014). Solution speciation of plutonium and americium at an Australian legacy radioactive waste disposal site. Environmental Science and Technology, Volume 48, pp. 10045–10053. https://doi.org/10.1021/es500539t.

18 Luo, X., Zhang, G., Xue W., et al. (2013). Research on a pellet co-precipitation micro-filtration process for the treatment of liquid waste containing strontium. Journal of Radio Analytical and Nuclear Chemistry, Volume 298, pp. 931–939. https://doi.org/10.1007/s10967-013-2495-x.

19 Bobrov, P.A., Slyunchev, O.M., and Semenova, T.A. (2015). Radionuclide removal from radioactively contaminated drainage water and groundwater by precipitation and sorption methods. Radiochemistry, Volume 57, pp. 537–541. https://doi.org/10.1134/S106636221505015X.

20 Rogers H., Bowers J., and Gates-Anderson, D. (2012). An isotope dilution–precipitation process for removing radioactive cesium from wastewater. Journal of Hazardous Materials, Volume 243, pp. 124–129. https://doi.org/10.1016/j.jhazmat.2012.10.006.

21 Kim, K.W., Baek, Y.J., Lee, K.Y., et al. (2016). Treatment of radioactive waste seawater by coagulation–flocculation method using ferric hydroxide and poly acrylamide. Journal of Nuclear Science and Technology, Volume 53, Issue 3, pp. 439–450. https://doi.org/10.1080/00223131.2015.1055313.

22 Oh, M., Lee, K., and Jeon, M.K. (2023). Chemical precipitation–based treatment of acidic wastewater generated by chemical decontamination of radioactive concrete. Journal of Environmental Chemical Engineering, Volume 11, Issue 5, p. 110306. https://doi.org/10.1016/j.jece.2023.110306.

23 Gäfvert, T., Ellmark, C., and Holm, E. (2002). Removal of radionuclides at a waterworks. Journal of Environmental Radioactivity, Volume 63, Issue 2, pp. 105–115. https://doi.org/10.1016/s0265-931x(02)00020-6.

24 Liu, Y., Gu, P., Jia, L., et al. (2016). An investigation into the use of cuprous chloride for the removal of radioactive iodide from aqueous solutions. Journal of Hazardous Materials, Volume 302, pp. 82–89. https://doi.org/10.1016/j.jhazmat.2015.09.045.

25 Rana, D., Matsuura, T., Kassim, M.A., et al. (2013). Radioactive decontamination of water by membrane processes-a review. Desalination, Volume 321, pp. 77–92. https://doi.org/10.1016/j.desal.2012.11.007.

26 International Atomic Energy Agency (2004). Application of Membrane Technologies for Liquid Radioactive Waste Processing, Technical Reports Series No. 431, IAEA, Vienna.

27 Zhao, J., Wang, T., Zhang, D., et al. (2007). Treatment of plutonium-containing wastewater by a combined flocculation-microfiltration process. Nuclear Radiochemistry, Volume 29, pp. 113–117.

28 Zhuang, S., and Wang, J. (2024). Cesium removal from radioactive wastewater by adsorption and membrane technology. Frontiers in Environmental Science and Engineering, Volume 18, Issue 3, p. 38. https://doi.org/10.1007/s11783-024-1798-1.

29 Ambashta, R.D., and Sillanp, M.E.T. (2012). Membrane purification in radioactive waste management: a short review. Journal of Environmental Radioactivity, Volume 105, pp. 76–84.

30 Ahmad, N.A., Tam, L.J., Goh, P.S., et al. (2023). Treatment of radionuclide-containing wastewater using thin film composite reverse osmosis membrane with spray coating-assembled titania nanosheets. Journal of Environmental Chemical Engineering, Volume 11, Issue 5, p. 110540. https://doi.org/10.1016/j.jece.2023.110540.

31 Yang, H.M., Park, C.W., Kim, I., et al. (2019). Hollow flower-like titanium ferrocyanide structure for the highly efficient removal of radioactive cesium from water. Chemical Engineering Journal, Volume 392, p. 123713. https://doi.org/10.1016/j.cej.2019.123713.

32 Karabayir, E., Ozdemir, A., Senkal, B.F., et al. (2019). A radioactively durable melamine-styrene based polymer: highly efficient removal of ^{90}Sr. Applied Radiation and Isotopes, Volume 149, pp. 96–103. https://doi.org/10.1016/j.apradiso.2019.04.024.

33 Ma, H., Shen, M., Tong, Y., et al. (2023). Radioactive wastewater treatment technologies: a review. Molecules, Volume 28, Issue 4, p. 1935. https://doi.org/10.3390/molecules28041935.

34 Dai, S., Ning W., Chenjia, Q., et al. (2019). Preparation of core-shell structure Fe_3O_4@C@MnO_2 nanoparticles for efficient elimination of U (VI) and Eu (III) ions. Science of the Total Environment, Volume 685, pp. 986–996. https://doi.org/10.1016/j.scitotenv.2019.06.292.

35 Nekrasova, N.A., Milyutin, V.V., Kaptakov, V.O., et al. (2023). Inorganic sorbents for wastewater treatment from radioactive contaminants. Inorganics, Volume 11, Issue 3, p. 126. https://doi.org/10.3390/inorganics11030126.

36 Hassan, R.S., Abass, M.R., Eid, M.A., et al. (2021). Sorption of some radionuclides from liquid waste solutions using anionic clay hydrotalcite sorbent. Applied Radiation and Isotopes, Volume 178, p. 109985. https://doi.org/10.1016/j.apradiso.2021.109985.

37 Fuks, L., and Oszczak-Nowińska, A. (2021). Sorption of selected radionuclides from liquid radioactive waste by sorbents of biological origin: the alkaline earth alginates. Nukleonika, Volume 66, Issue 4, pp. 153–158. https://doi.org/10.2478/nuka-2021-0023.

38 Moussa, S.I., Mekawy, Z.A., Dakroury, G.A., et al. (2023). Linear and nonlinear recognition for the sorption of ^{60}Co and $^{152+154}$Eu radionuclides onto bio CuO nanocomposite. Journal of Polymers and the Environment, Volume 31, pp. 2148–2165. https://doi.org/10.1007/s10924-022-02735-4.

39 Modi, M.K., Pattanaik, P., and Dash, N. (2015). Sorption of radionuclide. International Journal of Pharmaceutical Sciences Review and Research, Volume 34, Issue 1, pp. 122–130.

40 Xu, M., Cai, Y., Chen, G., et al. (2022). Efficient selective removal of radionuclides by sorption and catalytic reduction using nanomaterials. Nanomaterials, Volume 12, Issue 9, p. 1443. https://doi.org/10.3390/nano12091443.

41 Helal, A.S., Hémadi, M., Lomas, J.S., et al. (2025). Uranium removal from environmental water and nuclear waste: nanomaterial solutions and their environmental sustainability. Chemical Engineering Journal, Volume 507, p. 160298. https://doi.org/10.1016/j.cej.2025.160298.

42 Wang, L.L., Luo, F., Dang, L.L., et al. (2015). Ultrafast high-performance extraction of uranium from seawater without pretreatment using an acylamide- and carboxyl-functionalized metal–organic framework. Journal of Materials Chemistry A, Volume 3, pp. 13724–13730. https://doi.org/10.1039/C5TA01972A.
43 Chen, L., Bai, Z., Zhu, L., et al. (2017). Ultrafast and efficient extraction of uranium from seawater using an amidoxime appended metal-organic framework. ACS Applied Materials and Interfaces, Volume 9, pp. 32446–32451. https://doi.org/10.1021/acsami.7b12396.
44 Wang, X.F., Chen, Y., Song, L.P., et al. (2019). Cooperative capture of uranyl ions by a carbonyl-bearing hierarchical-porous Cu-organic framework. Angewandte Chemie International Edition, Volume 58, pp. 18808–18812. https://doi.org/10.1002/anie.201909045.
45 Yuan, Y., Feng, S., Feng, L., et al. (2020). Spatial structure bio-inspired nano-pocket for targeting uranyl capture. Angewandte Chemie International Edition, Volume 59, Issue 11, pp. 4262–4268. https://doi.org/10.1002/anie.201916450.
46 Cheng, G., Zhang, A., Zhao, Z., et al. (2021). Extremely stable amidoxime functionalized covalent organic frameworks for uranium extraction from seawater with high efficiency and selectivity. Science Bulletin, Volume 66, pp. 1996–2001. https://doi.org/10.1016/j.scib.2021.05.012.
47 Majeed, M.D., and Roushani, M. (2024). Uranium (VI) ion removal using a novel high adsorption COFs/GO/CH aerogel nanocomposite for wastewater treatment. Results in Chemistry, Volume 12, p. 101933. https://doi.org/10.1016/j.rechem.2024.101933.
48 Gu, Z., Wang, Y., Tang, J., et al. (2015). The removal of uranium (VI) from aqueous solution by graphene oxide-carbon nanotubes hybrid aerogels. Journal of Radioanalytical and Nuclear Chemistry, Volume 303, pp. 1835–1842. https://doi.org/10.1007/s10967-014-3795-5.
49 Zhao, G., Wen, T., and Yang, X. (2012). Preconcentration of U (VI) ions on few-layered graphene oxide nanosheets from aqueous solutions. Dalton Transactions, Volume 41, pp. 6182–6188. https://doi.org/10.1039/c2dt00054g.
50 Sun, Y., Yang, S., and Sheng, G. (2012). The removal of U (VI) from aqueous solution by oxidized multiwalled carbon nanotube. Journal of Environmental Radioactivity, Volume 105, pp. 40–47. https://doi.org/10.1016/j.jenvrad.2011.10.009.
51 Schierz, A., and Zänker, H. (2009). Aqueous suspensions of carbon nanotubes: surface oxidation, colloidal stability and uranium sorption. Environmental Pollution, Volume 157, pp. 1088–1094. https://doi.org/10.1016/j.envpol.2008.09.045.
52 Tan, L., Wang, Y., and Liu, Q. (2015). Song, enhanced adsorption of uranium (VI) using a three-dimensional layered double hydroxide/graphene hybrid

material. Chemical Engineering Journal. Volume 259, pp. 752–760. https://doi.org/10.1016/j.cej.2014.08.015.

53 Shao, M.Y., Jiang, D.Z., and Wang, X. (2009). Plasma induced grafting carboxymethyl cellulose on multiwalled carbon nanotubes for the removal of UO^{22+} from aqueous solution. The Journal of Physical Chemistry B, Volume 113, pp. 860–864. https://doi.org/10.1021/jp8091094.

54 Zhu, M., Cai, Y., Liu, S., et al. (2019). $K_2Ti_6O_{13}$ hybridized graphene oxide: effective enhancement in photodegradation of RhB and photoreduction of U (VI). Environmental Pollution, Volume 248, pp. 448–455. https://doi.org/10.1016/j.envpol.2019.02.025.

55 https://encyclopedia.pub/entry/46925 (Last accessed 24.03.25)

56 Liu, X., Du, P., Pan, W., et al. (2018). Immobilization of uranium (VI) by niobate/titanate nanofakes heterojunction through combined adsorption and solar-light-driven photocatalytic reduction. Applied Catalysis B: Environmental, Volume 231, pp. 11–22. https://doi.org/10.1016/j.apcatb.2018.02.062.

57 Li, P., Wang, J., and Wang, Y., (2019). Photoconversion of U (VI) by TiO_2: an efficient strategy for seawater uranium extraction. Chemical Engineering Journal, Volume 365, pp. 231–241. https://doi.org/10.1016/j.cej.2019.02.013.

58 Salomone, V.N., Meichtry, J.M., and Litter, M.I. (2015). Heterogeneous photocatalytic removal of U (VI) in the presence of formic acid: U (III) formation. Chemical Engineering Journal, Volume 270, pp. 28–35. https://doi.org/10.1016/j.cej.2015.01.118.

59 Salomone, V.N., Meichtry, J.M., Schinelli, G., et al. (2014). Photochemical reduction of U (VI) in aqueous solution in the presence of 2-propanol. Journal of Photochemistry and Photobiology A: Chemistry, Volume 277, pp. 19–26. https://doi.org/10.1016/j.jphotochem.2013.12.006.

60 Chen, J., Ollis, D.F., Rulkens, W.H., et al. (1999). Photocatalyzed deposition and concentration of soluble uranium (VI) from TiO_2 suspensions. Colloids and Surfaces A: Physicochemical and Engineering Aspects, Volume 151, pp. 339–349. https://doi.org/10.1016/S0927-7757(98)00506-8.

61 Guo, Y., Guo, Y., Wang, X., et al. (2017). Enhanced photocatalytic reduction activity of uranium (VI) from aqueous solution using the Fe_2O_3–graphene oxide nanocomposite. Dalton Transactions, Volume 246, pp. 14762–14770. https://doi.org/10.1039/C7DT02639K.

62 Lu, C., Zhang P., Jiang S., et al. (2016). Photocatalytic reduction elimination of UO22+ pollutant under visible light with metal-free sulfur doped g-C_3N_4 photocatalyst. Applied Catalysis B: Environmental, Volume 200, pp. 378–385. https://doi.org/10.1016/j.apcatb.2016.07.036.

63 Jiang, X., Xing, Q.J., Luo, X.B., et al. (2018). Simultaneous photoreduction of Uranium (VI) and photooxidation of Arsenic (III) in aqueous solution over g-$C3N_4$/TiO_2 heterostructured catalysts under simulated sunlight irradiation. Applied Catalysis B Environmental, Volume 228, pp. 29–38.

64 Wang, G., Zhen, J., Zhou, L., et al. (2015). Adsorption and photocatalytic reduction of U (VI) in aqueous TiO2 suspensions enhanced with sodium formate. Journal of Radio Analytical and Nuclear Chemistry, Volume 304, pp. 579–585.

65 Li, Z., He, Y., Sonne, C., et al. (2023). A strategy for bioremediation of nuclear contaminants in the environment. Environmental Pollution, Volume 319, p. 120964.

66 Francis, A.J., and Nancharaiah, Y.V. (2015). In situ and ex situ bioremediation of radionuclide-contaminated soils at nuclear and NORM sites. In: van Velzen L (ed) Environmental remediation and restoration of contaminated nuclear and norm Sites (PDF). Woodhead publishing series in energy. Elsevier. pp. 185–236. https://doi.org/10.1016/B978-1-78242-231-0.00009-0.

67 Song, W., Wang, X., and Sun, Y. (2019). Bioaccumulation and transformation of U (VI) by sporangiospores of Mucor circinelloides. Chemical Engineering Journal, Volume 362, pp. 81–88.

68 Zheng, X.Y., Shen, Y.H., Wang, X.Y., et al. (2018). Effect of pH on uranium (VI) biosorption and biomineralization by *Saccharomyces cerevisiae*. Chemosphere, Volume 203, pp. 109–116.

69 Ding, L., Tan, W.F., and Xie, S.B. (2018). Uranium adsorption and subsequent re-oxidation under aerobic conditions by Leifsonia sp. - coated biochar as green trapping agent. Environmental Pollution, Volume 242, pp. 778–787.

70 Saleh, H.M., Bayoumi, T.A., and Mahmoud, H.H., (2017). Uptake of cesium and cobalt radionuclides from simulated radioactive wastewater by Ludwigia stolonifera aquatic plant. Nuclear Engineering and Design, Volume 315, pp. 94–199. https://doi.org/10.1016/j.nucengdes.2017.02.018.

71 Zheng, X.Y., Wang, X.Y., and Shen, Y.H. (2017). Biosorption and biomineralization of uranium (VI) by Saccharomyces cerevisiae- crystal formation of chernikovite. Chemosphere, Volume 175, pp. 161–169. https://doi.org/10.1016/j.chemosphere.2017.02.035.

72 Lee, S.Y., Jung, K.H., and Lee J.E. (2014). Photosynthetic bio mineralization of radioactive Sr via micro algal CO_2 absorption. Bioresource Technology, Volume 172, pp. 449–452. https://doi.org/10.1016/j.biortech.2014.09.023.

73 Martins, M., Faleiro, M.L., and Chaves, S. (2010). Anaerobic bio-removal of uranium (VI) and chromium (VI): comparison of microbial community structure. Journal of Hazardous Materials, Volume 176, pp. 1065–1072. https://doi.org/10.1016/j.jhazmat.2009.11.149.

74 Carvajal, D.A., Katsenovich, Y.P., and Lagos L.E. (2012). The effects of aqueous bicarbonate and calcium ions on uranium biosorption by Arthrobacter G975 strain. Chemical Geology, Volume 330-331, pp. 1–59. https://doi.org/10.1016/j.chemgeo.2012.08.018.

75 Kang, D.J., Seo, Y.J., and Saito, T. (2012). Uptake and translocation of cesium-133 in Napier grass (*Pennisetum purpureum Schum.*) under hydroponic conditions. Ecotoxicology and Environmental Safety, Volume 82, pp. 122–126. https://doi.org/10.1016/j.ecoenv.2012.05.017.

76 Saleh, H.M. (2012). Water hyacinth for phytoremediation of radioactive waste simulate contaminated with cesium and cobalt radionuclides. Nuclear Engineering and Design, Volume 242, pp. 425–432. https://doi.org/10.1016/j.nucengdes.2011.10.023.

77 Katsoyiannis, I.A., Althoff, H.W., and Bartel, H. (2006). The effect of groundwater composition on uranium (VI) sorption onto bacteriogenic iron oxides. Water Research, Volume 40, pp. 3646–3652. https://doi.org/10.1016/j.watres.2006.06.032.

78 Soudek, P., Tykva, R., and Vaněk, T. (2004). Laboratory analyses of ^{137}Cs uptake by sunflower, reed and poplar. Chemosphere, Volume 55, pp. 1081–1087. https://doi.org/10.1016/j.chemosphere.2003.12.011.

79 Bahaj, A.S., Croudace, I.W., and James, P.A.B. (1998). Continuous radionuclide recovery from wastewater using magnetotactic bacteria. Journal of Magnetism and Magnetic Materials, Volume 184, pp. 241–244. https://doi.org/10.1016/S0304-8853 (97)01130-X.

80 Ferreira, R.V.D.P., Sakata, S., Dutra, F., et al. (2011). Treatment of radioactive liquid organic waste using bacteria community. Journal of Radioanalytical and Nuclear Chemistry. Volume 292, pp. 811–817.

81 Liu, X., Hu, W., Huang, X., et al. (2015). Highly effective biosorption of Sr (II) from low level radioactive wastewater. Water Science and Technology, Volume 71, pp. 1727–1733. https://doi.org/10.2166/wst.2015.156.

82 Tsezos, M., and Volesky, B. (2010). Biosorption of uranium and thorium. Biotechnological Bioengineering, Volume 23, pp. 583–604.

83 Sylvester, P., Milner T., and Jensen, J. (2013). Radioactive liquid waste treatment at Fukushima Daiichi. Journal of Chemical Technology and Biotechnology. Volume 88, pp. 1592–1596. https://doi.org/10.1002/jctb.4141.

84 https://www.frtr.gov/matrix/Electrokinetic-Enhanced-Remediation/ (Last accessed 14.3.25)

85 Yang, Q., Hou, L., and Wang, Y. (2007). Research progress in low- and medium-level radioactive wastewater treatment technology. Environmental Science Management, Volume 32, pp. 103–110.

86 International Atomic Energy Agency (2002). Application of ion exchange processes for the treatment of radioactive waste and management of spent ion exchangers, technical reports series no. 408. International Atomic Energy Agency, Vienna.

87 Yu, K., Shao, P., Meng, P., et al. (2020). Superhydrophilic and highly elastic monolithic sponge for efficient solar-driven radioactive wastewater treatment under one sun. Journal of Hazardous Materials, Volume 392, p. 122350. https://doi.org/10.1016/j.jhazmat.2020.122350.

88 Fang, X., Xu, Z., Luo, Y., et al. (2016). Removal of radionuclides from laundry wastewater containing organics and suspended solids using inorganic ion exchanger. Procedia Environmental Sciences, Volume 31, pp. 375–381. https://doi.org/10.1016/j.proenv.2016.02.053.

89 Youssefa, M.A., Attallaha, M.F., Lasheen, Y.F., et al. (2019). Assessment removal of tritium radionuclide from liquid waste using sequential ion exchange resin. Desalination and Water Treatment, Volume 160, pp. 50–56. https://doi.org/10.5004/dwt.2019.24374.

90 Figueiredo, B.R., Cardoso, S.P., Portugal, I., et al. (2017). Inorganic ion exchangers for cesium removal from radioactive wastewater. Separation and Purification Reviews, Volume 47, Issue 4, pp. 306–336. https://doi.org/10.1080/15422119.2017.1392974.

91 El-Naggar, M., El-Kamash, A., El-Dessouky, M., et al. (2008). Two-step method for preparation of NaA-X zeolite blend from fly ash for removal of cesium ions. Journal of Hazardous Materials, Volume 154, pp. 963–972. https://doi.org/10.1016/j.jhazmat.2007.10.115.

92 Galamboš, M., Paučová, V., Kufčáková, J., et al. (2010). Cesium sorption on bentonites and montmorillonite K10. Journal of Radioanalytical and Nuclear Chemistry, Volume 284, pp. 55–64. https://doi.org/10.1007/s10967-010-0480-1.

93 Adamson, D., Howe, A., and McCabe D. (2017). Preparation and evaporation of Hanford waste treatment plant direct feed low activity waste effluent management facility simulant. Savannah River Site, Aiken, SC, USA.

94 Asghar, K., Ngulimi, M.F., Kim, S., et al. (2024). Cobalt recovery from industrial and nuclear waste resources: a review. Chemical Engineering Journal Advances, Volume 20, p. 100668. https://doi.org/10.1016/j.ceja.2024.100668.

95 Bhattacharyya, A., Mohapatra, P.K., Pathak, P.N., et al. (2006). Cation-exchange separation of uranium from thorium in nitric acid medium. Journal of Radio Analytical and Nuclear Chemistry, Volume 268, pp. 323–328. https://doi.org/10.1007/s10967-006-0168-8.

96 Hou, C., Lin, W., Yang, L. et al. (2022). Experimental performance evaluation and model-based optimal design of a mechanical vapour recompression system for radioactive wastewater treatment. Energy Conversion and Management, Volume 252, p. 115087. https://doi.org/10.1016/j.enconman.2021.115087.

97 McCullough, G.E. (1951). Concentration of radioactive liquid waste by evaporation. Industrial and Engineering Chemistry, Volume 43, Issue 7, pp. 1505–1509. https://doi.org/10.1021/ie50499a017.

98 Zaki, A.A., Hala, A.A., and Othman, E.A. (2016). Performance improvement of a radioactive forced circulation evaporator system. Arab Journal of Nuclear Sciences and Applications, Volume 49, Issue 1, pp. 76–86.

99 IAEA (2003). Combined methods for liquid radioactive waste treatment. IAEA, Vienna.

100 Cao, J.G., Gu, P., Zhao, J., et al. (2010). Removal of strontium from an aqueous solution using co-precipitation followed by microfiltration (CPMF). Journal of Radioanalytical and Nuclear Chemistry, Volume 285, pp. 539–546. https://doi.org/10.1007/s10967-010-0564-y.

101 Arnal, J.M., Sancho, M., García-Fayos, B., et al. (2017). Declassification of radioactive water from a pool type reactor after nuclear facility dismantling. Radiation Physics and Chemistry, Volume 138, pp. 72–74. https://doi.org/10.1016/j.radphyschem.2017.03.027.

102 Sancho, M., Arnal J.M., and García-Fayos, B. (2013). Treatment of hospital radioactive liquid wastes from RIA (radioimmunoassay) by membrane technology. Desalination, Volume 321, pp. 110–118. https://doi.org/10.1016/j.desal.2013.03.032.

103 Zhang, C.P., Gu, P., Zhao, J., et al. (2009). Research on the treatment of liquid waste containing cesium by an adsorption-microfiltration process with potassium zinc hexacyanoferrate. Journal of Hazardous Materials, Volume 167, Issue (1-3), pp. 1057–1062. https://doi.org/10.1016/j.jhazmat.2009.01.104.

104 Zhang, X., Gu, P., and Liu, Y. (2019). Decontamination of radioactive wastewater: state of the art and challenges forward. Chemosphere, Volume 215, pp. 543–553. https://doi.org/10.1016/j.chemosphere.2018.10.029.

105 Lei, H., Muhammad, Y., Wang, K., et al. (2021). Facile fabrication of metakaolin/slag-based zeolite microspheres (M/SZMs) geopolymer for the efficient remediation of Cs^+ and Sr^{2+} from aqueous media. Journal of hazardous materials, Volume 406, p. 124292. https://doi.org/10.1016/j.jhazmat.2020.124292.

106 He, W., Ai, K., Ren, X., et al. (2017). Inorganic layered ion-exchangers for decontamination of toxic metal ions in aquatic systems. Journal of Materials Chemistry A, Volume 5, pp. 19593–19606. https://doi.org/10.1039/C7TA05076C.

107 https://www.opb.org/article/2024/10/12/hanford-radioactive-nuclear-richland-washington-plutonium-world-war-2-manhattan-project/ (Last accessed 28.01.25)

108 https://www.britannica.com/topic/Hanford-Engineer-Works (Last accessed 21.03.25)

109 https://darrp.noaa.gov/hazardous-waste/hanford-nuclear-site (Last accessed 28.01.25)

110 https://ecology.wa.gov/waste-toxics/nuclear-waste/hanford-cleanup/hanford-overview (Last accessed 28.01.25)

111 https://www.unscear.org/unscear/en/areas-of-work/fukushima.html (Last accessed 28.01.25)

112 https://www.britannica.com/event/Fukushima-accident (Last accessed 21.03.25)

113 Akleyev A.V., Degteva M.O., and Krestinina L.Y.(2022) Medical consequences of the Urals radiation accidents: comparative analysis. Environmental Advances, Volume 9, p. 100253. https://doi.org/10.1016/j.envadv.2022.100253.

114 https://www.greenpeace.org/static/planet4-hungary-stateless/2018/10/caeaa34d-caeaa34d-mayak_a-50-year-tragedy.pdf (Last accessed 28.01.25)

115 Akleev, A.V., and Kiselev, M.F. (eds) (2001). Muslyumovo: results of 50 years monitoring Ural Scientific Practical Center of Radiological Medicine. Cheljabinsk.

116 Report, Mayak: A 50-Year Tragedy, Greenpeace Russia, 2007

117 Trabalka, J.R. (1979). "Russian experience". Environmental decontamination: proceedings of the workshop, 4-5 December 1979, Oak Ridge, Tennessee. Oak Ridge National Laboratory. pp. 3–8. CONF-791234.

118 Kellerer, A. (2002). The Southern Urals radiation studies. A reappraisal of the current status. Radiation and Environmental Biophysics, Volume 41, Issue 4, pp. 307–316. https://doi.org/10.1007/s00411-002-0168-1.

119 https://www.britannica.com/event/Chernobyl-disaster (Last accessed 21.03.25)

120 Voitsekhovych, O.V, Kanivets, V., and Onishi, Y. (2007). The Chernobyl accident and its aquatic impacts on the surrounding area. In: Onishi Y, Voitsekhovich OV, Zheleznyak MJ (eds) Chernobyl–what have we learned? Environmental pollution, Volume 12. Springer. https://doi.org/10.1007/1-4020-5349-5_2.

121 Rajkhowa, S., Sarma, J., and Das, A.R. (2021). Chapter 15 - Radiological contaminants in water: pollution, health risk, and treatment. In: Ahamad A, Siddiqui SI, Singh P (eds) Contamination of water, health risk assessment and treatment strategies. Academic Press, pp. 217–236. https://doi.org/10.1016/B978-0-12-824058-8.00013-X.

122 https://www.world-nuclear-news.org/articles/south-korea-develops-novel-nuclide-separation-technology (Last accessed 24.03.25)

123 Azman, M.A.N., Goh, P.S., Ismail, A.F., et al. (2024). Forward osmosis as a contemporary treatment solution for mitigating radionuclide pollution in water bodies. Journal of Environmental Chemical Engineering, Volume 12, Issue 3, p. 112542. https://doi.org/10.1016/j.jece.2024.112542.

124 Qian, Z., Tao, Y., Peng, Y., et al. (2025). Capacitive deionization: an emerging electrosorption technology for efficient radioactive wastewater treatment. Journal of Environmental Chemical Engineering, Volume 13, Issue 1, p. 115104. https://doi.org/10.1016/j.jece.2024.115104.

125 Won, Y. (2023). Removal of Radionuclides using Capacitive Deionization. *LSU Doctoral Dissertations*. 6091. Louisiana State University and Agricultural and Mechanical College. https://repository.lsu.edu/gradschool_dissertations/6091

126 Alkhadra, M.A., Conforti, K.M., Gao, T., et al. (2020). Continuous separation of radionuclides from contaminated water by shock electrodialysis. Environmental Science and Technology, Volume 54, Issue 1, pp. 527–536. https://doi.org/10.1021/acs.est.9b05380.

127 Chaalal, O., and Islam, M.R. (2001). Integrated management of radioactive strontium contamination in aqueous stream systems. Journal of Environmental Management, Volume 61, Issue 1, pp. 51–59. https://doi.org/10.1006/jema.2000.0399.

128 Srinath, N., Vivek, P., Rajakumari, K., et al. (2025). Biotechnological innovations in radioactive waste management technologies. In: Kumar N (ed). Radioactive pollutant. Environmental science and engineering. Springer, Cham. https://doi.org/10.1007/978-3-031-73796-1_7.

129 Jianga, B., Lia F., and Zhao, X. (2019). Removal of trace Cs(I), Sr(II), and Co(II) in aqueous solutions using continuous electrodeionization (CEDI). Desalination and Water Treatment, Volume 155, pp. 175–182.

130 Seo, S.D., Truong-Lam, H.S., Jeon, C., et al. (2024). Simultaneous removal of multi-nuclide (Sr^{2+}, Co^{2+}, Cs^{+}, and I^{-}) from aquatic environments using a hydrate-based water purification process. Journal of Hazardous Materials, Volume 462, p. 132700. https://doi.org/10.1016/j.jhazmat.2023.132700.

131 Chakraborty, A., Pal, A., and Saha, B.B., (2022). A critical review of the removal of radionuclides from wastewater employing activated carbon as an adsorbent. Materials (Basel), Volume 15, Issue 24, p. 8818. https://doi.org/10.3390/ma15248818.

132 Khandaker, S., Toyohara Y., Kamida, S., et al. (2018). Adsorptive removal of cesium from aqueous solution using oxidized bamboo charcoal. Water Resources and Industry, Volume 19, pp. 35–46. https://doi.org/10.1016/j.wri.2018.01.001.

133 Xiao-Teng, Z., Dong-Mei, J., Yi-Qun, X., et al. (2019). Adsorption of uranium (VI) from aqueous solution by modified rice stem. Journal of Chemistry, pp. 1–10. https://doi.org/10.1155/2019/6409504.

134 Abu-Dalo, M.A., Nevostrueva, S., and Hernandez, M. (2020). Removal of radionuclides from acidic solution by activated carbon impregnated with methyl- and carboxy-benzotriazoles. Scientific Reports, Volume 10, p. 11712. https://doi.org/10.1038/s41598-020-68645-4.

135 Jiménez-Reyes, M., Almazán-Sánchez, P.T., and Solache-Ríos M. (2021). Radioactive waste treatments by using zeolites. A short review. Journal of Environmental Radioactivity. Volume 233, p. 106610. https://doi.org/10.1016/j.jenvrad.2021.106610.

136 Olguín, M.T., Solache-Ríos, M., Acosta, D., et al. (1999). Uranium sorption in zeolite X: the valence effect. Microporous and Mesoporous Materials, Volume 28, Issue 3, pp. 377–385. https://doi.org/10.1016/S1387-1811(98)00306-0.

137 Romanchuk, A.Y., Slesarev, A.S., Kalmykov, S.N., et al. (2012).Graphene oxide for effective radionuclide removal. Physical Chemistry Chemical Physics, Volume 15, pp. 2321–2327. https://doi.org/10.1039/C2CP44593J.

138 https://scitechdaily.com/graphene-oxide-removes-radioactive-material-from-water/ (Last accessed 28.01.25)

139 Xu, L., and Wang, J. (2017). The application of graphene-based materials for the removal of heavy metals and radionuclides from water and wastewater. Critical Reviews in Environmental Science and Technology. Volume 47, Issue 15, pp. 1042–1105. https://doi.org/10.1080/10643389.2017.1342514.

140 Zubair, M., Radkiany, R.S., Bilal M., et al. (2024) Innovative solutions for the removal of radionuclides from water – a review. Materials Science in Semiconductor Processing, Volume 178, p. 108450. https://doi.org/10.1016/j.mssp.2024.108450.

141 Ibrahim, Y., Meslam, M., and Eid, K. (2022). A review of MXenes as emergent materials for dye removal from wastewater. Separation and Purification Technology, Volume 282, p. 120083. https://doi.org/10.1016/j.seppur.2021.120083.

142 Mushtaq, S., Husnain, S.M., Kazmi, S.A.R., et al. (2023). MXene/AgNW composite material for selective and efficient removal of radioactive cesium and iodine from water. Scientific Reports, Volume 13, p. 19696. https://doi.org/10.1038/s41598-023-47075-y.

143 Wang, L., and Liu, Y.-L. (2023). Facile Synthesis of MXene/MnO_2 nanocomposites for efficient removal of radionuclide uranium. Crystals, Volume 13, Issue 5, p. 804. https://doi.org/10.3390/cryst13050804.

144 Liu, Y.-L., Li, D., Cao, P., et al. (2024). Advances in MXene-based composite materials for efficient removal of radioactive nuclides and heavy metal ions. Materials Today Physics, Volume 44, p. 101444. https://doi.org/10.1016/j.mtphys.2024.101444.

145 Sangeetha, J., David, M., Naik, J., et al. (2020). Removal of radioactive wastes using nanomaterials. In: Kharissova OV, Torres-Martínez LM, Kharisov BI (eds). Handbook of nanomaterials and nanocomposites for energy and environmental applications, Springer, Cham. https://doi.org/10.1007/978-3-030-36268-3_80.

146 Hao, M., Liu, Y., Wu, W., et al. (2023). Advanced porous adsorbents for radionuclides elimination. EnergyChem, Volume 5, Issue 4, p. 100101. https://doi.org/10.1016/j.enchem.2023.100101.

147 Yao, C., Dai, Y., Chang, S., et al. (2023). Removal of cesium and strontium for radioactive wastewater by Prussian blue nanorods. Environmental Science and Pollution Research, Volume 30, pp. 36807–36823. https://doi.org/10.1007/s11356-022-24618-w.

148 Lee, K.-Y., Kim, K.-W., Park, M., et al. (2016). Novel application of nanozeolite for radioactive cesium removal from high-salt wastewater. Water Research, Volume 95, pp. 134–141. https://doi.org/10.1016/j.watres.2016.02.052.

149 https://www.oecd-nea.org/jcms/c_12892/radioactive-waste-management (Last accessed 21.03.25)

150 Radioactive waste and spent fuel - European Commission. https://energy.ec.europa.eu/topics/nuclear-energy/radioactive-waste-and-spent-fuel_en (Last accessed 21.03.25)

151 Radioactive waste management: local communities need to have their say | EESC (2024). https://www.eesc.europa.eu/en/news-media/press-releases/radioactive-waste-management-local-communities-need-have-their-say (Last accessed 21.03.25)

152 Radioactive Waste Management | AERB - Atomic Energy Regulatory Board. https://aerb.gov.in/english/publication/codes-guides/general/radioactive-waste-management (Last accessed 21.03.25)

10

Microbes as Contaminants

Wastewater is one of the most ideal medium for the growth of microbes like bacteria, viruses, fungi, parasitic cysts, eggs, helminths, etc. Owing to the health hazards caused by these microorganisms, their identification and complete removal becomes essential for the safe use and reuse of water.

Microbial infestation and growth in the water bodies can lead to several detrimental effects, ranging from physical ailments to genetic disorders. Bacteria can cause typhoid, cholera, tuberculosis, while protozoans are responsible for dysentery. These put humans at risk when contaminated wastewater is discharged into the rivers, ponds, green spaces, and agricultural fields. This poses serious health implications, even if present in low concentrations. One such global effect has already been witnessed due to coronavirus 2 (SARS-CoV-2), where the deadly virus managed to enter the water and water bodies through the masks and stool of the infected persons [1]. The microbe was so powerful that it has almost affected the entire globe and led to several deaths. Due to the growing population and industrialization, the emergence of contaminants hassled to a greater risk of contamination of water bodies and sources by microbial species. Development of new strains and the re-emergence of microbial species are also of great concern.

Table 10.1 summarizes a few diseases caused due to pathogens associated with wastewater that come in contact with humans via water [2].

There are various methods for the detection of microbes in water and wastewater, such as culture-based techniques, next-generation sequencing, molecular approach, microscopy, polymerase chain reaction (PCR), multiplex PCR, nanotechnology, etc. [3–5]. Though microbes are an integral part of some biological treatment processes yet there could be adverse effects when this microbial presence is of pathogens. Also, due to diverse forms of microbial species, it is impractical to monitor them all to assess the treatment method. Thus, usually an indicator organism is chosen that helps to suggest the contamination level of the

Table 10.1 Diseases caused due to pathogens.

Contaminant	Pathogenic species	Diseases
Virus	*Enteroviruses*	Fever, meningitis, paralysis
	Caliciviruses	Hepatitis
	Astroviruses	Fever, diarrhea
	Coronavirus	Respiratory disorder
	Enterovirus	Respiratory tract infection, blurred vision
	Reovirus	Pneumonia, encephalitis, meningitis
	Parvovirus	Rashes
Bacteria	*Salmonella*	Typhoid and diarrhea
	Campylobacter	Foodborne diseases
	Escherichia coli	Renal failure
	Vibrio cholerae	Cholera
Cyanobacteria	*Microcystis, Anabaena,*	Diarrhea, liver damage
	Aphanizomenon	Dysentery
Protozoa	*Cryptosporidium parvum*	Cryptosporidiosis
	Giardia lamblia	Chronic diarrhea, abdominal discomfort
	Entamoeba histolytica	Amebic dysentery
Helminths	*Schistosoma mansoni*	Liver and intestinal damage
	Trichuris trichiura	Trichuriasis
	Ascaris lumbricoides	Ascariasis

pathogens. Widely used indicators include coliform bacteria, aerobic and anaerobic spores, helminth ova, and T4 coliphage viruses [6].

The possible presence of facial contamination and pathogens can be detected with the help of indicator organisms. Indicator organisms are those that are used as a surrogate to evaluate the pathogenic occurrence. These are non-pathogenic with minimal or no growth in water and have high detection ability. It must be available in abundance and should be easy to culture. The US Environmental Protection Agency mandates the use of *E. coli* as a fecal indicator to comply with the monitoring standards of the wastewater discharges. *Enterococci* have also been used as fecal indicators, but are a poor surrogate for viruses. Bacteriophages are potential virus fecal indicators. These indicators convey no information regarding the pathway of decontamination or inactivation of pathogens. Thus, organisms or their consortium should be developed that can suggest the pathogenic behavior like *E. coli* and F-RNA phage [7].

Apart from these, particle size distributions are also used as a surrogate for helminth ova analysis in wastewater [8]. Table 10.2 summarizes various indicator

Table 10.2 Various indicator organisms and their nature.

Pathogen	Indicator organism	Characteristic features
Bacteria	*E.coli* Enterococci	• *E. coli* is Gram-negative, rod-shaped organism with, 1.8 μm length and 0.8 μm diameter • Nonspore-forming • Present in the gut of warm-blooded animals • Ferment lactose
	Total coliforms	• Gram-negative • Nonspore-forming • Rod-shaped facultative anaerobic bacteria • Ferment lactose to acid and gas • No specific indicators of fecal pollution
	Fecal coliforms	• Thermotolerant coliforms • Ferment lactose to acid and gas
	Fecal streptococci	• Gram-positive • Nonspore-forming • Grow on bile agar at 45 °C, • Genera *Enterococcus* and *Streptococcus*
Viruses	Fecal bacteriophage F-RNA Coliphage MS2	• Resistant to disinfection • Structural and compositional • Present in abundance • Resemblance with the target • Specific to host infection
Protozoa Cryptosporidium Giardia	*Clostridium perfringens*	• Spore-forming • Gram-positive • Obligate anaerobic species • Abundant availability • Present in feces of warm-blooded animals or decaying matter

Source: [8]/Victoria University.

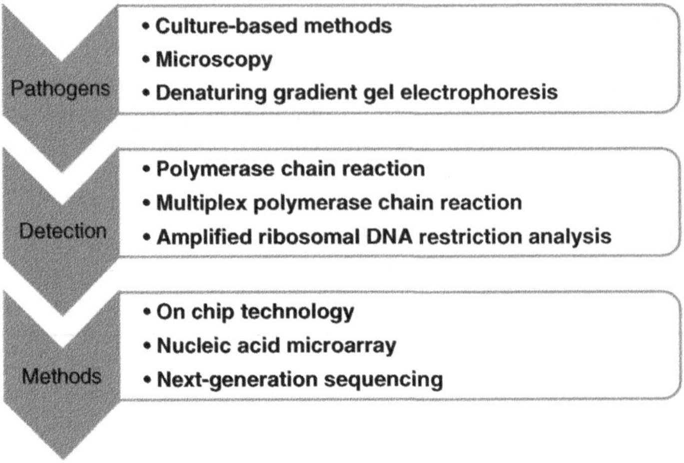

Figure 10.1 Pathogen detection methods.

organisms commonly employed for the removal of pathogens in wastewater treatment units.

The isolation and detection of pathogenic microbes can be done using various techniques [9] as mentioned in Figure 10.1.

10.1 Treatment Methods

As mentioned by the World Health Organization (WHO) [10], different methods which are used for the treatment of microorganisms present in water are shown in Figure 10.2.

The above-depicted processes are used for the treatment of drinking water. The situation might become complex when microbes are to be eliminated from contaminated wastewater.

Adoption of the correct treatment process becomes a matter of vital importance as different pathogens are to be removed, such as bacteria, protozoa, etc., effectively, giving due consideration to the cost involved. The types of microbes in wastewater differ from region to region depending on the kind of population [11, 12]. The treatment of microbe-loaded wastewater is carried out as per WHO guidelines [13]. The treatment method aims at obtaining water that can be reused in various activities. Reuse of wastewater for agricultural irrigation involves the attainment of a particular concentration of certain microbes in the treated water. The permissible concentration for reuse in agriculture is fecal coliforms ≤103–104/100 mL and helminth eggs ≤1/L in treated wastewater.

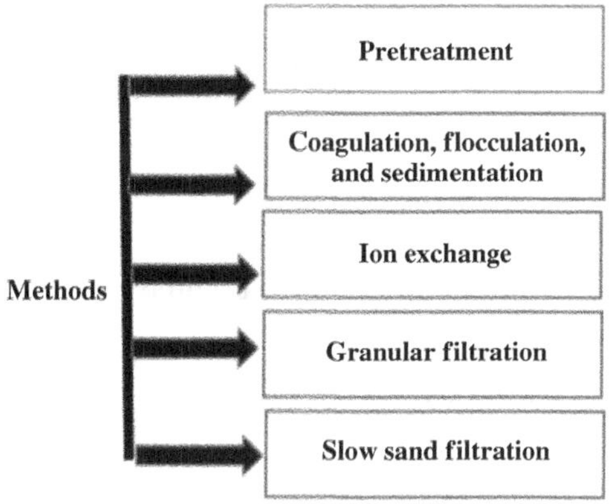

Figure 10.2 Various methods for the removal of microbes.

The treatment processes for wastewater comprise of primary, secondary, and tertiary steps. Bacteria, viruses, and helminth eggs are eliminated by natural treatment processes and stabilization ponds for wastewater. Other methods, such as ozonation, chlorination, and application of ultraviolet radiation, are also used for inactivation. Out of the mentioned microbes, helminth eggs are a more resistant species to the methods of chlorination, etc. Protozoa can be treated by sedimentation and other techniques [14]. The efficacy of the treatment method is usually measured in terms of log value (Equation 10.1).

$$\mathrm{LRV} = \log\left(\frac{C_i}{C_f}\right) \tag{10.1}$$

where

LRV is the log removal value

C_i is the initial pathogen concentration (influent)

C_f is the final pathogen concentration (effluent)

If the log removal value is unity, it implies a 90% reduction in microbial population. If the value is two, removal is 99%, while if the log value is 3, then 99.9% reduction of microorganisms occurs. The specified efficiency of *C. parvum* is 2, while 3 for *G. lamblia* and 4 for viruses according to the US Environmental Protection Agency [15]. The different treatment methods adopted for the removal of microbes are mentioned in Figure 10.3.

10.1.1 Primary Sedimentation

Wastewater is let into large settling tanks, wherein, with low turbulence, the organic matter is made to settle. Primary settlement and sedimentation can have varying effects on the microbial population depending on their shapes, sizes, etc.

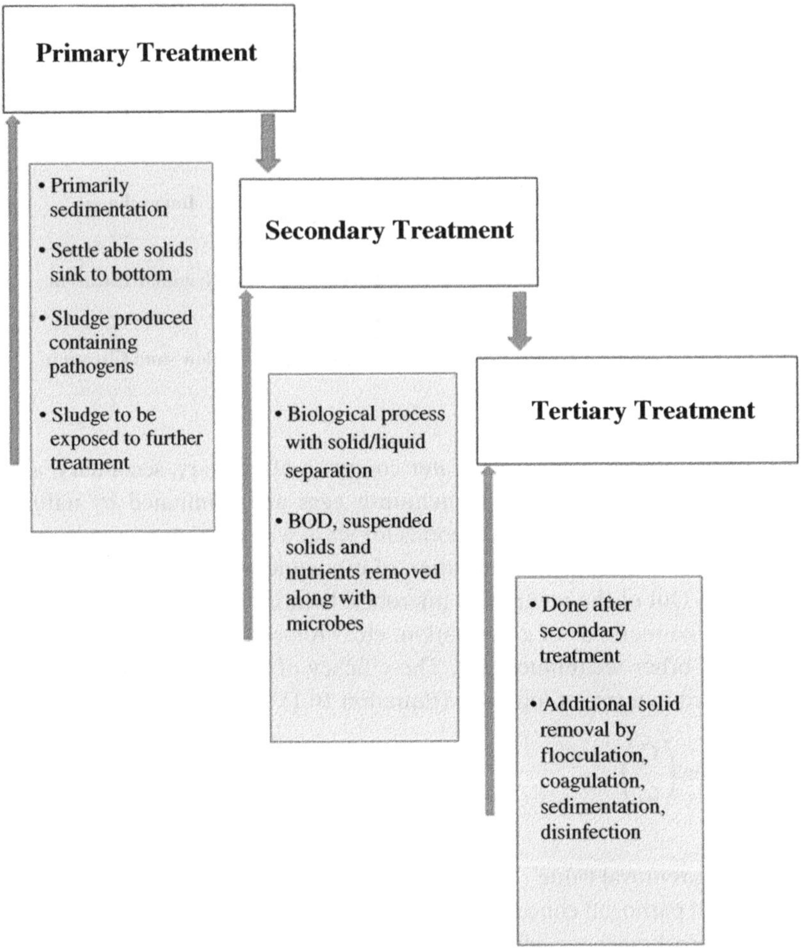

Figure 10.3 Treatment steps for the removal of microbes.

Some researchers have found significant changes in the microbial concentration after the sedimentation process in the sedimentation tank [16], whereas others have not reported substantial changes [17]. Studies have proven a significant variation in the microbial population in the wastewater of municipal wastewater treatment plants in Denmark [18].

10.1.2 Coagulation-flocculation

This method can be a reliable process for the elimination of several bacteria, as it removes all the suspended and soluble particles from wastewater. Organisms such as *Staphylococcus* sp. and *E. coli* have been successfully removed by this method, where diatomite and alum were used as coagulants [19]. Other coagulants such as ferric chloride ($FeCl_3$), polyaluminum chloride–PAC ferric sulfate [$Fe_2(SO_4)_3$] and aluminum sulfate [$Al_2(SO_4)_3$], etc., have also been used [20]. This method is not recommended for sewage wastewater treatment as it leads to the elimination of all types of microbial species present in the wastewater, which is required in the later stages of sewage treatment. But, in wastewater from nonsewage sources, the method is quite effective and adoptable.

10.1.3 Filtration

With the growing resistance of the microbes to the different chemicals, it becomes essential to explore other methods for the treatment of microbes [21]. This technology is associated with a few advantages and disadvantages. The procedure does not require the use of chemicals, in contrast to the chemical disinfection process, and can be scaled up. This method can eliminate microbes based on size, but has the disadvantage of letting small viruses pass through if the pore size of the filter which not large enough to capture them. The filter may also suffer from clogging and biofilm formation issues, which demand frequent cleaning. The process overall is a slow process, which could limit its application for time-sensitive processes.

Filtration techniques such as biofiltration and membrane filtration have proved to be advantageous [22]. Nanofibers incorporating metals and metal oxide [23], also have proved their worth in the filtration of microbes, where particles of silver [24], iron oxide [25], copper [24], zinc oxide [26], titanium dioxide [27], lanthanum oxide [28], etc., have been explored. Filtration by membranes has been effective in the elimination of a large variety of bacteria, protozoans, viruses, etc. The size exclusion ability of membranes shows the potential of pathogenic removal. Ultrafiltration is preferred over microfilters for the removal of viruses, as it operates on the principle of physical exclusion.

10.1.4 Conventional Disinfection

For the process of disinfection, the aspects of prime importance to be considered are the concentration of the disinfectant, temperature, pH, and contact time [10]. The disinfectant chosen should have the ability to successfully destroy the microbes, should be easy to store and handle, should not lead to any toxic or harmful products, and should be cost-effective. This process becomes very important when it is targeted to convert wastewater to potable water. Chlorine dioxide can be produced and utilized onsite for the elimination of algae from wastewaters. It also eliminates odor, taste, and heavy metals from water.

Similarly, the use of ultraviolet radiation prevents the replication of microbes and damages their building blocks.

10.1.5 Chlorination

Chlorine [29] is considered to be the most ideal disinfectant as it oxidizes the cellular material and destroys the microbes. It inactivates the microbe by destroying its cell membrane, terminating the vital activities like cell respiration and DNA activity of the organism. For chlorination, chlorine is applied in controlled doses in the form of a liquid, gas, or in the form of compounds of chlorine. The method of chlorination can be applied to a wide spectrum of microbes. It also oxidizes certain organic compounds, removes bad odors, and inorganic compounds. Chlorine, however, can prove to be toxic to aquatic life and needs dechlorination [30–32].

Chlorination by chlorine gas or chlorine-based disinfectants is one of the most proven processes of removing pathogens from water. It involves the rupture of their cell walls and the oxidation of their cellular components. It is a broadly used disinfectant but suffers from the disadvantage of unpleasant taste and odor, and ineffectiveness against cysts and protozoans.

Chloramines are also found to interact with the protein makeup of the microbes and thus prevent their growth and proliferation. The use of chloramines is not associated with the formation of any deleterious byproducts, but it has less potency for pathogen elimination and needs skilled monitoring. Algal removal can be facilitated by using chlorine dioxide, which is preferably used in the contamination zone and prevents distribution-related issues. The disinfecting agent is susceptible to light, pressure, and temperature and should be handled with care as it is associated with explosive characteristics at higher concentrations [33].

10.1.6 Ultraviolet (UV) Light

The wavelength of UV radiation is from 200 to 400 nm, comprising of UV-A, ranging from 315 to 400 nm, UV-B, ranging from 280 to 315 nm, and UV-C, ranging from 200 to 280 nm. UV-C constitutes the germicidal range, which deactivates microbes [34, 35]. UV radiation is lethal to microorganisms such as bacteria, viruses, protozoa, etc. [36]. UV radiations directly impact the nucleic acids, thereby preventing replication. The biggest advantage of UV radiation is that these leave no residue and prevents further contamination. The process exhibits some limitations, as sometimes low doses might not effectively treat microbes, and the presence of turbidity as total suspended solids (TSS), might interfere, affecting the efficiency of the process. Total coliform, *Fecal coliforms,* and *streptococci* microbial species were eliminated using a UV disinfection system at low pressure, though direct disinfection is not possible, which demands better effluent quality before disinfection [37]. Disinfection of *coliforms* and *salmonella* species can be effectively carried out by UV irradiation. The radiation destroys the genetic assembly of the bacterial species and obstructs its reproductive ability. The effectiveness of the process depends on effluent characteristics like concentration, UV radiation intensity, exposure time, etc. Reports have suggested the effective elimination of pathogenic bacteria and viruses using UV disinfection. Though the process is found not to be effective against *Clostridium* spp., *E. coli,* and *B. subtilis* may be due to the presence of high concentrations of particles in the effluent, which might have confined the microbes from radiation.

Ultraviolet, along with hydrogen peroxide, is employed as the agent to successfully remove *E. coli, Enterococcus* spp., and *G. Lambia* from wastewaters [33].

10.1.7 Ozonation

Ozone is a stronger oxidizing agent than chlorine, which works by breaking down the cell membranes of the pathogens. Ozone interferes with the enzymatic activity and kills the microbes. This oxidant not only kills the microbes but is also successful in eliminating several pollutants like microplastics, drugs, dyes, etc., from wastewater. Ozone has shown its potential to remove cysts and spores of microbial species. It successfully acts against the proliferation of *Giardia* and *Cryptosporidium* species. Though there is no formation of byproducts during the process, chances of recontamination exist. The process is expensive and requires post-treatment for organics generated during the process.

Ozone can destroy the cell wall of bacterial species, leading to their death. The pathogenic inactivation up to 98% is possible with five minutes of exposure to ozone having 10 mg/L. The effectiveness of ozonation for disinfection is influenced by ozone supply and its demand, the nature of the effluent, COD, and TDS of the wastewater. But, the process can lead to the generation of toxic byproducts like Bromate ions, peroxides, and sometimes the microbial species regrow due to assimilable organics [38].

The dissociation of O_2 molecules leads to the formation of oxygen atoms, which lead to the formation of ozone. These oxygen atoms subsequently collide with oxygen (O_2) molecules to form ozone (O_3). Usually, the generation of ozone is done onsite as it is unstable and leads to elemental oxygen.

Ozone is considered to be a very powerful viricide and oxidant. It results in the leakage of the cellular contents by leaking into the microbial cell wall, causing destruction of nucleic acids, and reacting with the byproducts. In case of bacteria, it causes cell wall disintegration, commonly known as lysis. Concentration of ozone, time of contact, and the sensitivity of microbes are the important factors governing the efficiency of the process.

The process of ozonation is more effective than chlorination; however, the application globally varies depending upon the wastewater characteristics and cost efficiency. The advantages offered are no residues, no re-infestation of microbes, onsite generation, and an increase in dissolved oxygen levels. The limiting factors in the applicability can be the ineffectiveness in case of low microbial concentration, complex process in comparison to chlorination, demanding the use of corrosion-resistant materials, as it is very corrosive and not cost-effective [39].

10.1.8 Waste Stabilization Ponds

Waste stabilization ponds (WSPs) [40] enable the treatment by microbes by storing water in large shallow basins [41]. Pathogens get eliminated due to adverse pond conditions such as sunlight, temperature, etc., causing natural death and sedimentation after getting adsorbed on particles [42]. Several factors, such as the geometry of the pond, depth, and time, are the factors governing the efficiency of the process [43]. Fecal coliforms and other pathogenic microbes are removed due to the effect of sunlight, leading to photo-oxidative deoxyribonucleic acid (DNA) damage, pH, temperature, and predation [44]. The removal of pathogens is ensured by either using anaerobic ponds or facultative ponds, followed by maturation ponds, depending upon the aeration conditions and microbial activity involved. The technique is associated with the drawbacks of the growth of a large amount of microalgae, which can help in the induction of cyanobacterial growth, associated with toxin production, reducing the waste treatment efficiency [45]. Microbial decontamination involving cysts and parasitic eggs was studied by using stabilization ponds. 100% efficiency was achieved. These systems have shown good acceptability due to simple operation and setup [46]. Figure 10.4 gives a representation of waste stabilization ponds.

10.1.9 Constructed Wetlands

Constructed wetlands have been employed for the elimination of nutrients and organic matter. These involve the utilization of the microorganisms as detoxification tools. The economics of wetlands and easy operation and maintenance encouraged researchers to study the removal of microbial agents using wetlands. *E. coli* showed a mean removal of 98.0% at the experimental scale. Factors like protozoans, bacteriophages governed the removal of coliform and other microbial agents. But the overall wetland did not seem to be a viable option for microbial decontamination [3].

Studies have shown that constructed wetlands can be effectively used for the 99% elimination of pathogens, along with other contaminants and nutrients [47]. Various designs of constructed wetlands are structured for efficient removal of contaminants, including microbes from wastewater (Figure 10.5a–c).

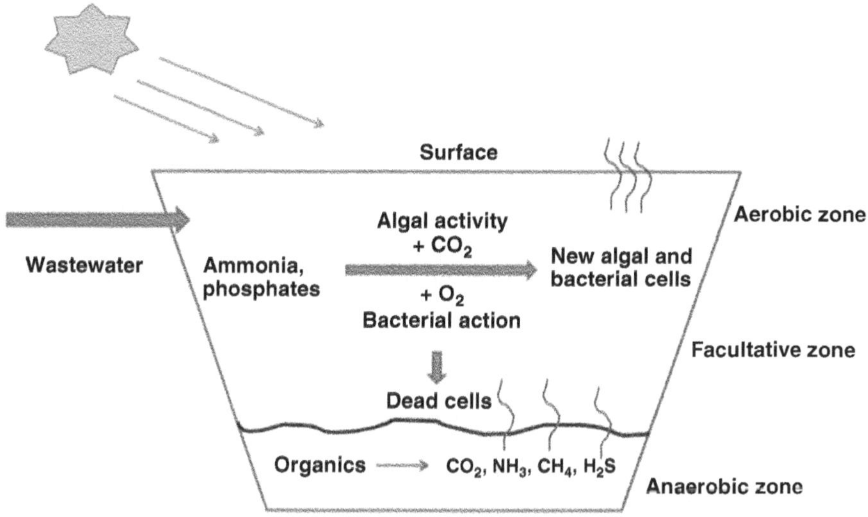

Figure 10.4 Waste stabilization pond.

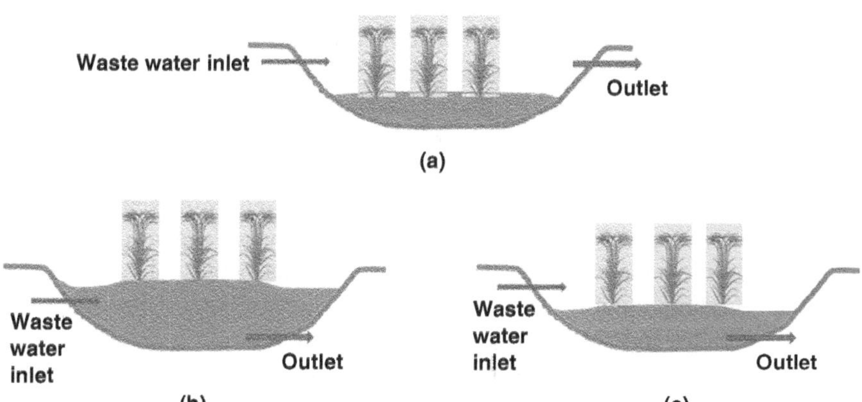

Figure 10.5 (a) Free water surface constructed wetland, (b) horizontal flow constructed wetland, and (c) vertical flow constructed wetland.

An integration of several treatment processes, such as biological, physical, and chemical, leads to the removal of pathogens from wastewater [48–50]. Physical processes such as sedimentation, filtration, chemical processes such as the use of UV light, adsorption, and biological treatment methods such as predation, retention on biofilms, natural die off, etc., have proved successful in elimination.

Sedimentation of the microbes can take place after accumulation on the gravel [51]. Mechanical filtration of the microbes can also occur [52]. The microbes present in the wastewater also get adsorbed onto the gravel and particles [51]. The presence or absence of oxygen also has an impact on the activity of microbes. Certain plants transfer the atmospheric oxygen to the roots, adversely affecting the growth of microbes that prefer anaerobic environments. Exposure of microbes in constructed wetlands to UV light or solar radiation helps eliminate pathogens such as coliforms [53]. Plant bactericides such as *Scirpus lacustris* can release certain chemicals from their roots, destroying microbes. Predation [54] of microbes by other microbes and natural die off [55] are also significant processes for pathogen removal [56]. Where predation depends upon the different targeted species characteristics, natural die-off of microbes takes place due to starvation of microbes.

The selection of the processes is available for the removal of pathogens, but the sensitivity of the microbes remains a crucial factor.

10.1.10 Novel Methods

10.1.10.1 Carbon Nanotubes

Filtering out pathogenic microbes can be effectively done by using nanotechnology. Antimicrobial nano-sized particles include silver and copper titanium dioxide. These particles generate reactive oxygen species and bring about the rupturing of the microbial cell wall. The large surface area to volume ratio enables nanotubes and nanofibers, along with nanocatalysts, to effectively remove microbial contaminants from water. Nanotechnology has enabled the development of nanosensors, which evolved to detect and track microbial contaminants in water, in real time. They have the advantage of high responsiveness to even very small concentrations of specific microbes, which facilitates better and accurate detection. Though the nanotechnology-based materials offer better potential for the detection and elimination of microbes, they face challenges like cost intensity, scalability limitations, and necessary research to facilitate widespread development.

Activated carbons, nanotubes have provided ample success in the elimination of microbes from the water in comparison to the conventional adsorbents due to their high surface area to volume ratio, good porosity, and presence of functional groups on their surface. Activated carbon incorporated with metals like silver, iron, and copper has shown better spore adsorption capacity and good bacterial inactivation competence. The evaluation of such metal composites was carried out against Gram-positive and Gram-negative bacteria's and results indicated high inhibition characteristics by the nanohybrids. This could be due to the destruction of the bacterial cell wall by the free radical species generated during the interaction with the suspension agents. The growth of species like *E. coli* and *S. aureus* is restricted due to the use of bimetallic carbon fiber composites. Similarly, activated carbon functionalized with quaternary ammonium salts was also established to be effective against *E. coli* contamination. The carbon nanotubes are reported to have the characteristics of targeting the bacterial cell covering and neutralization of bacteria and viruses. These can be employed to effectively filter off the microbial contaminants from the water. Destructive phospholipid extraction was prominently involved in the degradation of the cell wall of *E. coli*, as per reports [57]. Recently, graphene and its derivatives, like graphene oxide and reduced graphene oxide, have gained noteworthy consideration, as they possess remarkable antibacterial properties. In one of the investigations, the minimum inhibitory concentration of nanosheets of graphene was evaluated for the decontamination of bacterial pathogens. It was reported that graphene morphology enabled leakage in the intracellular components of the microorganism, leading to disturbance in the RNA sequencing and protein makeup and ultimately destruction of microbes [58].

10.1.10.2 Membrane Bioreactor

A membrane bioreactor is a wastewater treatment plant that has shown good abatement of microbial species. However, the effluents treated with ClO_2 showed contamination with *E. coli* and *Salmonella* species, focusing on more attention to be given for such wastewater. Also, *Legionella* species and *Leg. pneumophila* presence highlighted the need to further carry out investigations for better results [59].

An investigation conducted on the removal of total coliform from wastewater using an activated sludge system resulted in effluent with a permissible limit of microbial population, which can be discharged into the water streams [60].

However, in another study, the method was not sufficient to achieve the desirable remediation limit may be due to improper management factors and production of sludge waste in bulk. But, the method is considered one of the best methods due to its high efficiency in reducing the BOD [61].

The membrane bioreactor was found to give outstanding performance for the removal of *E. coli* and fecal coliforms. The efficiency of aluminum oxide–titanium oxide ceramic was investigated to reclaim wastewater, and complete removal of the heterotrophic bacterial population was achieved. The removal of viruses was also possible using a membrane bioreactor, though the removal was inconsistent. Few reported log removal value of 6.3 and 6.8 for human adenovirus and enterovirus, respectively [62].

A microbial evaluation of effluent treated by membrane bioreactor suggested acceptable virus removal efficiency, good coliphage elimination potential irrespective of high feed variation, and superior *E. coli* reduction with a log removal value of five attainments. Membrane bioreactor efficiency is influenced by the membrane material, its pore size, and membrane flux. The biomass processes regulate the pathogenic removal and include aggregation, biosorption, and biodegradation. Biofilms act as a barrier to microbes and reduce he permeability of the membranes. Thus, the efficiency of the membrane bioreactor can be reclaimed by backwashing and other cleaning methods. The physical cleaning methods can affect virus elimination due to damage to biofilms, and infiltration of pathogens can take place due to membrane breach. Thus, membrane bioreactor-treated effluent is recommended to be subjected to a disinfection process [62].

10.1.10.3 Biofilters

Biofilters are a very economical and promising method of disinfection of wastewaters. These include trickling filters, slow sand filters, rapid sand filters, storm water filters, and microbial-earthworm ecofilters. The above-mentioned methods were able to effectively remove *Giardia* sp., *Salmonella, E. coli, Coliforms*, etc. from wastewaters [6].

Slow sand filters consist of a reinforced concrete structure having fine sand placed over the support of gravel. The filtration mechanism involves gravity. The active biological species for wastewater treatment remain attached to the sand surface and are known as Schmutzdecke (German). The filtration rate is usually 0.08–0.4 m/hour. Slow sand filter working is based on the biological activity occurring in the ecosystem comprising of bacteria, ciliates, rotifers, aquatic worms, etc. Slow sand filters effectively remove waterborne pathogens like *Giardia* cysts, *Cryptosporidium enteroparasites* (99.9%). The oligotrophic bacterial population existing in the biofilm (Zooglea) eliminates viruses better in the slow sand filter. Certain autochthonous bacteria secrete toxic chemicals to kill pathogens. Various investigations revealed substantial removal for *E. coli*, enterococci, and coliphages, with log values of $\leq$4.7, $\leq$2.4, and $\leq$2.8, respectively. The mechanism involved in slow sand filters includes both physical and biological interactions. Physical processes involved are straining and adsorption. The transportation mechanisms, like impaction, sedimentation, and diffusion, predominantly occur in the lower zone of the sand filter. Biological dominance is observed in the upper layer of a slow sand filter, and a slow rate of filtration facilitates effective interaction and predation. This type of treatment is effective for low-demand situations. It is associated with the disadvantages of pre-sedimentation requirement, disinfection necessity, and the issue of clogging of the filtration bed due to turbid influent [6].

Rapid sand filter involves high-speed operation and timely backwashing. The addition of coagulants can modify the efficiency of the filtration process, but the method is found effective only for the removal of turbidity and not for bacteria, viruses, or organic matter. It is associated with high capital and operational costs, including backwashing, making it less suitable for microbial decontamination. Though it is a commonly employed process in municipal wastewater for the removal of a number of micro pollutants [63].

Microbial earthworm filters are an eco-friendly method of wastewater treatment. These filters are a passive but natural wastewater treatment system involving a symbiotic association between the earthworms and microbes, which jointly function for the removal of pathogens from wastewaters. The process involved includes sedimentation, filtration, precipitation, adsorption, and biochemical degradation. Earthworms function as a conditioning agent, improving the performance of microbes for pathogen removal. Pathogenic removal efficiency of the system was successfully investigated using riverbed material. It has been successful for domestic, urban, and wastewaters from the petroleum industry with proficient removal of COD, BOD, and coliforms.

Eisenia fetida is one of the widely employed epigeic earthworm species in microbial earthworm filters due to its favorable enzymatic activity and associated biochemical benefits (Figure 10.6). Microbial earthworm filters are a joint system involving passing of wastewater through a container tank with filtering media, followed by percolation through an earthworm-loaded layer of soil for phytoremediation of the effluent. Vermifiltration has no drawback of foul odor and requires a lower cost investment, both in terms of construction and sludge disposal. It involves low methane emissions in comparison to those from an anaerobic lagoon facility. It minimizes the environmental impact related to the wastewater disposal. The effectiveness of Microbial Earthworm filters is influenced by environmental conditions like temperature and humidity. The thriving temperature for earthworms is between 15 and 25 °C. The temperature above or below makes the earthworm's activity unavailable. Also, toxic constituents of the influent can be fatal to earthworms. It is thus required that pretreated water must be subjected to the Microbial Earthworm filter system. As the treatment system is based on a filtration

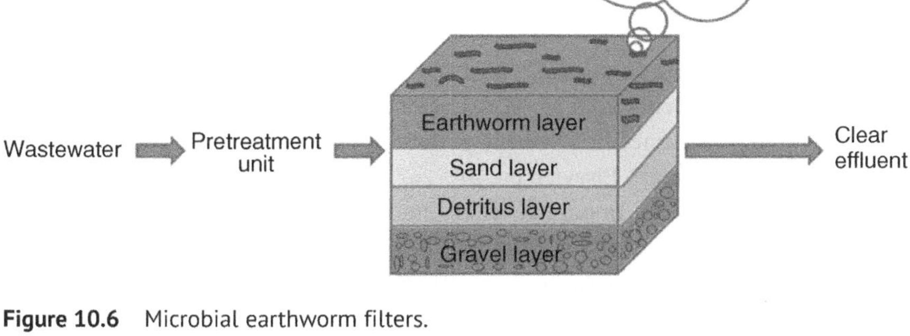

Figure 10.6 Microbial earthworm filters.

mechanism, it may get clogged due to solids and organic matter, which can reduce the efficiency of the system and can escalate the maintenance costs [6].

10.1.10.4 Advanced Oxidation Process

The advanced oxidation process involves a combination of processes to improve the removal performance. Hydrogen peroxide, ozone, and oxygen active radicals and energy radiations are utilized for microbial elimination. The disturbances in the structural makeup of the organism occur rapidly with a combined process. Combined processes like hydrogen peroxide and ultraviolet have been successfully utilized to remove pathogens like *E. coli, Enterococcus*, and *G. lambia* from the water bodies. Ozone-based advanced oxidation processes have an impressive performance in the removal of waterborne pathogens like Rotavirus, Hepatitis virus, *C. parvum* cyst, *Giardia* cyst, etc.

Similarly, the photocatalytic advanced oxidation process utilizes titanium oxide, which produces hydroxyl radicals that have the ability to damage the internal structure of the pathogenic microbes present in the wastewater. *S. enterica, V. cholerae*, Adenovirus, and Norovirus have been eliminated from water using the photocatalytic oxidation process [33].

References

1 Tran, H.N., Le, G.T., Nguyen, D.T., et al. (2021). SARS-CoV-2 coronavirus in water and wastewater: a critical review about presence and concern. Environmental Research, Volume 193, p. 110265. https://doi.org/10.1016/j.envres.2020.110265.

2 Gilbride, K., Lee D.Y., and Beaudette, L. (2006). Molecular techniques in wastewater: understanding microbial communities, detecting pathogens, and real-time process control. Journal of Microbiological Methods, Volume 66, pp. 1–20.

3 Ramirez-Castillo, F.Y., Loera-Muro, A., Jacques, M., et al. (2015). Waterborne pathogens: detection methods and challenges. Pathogens, Volume 4, pp. 307–334.

4 Singh, P., Yadav, P., Gupta, R.K., et al. (2023). Chapter 14: Pathogenic microbes in wastewater: identification and characterization. In: Ferreira LFR, Kumar A, Bilal M (eds) Advances in chemical pollution, Environmental management and protection, Volume 9, Elsevier, pp. 247–262.

5 Shen, Y., Giang, T.L., Dong, T.N., et al. (2021). Recent advances in nanotechnology for simultaneous detection of multiple pathogenic bacteria. Nano Today, Volume 38, p. 101121. https://doi.org/10.1016/j.nantod.2021.101121.

6 Maurya, A., Singh, M.K., and Kumar, S. (2020). Biofiltration technique for removal of waterborne pathogens. In: Narasimha M, Grobelak A (eds) Waterborne pathogens, Butterworth-Heinemann, pp. 123–141. https://doi.org/10.1016/B978-0-12-818783-8.00007-4.

7 Bosch, A. (2010). Human enteric viruses in the water environment: a mini review. International Microbiology, Volume 1, pp. 191–196.

8 Monis, P.T., Lau, M., Reeves, P., et al. (2015). National validation guidelines for water recycling: methods for pathogen isolation, culture, detection and enumeration. Brisbane, QLD: Australian Water Recycling Centre of Excellence.

9 Yergeau, E., Masson, L., Elias, M., et al. (2016). Comparison of methods to identify pathogens and associated virulence functional genes in biosolids from two different wastewater treatment facilities in Canada. PLoS One, Volume 11, pp. 1–20. https://doi.org/10.1371/journal.pone.0153554.

10 LeChevallier, M.W., and Kwok-Keung, A. (eds) (2004). Water treatment and pathogen control: process efficiency in achieving safe drinking water, World Health Organization, IWA Publishing.

11 Jiménez, B., Mara, D., Carr, R., et al. (2010). Wastewater treatment for pathogen removal and nutrient conservation: suitable systems for use in developing countries. IWMI Books, Reports H042608, International Water Management Institute.

12 Jiménez, B. (2003). Health risks in aquifer recharge with recycled water. In: Aertgeerts R, Angelakis A (eds) Aquifer recharge using reclaimed water, WHO Regional Office for Europe, Copenhagen, pp. 54–172.

13 Jiménez, B. (2007). Helminth ova control in sludge: a review. Water Science and Technology, Volume 56, Issue 9, pp. 147–155. https://doi.org/10.2166/wst.2007.713.

14 Jiménez, B. (2007). Helminth ova removal from wastewater for agriculture and aquaculture reuse. Water Science and Technology, Volume 55, Nos 1–2, pp. 485–493.

15 Antony, A., Blackbeard, J., and Leslie, G. (2011). Removal efficiency and integrity monitoring techniques for virus removal by membrane processes. Critical Reviews in Environmental Science and Technology, Volume 42, pp. 891–933. https://doi.org/10.1080/10643389.2011.556539.

16 Shin, J., Choi, S., Park, C.M., et al. (2022). Reduction of antibiotic resistome in influent of a wastewater treatment plant (WWTP) via a chemically enhanced primary treatment (CEPT) process. Chemosphere, Volume 286, Article 131569. https://doi.org/10.1016/j.chemosphere.2021.131569.

17 Lambirth, K., Tsilimigras, M., Lulla, A., et al. (2018). Microbial community composition and antibiotic resistance genes within a North Carolina urban water system. Water, Volume 10, Article 1539.

18 Riisgaard-Jensen, M., Dottorini, G., Nierychlo, M., et al. (2023). Primary settling changes the microbial community of influent wastewater to wastewater treatment plants. Water Research, Volume 244, p. 120495. https://doi.org/10.1016/j.watres.2023.120495.

19 Shaarani, S., Fitriah, A.S.N., MdAkhir, F.N., et al. (2019). Removal efficiency of Gram-positive and Gram-negative bacteria using a natural coagulant during coagulation, flocculation, and sedimentation processes. Water Science and Technology, Volume 80, Issue 9, pp. 1787–1795. https://doi.org/10.2166/wst.2019.433.

20 Rana, S., and Suresh, S. (2017). Comparison of different coagulants for reduction of COD from textile industry wastewater. Material Today Proceedings, Volume 4, Issue 2, pp. 567–574. https://doi.org/10.1016/j.matpr.2017.01.058.

21 Fahimirad, S., Abtahi, H., Razavi, S.H., et al. (2017). Production of recombinant antimicrobial polymeric protein beta casein-E 50-52 and its antimicrobial synergistic effects assessment with thymol. Molecules, Volume 22, Issue 6, p. 822. https://doi.org/10.3390/molecules22060822.

22 Zodrow, K.R., Tousley, M.E., and Elimelech, M. (2014). Mitigating biofouling on thin-film composite polyamide membranes using a controlled-release platform. Journal of Membrane Science, Volume 453, pp. 84–91.

23 Fahimirad, S., Fahimirad, Z., and Sillanpää, M. (2021). Efficient removal of water bacteria and viruses using electro spun nanofibers. Science of the Total Environment, Volume 751, p. 141673. https://doi.org/10.1016/j.scitotenv.2020.141673.

24 Taner, M., Sayar, N., Yulug, I.G., et al. (2011). Synthesis, characterization and antibacterial investigation of silver–copper nanoalloys. Journal of Material Chemistry, Volume 21, Issue 35, pp. 13150–13154.

25 Mortimer, C.J., and Wright, C.J. (2017). The fabrication of iron oxide nanoparticle-nanofiber composites by electrospinning and their applications in tissue engineering. Biotechnology Journal, Volume 12, Issue 7, p. 1600693.

26 Tiwari, V., Mishra, N., Gadani, K., et al. (2018). Mechanism of anti-bacterial activity of zinc oxide nanoparticle against carbapenem-resistant Acinetobacter baumannii. Frontiers in Microbiology, Volume 6, Issue 9, p. 1218.

27 de Dicastillo, C.L., Correa, M.G., Martínez, F.B., et al. (2020). Antimicrobial effect of titanium dioxide nanoparticles. In: Mareş M, Lim SHE, Lai K-S, et al. (eds) Antimicrobial Resistance - A One Health Perspective.

28 He, J., Wang, W., Shi R., et al. (2018). High speed water purification and efficient phosphate rejection by active nanofibrous membrane for microbial contamination and regrowth control. Chemical Engineering Journal, Volume 337, pp. 428–435.

29 Pang, Y.C., Xi, J.Y., Xu, Y., et al. (2016). Shifts of live bacterial community in secondary effluent by chlorine disinfection revealed by Miseq high-throughput sequencing combined with propidium monoazide treatment. Applied Microbiology and Biotechnology, Volume 100, pp. 6435–6446. https://doi.org/10.1007/s00253-016-7452-5.

30 https://www3.epa.gov/npdes/pubs/chlo.pdf (Last accessed 19.09.23)

31 https://www.ncbi.nlm.nih.gov/pmc/articles/PMC7832327/ (Last accessed 19.09.23)

32 https://www3.epa.gov/npdes/pubs/pwm_2010.pdf (Last accessed 19.09.23).

33 Oluwatobi, V.O., Damilare, C.O., Olumide, O., et al. (2024). Exploring emerging water treatment technologies for the removal of microbial pathogens. Current Research in Biotechnology, Volume 8, p. 100252. https://doi.org/10.1016/j.crbiot.2024.100252.

34 Francy, D.S., Stelzer, E.A., Bushon, R.N., et al. (2012). Comparative effectiveness of membrane bioreactors, conventional secondary treatment, and chlorine and UV disinfection to remove microorganisms from municipal wastewaters. Water Research, Volume 46, pp. 4164–4178.

35 Turtoi, M. (2013). Ultraviolet light potential for wastewater disinfection, Annals. Food Science and Technology, Volume 14, Issue 1, pp 153–164.

36 Hijnen, W.A.M., Beerendonk, E.F., and Medema, G.J. (2006). Inactivation credit of UV radiation for viruses, bacteria and protozoan (oo) cysts in water: a review. Water Research, Volume 40, Issue 1, pp. 3–22.

37 Aghalari, Z., Dahms, H.U., Sillanpää, M., et al. (2020). Effectiveness of wastewater treatment systems in removing microbial agents: a systematic review. Global Health, Volume 16, Issue 13. https://doi.org/10.1186/s12992-020-0546-y.

38 Al-Gheethi, A.A., Efaq, A.N., Bala, J.D., et al. (2018). Removal of pathogenic bacteria from sewage-treated effluent and biosolids for agricultural purposes. Applied Water Science, Volume 8, pp. 74. https://doi.org/10.1007/s13201-018-0698-6.

39 Nasuhoglu, D., Isazadeh, S., Westlund, P., et al. (2018). Chemical, microbial and toxicological assessment of wastewater treatment plant effluents during disinfection by ozonation. Chemical Engineering Journal, Volume 346, pp. 466–476. https://doi.org/10.1016/j.cej.2018.04.037.

40 Wastewater technology fact sheet ozone disinfection. United States environmental protection agency, EPA 832-F-99-063 September 1999. https://www3.epa.gov/npdes/pubs/ozon.pdf.

41 https://www.ncbi.nlm.nih.gov/pmc/articles/PMC7126130/ (Last accessed 19.09.24).

42 Alexiou, G., and Mara, D. (2003). Anaerobic waste stabilization ponds. Applied Biochemistry and Biotechnology, Volume 109, pp. 241–252.

43 Campos, C., Guerrero, A., and Crdenas, M. (2002). Removal of bacterial and viral faecal indicator organisms in a waste stabilization pond system in

Choconta, Cundinamarca (Colombia). Water Science and Technology, Volume 45, pp. 61–66.
44 Von Sperling, M. (2005). Modelling of coliform removal in 186 facultative and maturation ponds around the world. Water Research, Volume 39, pp. 5261–5273.
45 King, B., Fanok, R., Phillips, R., et al. (2017). Cryptosporidium attenuation across the wastewater treatment train: recycled water fit for purpose. Applied and Environmental Microbiology, Volume 83. https://doi.org/10.1128/AEM.03068-16.
46 Fazelipour, M., Takdastan, A., and Sekhavatjo, M. (2011). Survey on chlorine application in sequencing batch reactor waste sludge in order to sludge minimization. Asian Journal of Chemistry, Volume 23, Issue 6, pp. 2994–2998.
47 Gláucia, E., Gama, V., and Danilo Gualberto, Z. (2024). Chapter 5 - FAME analysis of microalgal biomass from waste stabilization ponds for biofuel purposes. In: Sangeetha J, Thangadurai D (eds) Woodhead series in bioenergy, microalgal biomass for bioenergy applications, Woodhead Publishing, pp. 99–112. https://doi.org/10.1016/B978-0-443-13927-7.00016-5.
48 Alexandros, S.I., and Akratos, C.S. (2016). Removal of pathogenic bacteria in constructed wetlands: mechanisms and efficiency. In: Ansari, A, Gill, S, Gill, R, et al. (eds) Phytoremediation. Springer. https://doi.org/10.1007/978-3-319-41811-7_17.
49 Decamp, O., and Warren, A. (2000). Investigation of Escherichia coli removal in various designs of subsurface flow wetlands used for wastewater treatment. Ecological Engineering, Volume 14, pp. 293–299.
50 Karathanasis, A.D., Potter, C.L., and Coyne, M.S. (2003). Vegetation effects on faecal bacteria, BOD, and suspended solid removal in constructed wetlands treating domestic wastewater. Ecological Engineering, Volume 20, Issue 2, pp. 157–169.
51 Karim, M.R., Manshadi, F.D., Karpiscak, M.M., et al. (2004). The persistence and removal of enteric pathogens in constructed wetlands. Water Research, Volume 38, Issue 7, pp. 1831–1837.
52 Sengupta, M.E., Thamsborg, S.M, Andersen, T.J., et al. (2011). Sedimentation of helminth eggs in water. Water Research, Volume 45, Issue 15, pp. 4651–4660.
53 Boutilier, L., Jamieson, R., Gordon, R., et al. (2009). Adsorption, sedimentation and inactivation of E. coli within wastewater treatment wetlands. Water Research, Volume 43, pp. 4370–4380.
54 Zdragas, A., Zalidis, G.C., Takavakoglou, V., et al. (2002). The effect of environmental conditions on the ability of a constructed wetland to disinfect municipal wastewater. Environmental Management, Volume 29, Issue 4, pp. 51048–51548.

55 Stott, R., May, E., Matsushita, E., et al. (2001). Protozoan predation as a mechanism for the removal of Cryptosporidium oocysts from wastewaters in constructed wetlands. Water Science and Technology, Volume 44, pp. 191–198.

56 Stevik, T.K., Aa, K., Ausland, G., et al. (2004). Retention and removal of pathogenic bacteria in wastewater percolating through porous media: a review. Water Research, Volume 38, pp. 1355–1367.

57 Haijiao, L., Wang, J., Stoller, M., et al. (2016). An overview of nanomaterials for water and wastewater treatment. Advances in Material Science. https://doi.org/10.1155/2016/4964828.

58 Tasrin, S., and Gasim, H. (2023). Exploring graphene's antibacterial potential for advanced and sustainable solutions in water treatment. Journal of Water Process Engineering, Volume 56, p. 104530. https://doi.org/10.1016/j.jwpe.2023.104530.

59 Bonetta, S., Pignata, C., Gasparro, E., et al. (2022). Impact of wastewater treatment plants on microbiological contamination for evaluating the risks of wastewater reuse. Environmental Sciences Europe, Volume 34, pp. 20. https://doi.org/10.1186/s12302-022-00597-0.

60 Ghoreishi, B., Aslani, H., Dolatkhah, A., et al. (2016). Evaluation of microbial quality in biosolids generated from municipal wastewater treatment plants. Iranian Journal of Health Environment, Volume 9, Issue 1, pp. 81–90.

61 Paiva, M.C., Ávila, M.P., Reis, M.P., et al. (2015). The microbiota and abundance of the class 1 Integron-Integrase Gene in tropical sewage treatment plant influent and activated sludge. PLoS One, Volume 10, Issue 6. https://doi.org/10.1371/journal.pone.0131532.

62 Hai, F.I., Riley, T., Shawkat, S., et al. (2014). Removal of pathogens by membrane bioreactors: a review of the mechanisms, influencing factors and reduction in chemical disinfectant dosing. Water, Volume 6, pp. 3603–3630. https://doi.org/10.3390/w6123603.

63 Seeger, E.M., Braeckevelt, M., Reiche, N., et al. (2016). Removal of pathogen indicators from secondary effluent using slow sand filtration: optimization approaches. Ecological Engineering, Volume 95, pp. 635–644. https://doi.org/10.1016/j.ecoleng.2016.06.068.

11

Pharmaceutics as Emerging Contaminants in Wastewater: Hazards and Removal Technologies

Increasing technological advancement is associated with an elevation of health-related issues. A comfortable and handy lifestyle has resulted in making a less laborious, more machine-dependent and lazy human race. As a result human population is now more prone to a number of health disorders. These could either be genetically inherited or environmentally induced. Researchers, scientists, medical practitioners, and pharmaceutical companies are continuously making efforts to develop drugs that make life disease-free and comfortable. The ever-increasing world population has made drug and medical-care products more prevalent. A huge number of pharmaceutical setups have thus been established, and so is the pollution load created by them. Life cannot be imagined now without drugs, which gives rise to related pollution issues, which cannot be ignored. Pharmaceutical contaminants are one of the potential categories of biologically active pollutants, classified as emerging contaminants, obtained from pharmaceutical setups.

Pharmaceutical wastes are the discarded end products or chemicals from the pharmaceutical industries, or items containing such substances. These encompass all solid or liquid expired drugs, waste medical items and even wastewater from the pharmaceutical sector. Pharma wastewater contains residues like unwanted stocks, sharps, scalpels, lancets and syringes, gloves, aprons, masks, bandages, intravenous bags and tubing, vials, medicine bottles, pill containers, medicine tubes, inhalers, nebulizing setups, etc.

11.1 Sources of Pharmaceutical Waste

The sources of pharma-related waste are not limited to a single point origin, but several diverse sources contribute to the introduction of these wastes to the environment. These are thus, diversified in their composition and consistency and

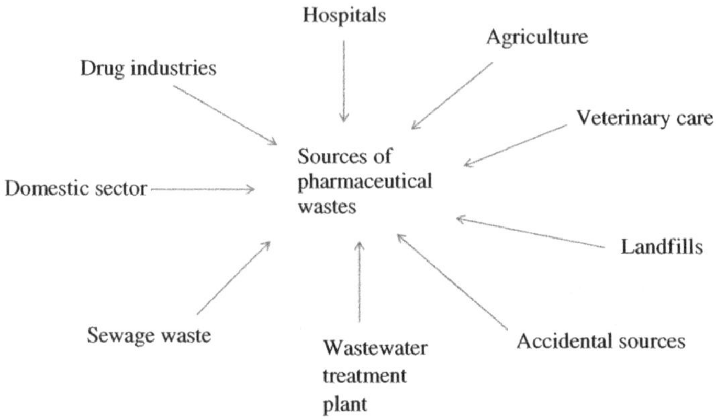

Figure 11.1 Sources of pharmaceutical wastes.

related effects. Pharmaceutical waste can severely impair the wellbeing of our environment and living organisms (Figure 11.1) [1].

11.1.1 Pharmaceutical Manufacturing Facilities

These contribute to physical and chemical waste generated during the processing of drugs and are the main sources of water body contamination. Physical waste refers to the unused products, spent chemical containers, cleansing items, etc. The water from the cleansing activity of the manufacturing units also gets contaminated with the pharmaceutical residues. This water when discarded into the water bodies poses serious threat to aquatic ecosystem.

11.1.2 Healthcare Facilities

Healthcare institutions and extended care facilities produce wastes like expired drugs, used syringes, medicine bottles and other refused items. These lead to the generation of huge amount of solid toxic waste. Lack of improper drug management policies also results in waste production.

11.1.3 Personal Care Products

Cosmetic industries generate several personal care products, such as perfumes, creams, aroma, lotions, etc., and also contribute to waste generation and accumulation. These find their way into the environment and contaminate the water bodies. These are one of the ignored wastes requiring special disposal methods.

11.1.4 Veterinary Institutions

Pharmaceutical waste generated from the veterinary facilities includes expired drugs, used gloves, syringes, tubes, and devices, which are in direct contact with the drugs.

11.1.5 Wastewater Treatment Plants

The waste from domestic sector and other sources is directed to wastewater treatment facilities. Most of the wastewater treatment plants still follow conventional treatment methods which are insufficient to treat the wastewater of varying pharmaceutical composition. Thus, inefficient treatment leads to the building up of contaminants in the water bodies.

11.1.6 Landfills

Disposal of waste through the landfills is one of the most commonly used methods for waste treatment. The defective landfill procedures could cause leaching of the toxic waste which can contaminate soil and ground water.

11.1.7 Other Sources

The floor and lab washing wastes also contribute to water contamination substantially. Considering the various sources of the inlet of pharmaceutical toxicants, it is essential to have a more trusted and practical waste disposal methods to ensure environmental safety. Figure 11.2 portrays the various sources and pathways of the contact of pharmaceuticals with humans.

11.2 Types of Pharmaceutical Waste

In order to understand the impact the pharmaceutical contaminants, these toxicants can be broadly classified into hazardous or nonhazardous wastes, and chemo or controlled wastes [2, 3].

Hazardous wastes are further subcategorized into listed wastes and characteristic wastes. A listed waste refers to hazardous waste given the label as F, K, P, and U. The notation F refers to "nonspecific source wastes," which are mostly the wastes from the spent solvent; K refers to "specific source wastes," which are generally generated wastes from industries; and P and U categories are for the unused wastes with one ingredient active such as residues from containers, spills, etc.

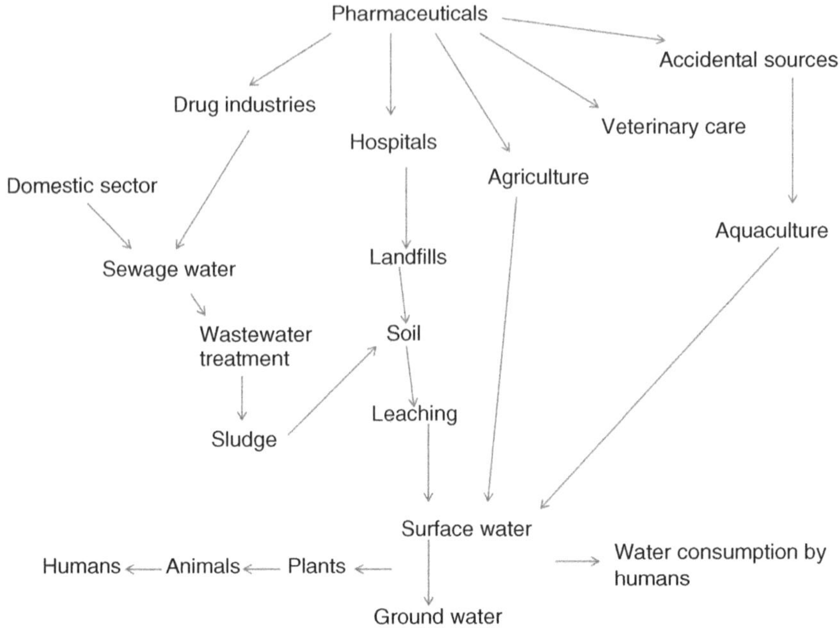

Figure 11.2 Pathways of pharmaceutical waste.

Characteristic wastes are given attention as these depict hazardous traits such as ignitability, reactivity, toxicity, and corrosion tendency. Wastes can also be classified as infectious and controlled substances. Infectious wastes are generally the biomedical waste from microbiological and biotechnological facilities. Controlled substances need special disposal methods. These include cytotoxic waste, narcotics, psychotropic, anti-infective drugs, antiseptics, disinfectants, etc. (Figure 11.3) [4].

The discharge of treated and untreated wastewater, with pharmaceuticals, is the major path for the entry of pharmaceuticals into the water bodies. Leaching from landfills is also one of the key sources of entry of drugs and pharmaceutics into water bodies. An improper land disposal and percolation of rain water, dissolves the toxins in it. Animal farms and crop fields are crucial pathways as they employ pharmaceuticals for inducing disease resistance and enhancing growth characteristics. Thus, agricultural runoff acts as a carrier of residual pharmaceuticals into the water bodies. A sound knowledge of the pathways involved in the migration of these wastes is essential to curtail their impact and to ensure safety of the aqueous bodies. Published research and findings suggest significant and deleterious effect on water health due to drug dissolution.

The effluent generated from diverse processes during the production of pharmaceutical medicines, includes an extensive range of compounds. The rate of

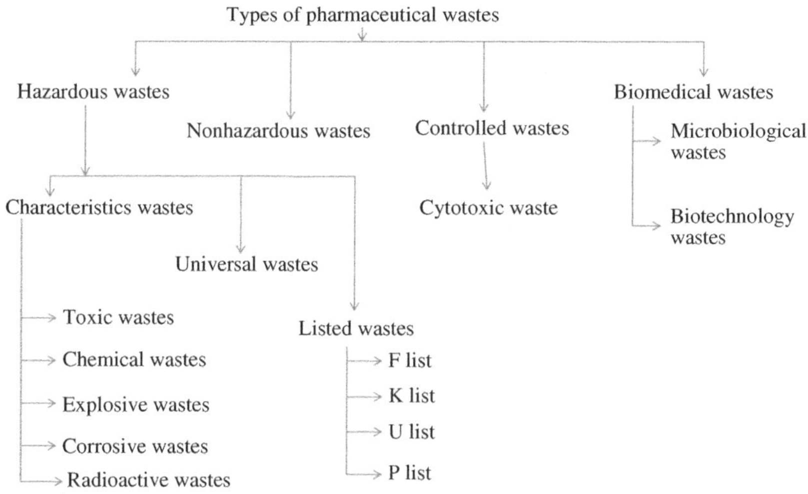

Figure 11.3 Types of pharmaceutical wastes.

production, activities generating waste and many other factors are the decisive factors for the composition of wastewater. The composition of pharmaceutical wastewaters depends upon the source of the contaminants. For instance, pharmaceutical compounds are reported in potable water mainly from the utilization of surface water by village community for drinking purpose. The nearby surface water bodies are the simplest points for dumping of the processed water from different processes running in the pharmaceutical industries. It is necessary to understand the composition of wastewater to decide the efficient treatment methodologies for wastewater management [5]. This is a crucial part for the reuse and recovery of wastewater for further utilization in the industrial setups, or discharging in the water sources [6].

The composition of pharmaceutical wastewater is quite variable and complex, with high organic matter and a recalcitrant nature. A knowledge of the type of pharmaceuticals present in wastewater helps devise a proper treatment strategy. Pharmaceuticals are of different types depending upon the mechanism operative in the drug action, the type of diseases against which these are being used, and their chemical structure. Analgesics and anti-inflammatories, antidepressants, antibiotics, antivirals, anticoagulants, sedatives, and cardiovascular medications are a few classes of pharmaceuticals. Analgesics and anti-inflammatories are one of the major contributors of contaminants in wastewater. Despite negligible concentrations of the drug in pharmaceutical wastewater, nonsteroidal anti-inflammatory drugs have proven to have prolonged health impacts. Such wastewater is odorous with high salt content, along with microbial toxicity. The carbon–nitrogen ratio is

Table 11.1 Composition of biopharmaceutical and pharmaceutical wastewater.

Parameters	Concentration in biopharmaceutical wastewater	Concentration in pharmaceutical wastewater
COD (mg/L)	2000–10000	1000–10000
BOD (mg/L)	1000–2500	500–2500
Temperature (°C)	25–80	25–80
pH	4–8	1–6
Suspended solids (mg/L)	200–500	200–500
Total phosphorus (mg/L)	50–250	50–250
Total nitrogen (mg/L)	500–1500	500–1500
Sulfate (mg/L)	5000–25000	5000–25000
Metal ion (mg/L)	1000–5000	1000–5000
Salts (mg/L)	10000–25000	10000–25000
Antibiotics (μg/L)	1–100	1–100

low for most of the biopharmaceuticals and pharmaceutical wastewaters. It contains solvents, ammonia–nitrogen, suspended solids, different catalysts, additives, intermediate compounds, etc. The typical composition of biopharmaceutical and pharmaceutical wastewater is presented in Table 11.1 [7].

According to one of the investigations, the mounting usage of drugs corresponds to a high drug wastage. In the United States alone, results suggested that drug wastage accounted for 2–3% of the entire drug costs, with greater than 1 billion dollar of drug wastage. In India, healthcare units generate 4057 tons of waste per day, as per a report by the Central Pollution Control Board [8–10]. A report from the US Geological Survey suggested pollution levels downstream from two pharmaceuticals manufacturing plants in New York State were ten to a thousand times higher than those at comparable units around the country [11]. Another report revealed that the Yangtze River in China, which introduces sewage water from 450 million people into the sea water, discharges 172 tons of pharmaceuticals each year [12].

11.3 Toxicity Assessment

One of the chief troubles associated with the release of pharmaceuticals into water bodies is the dependence of people living in villages on contaminated water sources for consumption and household activities. About the same, a report from

Spain highlighted the difficulty of reutilization of recovered wastewater for agricultural application, which led to contamination of water bodies in a few regions [13, 14]. Researchers have reported that in case a veterinary medicine is found to be associated with the water body, it can lose around 70% of the administered curative dosage to the immediate surroundings [15].

Pharmaceutical industries, wastewater treatment plants, hospitals, agricultural fields, and landfills contaminate the aquatic environment. Wastewater is an outcome of processed water effluent from diverse facilities like municipalities, industries, and agricultural fields, among which the flux from drug manufacturing plants is of key concern, due to the presence of physiologically and toxicologically potent compounds [16]. Wastewater is seen as a resource for the reclamation of usable materials, but it can lead to acute health risks. Overwhelming water demand and limited geographical freshwater resource distribution are an alarming situation and a challenge for sustainable development.

Wastewater from various sources should be collected, treated, and then reused; otherwise, the untreated liquid may introduce all the unknown and biologically active pollutants into the water bodies, which directly or indirectly can interact with the human body [17]. Wastewater from the pharmaceutical industry contains solvents, additives, chemicals, and especially active components of pharmaceutical composition. Such wastewater has a high chemical oxygen demand to biological oxygen demand (COD : BOD) ratio.

Metabolized drugs and their derivatives are often excreted by humans and animals [18]. Investigations have revealed that municipal wastewater facility usually contains human excreta, containing transformed or unchanged pharmaceutical ingredients. These may undergo cleavages and may reform the original pharmaceutical moieties. Such wastewater releases these toxic components into the environment. The nonjudicious consumption and unmonitored disposal of pharmaceuticals into water bodies impose deleterious effects on the health of the water ecosystem and humans. Also, the currently used wastewater treatment plants are not efficient enough to eradicate these from wastewater [19].

Pharmaceutical manufacturing processes include fermentation, isolation, synthesis, and recovery. Though recovery assemblies are associated with the manufacturing plants but still 100% solvent recovery is still not achievable. The chemical compounds are not efficiently filtered out during the synthesis process, and they readily dribble into the water system. Also, pharmaceutical wastewater is discharged into the open fields. Thus, pollution load due to pharmaceuticals occurs not only during manufacturing but also due to usage and improper disposal measures. Pharmaceuticals have been detected in drinking water, wastewater, sewage sludge, and soils. They enter the aquatic environment and, even if present in trace amount, can raise concerns over related human health issues. Pharmaceuticals can jeopardize the well-being of the aquatic ecosystem.

Pharmaceuticals in the water bodies are associated with the destruction of aquatic species, as well as biomagnification in the terrestrial ecosystem. These substances are fabricated to trigger biological activity, but if they are persistent in the environment, they can aggravate negative effects in non-targeted species [20, 21]. Endocrine disruption also occurs due to pharmaceutical wastes which seriously affect the biological species, as these tend to bioaccumulate in their bodies, as mentioned by Kayode-Afolayan et al. [22].

Pharmaceutical waste is incalcitrant, persists in water and gets bioaccumulated. Pharmaceutical waste, on reaching the biotic system, can lead to serious health issues both in animals and human. Fishes are very sensitive to these pollutants as they are subjected to these pollutants in an environment which is continuously mobile. Diclofenac and gemfibrozil drugs are found to directly influence the organ system of the fishes [23]. They result in impairment of reproduction in the exposed fishes. A number of studies have reported that estrogen related drugs or post-menopausal treatments have feminizing effects on male fish and alter male-female ratio [24]. The occurrence of pharmaceuticals like phenelzine and venlafaxine, including norfluoxetine and naltrexone metabolites have been found in the fish population [25].

Few researches have also unveiled the concentration of a few antidepressants in the cerebral tissues of fishes downstream from wastewater facilities [11]. The concentration of estrogens in potable water has a negative impact on human male fertility too. In addition, these raise the chances of occurrence of malignancy in breast and testicles [26, 27]. Jukosky et al. revealed about the induction of vitellogenesis in male *Oryzias latipes* by estrogen which gets introduced due to remnants of birth control pills and postmenopausal hormone treatments medicines, in the water bodies [28].

Another investigation related to the failure of the renal system in vultures, revealed the harm posed due to a nonsteroidal drug called diclofenac [29]. A similar observation was made by a team of researchers who confirmed the deleterious effect of diclofenac, which was found accumulated in the prey of vultures. Bioaccumulation led to detrimental effects on the vulture population [30].

Pharmaceuticals present in hospital wastewaters contain compounds that are estimated to have about 5–15 times greater toxic impacts than urban wastewater. A data published mentions the occurrence of different types of pharmaceuticals in the wastewater including antibiotics, antiepileptics, anticoagulants, analgesics and anti-inflammatory steroids, cosmetics, caffeine, paraxanthine, and antidepressants. They contribute in the generation of pathogenic microbes which are highly resistant to antibiotics and illustrate destructive effects on the aquatic ecosystem [31]. Resistance against antibiotics can lead to global health emergencies and can cascade effect on the ecological balance. The menace of resistance against antibiotics gets exacerbated and drastically toxifies the aquatic species. A study on Pacu fish in Argentina confirmed the presence of pharmaceuticals and their metabolites in the fish body [32]. This facilitated the swapping of resistant traits between the microbial populations, which further lead to spreading of antibiotic resistance among the aquatic species [33].

The decomposition of the organics present in sewage water, in the sewage treatment plants, is seriously affected due to the antibiotics present, which have an inhibiting effect on the activities of sewage microbial population. Pharmaceutical wastewater embodies substances which can persuade toxicity at acute and chronic levels, induce carcinogenicity, and can result into reproductive disorders, growth issues, and neurotoxic effect. The deleterious impact of little but unremitting revelation to pharmaceuticals wastes and personal care goods via wastewater are alarming (Figure 11.4) [34, 35].

The existence of pharmaceutical left over like antibiotics in the hydrosphere and lithosphere hampers the functions and assemblies of microbial populations. Researchers have observed changes in the microbial distribution nexus in a river with pharmaceutical contaminants [36, 37]. Drug remnants and their metabolites can seep into the aquifers, making water hazardous for flora, aquatic species and even humans [38]. Few investigations have revealed that presence of endocrine disrupting compounds in wastewater which can disrupt the human endocrine system [39]. Antibiotic resistive microbial species developed in the water bodies can influence the drifting characteristics of aquatic beings substantially [40, 41].

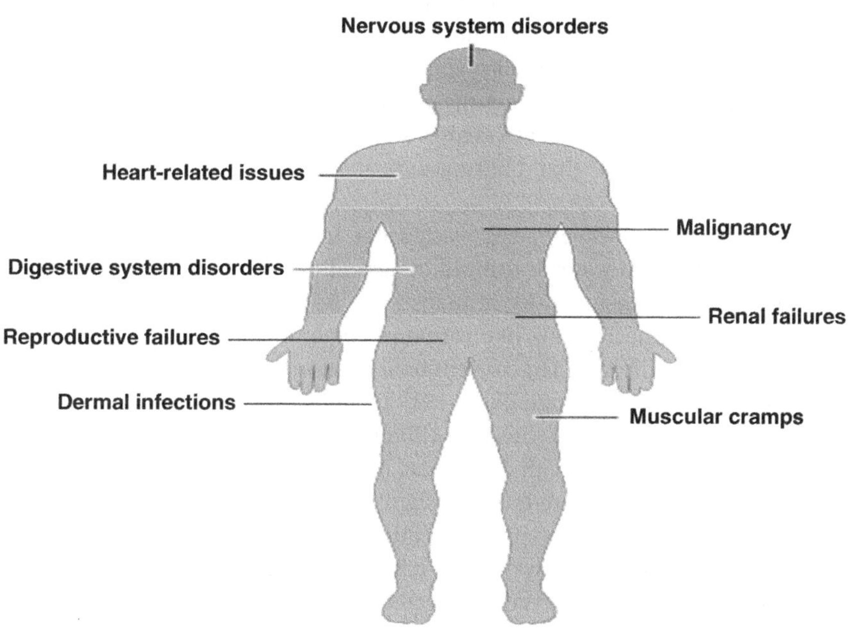

Figure 11.4 Harmful effects on vital organs of pharmaceutical wastewater.

Several pharmaceuticals simply find their way into the human body via food chain. Pharmaceutical contaminants can be partially eliminated using conventional wastewater treatment processes, which alarm the researchers about the fate of the unrecovered toxic waste [42].

Potential toxicity portrayed by these contaminants (antibiotics) was highlighted by Kovalakova and team, who suggested that the presence of antibiotics in the aqueous environment can lead to antibiotic-resistant bacteria (ARB) and antibiotic-resistant gene (ARG) evolution, with a possible threat of transfer to the human population [43]. Other researchers have also reported the development of resistance in several pathogenic bacterial strains due to relic antibodies. Pharmaceuticals also affect population sizes and the reproductive behavior [44]. The growth and development of *Rana pipiens* (Northern Leopard Frog) was greatly influenced due to the altered gene expression and metabolism by fluoxetine, a pharmaceutical present in water bodies [45]. Nontargeted Zebrafish embryo species showed malformations, apoptosis and decreased female population due to accumulated nonsteroidal anti-inflammatory drugs [46]. Organism like fishes and sea mammals which feed on the available food and fecal matter in aquatic environment, unintentionally disturb the transformation products of the pharmaceuticals [47]. The growth of algae is essential as they are phototropic which regulated the energy cycle and nutrients cycle in water bodies. The suspended waste matter of the drugs makes sunlight inaccessible to the aquatic algae and cyanobacteria, which disturbs the food production balance of the aquatic ecosystem [48].

Algal chloroplast is highly affected due to chronic exposure to pharma substances such as carbamazepine, sulfamethoxazole and diclofenac. This leads to algal toxicity and interferes with the food synthesis process. A reduced performance of the beneficial algal population greatly disturbs the aquatic flora and fauna [49, 50]. Another study confirms that pharmaceutical residues distort the growth of phytoplanktons [51].

Most of the cities lack efficient wastewater treatment plans and untreated pharmaceutical loaded water is utilized in agricultural fields to grow vegetables, which are directly consumed by humans [52]. According to the World Health Organization, pharmaceuticals like carbamazepine, diclofenac, and gabapentin are not completely eliminated by conventional wastewater treatment processes and reach to human system through plants [53]. In an investigation focusing on pharmaceutics risk assessment learning, on effluent from the hospitals, drugs such as acetaminophen, clarithromycin, diclofenac, naproxen, propranolol, ofloxacin, trimethoprim, and ibuprofen are found to significantly affect the aquatic organisms [54]. A widespread use of nano pharmaceuticals results in mounting quantities of used nanomaterials in the ecosystem, which can have serious implications [55].

Active pharmaceutical materials are used to fight against many life-threatening diseases and human health damages such as, COVID-19, diabetes, influenza, and goiters, allergies, and tumors [56, 57]. Novel pharmaceutics serve as future solutions for health problems, but their unidentified health implications, metabolic fortune and accumulation raises concerns in health related assessment [58].

Rathi and team presented a review report on the assessment and monitoring of the amount of pharmaceutical wastes in aqueous bodies [59]. It was recommended that utilization of active and passive integrated technologies should be done for better understanding of the occurrences and potential toxicity related to pharmaceuticals present in wastewater [60]. This can facilitate researchers to formulate, adopt and implement appropriate actions for strengthening public health [61].

Safe wastewater reuse is hampered due to technological limitations, becoming major limiting factors, as new chemical species continuously get introduced in the system, but their monitoring and treatment strategies are not designed at the same rate [62]. Thus, it is required to continuously explore and implement management strategies so that the disposal of pharmaceutical waste can be taken care of and also the effects of pharmaceuticals can be minimized. Still many pharmaceuticals and their metabolites need to be explored for studying their toxicity and carcinogenetic nature. The upcoming section of the chapter covers details of the treatment processes adopted for the treatment of pharmaceutical loaded wastewaters (Figure 11.5).

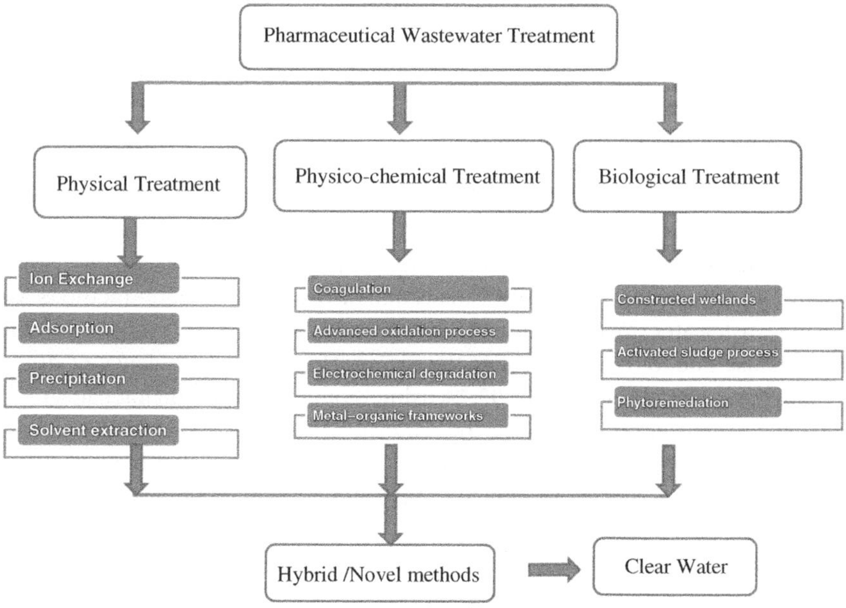

Figure 11.5 Popularly used techniques for pharmaceutical remediation.

11.4 Treatment Methods

11.4.1 Physical Treatment Methods

11.4.1.1 Autoclaving

An autoclave is a machine that operates at a very high pressure, sufficient for killing the pathogenic bacteria. Sterilization occurs at 121 °C for a specific time period. Rules for the biomedical waste are followed for deciding the temperature, pressure, and time for autoclaves. The time of exposure depends on the total load and volume of the pharmaceutical waste. This process helps disinfect the waste before its disposal, but the wastewater stream generated during the process needs a careful supervision and proper disposal. This method is not a promising way of controlling the impact of pharmaceuticals [63].

11.4.1.2 Precipitation

Precipitation is the method of letting the dissolved or suspended contaminants in solution to settle as an insoluble precipitates. The precipitate can be then filtered or centrifuged from the liquid. The method is largely used as a wastewater treatment method with advanced equipment and chemicals. The process offers the advantage of operating with variations in the flow rates and availability of a number of resins, having capacity to exchange specific ions [64]. In a report, chemical precipitation with iron chloride and calcium chloride was found to be effective in the pretreatment method for reducing processing cost and COD, along with total phosphorus. Though, investigations have revealed that overdosing of chemical and parallel reactions decrease the efficacy of the process [65].

11.4.1.3 Ion Exchange

This is a water treatment method where the unwanted dissolved ions are exchanged for ions that have specific charges. It includes softening, deionization, and demineralization by using insoluble organic polymers having high molecular weight and ion-exchange capacity. This is one of the most common methods for pharma wastewater treatment due to easy operation and less maintenance, but it limits its use due to the high cost of the resins. The removal efficiency of drugs like ibuprofen was observed to be 92.7% using Biomax resin, for diclofenac 89.5%, onto bioactivated carbon. Similarly, the removal efficiency for clarithromycin was 86.3% using conventional activated sludge. The results suggested biological ion exchange process was effective in degrading drugs by a simultaneous biodegradation and ion exchange process. Similarly, a mixed cationic and anionic exchange solid phase extraction method was successfully employed for the removal of various basic, acidic and neutral drugs present in wastewater [66].

11.4.1.4 Coagulation

It is a process of adding chemicals to wastewater, and converting stable pollutants into precipitated forms. As evident from literature, coagulation employing colloidal particles as contaminant carriers has also been investigated. Similarly, reports have shown the efficiency of coagulation-flocculation method for eradicating pharmaceuticals. The coagulation process is quite complex for both the natural and chemical coagulant. This involves formation of large flocs by destabilization followed by settling of flocs by precipitation. The mechanism dominantly involved is charge neutralization [67]. The process depends upon operating conditions like coagulant dose, settling time, rate of mixing, pH, etc. [68]. The process improves the biodegradability of the pharmaceutical waste and is effective in eliminating color, suspended solids, chromaticity, turbidity, and COD from wastewater.

Natural coagulants extracted from *Moringa oleifera* seeds, *H. Sabdariffa, J. Curcas, Azadirachta indica, A. Juss*, and aluminum sulfate were investigated by the researchers for their pharmaceutical removal potential. They revealed excellent suspended solids, COD, color, and turbidity removal, with low-cost investment in the process [69–71]. The use of natural coagulants for pharmaceutical removal proves to be more efficient than that of chemical coagulants, as it offers an advantage of less sludge formation, low cost investment, no special managerial skill, and environmental sustainability. The availability and efficiency of the process are affected by seasonal productivity and storage losses [72, 73]. Coagulation is usually followed by the sedimentation process. Both processes are easy to operate and have developed technology, though not very effective in removal of dissolved impurities.

11.4.1.5 Floatation Process

This is usually employed for the removal of suspended contaminants from the secondary wastewater effluent. It involves the production of large number of bubbles through aeration which form floating floc, with contaminants adhering to the bubbles, and can be thus separated from water. This method is preferred for wastewater having large amount of suspended impurities with less investment in terms of cost, labor and maintenance, though, being only a primary treatment process. Santosh and coworkers designed a coagulation-floatation method by employing oppositely charged surfactants and polyelectrolytes for quick eradication of antibiotics from water [74]. Air floatation process was used for removal of ibuprofen and diclofenac using graphene oxide based nanocomposite. Optimum pH for ibuprofen elimination was found to be 5, with a removal efficiency of 99.29%, while for diclofenac drug 95.31% efficiency was recorded at an optimum pH of 4, even at low concentration in wastewaters.

11.4.1.6 Solvent Extraction Process

This method is also known as the liquid extraction technique and is a method of separation of two immiscible liquids based on the difference in their relative solubility. The aqueous solute phase and organic liquid phase are usually separated. This method is quite simple with high operational ability. A high efficiency has been associated with this method [75]. Terpenoids, ketones, and diisopropyl ether have been employed for the preparation of eutectic solvents and were used for the removal of pharmaceuticals from aqueous solutions. The pharmaceuticals under investigation were analgesics, antidepressants, antibiotics, and blood pressure regulators. Hospital wastewater from a hospital care unit was collected, which was reported to have a chemical oxygen demand of 332 mg/L, total organic carbon as 111 mg/L, total nitrogen as 90.1 mg/L, and conductivity of 2360 μS/cm. The simulation of the extraction process was also carried out to understand the feasibility of complete removal of pharmaceuticals from hospital wastewater. The investigation revealed the best solvents as carvacrol and dodecanoic acid. pH was found to influence the drug removal process and the highest extraction yield with carvacrol solvent was obtained at pH 4. Drug removal using the carvacrol solvent was 100% using counter-current extraction. The utilization of hydrophobic solvents of natural terpenoids proved the sustainability of the process for drug elimination from pharmaceutical-loaded wastewater [76]. A team of researchers performed solvent extraction of an antibiotic called AMOX from liquid aqueous phase into 1-decanol using trialkyl methylammonium chloride, di-(2-ethylhexyl)-phosphoric acid, and tributyl phosphate as extractants [77].

11.4.1.7 Evaporation

Pretreatment of wastewater involves the use of an evaporator wherein the removal of water from water-based contaminants occurs by removing water and leaving behind the pollutants. It mainly reduces the volume of the wastewater. It is an effective and reliable method for removing contaminants. The system works on thermodynamic and mass transfer principles. Salts, heavy metals, and other hazardous materials can be removed by this method. It provides high-quality distillate, and possibility of water reuse along with the recovery of precious and dangerous materials through wastewater flow. The evaporators include falling film evaporators, rising film evaporators, and forced circulation evaporators, which involve heating the wastewater and collection the produced steam. The conventional evaporator technique, when replaced with a novel multi-effect evaporator, has proved to be an efficient method for pharmaceutical wastewater treatment [78].

11.4.2 Physio-chemical Methods

11.4.2.1 Advanced Oxidation Process

Advanced oxidation processes have enriched the anti-pollutant methodology for wastewater treatment. These are characterized by the in situ formation of reactive oxygen species like hydroxyl radicals, hydrogen peroxides, superoxide radicals, etc. which can effectively decontaminate the pharmaceutical water by complete conversion to carbon dioxide and water. These have gained attention in the last few years due to availability of various oxidative agents which form products that are susceptible to bioremediation. But, their high operating cost still is a challenge for the research community [79]. Advanced oxidation methods include wet oxidation, Fenton process, supercritical oxidation, UV-photocatalytic oxidation, ultrasound oxidation, electrochemical oxidation, and ozone oxidation. Among various available advance oxidation processes, Fenton process is one of the most promising methods for sequestrating pharmaceutical active materials from wastewater [80]. However, it suffers from the limitations of self-decomposition and iron mud formation [81]. Wet air oxidation process was formulated in an attempt to treat black liquor of the paper industry. It involved the decomposition of the organics into inorganics at elevated temperature of around 400 °C and high pressure. It effectively removed COD and had low energy consumption and easy operation. In another technique chemical reaction was carried out between organic load and dissolved oxygen in supercritical water at 24 Mpa pressure and 400 °C temperature. The method is known as supercritical water oxidation. The process speeds up with an elevation in temperature, but involves high cost input in terms of energy [7]. A number of types of pharmaceutical contaminants can be removed from wastewater by using photo-Fenton method. Fenton process executed using ferrioxalate showed excellent results for a pharmaceutical plant wastewater. The results were the outcome of ready generation of intermediate hydroxyl radical, hydroperoxyl radials, and the superoxide radical anion having reactive nature.

11.4.2.2 Photocatalytic Degradation

It is a combination of a catalyst and ultraviolet (UV) radiation. The oxidative capacity of the oxidants increases in the presence of a light source of 100–400 nm. Photocatalytic oxidation uses titanium and tungsten oxide as catalysts. UV light-mediated iron-doped titanium oxide nanomaterials were investigated for their degradation efficiency for metronidazole aqueous solution. The highest removal performance was achieved at alkaline pH. The method has been acknowledged for the removal of drugs and their derivatives; however, the process lacks recovery of catalytic nanoparticles and has a narrow light energy utilization. It requires an additional cooling water system to maintain the temperature in the entire process. Also, contact with a UV light source could lead to cancerous cell growth [82]. Techniques involving doping of metals and nonmetals onto titanium oxide are being developed to fabricate nanophotocatalysts that ease the charge carrier separation and improve degradation performance under a light source. The improved efficacy of the photo catalyst is due to narrowing of the band gap due to metal and nonmetal incorporation. It was observed that silver doped titanium oxide showed better degradation of pharmaceuticals due to surface Plasmon resonance effect. The carbon doped nanomaterials were found to show acceptable stability after recycling. The procedure adopted to prepare the photocatalyst also influenced the performance of the catalyst against contaminants. Hollow nanomaterials fabricated via sacrificial core agent in sol–gel technique were superior to solid nanomaterials. Also, microwave assisted sol–gel synthesis generated better nanomaterials.

11.4.2.3 Electrochemical Removal

Electrochemical methods are used both for the detection as well as for elimination of the pharmaceuticals from wastewaters. The main requirement for this process to be executed is to remove any macro particles from the water prior to subjection to electrochemical method. This method is of preferred choice due to its low operating cost and applicability to a large type of drug contaminants along with other organic, inorganic and ionic species. The process involves either the electrochemical conversion of recalcitrant drugs into biodegradable compounds or electrochemical combustion of the pollutants into inorganic molecules. Electrochemically generated chlorine in a setup with titanium oxide at anode and zirconium at cathode, along with sodium chloride salt, showed 90% degradation efficiency.

Electrochemical methods are used to eliminate toxic components from water and involve electrolytic recovery, electrochemical oxidation and air flotation method, mostly carried out at normal temperatures, showing high efficiency [66]. Norfloxacin drug when subjected to this process for 60 minutes, revealed an elimination efficiency of 85.29%. Similarly, degradation of acetaminophen was achieved within 90 minutes using electro catalytic anode [83]. In an investigation 100% degradation of a drug amoxicillin was achieved by hydroxyl radicals generated electrochemically using carbon felt fiber. Wastewater carrying pharmaceutical waste could be successfully treated using electrochemical methods.

11.4.2.4 Ozonation

This method involves the use of ozone for the removal of pharmaceuticals from wastewater. Ozonation carried out at pH greater than 8 is also a type of advanced oxidation process leading to the generation of greater amount of hydroxyl radicals. Ozone selectively attacks carbon–carbon linkages, aromatic moieties, sulfur, and phosphorus containing organic compounds, while hydroxyl radicals are involved in hydrogen abstraction, electrophilic addition and electron transfer process, thus resulting in 100% degradation of the drug contaminants. Additionally, ozone also eliminates pathogenic microbes associated with the wastewater. Ozonation of wastewaters from drug formulation industries can be conducted after pretreatment of the wastewater for pH adjustment. This process is considered as part of tertiary or disinfection stage of water.

Ozone has a good oxidizing ability with a redox potential of 2.07 eV. The use of ozone as oxidant is promising in decomposing the pharmaceuticals, as it has no secondary pollution. It can be performed over a wide range of pH and can be generated in situ. Ozone has the capacity to oxidize and decompose most of the organics present in the wastewater. During the process the contaminants undergo a series of oxidations, reductions and spontaneous transformations. Also, it is capable of eliminating turbidity and curbing of microbes in wastewater. Ozone in combination with other treatment technologies (O_3/H_2O_2, O_3/UV), provides effective outcomes. The ozonation process is influenced by a number of factors like amount of ozone, pharmaceutical load, other contaminants present in wastewater, along with the mode of operation. In an investigation for degradation of pharmaceuticals and personal care drugs it was observed that the degradation required a 30 minutes of ozonation period for 100% removal of the drugs under study [83].

11.4.2.5 Adsorption

The adsorption process has been proven to be an efficient process for pharmaceutical wastewater remediation due to its simple operation, low energy and cost consumption, and greater efficiency. Activated carbon has been used on a large industrial scale for this purpose, due to the presence of large specific surface area, high contaminant retaining capacity, porous structure and stability. Adsorption using activated carbon occurs either by reversible and nonselective physical adsorption or by chemical adsorption, which specifically targets the adsorbate and is difficult to desorb. The high cost involved in the use of activated carbon, low regeneration ability and ill-defined structure has however hindered its long-term utilization. Researchers are constantly exploring alternative adsorbents which are potentially effective in eliminating pharmaceuticals, are cost effective, eco-friendly and have the ability to regenerate [84]. In the search for the same a large number of low cost adsorbent like clay [85–87], biochars [88–93], metal–organic frameworks, carbon nanotubes [94], chitosan [95], etc. have been addressed in many reports for drug uptake from wastewaters.

The structure of clay offers an opportunity for degrading pharmaceuticals, as it can be used in conjunction with semiconductors or can be modified with acids or surfactants for adsorbing target species [93]. Among the clay minerals, montmorillonite and kaolinite have shown significant adsorption abilities [94]. Clays are naturally available, cost effective, have mechanical as well as chemical stability, porous structures, and good ion exchange capacities.

Biochar is another promising adsorbent for pharmaceutics elimination due to its physicochemical attributes [93]. It provides inexhaustible opportunities for the elimination of pharmaceuticals. It is a carbon rich substance obtained by pyrolysis, where the thermal decomposition of the natural waste occurs in the absence of oxygen at 300–1000 °C [89]. Biochar prepared from banana pseudo stem fibers was tested for the elimination of Amoxicillin. Biochar was impregnated with cobalt iron oxide nanocomposites, which generated mesoporous structure and excellent adsorption capacity at neutral pH for the drug [96]. Corncob based biochar was prepared by pyrolysis at 600 °C for 2 hours, and was observed to show a removal capacity of 399.6 μg/g [91]. Similarly, tea leaves biochar derived at 450 °C revealed exceptional ciprofloxacin absorption ability at pH 6 and 40 °C. Researchers reported the adsorption mechanism to be π–π interactions, hydrogen bonding, and electrostatic attractions [92].

Chitin is one of the most abundantly available and explored biopolymer worldwide. Its de-acetylation leads to the formation of chitosan, which is a polycationic polymer. It has an excellent prospective as adsorbent due to its biodegradability, compatibility and capability to form a number of derivatives. The amino and hydroxyl (–OH) groups on the structure of the adsorbent enhance the chemical modification of chitosan [96]. Modified chitosan has a better surface area, mechanical strength and functional groups for binding with the drugs. Chitosan is employed either as nanoparticle or as a composite for effectively removing pharmaceuticals form wastewaters [97]. Kyzas et al. reported a high percentage removal of pramipexole (307.18 mg/g) from wastewater, at pH 10, using grafted chitosan material. The modified chitosan showed better removal due to the creation of oppositely charged centers for binding [98].

Diclofenac, carbamazepine, and clofibric acid were eliminated from water using modified magnetic chitosan. This novel material was characterized by X-ray diffraction, scanning electron microscopy (SEM), transmission electron microscopy (TEM), and magnetometeric studies. The sorption capacities of clofibric acid and diclofenac obtained were 191.2 mg/g and 57.5 mg/g, respectively, but no noticeable adsorption occurred for carbamazepine [99]. In another investigation, ciprofloxacin was adsorbed onto a magnetite-based chitosan nanocomposite at an optimum pH of 6. With a decrease in pH, the adsorption capacity lowered, probably due to the existence of cationic drug and positive charge of the modified chitosan, along with the existence of protons with smaller sizes than adsorption sites of the composite [100].

Another nanocomposite of chitosan-terephthalaldehyde was employed to examine the removal of cefotaxime. A large variation in the existence of ionic forms of the drug was noticed with changes in the pH. The zero point charge (ZPC) of the adsorbent was found to be 8.6. Hydrogen bonding and π–π interaction also governed the adsorption of the antibiotic onto the composites [101]. Norfloxacin was observed to have a high adsorption capacity of 294 m^2/g due to high surface area of magnesium oxide-chitosan-graphene oxide nanocomposite [102]. The removal of an antibiotic diclofenac was tried by synthesizing graphene oxide–chitosan and amine based adsorbents using the method of mechanical mixing as a facile method. The fabricated adsorbents were characterized through Fourier transform infrared spectroscopy (FTIR), Brunauer–Emmett–Teller (BET), X-ray diffraction (XRD), TEM, SEM, Raman spectroscopy, and thermogravimetric analysis (TGA) techniques. At pH 5, greater than 90% removal was achieved by the hydrogel adsorbents. The isothermal investigations revealed monolayer, and homogeneous adsorption, while thermodynamics suggested adsorption to be exothermic with higher removal efficiency at lower temperatures [103].

11.4.2.6 Metal–Organic Frameworks

These are porous crystalline hybrids of metal ions with organic ligands. They are well known for their various potentials [104]. They are of key interest in wastewater treatment due to their easy tunable pore size and change in shapes from microporous to mesoporous. Their well-defined dimensions favor the interaction between the framework and the targeted pharmaceuticals. Metal–organic frameworks were first reported to be utilized for the removal of a nonsteroidal drug named Naproxen, which is widely prescribed for pain fever and stiffness of body [105]. The adsorption mechanism was electrostatic interaction between the drug and the adsorbent. In another investigation stable zirconium-ellagate, a plant based metal framework was employed to remove pharmaceutical pollutant from real wastewater. The efficiency of removal was in the range of 79–96.6% for most of the targeted species [106]. Metal–organics are not only limed to use for wastewater remediation, but their superficial surface modification enables them to be utilized for storage, sensing and catalysis as well [107]. Azhar et al. synthesized a submicron sized metal–organic framework for catalytic degradation of water contaminants via adsorption for sulfachloropyridazine and a personnel care product named *p*-hydroxybenzoic acid [108].

Nowadays, the wastewater remediation techniques use modifications of the composite materials along with nanostructured substances, due to their surface area, size flexibility, availability of reactive sites and regeneration abilities. These are the hexagonal plates of carbon stacked in the form of helicities, either as multiwalled or as single-walled carbon nanotubes [109]. Carbon nanotubes have good mechanical strength, thermal conductivity and very low energy loss. Thus, these vary in their potential applications. In last few decades they have been widely used for exclusion of pharmaceutical, and cosmetic related products from wastewater due to a large surface availability along with a porous morphology [94, 110].

Yu et al. sequestrated tetracycline and ciprofloxacin by multi-walled carbon nanotubes modified with oxygen contents [111]. In another investigation triclosan and paracetamol were removed with their excellent adsorption capacity onto carbon nanotubes, from a wide range of solutions with pH ranging from four to ten [112].

11.4.3 Biological Treatment

Biological treatment plays a very important part in conventional water treatment for the elimination of pharmaceutical waste from wastewaters (Figure 11.6).

11.4.3.1 Constructed Wetlands

The use of constructed wetlands is a biological treatment process showing effectiveness both in terms of cost and performance in comparison to various physical and chemical processes. Researchers have confirmed better removal of antibiotics through constructed wetlands [113]. The microbial interaction of bacteria, actinomycetes, yeasts, and protozoans brings about the degradation of the antibiotics. Though temperature, pH, and concentration of the load are the factors affecting the degradation, bacterial performance has been found to be better than any other species for the treatment of enrofloxacin and tetracycline.

Nitrifying and protein degrading bacteria better perform at high temperatures with a greater reduction ability. Also, integrated wetlands show an elimination of 99% of the antibiotics [114]. Efficiency greater than 70% was observed for vertical subsurface type constructed wetlands [115]. The technique still needs to be explored for removal of antibiotics like macro-lactone, chloramphenicol, sulfonamide, polyether and β-lactam, etc. from wastewaters [116].

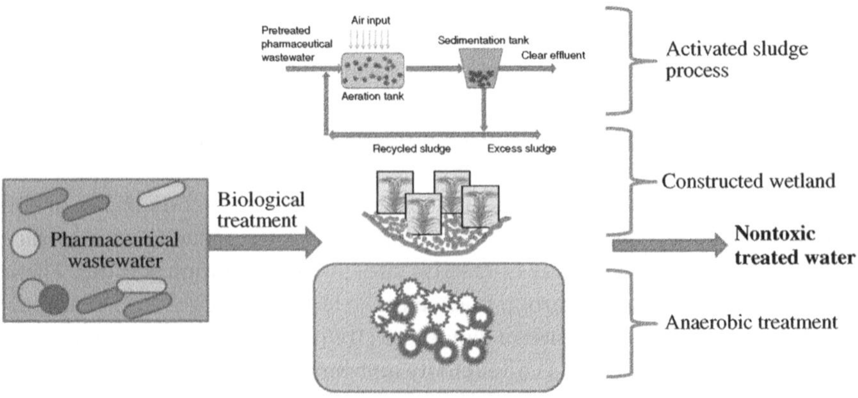

Figure 11.6 Biological treatment process for pharmaceutical wastewaters.

In a similar development, the membrane bioreactor has been seen as a more efficient bio-degradation wastewater treatment process. The setup takes up less space, and the amount of sludge production is also less. A COD removal of 96% and BOD removal of about 96% was observed for pharmaceutical wastewater [117]. The removal efficiency of membrane bioreactors was confirmed suggesting 82% removal of hydroxyl-bupropion, 98% elimination of ibuprofen, 95% and 92% for metformin drug and valsartan drug, respectively [118].

11.4.3.2 Anaerobic Treatment

Pharmaceutical wastewater is effectively treated by anaerobic treatment method. The anaerobic methods are projected to be sustainable and low-cost cost effective methods for drug degradation. These include anaerobic digesters, membrane bioreactors, anaerobic sludge beds, and sequencing batch reactors. Anaerobic digestion is seen as a promising method for drug degradation due to its efficiency of recovery carbon as biogas. It is observed that electrically stimulated anaerobic treatment ensures rapid degradation of antibiotics and improves renewable methane production.

In an investigation, COD removal efficiency was observed to be 50% for high-sulfate containing pharmaceutical wastewater. The efficiency was found to increase by using an anaerobic suspended film contact reactor [119]. A fixed-film reactor was employed for detoxifying herbal-based wastewater [120]. Saravanane et al. investigated the performance of the up-flow anoxic fluidized bed technique for remediation of cephalexin-loaded pharmaceutical effluent [121].

Anaerobic degradation method, however, suffers from the limitation of slow start-up time, slow growth, and poor biomass retention. Researchers have proposed the use of a combined anaerobic wastewater treatment process, like anaerobic treatment using microbial fuel cells, sludge blanket process with bio-electrocatalysis, upflow anaerobic sludge blanket process combined with constructed wetlands to enhance the environmental sustainability, and recommend a green biofuel production pathway.

11.4.3.3 Activated Sludge Process

This is one of the most conventionally adopted methods of treating drug-loaded wastewaters. The process converts the pollutants into inorganics along with sludge production which can be safely disposed off. The cost of investment in the process is less as compared of other biological methods. But it suffers from the disadvantage of energy consumption, the sludge disposal issue, and foaming problems. The efficiency of the activated sludge process for pharmaceutical elimination is governed by wastewater retention time, temperature, pH, oxygen availability, organic load, and the presence of microbial population. Reports suggest that Amoxicillin was removed completely using the process and the chemical

Table 11.2 Pharmaceutical wastewater treatment technologies [63, 74, 91].

Treatment method	Drug	Initial concentration (mg/L)	Contact time (min)	Removal (%)	pH
Ozonation	Tetracycline	0.05	20	95	3
	Sulfamethizole	1	2	95	7
	Diclofenac	0.67–1.2	10	99	–
	Ibuprofen	0.76–2	10	90	–
	Estrone	0.1	1	99.7	–
Photocatalysis	Ofloxacin	0.0002	360	77.8	8.5
	Azithromycin	0.1	180	100	7.2
	Ibuprofen	10	–	69	–
	Estrone	1	180	100	6–8
	Levonorgestrel	50	30	97	5.5
Fenton process	Ampicillin	0.1	20	100	–
	Tetracycline	0.1	20	100	–
	Amoxicillin	138	1	100	–
	Sulfamethizole	25	15	100	6
	Diclofenac	5	120	99	3.5
	Ibuprofen	20	180	100	3.0
Adsorption	Ciprofloxacin	–	80	91.75%	8

oxygen demand of the effluent was also successfully stabilized. A good decrease in the bacterial population resistant against antibiotics was observed. The process also enabled low environmental risk of the drugs as was confirmed by the risk quotients. Activated sludge process in combination with bioaugmentation helped increase the mineralization rate. Caffeine, Ibuprofen, acetaminophen and metronidazole were efficiently degraded using the process. Ibuprofen and paracetamol were eliminated using a benchtop reactor with activated sludge in a synthetic medium. The removal efficiency for these drugs was found to be 99.1% and 93.3%, respectively. Table 11.2 gives a list of different drugs removed by several methods.

11.4.4 Novel Technologies

11.4.4.1 Membrane Separation Technology

This technology has proven itself in the pharmaceutical industrial effluent treatment. Membrane separation is the process where component present in water selectively permeates through the selective membrane under some driving force. This way, pharmaceutical species can be selectively removed from the wastewater. It is a technique proving to be a feasible option against conventional treatment methods. Membrane separation techniques include microfiltration, ultrafiltration, and nanofiltration, which differ from one another in their pore sizes, which are 0.1, 0.01, and 0.001 μm, respectively. Membrane separation is based on the mechanisms of size exclusion, electrostatic repulsion, and adsorption. The extent of drug elimination is governed by various factors, including the drug's physicochemical attributes and the pore size, weight, potential, and driving force of the membranes. Micro filters and ultrafilters are less effective in eliminating the drugs, due to their large pore sizes [122].

In an investigation, 10 different categories of negatively charged (e.g., gemfibrozil) and neutral (e.g., triclosan) drugs were monitored for their removal using nanofiltration [123]. In a similar study, amoxicillin was tested with the same techniques, using a polyamide spiral, where the researchers reported a steady infiltration flux of 1.5 $L/min{\cdot}m^2$ and high refutation of amoxicillin (40%). The efficiency of the membrane was evaluated under various factors, including flow rate, pressure concentration, and COD in the feed [124]. The membrane process suffers from a limitation of fouling, thus requiring periodic cleaning. Also, the concentrate eliminated during the process was more concentrated and required critically monitored disposal strategies [122].

A composite membrane framework was designed for the removal of drugs like ibuprofen, diclofenac, ketoprofen, sulfamethoxazole, and naproxen. Results revealed the efficiency of the covalent organic framework with stable and high permeation ability of 0.75 $L/m^2/h/bar\,\mu m$. The separation mechanism was mainly influenced by steric and electrostatic interactions [125].

Researchers are now focusing on the development of hybrid technologies that involve the combination of two or more conventional or novel methods for the complete eradication of the toxic and persistent drugs from wastewaters. This need is due to drawbacks related to the conventional physical or chemical, or biological processes employed to date, for the eradication of the recalcitrant compounds. Several studies have been carried out in this regard in the last few years.

A combined technical setup was investigated for the elimination of real pharmaceuticals, which included a combination of up-flow anaerobic sludge bed with anaerobic tank and other advanced oxidation methods. The output supported that combined technologies can better eliminate antibiotics from wastewaters. The success of the technique was supported by another study in which pharmaceutical wastewater on treatment with TiO_2-photocatalysis and biological treatment, contributed to 96% of degradation, much greater than that provided by individual methods [126, 127].

11.4.4.2 Membrane Bioreactor System

Membrane bioreactors combine membrane and biological treatment processes and involve the principle of physical retention and selective permeation. Initially, membrane biological reactors were employed to treat hydrophobic pollutants, but the process, when coupled with other techniques, showed better performance with clearly monitored membrane specification and energy demand. The system, coupled with nanofiltration, was effective for sequestrating spiramycin and a decline in the acute toxicity was observed. Another study showed a fall in the membrane fouling. Membrane bioreactors with constructed wetland exhibited complete purging of drugs like ibuprofen, carbamazepine, frusemide, and ofloxacin, from hospital effluent [128].

The application of a single treatment method for wastewater treatment is usually not very fruitful and is limited in use due to the associated cost. To overcome the issue, an integration of techniques like biodegradation, adsorption, ozonation, membrane method, etc. is done. In this regard, a combined approach of ozone-based oxidation process and adsorption was carried out for a few anti-cancer drugs, including pain killers. With the combined method, COD was reduced to 85.4–92.7% at a pH range of 5–11 in 2.5 hours [129]. Pharmaceuticals like sulfamethoxazole, diclofenac, clofibric acid, and ibuprofen were removed from wastewater by using hybrid ceramic membranes having layer-by-layer coating at low transmembrane pressure, resulting in greater than 80% of the removal for drugs and other organics. This aids in the advanced treatment of water, which can be applied at a larger scale [130]. Similarly, the elimination of veterinary antibiotics such as tylosin tartrate and enrofloxacin from wastewater was done in a novel way using a hybrid of electroflocculation, catalytic ozonation, a catalyst Ni-Co-zeolite, and ceramic membrane filtration methods. It was revealed that the treatment method removed 97% of enrofloxacin, 98% of tylosin tartrate, 88% COD, 85% BOD_5, and 93% turbidity [131]. The success of the hybrid methods depends on the execution of the technique using real wastewater. Pharmaceutical industries can themselves help in achieving the aim of zero liquid discharge. A pilot study was carried out to assess the performance of a zero liquid discharge system installed in a pharmaceutical industry. The effluents were subjected to a stripper system, multi-effect evaporator, and agitated thin film dryer, aerobic biological process, followed by a reverse osmosis system. It was observed that the tubular filter membrane removed about 95% of the suspended solids from wastewaters. The removal facilitated biofouling of the membranes, and improvement in the efficiency was recorded [132].

Thus, several techniques are available for the elimination of drugs from wastewaters; however, their large-scale application is limited due to the associated weakness observed in their sequestration process. One of the evaluation studies concluded reverse osmosis as one of the best methods of eliminating drugs like ofloxacin, sulfamethoxazole, and diclofenac from sewage wastewater, with a removal efficiency of 99%, while the activated sludge process was efficient in treating wastewater received from hospital facilities. The deficiencies of the individual treatment methods can be made up only by a combination of techniques to remove drugs from wastewater.

11.5 Legislative Framework of Pharmaceuticals

The increasing prevalence of various diseases and the potential influence of drug residues on the ecosystem have led several nations to formulate and implement environmental risk assessment for drug discharge control. The effluent guidelines are a complex set of essentials that cover many aspects of pharmaceutical manufacturing and processing. In the United States, the National Pollutant Discharge Elimination System (NPDES) is a regulatory agency aiming at protecting water quality. The agency performs administrative and enforcement tasks for discharges to water bodies, covered under the Clean Water Act. The disposal standards and requirements for sewage wastewater have been covered under it. The formulation of pretreatment requirements is intended to protect publicly owned treatment systems and other treatment setups for the protection of the aqueous environment. It includes general and specific rules and limits for effluent discharges, considering various pollutants and chemicals. The European Nations, the United States, and Canada follow an assessment system for most of their medicines. European Union under 2008/105/EC directive sets guidelines for several drug acceptable concentrations in the aqueous environment, which includes Ibuprofen and Ciprofloxacin.

The Environmental Protection Agency in the United States is the authority for the regulation of several contaminants. These regulations include checks for pharmaceutical contaminants, in addition to other categories of contaminants, under the Safe Drinking Water Act (SDWA) and Clean Water Act (CWA), for drinking water and wastewater, respectively. The states can adopt standards, as given by CWA, through the National Pollutant Discharge Elimination System (NPDES), which can be enforced. Also, in Sections 306, 304, 302, and 301 of the Clean Water Act (CWA), guidelines for effluent, based on the technology of treatment, have been issued, which can be adopted.

Pharmaceutical Manufacturing Effluent Guidelines and Standards (40 CFR Part 439) were disseminated by the Environmental Protection Agency in 1976 for direct and indirect discharges of wastewater from such industries, which were then revised. In 1976, the EPA formulated "Best Practicable Control Technology, Currently Available" effluent limit guidelines (41 Federal Register 50676) for effluent limitations "Pharmaceutical Manufacturing Point Source Category."

The same was amended in 1983 (48 Federal Register 49808) and further in 1998 (63 Federal Register 50388). Pharmaceuticals are addressed in 40 CFR part 439. Section 439.12 recommends a maximum monthly average limit of BOD not less than 90%. The COD limitation is 1675 ppm, maximum daily load, and 856 ppm, as maximum monthly average load [133].

According to the United Kingdom, legislation for the maximum permitted concentration of BOD for five days, at 20 °C, is 50 mg/L, while for COD, it is 250 mg/L.

In China, pharmaceutical producing facilities need to follow the national discharge standards, GB8978, to discharge wastewater. These provide a maximum allowable discharge concentration for about 69 water contaminants.

Morocco has implemented various environmental reforms committed to sustainable development. Moroccan law enacted in 2016 focuses on the reuse of wastewater and sewage, simultaneously fulfilling the water quality standards. The National Wastewater Reuse Plan of Morocco is committed to the installation of several wastewater treatment plants by 2030, which will open up new ventures for water recycling and its reuse. Its plan would comprise of joint efforts by the public and private sectors to launch projects for water purification, and to achieve the environmental quality goal.

Pharmaceuticals are eliminated from wastewater by several conventional treatment methods. Physical processes are found to be highly efficient, but they produce concentrated wastewater that needs further treatment, followed by proper disposal. Membrane processes have limitations of cost and fouling, while adsorption poses an issue of loss of adsorbent and the costs involved in the preparation of the adsorbent. Biological processes are simple to operate and execute, but they are influenced by operational conditions and have low removal efficiency for biologically stable compounds. Thus, combined processes could be ideal alternatives for the destruction of both the drug and its transformation products. The installation of a hybrid wastewater treatment plant has been seen as a promising option for the reduction of drugs and their transformative products from wastewaters, before these enter the water body or ecosystem. Though the optimization of the treatment method is based on the load matrix, quality, and quantity of the contaminated water and their residues and impacts. Hybrid systems can produce bio-energy, which reduces the cost of operations or handling of wastewater treatment utilities. The effectiveness of the hybrid process depends on the characteristics of the wastewater, which should be pre-evaluated to achieve the greatest output. The process has the advantage of low energy input, competent degradation, and nontoxic by-product generation. Treatment of contaminated wastewater is necessary for the quality of life for humans and other living organisms. An appropriate treatment technique is necessary to be adopted that can meet the requirements of the quality of dischargeable water. More information is required to completely assess and mitigate the effects of pharmaceuticals on water systems. It is crucial to conduct investigations for a deeper perception of the interactions of these drugs with the ecosystem. An evaluation of the complex behavior of pharmaceuticals and their derivatives can help mitigate the negative effects and ensure the sustainability of water systems, as well as health. Also, focus on drinking water treatment technologies is currently imperative, as existing processes are not effective to purge the contaminants entirely. In this regard, collaborative efforts by researchers, practitioners, industrialists, policy makers, and governmental agencies are crucial.

11.6 Case Studies

11.6.1 Sewerage Treatment Plant, Ghana

The wastewater treatment plant, Mudor, Accra, Ghana, was established in 2000, but some management and finance-related issues led to the shutdown of the facility. The sewage treatment plant consisted of screens, a grit chamber, a trickling filter, a sedimentation tank, and up-flow anaerobic sludge blanket units. The increasing sewage water load, requirement of treatment plants, and revision in the previously formulated strategies, the treatment plant was again made functional in 2017. Places like Osu, Korle Bu, and High streets were observed to be the major sites, contributing to the sewage system, having a discharge flow rate of 16000–180000 m^3/day.

The samples of wastewater from a treatment plant in Ghana were investigated for pharmaceutically active compounds like diclofenac, paracetamol, ibuprofen, and aspirin.

The study confirmed that influent wastewater directed to the treatment facility contained an average maximum concentration of 121.31 μg/mL of diclofenac, 65.54 μg/mL of paracetamol, 19.54 μg/mL, ibuprofen, and 0.27 μg/mL of aspirin. The wastewater showed high concentrations of drugs, which can be linked with the release of drug waste from the Korle Bu hospital, which discharged the high est amount of diclofenac. In a 16 days of sample collection time, 121.16 μg/mL diclofenac was observed, while the Aspirin concentration was merely obtained as 0.27 μg/mL, which suggested that the country community was using Aspirin in relatively lesser amounts. The investigation samples were collected at the time of maximum load received by the wastewater treatment plant. The study included an exploration of the ability of the trickling filter for the removal of these pharmaceutical contaminants. The efficiency of the trickling filter was also assessed by analyzing the grab samples for two days. The average removal efficiency of the pharmaceutical contaminants was obtained as 74%, 93%, 98%, and 99% for diclofenac, aspirin, paracetamol, and ibuprofen, respectively. The trickling filter was found to reduce the concentration of the drugs to a large extent. The removal efficiency of the wastewater treatment plant for COD and BOD was 94.4% and 98%, respectively. The nitrate concentration was recorded as 22.1 mg/L, which exceeded the EPA guidelines for release concentration, which was 1.5 mg/L. The effluent from the treatment unit was suitable for irrigation purpose as the composition could help enhance the soil fertility. The final effluent concentration of pharmaceutical wastewater from Ghana was in agreement with the EPA standards of 2 mg/L. The removal efficiency of diclofenac was 74%, which was low in comparison to other pharmaceutical contaminants. The overall removal efficiency for turbidity was only 68.8%, possibly due to the limited retention time provided to the wastewater.

Thus, it was concluded that the wastewater treatment plant needed to upgrade the treatment units to increase the removal efficiency for drugs, especially contaminants like diclofenac. The incorporation of photo transformation facilitated the drug elimination, due to photolytic degradation. Also, the sorption of pharmaceutics to the membrane bioreactor sludge could be a potential way of enhancing the performance of the treatment plants. Research suggested that the inclusion of photo transformation and adsorption by the treatment plant could help achieve a higher removal efficiency for the pharmaceutical pollutants [134].

11.6.2 Pharmaceutical Wastewater Treatment Using High-rate Algal Pond

A study was carried out for investigating the efficiency of high-rate algal ponds for the removal of pharmaceutical contaminants from domestic wastewater in the city of Santiago de Cali, Colombia.

High-rate algal pond is a shallow biomass reactor of microalgae and bacterial symbiotic association, efficient enough to eradicate the pharmaceutical pollutants. High-rate algal pond involves both biotic and abiotic factors contributing to reducing pollutants by biodegradation, light-assisted degradation, and sorption. All these processes together convert the complex organic compounds into simple ones. The performance of the process is affected by the presence of other contaminants, for target species, which may preferentially adsorb on the adsorption site.

High-rate algal pond was functional under a volume of 0.6 m^3, with a water flow of 0.2 m^3/day, and a retention time of three days. Removal efficiency was greater than 70% for ibuprofen and paracetamol, 30–70% for diclofenac and naproxen, and less than 30% for carbamazepine and gemfibrozil. The advantageous aspect of using this technique was good performance in a short retention time. A decrease in the algal community, decreased the removal efficacy for a few drugs, while for a few, like gemfibrozil and diclofenac, the removal efficiency was found to increase due to the direct photolysis process. The investigation confirmed that photo degradation was a favored process under tropical environmental conditions due to favored growth and action of microalgae. The other processes involved were biodegradation and sorption. Various microalgae species were identified in the ponds, with *Chlorophyta* as the most abundant of them all. The case study also included the application of the hazard quotient to measure ecological hazard before and after the high-rate algal pod treatment. Hazard quotient is the ratio of measured environmental concentration to the estimated no-effect amount of the pollutant. A hazard quotient less than 0.1 suggests an insignificant hazard, obtained for paracetamol in the investigated sample. Medium scale hazard for diclofenac was observed, as the Hazard Quotient (HQ) for this drug was in the range 1 to 10. High hazard was reported for gemfibrozil, ibuprofen, and naproxen as the HQ value was found to be greater than 10. After subjecting the wastewater to high-rate algal pond, no significant change occurred in the gemfibrozil and ibuprofen as the HQ for gemfibrozil and ibuprofen at influent was 52.30 and 318.13, while at effluent was merely 35.12 and 74.91, respectively.

These results suggested the requirement for a complementary treatment process. It was required to have an additional treatment process to reduce the health risk related to the pharmaceutical pollutants. The ecological hazard associated with the pharmaceutical waste was reduced by more than 50% [135].

References

1 Sreekanth, K., Gupta, N.V., Raghunandan, H.V., et al. (2014). A review on managing of pharmaceutical waste in industry. International Journal of Pharma Technological Research, Volume 6, Issue 3, pp. 899–907.

2 Kadam, A., Patil, S., Patil, S., et al. (2016). Pharmaceutical waste management an overview. Indian Journal of Pharmacy Practice, Volume 9, Issue 1, pp. 2–8.

3 https://www.vlses.com/2022/10/31/types-of-pharmaceutical-waste/ (Last accessed 23.09.24).

4 Shinde, A.R., Patil, A.P., Patil, T.M., et al. (2020). Pharmaceutical waste management: significant for health of nature and future. International Journal of Pharmaceutical Sciences Review and Research, Volume 65, Issue 2, pp. 68–74.

5 Gadipelly, C., Pérez-González, A., Yadav, G.D., et al. (2014). Pharmaceutical industry wastewater: review of the technologies for water treatment and reuse. Industrial and Engineering Chemistry Research, Volume 53, Issue 29, pp. 11571–11592.

6 Zhang, R., Wang, Z., Zhou, Z., et al. (2019). Highly effective removal of pharmaceutical compounds from aqueous solution by magnetic Zr-based MOFs

composites. Industrial and Engineering Chemistry Research, Volume 58, Issue 9, pp. 3876–3884. https://doi.org/10.1021/acs.iecr.8b05244.

7 Guo, Y., Qi, P.S., and Liu, Y.Z. (2017). A review on advanced treatment of pharmaceutical wastewater. IOP Conference. Series: Earth and Environmental Science, Volume 63, p. 012025. https://doi.org/10.1088/1755-1315/63/1/012025.

8 Scenario of pharmaceutical market in India. http://www. smetimes.in/smetimes/in-depth/2013/May/27/scenario-of-pharmaceuticalmarket-in india629370 (Last accessed 23.09.24).

9 Seeheusen, D.A., and Edwards, J. (2006). Patient practices and beliefs concerning disposal of medications. American Board of Family Medicine, Volume 19, Issue 6, pp. 542–547.

10 Over 1 million tones medical waste headache for India. http://www.thepharmatimes.in/./237-over-1-million-tonnes-medicalwaste-he. (Last accessed 23.09.2024).

11 https://www.health.harvard.edu/newsletter_article/drugs-in-the-water (Last accessed 23.09.24).

12 Ciara, B., Gokulajri, S., and Suresh, C. (2018). Recent advances in photocatalysis for environmental applications. Journal of Environmental Chemical Engineering, Volume 6, pp. 3531–3555.

13 Chen, Y., Yu, G., Cao, Q., et al. (2013). Occurrence and environmental implications of pharmaceuticals in Chinese municipal sewage sludge. Chemosphere, Volume 9, pp. 1765–1772.

14 Rodriguez, N.C., Bjorklund, E., Bak, M., et al. (2013). Pollution pathways of pharmaceutical residues in the aquatic environment on the Island of Mallorca Spain. Reviews of Environmental Contamination and Toxicology, Volume 65, pp. 56–66.

15 Tian, M., He, X., Feng, Y., et al. (2021). Pollution by antibiotics and antimicrobial resistance in livestock and poultry manure in China, and counter measures. Journal of Antibiotics, Volume 10, Issue 5, p. 539.

16 Bahadar, H., Abdollahi, M., Maqbool, F., et al. (2014). Mechanistic overview of immune modulatory effects of environmental toxicants. Inflammation and Allergy Drug Targets, Volume 13, pp. 382–386. https://doi.org/10.2174/1871528114666150529103003.

17 Touraud, E., Roig, B., Sumpter, J.P., et al. (2011). Drug residues and endocrine disruptors in drinking water: risk for humans? International Journal of Hygiene and Environmental Health, Volume 214, Issue 6, pp. 437–441. https://doi.org/10.1016/j.ijheh.2011.06.003.

18 Patel, M., Kumar, R., Kishor, K., et al. (2019). Pharmaceuticals of emerging concern in aquatic systems: chemistry, occurrence, effects, and removal methods. Chemical Reviews Journal, Volume 119, pp. 3510–3673.

19 AL Falahi, O.A., Abdullah, S.R.S., Hasan, H.A., et al. (2021). Simultaneous removal of ibuprofen, organic material, and nutrients from domestic wastewater through a pilot-scale vertical sub-surface flow constructed wetland with aeration system. Journal of Water Process Engineering, Volume 43, p. 102214.

20 Rivera-Utrilla, J., Sánchez-Polo, M., Ferro-García, M.A., et al. (2013). Pharmaceuticals as emerging contaminants and their removal from water. A review. Chemosphere, Volume 93, Issue 1, pp. 268–1287.

21 Gwenzi, W., and Chaukura, N. (2018). Organic contaminants in African aquatic systems: current knowledge, health risks, and future research directions. Science of the Total Environment, Volume 619–620, pp. 1493–1514. https://doi.org/10.1016/j.scitotenv.2017.11.121.

22 Kayode-Afolayan, S.D., Ahuekwe, E.F., and Nwinyi, O.C. (2022). Impacts of pharmaceutical effluents on aquatic ecosystems. Scientific African, Volume 17, p. e01288.

23 Beiras, R. (2021). Chapter 6: Environmental risk assessment of pharmaceutical and personal care products in estuarine and coastal waters. In: Juan C, Álvarez D, Jiménez-Cisneros B (eds) Pharmaceuticals in marine and coastal environments. Estuarine and coastal sciences series, Volume 1. Elsevier, pp. 195–252.

24 Larsson, D.G.J., Adolfsson-Erici, M., Parkkonen, J., et al. (1999). Ethinyloestradiol- an undesired fish contraceptive? Aquatic Toxicology, Volume 45, Issues 2–3, pp. 91–97. https://doi.org/10.1016/S0166-445X(98)00112-X.

25 Fatta-Kassinos, D., and Kümmerer, K. (2010). Pharmaceuticals in the environment: sources, fate, effects, and risks. Environmental Science and Pollution Research, Volume 17, pp. 519–521.

26 Mahapatra, S., Ali, M.H., and Samal, K. (2022). Assessment of compost maturity-stability indices and recent development of composting bin. Energy Nexus, Volume 6, Article 100062. https://doi.org/10.1016/j.nexus.2022.100062.

27 Schaider, L.A., Rudel, R.A., Ackerman, J.M., et al. (2014). Pharmaceuticals, perfluoro surfactants, and other organic wastewater compounds in public drinking water wells in a shallow sand and gravel aquifer. Science of the Total Environment, Volume 468–469, pp. 384–393.

28 Jukosky, J.A., Watzin, M.C., and Leiter, J.C. (2008). The effects of environmentally relevant mixtures of estrogens on Japanese medaka (*Oryzias latipes*) reproduction. Aquatic Toxicology, Volume 86, Issue 2, pp. 323–331.

29 Green, R.E., Ian, N., Susanne, S., et al. (2004). Diclofenac poisoning as a cause of vulture population declines across the Indian subcontinent. Journal of Applied Ecology, Volume 41, Issue 5, pp. 793–800. https://doi.org/10.1111/j.0021-8901.2004.00954.x.

30 Arockianathan, S., and Balasundaram, R. (2016). Observation of a population of Egyptian vulture *Neophron percnopterus* in Ramanagaram hills, Karnataka, southern India, Vulture News, Volume 71, pp. 1–14. https://doi.org/10.4314/vulnew.v71i1.3.

31 Escher, B.I., Baumgartner, R., Koller, M., et al. (2011). Environmental toxicology and risk assessment of pharmaceuticals from hospital wastewater. Water Research, Volume 45, pp. 75–92.

32 Carrizo, J.C., VoDuy, S., Munoz, G., et al. (2022). Suspect screening of pharmaceuticals, illicit drugs, pesticides, and other emerging contaminants in Argentinean *Piaractus mesopotamicus*, a fish species used for local consumption and export. Chemosphere, Volume 309, p. 136769.

33 Cui, C., Han, Q., Jiang, L., et al. (2018). Occurrence, distribution, and seasonal variation of antibiotics in an artificial water source reservoir in the Yangtze river delta, East China. Environmental Science and Pollution Research, Volume 25, pp. 19393–19402.

34 Sharma, S.K., and Sanghi, R. (2012). Wastewater reuse and management. Springer Science and Business Media, Berlin.

35 Samal, K., and Dash, R.R. (2021). Modelling of pollutants removal in Integrated Vermifilter (IVmF) using response surface methodology. Clean Engineering and Technology, Volume 2, p. 100060. https://doi.org/10.1016/j.clet.2021.100060.

36 Li, B., Xu, D., Zhou, X., et al. (2023). Environmental behaviors of emerging contaminants in freshwater ecosystem dominated by submerged plants: a review. Environmental Research, Volume 227, p. 115709.

37 Mastrángelo, M.M., Valdés, M.E., Eissa, B., et al. (2022). Occurrence and accumulation of pharmaceutical products in water and biota of urban lowland rivers. Science of Total Environment, Volume 828, p. 15430322.

38 Petrie, B., Barden, R., and Kasprzyk-Hordern, B. (2015). A review on emerging contaminants in wastewaters and the environment: current knowledge, understudied areas, and recommendations for future monitoring. Water Research, Volume 72, pp. 3–27.

39 Gaffney, V.J., Almeida, C.M., Rodrigues, A., et al. (2015). Occurrence of pharmaceuticals in a water supply system and related human health risk assessment. Water Research, Volume 72, pp. 199–208.

40 Kummerer, K. (2001). Drugs in the environment: emission of drugs, diagnostic aids and disinfectants into wastewater by hospitals in relation to other sources: a review. Chemosphere, Volume 45, pp. 957–969.

41 Orlando, E.F., Kolok, A.S., Binzcik, G.A., et al. (2004). Endocrine-disrupting effects of cattle feedlot effluent on an aquatic sentinel species, the Fathead Minnow. Environmental Health Perspectives, Volume 112, pp. 353–358.

42 Ternes, T., Janex-Habibi, M., Knacker, T., et al. (2006). Assessment of technologies for the removal of pharmaceuticals and personal care products in sewage and drinking water facilities to improve the indirect potable water reuse. Environmental Science Engineering.

43 Kovalakova, P., Cizmas, L., McDonald, T.J., et al. (2020). Occurrence and toxicity of antibiotics in the aquatic environment: a review. Chemosphere, Volume 251, p. 126351.

44 He, W.C., Lin, J., Bing, Yu., et al. (2021). Reducing the impact of antibiotics in wastewaters: increased removal of mitoxantrone from wastewater by biosynthesized manganese nanoparticles. Journal of Cleaner Production, Volume 293, p. 126207.

45 Foster, H.R., Burton, G.A., Basu, N., et al. (2010). Chronic exposure to fluoxetine (prozac) causes developmental delays in Rana pipiens larvae. Environmental Toxicology and Chemistry, Volume 29, pp. 2845–2850.

46 Bereketoglu, C., Pradhan, A., and Olsson, P.E. (2020). Nonsteroidal anti-inflammatory drugs (NSAIDs) cause male-biased sex differentiation in zebrafish. Aquatic Toxicology, Volume 223, p. 105476.

47 Nakagawa, S., Saito, H., Tame, A., et al. (2017). Microbiota in the coelomic fluid of two common coastal starfish species and characterization of an abundant helicobacter-related taxon. Scientific Reports, Volume 7, Issue 1, p. 8764.

48 Ogueji, E., Nwani, C., Iheanacho, S., et al. (2018). Acute toxicity effects of ibuprofen on behaviour and hematological parameters of African catfish *Clarias gariepinus*. African Journal of Aquatic Science, Volume 43, p. 293.

49 Harshkova, D., Majewska, M., Pokora, W., et al. (2021). Diclofenac and atrazine restrict the growth of a synchronous chlamydomonas reinhardtii population via various mechanisms. Aquatic Toxicology, Volume 230, p. 105698.

50 Mojiri, A., Zhou, J.L., Ratnaweera, H., et al. (2022). Pharmaceuticals and personal care products in aquatic environments and their removal by algae-based systems. Chemosphere, Volume 288, p. 132580.

51 Ekta, B., and Pammi, G. (2018). Impact of antibiotics on plants. International Journal of Pharmaceutical Sciences Review and Research, Volume 52, pp. 49–53.

52 Qishlaqi, A., Moore, F., and Forghani, G. (2008). Impact of untreated wastewater irrigation on soils and crops in Shiraz suburban area, SW Iran. Environmental Monitoring Assessment, Volume 141, Issue 1-3, pp. 257–273. https://doi.org/10.1007/s10661-007-9893-x.

53 World Health Organization (2012). Pharmaceuticals in drinking-water. WHO, Geneva.

54 Mendoza, A., Aceña, J., Pérez, S., et al. (2015). Pharmaceuticals and iodinated contrast media in a hospital wastewater: a case study to analyse their presence and characterize their environmental risk and hazard. Environmental Research, Volume 140, pp. 225–241. https://doi.org/10.1016/j.envres.2015.04.003.

55 O'Brien, N., and Cummins, E. (2010). Ranking initial environmental and human health risk resulting from environmentally relevant nanomaterials. Journal of Environmental Science and Health A, Volume 45, Issue 8, pp. 992–1007. https://doi.org/10.1080/10934521003772410.

56 Chaubey, A.K., Patel, M., and Mohan, D. (2021). Synthesis of engineered biochar for aqueous pharmaceutical removal. EGU General Assembly 2021. https://doi.org/10.5194/egusphere-egu21-9358.

57 Dehkordi, S.K., Paknejad, H., Blaha, L., et al. (2021). Instrumental and bioanalytical assessment of pharmaceuticals and hormone-like compounds in a major drinking water source-wastewater receiving Zayandeh Rood River, Iran. Environmental Science and Pollution Research, Volume 29, pp. 9023–9037.

58 Prasse, C., Stalter, D., Schulte-Oehlmann, U., et al. (2015). Spoilt for choice: a critical review on the chemical and biological assessment of current wastewater treatment technologies. Water Research, Volume 87, pp. 237–270. https://doi.org/10.1016/j.watres.2015.09.023.

59 Rathi, B.S., Kumar, P.S., and Vo, D.V.N. (2021). Critical review on hazardous pollutants in water environment: occurrence, monitoring, fate, removal technologies and risk assessment. Science of the Total Environment, Volume 797, p. 149134.

60 Praveena, S.M., Mohd Rashid, M.Z., Mohd Nasir, F.A., et al. (2019). Occurrence and potential human health risk of pharmaceutical residues in drinking water from Putrajaya (Malaysia). Ecotoxicology and Environmental Safety, Volume 180, pp. 549–556.

61 Gallego-Ríos, S.E., Peñuela, G.A., and Martínez-López, E. (2021). Updating the use of biochemical biomarkers in fish for the evaluation of alterations produced by pharmaceutical products. Environmental Toxicology and Pharmacology, Volume 88, p. 10375611.

62 Matamoros, V., Rodríguez, Y., and Albaigés, J. (2016). A comparative assessment of intensive and extensive wastewater treatment technologies for removing emerging contaminants in small communities. Water Research, Volume 88, pp. 777–785. https://doi.org/10.1016/j.watres.2015.10.058.

63 Jaseem, M., Kumar, P., and John, R.M. (2017). An overview of waste management in pharmaceutical industry. The Pharma Innovation Journal, Volume 6, Issue 3, pp. 158–161.

64 Venkateswara Reddy, B., Sandeep, P., Ujwala, P., et al. (2014). Water treatment process in pharma industry: a review. International Journal of Pharmacy and Biological Sciences, Volume 4, Issue 2, pp. 7–18.

65 Wenquan, S., Genchao, M., Yongjun, S., et al. (2017). Effective treatment of high phosphorus pharmaceutical wastewater by chemical precipitation. The Canadian Journal of Chemical Engineering, Volume 95, Issue 8, pp. 1585–1593. https://doi.org/10.1002/cjce.22799.

66 Shearer, L., Pap, S., and Gibb, S.W. (2022). Removal of pharmaceuticals from wastewater: a review of adsorptive approaches, modelling and mechanisms for metformin and macrolides. Journal of Environmental Chemical Engineering, Volume 10, Issue 4, p. 108106. https://doi.org/10.1016/j.jece.2022.108106.

67 Ahmad, A., Abdullah, S.R.S., Hasan, H.A., et al. (2021). Plant-based versus metal-based coagulants in aquaculture wastewater treatment: effect of mass ratio and settling time. Journal of Water Process Engineering, Volume 43, p. 102269.

68 Sulaiman, M., Zhigila, D.A., Umar, D.M., et al. (2017). *Moringa oleifera* seed as alternative natural coagulant for potential application in water treatment: a review. Journal of Advanced Research, Volume 30, pp. 1–11.

69 Nonfodji, O.M., Fatombi, J.K., Ahoyo, T.A., et al. (2022). Performance of *Moringa oleifera* seeds protein and *Moringa oleifera* seeds protein-polyaluminum chloride composite coagulant in removing organic matter and antibiotic-resistant bacteria from hospital wastewater. Journal of Water Process Engineering, Volume 33, p. 101103.

70 Thirugnanasambandham, K., and Karri, R.R. (2021). Preparation and characterization of *Azadirachta indica A. Juss.* plant based natural coagulant for the application of urban sewage treatment: modelling and cost assessment. Environmental Technology Innovations, Volume 23, Issue 3, p. 101733.

71 Sibartie, S., and Ismail, N. (2018). Potential of *Hibiscus Sabdariffa* and *Jatropha Curcas* as natural coagulants in the treatment of pharmaceutical wastewater. MATEC Web Conference, Volume 152, p. 01009.

72 Awolola, G., Oluwaniyi, O., Solanke, A., et al. (2010). Toxicity assessment of natural and chemical coagulants using brine shrimp (*Artemia salina*) bioassay. International Journal of Biological and Chemical Sciences, Volume 4, pp. 633–641.

73 Formicoli, T.K., Freitas, S., Almeida, A., et al. (2018). Review of utilization plant-based coagulants as alternatives to textile wastewater treatment. In: Muthu S (ed). Detox fashion, Springer, Singapore, pp. 27–79.

74 Saitoh, T., Shibata, K., Fujimori, K., et al. (2017). Rapid removal of tetracycline antibiotics from water by coagulation-flotation of sodium dodecyl sulfate and poly (allylamine hydrochloride) in the presence of Al (III) ions. Separation and Purification Technology, Volume 187, pp. 76–83.

75 Xu, Y., Souza, M., Ribeiro-Pontes, M.Z., et al. (2001). Liquid liquid extraction of pharmaceuticals by aqueous two-phase systems. Brazilian Journal of Pharmaceutical Science, Volume 37, Issue 3, pp. 305–320.

76 Gu, T. (2000). Liquid-liquid partitioning methods for bio separations. Separation Science and Technology, Volume 2, pp. 329–364.

77 Homsirikamol, C., Sunsandee, N., Pancharoen, U., et al. (2016). Synergistic extraction of amoxicillin from aqueous solution by using binary mixtures of

Aliquat 336, D2EHPA and TBP. Separation and Purification Technology, Volume 162, pp. 30–36.

78 Periyannan, R., Prasad, A., Arulmozhi, J.S., et al. (2022). Pharmaceutical effluent treatment using multi-effect evaporator process. Advances in Materials Science and Engineering. https://doi.org/10.1155/2022/5238033.

79 Pandis, P.K., Kalogirou, C., Kanellou, E., et al. (2022). Key points of advanced oxidation processes (AOPs) for wastewater, organic pollutants and pharmaceutical waste treatment: a mini review. Chemical Engineering, Volume 6, Issue 1, p. 8. https://doi.org/10.3390/chemengineering6010008.

80 Pignatello, J.J., Oliveros, E., and MacKay, A. (2006). Advanced oxidation processes for organic contaminant destruction based on the Fenton reaction and related Chemistry. Critical Reviews in Environmental Science and Technology, Volume 12, pp. 1–84.

81 Bokare, A.D., and Choi, W. (2014). Review of iron-free Fenton-like systems for activating H_2O_2 in advanced oxidation processes. Journal of Hazardous Materials, Volume 275, pp. 121–135.

82 Yang, J., Xiao, J.D., Cao, H.B., et al. (2018). The role of ozone and influence of band structure in WO_3 photocatalysis and ozone integrated process for pharmaceutical wastewater treatment. Journal of Hazardous Materials, Volume 360, pp. 481–489. https://doi.org/10.1016/j.jhazmat.2018.08.033.

83 Jiao, J., Li, Y., Song, Q., et al. (2022). Removal of pharmaceuticals and personal care products (PPCPs) by free radicals in advanced oxidation processes. Materials, Volume 17, Issue 15, p. 228152. https://doi.org/10.3390/ma15228152.

84 Andrade, J.R., de, Oliveira, M.F., da Silva, M.G.C., et al. (2018). Adsorption of pharmaceuticals from water and wastewater using nonconventional low-cost materials: a review. Industrial and Engineering Chemistry Research, Volume 57, Issue 9, pp. 3103–3127. https://doi.org/10.1021/acs.iecr.7b05137.

85 Fraiha, O., Hadoudi, N., Zaki, N., et al. (2024). Comprehensive review on the adsorption of pharmaceutical products from wastewater by clay materials. Desalination and Water Treatment, Volume 317, p. 100114. https://doi.org/10.1016/j.dwt.2024.100114.

86 Mohapi, M., Sefadi, J.S., Mochane, M.J., et al. (2020). Effect of LDHs and other clays on polymer composite in adsorptive removal of contaminants: a review. Crystals, Volume 10, pp. 1–39. https://doi.org/10.3390/cryst10110957.

87 Wan, M., Li, Z., Hong, H., et al. (2013). Enrofloxacin uptake and retention on different types of clays. Journal of Asian Earth Sciences, Volume 77, pp. 287–294. https://doi.org/10.1016/j.jseaes.2013.02.032.

88 Kang, Z., Jia, X., Zhang, Y., et al. (2022). A review on application of biochar in the removal of pharmaceutical pollutants through adsorption and persulfate-based AOPs. Sustainability, Volume 14, Issue 16, p. 10128.

89 Martín-Pozo, L., de Alarcón-Gómez, B., and Rodríguez-Gómez, R. (2019). Analytical methods for the determination of emerging contaminants in sewage sludge samples. A review. Talanta, Volume 192, pp. 508–533. https://doi.org/10.1016/j.talanta.2018.09.056.

90 Chakhtouna, H., Benzeid, H., Zari, N., et al. (2021). Functional $CoFe_2O_4$-modified biochar derived from banana pseudo stem as an efficient adsorbent for the removal of amoxicillin from water. Separation and Purification Technology, Volume 266, pp. 1383–5866. https://doi.org/10.1016/j.seppur.2021.118592.

91 Dang, B.T., Gotore, O., Ramaraj, R., et al. (2021). Sustainability and application of corncob-derived biochar for removal of fluoroquinolones. Biomass Conversion and Biorefinery, Volume 12, pp. 913–923. https://doi.org/10.1007/s13399-020-01222-x.

92 Li, J., Yu, G., Pan, L., et al. (2018). Study of ciprofloxacin removal by biochar obtained from used tea leaves. Journal of Environmental Sciences, Volume 73, pp. 20–30. https://doi.org/10.1016/j.jes.2017.12.024.

93 Chauhan, S., Shafi, T., Dubey, B.K., et al. (2023).Biochar-mediated removal of pharmaceutical compounds from aqueous matrices via adsorption. Waste Disposal and Sustainable Energy, Volume 5, pp. 37–62. https://doi.org/10.1007/s42768-022-00118-y.

94 Duarte, E.D.V., Oliveira, M.G., Spaolonzi, M.P., et al. (2022). Adsorption of pharmaceutical products from aqueous solutions on functionalized carbon nanotubes by conventional and green methods: a critical review. Journal of Cleaner Production, Volume 372, p. 133743. https://doi.org/10.1016/j.jclepro.2022.133743.

95 Abd El-Monaem, E.M., Eltaweil, A.S., Elshishini, H.M., et al. (2022). Sustainable adsorptive removal of antibiotic residues by chitosan composites: an insight into current developments and future recommendations. Arabian Journal of Chemistry, Volume 15, Issue 5, p. 103743. https://doi.org/10.1016/j.arabjc.2022.103743.

96 Liakos, E.V., Lazaridou, M., Michailidou, G., et al. (2021). Chitosan adsorbent derivatives for pharmaceuticals removal from effluents: a review. Macromolecules, Volume 1, Issue 2, pp. 130–154. https://doi.org/10.3390/macromol1020011.

97 Ashraf, M.H., Hussain, N., Muneer, M.A., et al. (2024). Chapter 4: Chitosan-based nanomaterials for pharmaceutical waste remediation. In: Kumar A, Bilal M, Ferreira LFR (eds) Advances in chemical pollution, Environmental management and protection, Elsevier, Volume 10, pp. 83–116. https://doi.org/10.1016/bs.apmp.2023.09.001.

98 Kyzas, G.Z., Kostoglou, M., Lazaridis, N.K., et al. (2013). Environmental friendly technology for the removal of pharmaceutical contaminants from wastewaters using modified chitosan adsorbents. Chemical Engineering Journal, Volume 222, pp. 248–258.

99 Zhang, Y., Shen, Z., Dai, C., et al. (2014). Removal of selected pharmaceuticals from aqueous solution using magnetic chitosan: sorption behavior and mechanism. Environmental Science and Pollution Research, Volume 21, pp. 12780–12789. https://doi.org/10.1007/s11356-014-3212-1.

100 Rasoulzadeh, H., Mohseni-Bandpei, A., Hosseini, M., et al. (2019). Mechanistic investigation of ciprofloxacin recovery by magnetite–imprinted chitosan nanocomposite: isotherm, kinetic, thermodynamic and reusability studies. International Journal of Biological Molecules, Volume 133, pp. 712–721.

101 Kentsa, E., Abi, C.F., Ngomo, H.M., et al. (2020). Characterization of Akilbenza clay from Cameroon and its performance for the removal of copper (II) ions from aqueous solution. Environmental Science and Pollution Research, Volume 27, pp. 36487–36497.

102 Nazraz, M., Yamini, Y., and Asiabi, H. (2019). Chitosan-based sorbent for efficient removal and extraction of ciprofloxacin and norfloxacin from aqueous solutions. Mircochimica Acta, Volume 186, pp. 1–9.

103 Mahmoodi, H., Fattahi, M., and Motevassel, M. (2021). Graphene oxide-chitosan hydrogel for adsorptive removal of diclofenac from aqueous solution: preparation, characterization, kinetic and thermodynamic modelling. RSC Advances, Volume 11, Issue 57, pp. 36289–36304. https://doi.org/10.1039/d1ra06069d.

104 Kumar, P., Bansal, V., Kim, K.H., et al. (2018). Metal-organic frameworks (MOFs) as futuristic options for wastewater treatment. Journal of Industrial and Engineering Chemistry, Volume 62, pp. 130–145. https://doi.org/10.1016/j.jiec.2017.12.051.

105 Hasan, Z., Jeon, J., and Jhung, S.H. (2012). Adsorptive removal of naproxen and clofibric acid from water using metal-organic frameworks. Journal of Hazardous Materials, Volume 209–210, pp. 151–157. https://doi.org/10.1016/j.jhazmat.2012.01.005.

106 Grape, E.S., Chacón-García, A.J., Rojas, S., et al. (2023). Removal of pharmaceutical pollutants from effluent by a plant-based metal–organic framework. Nature Water, Volume 1, pp. 433–442. https://doi.org/10.1038/s44221-023-00070-z.

107 Ahmad, S., Abbasi, B.B.K., Nazir, M.S., et al. (2022). Metal organic frameworks (MOFs) as formidable candidate for pharmaceutical wastewater treatment. In: Lichtfouse E, Muthu SS, Khadir A (eds) Inorganic-organic composites for water and wastewater treatment. Environmental footprints and eco-design of products and processes. Springer, Singapore. https://doi.org/10.1007/978-981-16-5928-7_2.

108 Azhar, M.R., Vijay, P., Tadé, M.O., et al. (2018). Submicron sized water-stable metal organic framework (bio-MOF-11) for catalytic degradation of pharmaceuticals and personal care products. Chemosphere, Volume 196, pp. 105–114. https://doi.org/10.1016/j.chemosphere.2017.12.164.

109 Peng, J., He, Y., Zhou, C., et al. (2021). The carbon nanotubes-based materials and their applications for organic pollutant removal: a critical review. Chinese Chemical Letters, Volume 32, Issue 5, pp. 1626–1636.

110 Lavagna, L., Nisticò, R., Musso, S., et al. (2021). Functionalization as a way to enhance dispersion of carbon nanotubes in matrices: a review. Materials Today Chemistry, Volume 1, Issue 20, p. 100477.

111 Yu, F., Sun, S., Han, S., et al. (2016). Adsorption removal of ciprofloxacin by multi-walled carbon nanotubes with different oxygen contents from aqueous solutions. Chemical Engineering Journal, Volume 285, pp. 588–595.

112 Wang, Y., Zhu, J., Huang, H., et al. (2015). Carbon nanotube composite membranes for microfiltration of pharmaceuticals and personal care products: capabilities and potential mechanisms. Journal of Membrane Science, Volume 479, pp. 165–174.

113 Samal, K., Kar, S., Trivedi, S., et al. (2021). Assessing the impact of vegetation coverage ratio in a floating water treatment bed of *Pistia stratiotes*. SN Applied Science, Volume 3, pp. 1–8. https://doi.org/10.1007/s42452-020-04139-2.

114 Chen, J., Liu, Y.S., Su, H.C., et al. (2015). Removal of antibiotics and antibiotic resistance genes in rural wastewater by an integrated constructed wetland. Environmental Science and Pollution Research, Volume 22, Issue 3, pp. 1794–1803.

115 Liu, X, Guo, X., Liu, Y., et al. (2019). A review on removing antibiotics and antibiotic resistance genes from wastewater by constructed wetlands: performance and microbial response. Environmental Pollution, Volume 254, Issue Part A, Article 112996. https://doi.org/10.1016/j.envpol.2019.112996.

116 Samal, K., Bandyopadhyay R., and Dash, R.R. (2022). Biological treatment of contaminants of emerging concern in wastewater: a review. Journal of Hazardous, Toxic and Radioactive Wastes, Volume 26, Article 04022. https://doi.org/10.1061/(ASCE)HZ.2153-5515.000068.

117 Chang, C.Y., Chang, J.S., Vigneswaran, S., et al. (2008). Pharmaceutical wastewater treatment by membrane bioreactor process case study in southern Taiwan. Desalination, Volume 234, Issue (1–3), pp. 393–401.

118 Mousel, D., Bastian, D., Firk, J., et al. (2021). Removal of pharmaceuticals from wastewater of health care facilities. Science of the Total Environment, Volume 751, Article 141310.

119 Mohan, S.V., Prakasham, R.S., Satyavathi, B., et al. (2001). Biotreatability studies of pharmaceutical wastewater using an anaerobic suspended film contact reactor. Water Science and Technology, Volume 43, Issue 2, pp. 271–276.

120 Nandy, T., and Kaul, S.N. (2001). Anaerobic pre-treatment of herbal-based wastewater using fixed–film reactor with recourse to energy recovery. Water Research, Volume 35, Issue 2, pp. 351–362.

121 Saravanane, R., Murthy, D.V.S., and Krishnaiah, K. (2001). Bioaugmentation and treatment of cephalexin drug-based pharmaceutical effluent in an upflow anaerobic fluidized bed system. Bioresource Technology, Volume 76, Issue 3, pp. 279–281.

122 Loganathan, P., Vigneswaran, S., Kandasamy, J., et al. (2023). Treatment trends and combined methods in removing pharmaceuticals and personal care products from wastewater–a review. Membranes, Volume 13, Issue 2, p. 158. https://doi.org/10.3390/membranes13020158.

123 Jamil, S., Loganathan, P., Khan, S.J., et al. (2021). Enhanced nanofiltration rejection of inorganic and organic compounds from a wastewater-reclamation plant's micro-filtered water using adsorption pre-treatment. Separation and Purification Technology, Volume 260, p. 11820.

124 Shahtalebi, M.H., Sarrafzadeh, M.M., and Montazer, R. (2011). Application of nanofiltration membrane in the separation of amoxicillin from pharmaceutical wastewater. Iranian Journal of Environmental Health, Science and Engineering, Volume 8, Issue 2, pp. 109–116.

125 Banjerdteerakul, K., Peng, H., and Li, K. (2023). COF-based nanofiltration membrane for effective treatment of wastewater containing pharmaceutical residues. Journal of Membrane Science, Volume 681, p. 121780. https://doi.org/10.1016/j.memsci.2023.121780.

126 Talwar, S., Sangal, V.K., and Verma, A. (2018). Feasibility of using combined TiO_2 photocatalysis and RBC process for the treatment of real pharmaceutical wastewater. Journal of Photochemistry and Photobiology A: Chemistry, Volume 353, pp. 263–270. https://doi.org/10.1016/j.jphotochem.2017.11.01.

127 Ghazal, H., Koumaki, E., Hoslett, J., et al. (2022). Insights into current physical, chemical and hybrid technologies used for the treatment of wastewater contaminated with pharmaceuticals. Journal of Cleaner Production, Volume 361, p. 132079. https://doi.org/10.1016/j.jclepro.2022.132079.

128 Dhangar, K., and Kumar, M. (2020). Tricks and tracks in removal of emerging contaminants from the wastewater through hybrid treatment systems: a review. Science of the Total Environment, Volume 738, p. 140320. https://doi.org/10.1016/j.scitotenv.2020.140320.

129 Patel, S., Mondal, S., Majumder, S.K., et al. (2020).Treatment of a pharmaceutical industrial effluent by a hybrid process of advanced oxidation and adsorption. ACS Omega, Volume 5, Issue 50, pp. 32305–32317. https://doi.org/10.1021/acsomega.0c04139.

130 Vergara-Araya, M., Oeltze, H., Radeva, J., et al. (2022). Operation of hybrid membranes for the removal of pharmaceuticals and pollutants from water and wastewater. Membranes, Volume 12, Issue 5, p. 502. https://doi.org/10.3390/membranes12050502.

131 Masood, Z., Ikhlaq, A., Aziz, H.A., et al. (2022). A novel hybrid treatment for pharmaceutical wastewater implying electro flocculation, catalytic ozonation with Ni-Co Zeolite 5A° catalyst followed by ceramic membrane filtration. Journal of Water Process Engineering, Volume 51, p. 103423. https://doi.org/10.2139/ssrn.4225461.

132 Yilma, B., and Eranna, B.C. (2023). Performance evaluation and modification of zero liquid discharge plant in pharmaceutical industry. European Chemical Bulletin, Volume 12, Issue 9, pp. 2114–2132. https://doi.org/10.48047/ecb/2023.12.9.196.

133 Samal, K., Mahapatra, S., and Ali, M.H. (2022). Pharmaceutical wastewater as emerging contaminants (EC): treatment technologies, impact on environment and human health. Energy Nexus, Volume 6, p. 100076. https://doi.org/10.1016/j.nexus.2022.100076.

134 Kodom, K., Attiogbe, F., and Kuranchie, F.A. (2021). Assessment of removal efficiency of pharmaceutical products from wastewater in sewage treatment plants: a case of the sewerage systems Ghana limited, Accra. Heliyon, Volume 7, Issue 11, p. e08385. https://doi.org/10.1016/j.heliyon.2021.e08385.

135 Jiménez-Bambague, E.M., Madera-Parra, C.A., Ortiz-Escobar, A.C., et al. (2020). High-rate algal pond for removal of pharmaceutical compounds from urban domestic wastewater under tropical conditions. Case study: Santiago de Cali, Colombia. Water Science and Technology, Volume 82, Issue 6, pp. 1031–1043. https://doi.org/10.2166/wst.2020.362.

12

Surfactants: Behavior and Remediation Process

Surfactants are the most widespread and commercially used chemicals in various industrial sectors due to their cleansing ability. Surfactants are chemical compounds with the ability to decrease the surface tension between two liquids, a liquid and a gas, or a liquid and a solid. They are also termed as surface-active agents. They are used in applications related to foaming, cleansing, emulsifying, and solubilization. These are high molecular weight molecules having both hydrophilic and hydrophobic parts with natural alignment at the air–water interface. The hydrophobic part is exposed to the air while the hydrophilic part is in water. This alignment leads to a reduction in the surface tension [1].

Surfactants constitute an important part of the detergents, along with the builders and additives. The builders present in detergents are the carbonates, silicates, and phosphates, while the additives are the brighteners, fresheners, agents with antimicrobial properties, etc. They are added to get rid of dirt or oil from textiles, household utilities, or skin. Surfactants have a widespread applicability not only as a household cleansing agent but also for various other applications in industries and the medical field. They are frequently used in laundry facilities, fabric softeners, and personal care products. They mainly have broad-scale releases in the wastewater streams and the environment, from the waste outlets of kitchens and bathrooms. A laundry industrial operation utilizes about 15 L of water and discharges about 400 m^3 of wastewater daily, having high surfactant content along with other pollutants of concern [2]. It is reported that laundry wastewater can have chemical oxygen demand (COD) of 1200 up to 20000 mg/L [3]. The outbreak of the COVID-19 pandemic increased the consumption of surfactants and their concentration in the water bodies, which posed a threat to the environment.

Hazards and Treatment of Wastewater Contaminants, First Edition. Dipika Jaspal and Arti Malviya. Published 2026 by John Wiley & Sons, Inc.

12.1 Sources of Surfactants

Surfactants are either produced from naturally existing biomass like sugar, vegetable oils, or are synthesized in a laboratory from petroleum products. One of the earliest surfactants prepared is known as soap. The process of formation of soaps is termed as saponification. These are the saponified products obtained from the hydrolysis of triglycerides and an alkali, like sodium hydroxide. The manufacturing of surfactants on a large scale is done using saturated and unsaturated carboxylic acids. A synthetic surfactant has an advantage over a soap surfactant as no scum formation occurs during their application. Most of the commonly used anionic surfactant, namely linear alkylbenzene sulfonate, is prepared by chloroalkane in the presence of an acidic promoter like aluminum chloride or hydrofluoric acid. Most of the liquid detergents containing alkyl sulfates are manufactured using esters of alcohols and sulfuric acids. Another anionic surfactant, alkyl ether sulfate, is prepared by ethoxylation of primary alcohol with epoxyethane, followed by sulfonation using sulfur trioxide and neutralization by alkali. These are chiefly employed in dishwashing liquids, gels, shampoos, etc. Among the cationic surfactants are monoalkyl quaternary systems and esterquats.

A major group of nonionic surfactants is the ethoxylates, prepared by the condensation of long-chain alcohols with ethylene oxide. These surfactants do not bear any charge and are often used in conjunction with the anionic surfactants. Nonionic surfactants are better emulsifiers than anionic surfactants at the same concentration. They foam less, but are more efficient in the removal of oil and dirt, in comparison to the anionic surfactant. Nonionic surfactants are employed in both powder and liquid forms, as surface cleansing agents, fabric detergents, in many industrial emulsion polymerization processes, and for agro-based formulations. Amphoteric surfactants are one of the pH-balanced surfactants. They are very mild in action and thus widely used in shampoos and cosmetics [4].

Natural surfactants are those which synthesized from natural raw materials such as fruits, vegetables, plants, and microorganism. These surfactants are naturally obtained from soap nuts, algal oil, inulin, sugar beet, coconut oil, palm oil, and even from the secretion of the caterpillars. Bio surfactants are generally synthesized by microorganisms such as bacteria and other microbes. Surfactants are now produced from renewable biomass like vegetable oils, biofuel by-products, sugarcane, etc. Coco glucoside, decyl glucoside, lauryl glucoside, and cocobetaine are a few naturally obtained nonionic surfactants. The biodegradability of natural surfactants makes them superior to most of the synthetic surfactants, which are nonbiodegradable. Among the natural surfactants, microbial surfactants are better than plant-originated surfactants, due to their large-scale scale quick production, and diverse characteristics. Plant-based natural surfactants are costly and suffer from the disadvantage of solubility [5].

12.2 Types of Surfactants

They are divided into various types depending upon the polar head group of the surfactant molecules [4, 5]. Surfactants can be:

1. **Nonionic Surfactants:** These are the surfactants characterized by the presence of hydrophilic and hydrophobic groups with no charge on them. These types of surfactants are effectively capable of binding with the wetting agents and thus have better spreading and wetting features. This makes them efficient emulsifiers and foaming agents. They find a wide utility in the applications, ranging from personal care cosmetics to various industries. Nonionic surfactants are divided into two main types depending upon the nature of the hydrophilic group, viz, polyethylene glycol type and polyhydric alcohol type. Polyethylene glycol-type nonionic surfactants are synthesized by mixing ethylene oxide as a hydrophilic group with a hydrophobic base chemical. In polyhydric alcohol type nonionic surfactants, the hydrophobic group is linked to a polyhydric alcohol. The hydrophobic groups include glycerol, pentaerythritol, sorbitol, etc., which have many hydroxyl groups attached to them, thereby imparting hydrophilic characteristics (Figure 12.1).
2. **Anionic Surfactants:** Anionic surfactants are characterized by the presence of a negative charge on the polar head groups (Figure 12.2), which is the hydrophilic end. These negatively charged parts of the molecules are usually carboxylate, sulfonate, sulfate, phosphate, and ester which are neutralized by positively charged metal cations, such as sodium or potassium. They are one of the most widely used surfactants, mainly in the detergent sector. Sodium lauryl sulfate, polyoxyethylene alkyl ether carboxylate, sodium alkylbenzene sulfonate, and sodium laureth sulfate are a few commonly used anionic surfactants. Anionic surfactants are widely used in detergents due to their easy method of synthesis and providing better performance. Thus, the detection

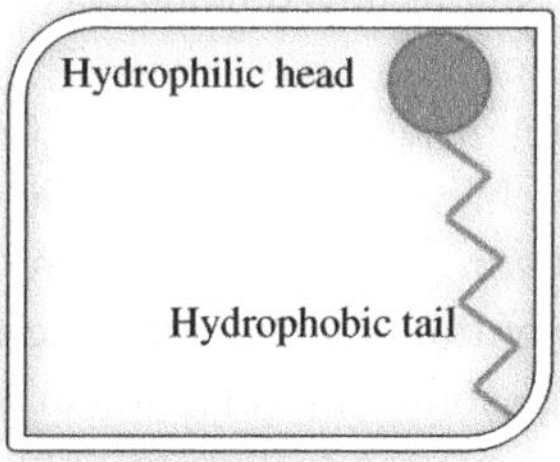

Figure 12.1 Hydrophilic and hydrophobic components in nonionic surfactant structure.

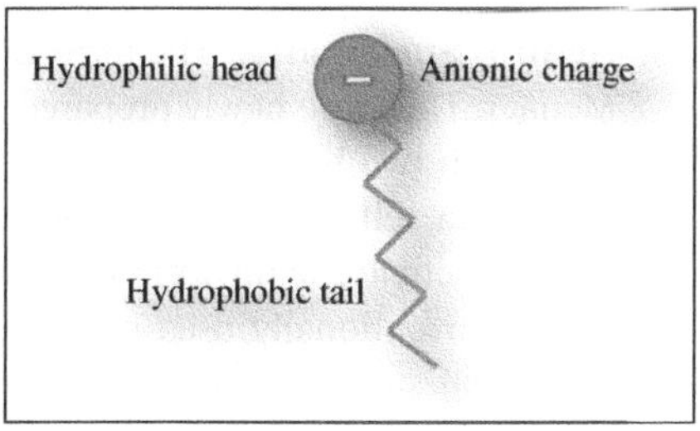

Figure 12.2 Anionic surfactant.

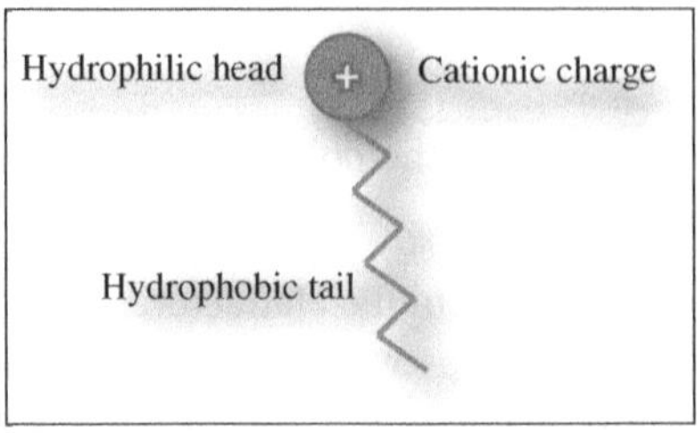

Figure 12.3 Cationic surfactant.

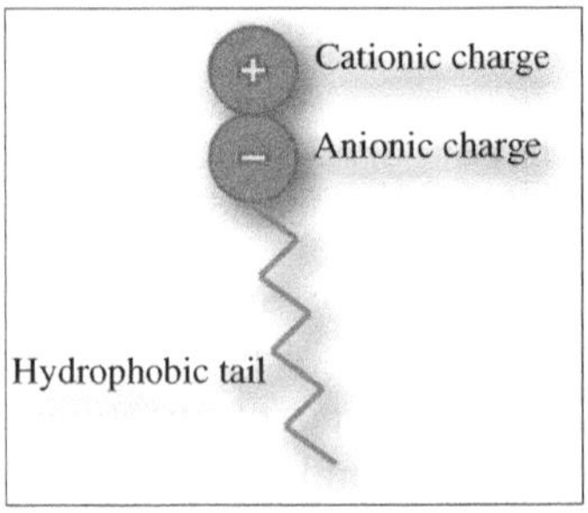

Figure 12.4 Zwitterion surfactant.

possibility of anionic surfactants in the wastewater is more than any other category [6].

3. **Cationic Surfactants:** The structural makeup of cationic surfactants is opposite to that of the anionic surfactants (Figure 12.3). These are the surfactants with a positive charge on the polar head. They are sometimes also known by the term *Invert soaps*. Soromine A, a cationic textile softener, belongs to this category. Such surfactants also possess disinfecting characteristics. These do not have good foaming property, and thus are mainly limited to their utilization in shampoos and conditioners.
4. **Zwitterion Surfactants:** These are also called amphoteric surfactants since they contain both types of charges, negative and positive, within the same molecule (Figure 12.4). The cationic part is either primary, secondary, or tertiary amines or quaternary ammonium cations, while the anionic part includes sulfonates. They are resistant to hard water and have antibacterial characteristics, due to which they can be employed well in shampoos and microbicidal agents, as a component. Zwitterion surfactants are employed as antistatic agents for synthetic fibers, as well as anticorrosive agents. They can be dried easily and are employed on a large scale in latex preparation in the paint industry. They exhibit low-to-moderate toxicity and can be easily removed using wastewater treatment. They have low bioaccumulation potential.
5. **Gemini Surfactants:** The term *Gemini* surfactant was coined by Fredric M. Menger. These are newly prepared surfactants characterized by dimeric structures with two hydrophobic chains and two hydrophilic heads (Figure 12.5). These can self-assemble at much lower concentrations and are superior in surface activity, foam formation, and wetting ability in comparison to the other conventionally used surfactants. These have the feature to reduce the interfacial surface tension 10–100 times. These surfactants can have different categories of spacer groups (groups which link the hydrophilic head and hydrophobic tail), such as flexible, rigid, or semi-flexible.

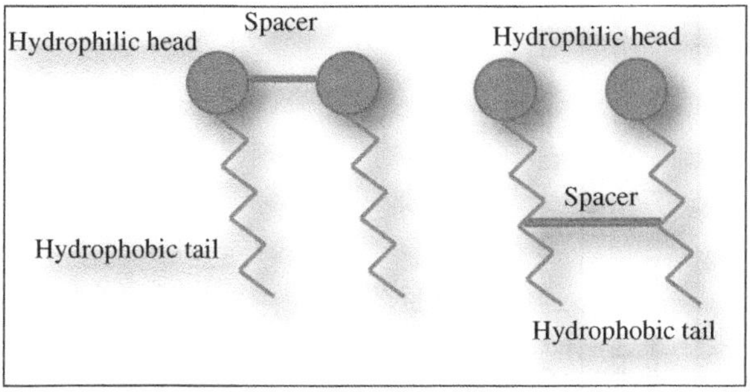

Figure 12.5 Gemini surfactant.

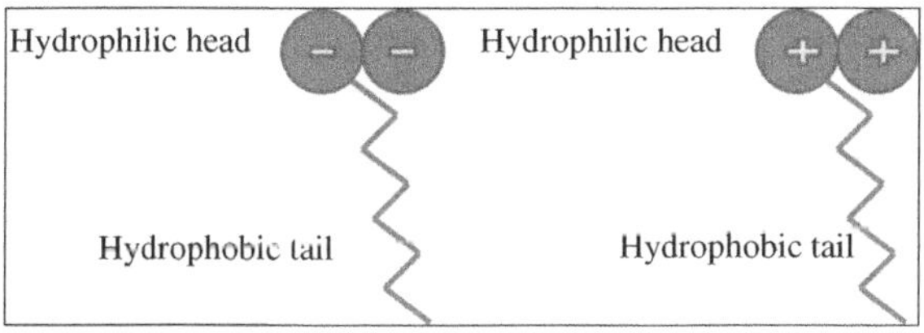

Figure 12.6 Bicephalous surfactants.

6. **Bicephalous Surfactants:** These are newly designed surfactants having two hydrophilic heads, a hydrophobic tail, and a spacer (Figure 12.6). These are mainly synthesized for their special application in the medical field.
7. **Biosurfactants:** These are the biodegradable surfactants obtained from microbial sources. Some examples of biosurfactants are glycolipids like rhamnolipid, emulsan from *Acinetobacter calcoaceticus*, and lipopeptides like viscosin.

Table 12.1 summarizes a few characteristics of various types of surfactants [4, 5].

12.3 Toxicity Assessment of Surfactants

Surfactants have a positive response to the adsorption characteristics in eradicating pollutants from wastewater, but at the same time, are themselves prominent water contaminants. The presence of these in the aqueous environment can lead to toxic impacts at the cellular or molecular level in living organisms. If their concentration in the water ranges from 0.4 to 40 mg/L, they have significant chronic lethal impacts on the aquatic species, in addition to carcinogenic effects [7].

Surfactants in the present scenario, used in various industrial sectors or at the domestic level, are mainly synthetic surfactants, which are xenobiotics. Surfactants are considered to be relatively biodegradable, but the presence of ring moiety or aryl groups makes them resistant to degradation. Also, the biodegradability of

Table 12.1 Types of surfactants and their characteristics [4, 5].

Parameters	Nonionic surfactant	Anionic surfactant	Cationic surfactant	Zwitterion surfactant	Gemini surfactant	Bicephalous surfactants	Biosurfactants
Recognition	These surfactants have no charge on the head group.	Anionic surfactants have a negative charge on the head group.	These are the surfactants with a positive charge on the head group.	Zwitterion surfactants possess both positive and negative charges.	These are newly prepared surfactants characterized by dimeric structures with two hydrophobic chains and two hydrophilic heads.	These are also newly designed surfactants and have two hydrophilic heads, a hydrophobic tail, and a spacer.	These are the surfactants obtained from microbial sources.
Characteristics	They are resistant to water hardness.	These are less expensive than cationic surfactants.	They are more expensive than anionic surfactants.	These are amphoteric.	The two main structures are linked by a spacer at or near the head groups.	These can be dicationic and dianionic surfactants.	These are known for their biodegradability, as they are produced from renewable sources.
		Possess excellent foaming and cleaning properties and are temperature insensitive.	Known for antimicrobial activity and conditioning effects.	Expensive surfactants.	These have the feature to reduce the interfacial surface tension 10–100 times.	Dicationic surfactants have a hydrophobic tail, a spacer, and two positively charged heads.	They have low toxicity impacts on the environment.
		Excellent emulsification and dispersability.		They have high solubility and low irritation potential on living organisms.		Dianionic surfactants contain heads with negative charges, along with spacers and tails.	Functionally, under extreme pH conditions, temperature, and are quite stable.

Uses	They are mainly used as wetting agents, as emulsifiers and as industrial cleaners.	These are one of the most widely used surfactants, mainly in the detergent sector.	Employed in fabric softeners, disinfectants, and as preservatives.	Employed for emulsion preparations and oil recovery, textile processing, and cosmetic industries also make use of these surfactants.	They have catalytic applications and adsorption-related uses in pollutant eradication. They also find their uses in the medical industry.	These are mainly synthesized for special applications in medicinal applications.	Have therapeutic and biomedical potential and are thus employed for pharmaceutical-related products and their processing or uses.
Examples	Tween 20 and triton X-100 are nonionic types of surfactants.	Sodium Lauryl Sulfate Sodium Laureth Sulfate	These include benzalkonium chloride and cetyltrimethyl-ammonium bromide.	Cocamidopropyl betaine and sulfobetaine are a few zwitterionic surfactants.	Amino acid-based and sugar-based, gemini surfactant.	Propyl esters and propyl ester sodium salt.	Glycolipids like rhamnolipid, Lipopeptides like viscosin are examples of biosurfactants.

the surfactants is dependent upon the type of surfactant, hydrophilic and hydrophobic groups present in the surfactants, chain structure, and type of treatment process employed for their eradication. The distribution and fate of surfactants is influenced by pH, temperature, salinity, and biodegradation and sorption characteristics.

Surfactants with complex structures are not removed by conventional wastewater treatment methods, and have been reported to be present even in treated wastewater, along with their degraded products. Conventional physicochemical and biological methods are not enough to completely eliminate surfactants from wastewater, especially if present in high concentrations. These surfactants when present in wastewater pose a high chemical oxygen demand.

The toxicity level of surfactants depends upon various physicochemical parameters of wastewater, including the biological character. The interaction between the surfactants and the living species is regulated by the pH, hardness of water, temperature, and presence of polar compounds in wastewater. The types of aquatic organisms, age of the species, and sensitivity level of the organisms in the aquatic system determines the extent of the impact of surfactants on them. This is however, depends on the nature of the surfactants, their chain length, and the presence of characteristic groups and overall structural makeup of the surfactant. Surfactants can have different structural characteristics due to different chain lengths, degrees of unsaturation, and the presence of polar groups. The toxicity level of surfactants depends upon the hydrocarbon chain length and degree of linearity. However, it is observed that the toxicity of anionic surfactants is greater than that of nonionic surfactants.

12.3.1 Surfactants and the Abiotic Environment

Investigations have reported the presence of surfactants in domestic and manufacturing wastewater effluent ranging from 1 to 10 mg/L and 1 to 300 mg/L, respectively [6]. A surfactant treatment plant leaves an appreciable amount of these in the effluent, creating an environmental issue. They lead to short as well as long-term influences on the ecosystem and poses risks to aquatic species and water quality. Surfactants degrade the aesthetic quality of water bodies and cause foaming problems. Leaching of surfactants from industrial waste and municipal sewage facilities contaminates the groundwater resources. Surfactants and related products pose negative effects on water bodies. Further, their adsorption over the solid substances reduces the natural sedimentation of the impurities, affecting the water quality. Their presence also leads to a reduced air transfer and oxygenation rate, deteriorating the self-cleansing ability of water systems, as the microbial population is affected at the cellular level. The ecological water cycle gets disturbed due to a decrease in the evaporation process.

Anionic surfactants help increase the solubility of various pollutants and increase the chances of eutrophication and toxicity. Surfactants can result in increased nitrogen and phosphate in the water, which can cause excessive eutrophication, algal blooms, and the utilization of oxygen by bacteria for decomposition of dead organic matter. The oxygen consumption during the natural decomposition process depletes the oxygen present in water bodies. This endangers the life of aquatic species as well as disturbs the ecological balance. Anionic surfactants get associated with the bioactive macromolecules (proteins) and alter their polypeptide chain arrangement, leading to an alteration in the surface charge of the molecule. This results in the modification of the biological activity of the organisms.

Surfactants are persistent contaminants and can pass through the wastewater treatment process and pollute drinking water sources, posing a risk to living species. These compounds are difficult to remove from water as they are stabilized in water due to dispersion and micelle formation.

12.3.2 Surfactants in the Biotic Environment

Anionic surfactants like alkylbenzene detergents cause physiological as well as pathological impacts on terrestrial as well as aquatic ecosystems. These are responsible for skin irritation and respiratory problems. Surfactants can be related to mortality, inhibition of mobility of aquatic species, interference with reproductive activity, and issues of reduced respiration in the aqueous environment. Figure 12.7 illustrates the impacts on aquatic species from exposure to surfactants.

Quaternary ammonium cationic surfactants can bind to the inner cytoplasmic membranes and disassemble them.

Surfactants lead to changes in the physiological and biochemical activities of aquatic species and influence their metabolic functions, decrease the growth patterns, damage cell structures or membranes, and also lead to damage to chlorophyll and proteins. It is reported that most of the Pisces population cannot survive if the concentration of the surfactants in wastewater is 15 mg/L, while their eggs and reproductive cycles are disturbed if the concentration is around 5 mg/L. Surfactants have also shown toxic impacts on vertebrates. A study reported that 0.4–40 mg/L of the surfactants can damage the gills of the aquatic species, delay their growth, and also alter their feeding characteristics. For invertebrates, a detergent concentration less than 0.01 mg/L proves to be toxic and can affect their growth and development.

Lethal concentration (50 LC_{50}) refers to the concentration of a chemical that kills 50% of the sample population. The European community Environmental legislation count for L_{50} is observed between 10 and 100 mg/L for most of the anionic surfactants, while nonionic surfactants have LC_{50} between 1 and 10 mg/L.

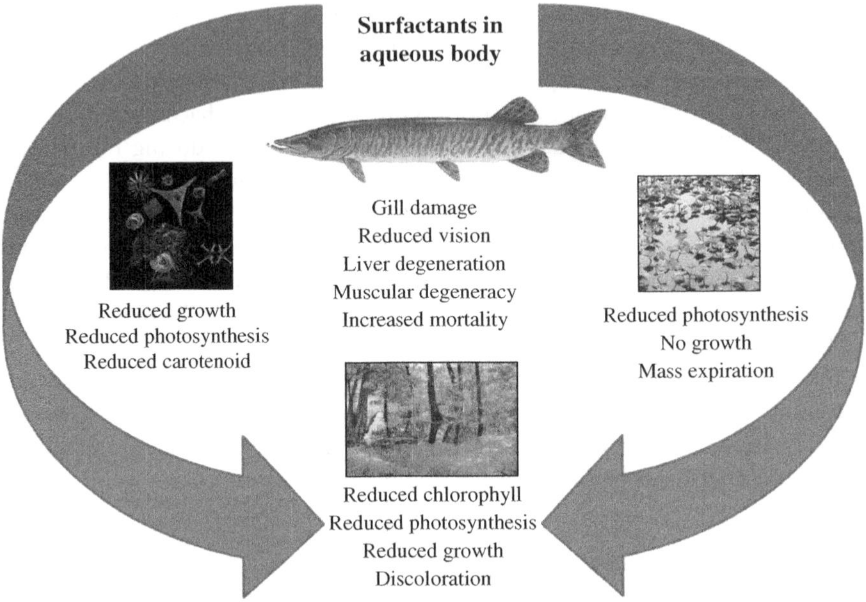

Figure 12.7 Adverse effects of surfactants on aquatic species. Adapted from [4, 5].

The nonionic surfactants are capable of exerting antimicrobial influence as they can bind with the phospholipid structures and protein chains. This ruptures the cell membranes and results in the death of the organism due to loss of ions and proteins [8].

Gheorghe et al. have reported that surfactants can retard the growth of the microbial population and increase the chances of mutations. Anionic surfactants can disrupt the internal structures of microbial cells and disturb their reproductive behavior. These decrease the absorption of nutrients by depolarization of the microbial cell membranes and can lead to the releases of toxic metabolites [9].

Some (like linear alkyl benzene sulfonates) have shown the ability to damage plant roots, which vitiate the transpiration and transfer of essential nutrients and water in them. Also, the transformed products of surfactants also are of great concern for the environment. Investigators have revealed a greater toxicity of the transformed products of surfactants than their precursors [10, 11]. Severe health impacts have been observed for humans, soil, microbes, and even terrestrial plants.

In an investigation, 4-nonylphenol and 4-tertoctylphenol were observed to exhibit estrogenic activity. These were capable of inducing vitellogenin in male fish in the estrogenic response. Similarly, *Bombyx mori*, the common name domestic silk moth, when exposed to 0.05–0.2 g/kg of nonylphenol, the vitellogenin expression of female insects declined, but was observed in male species. In another investigation, nonylphenol was found to influence the functioning of the adrenal gland when about 1.9 μg/L of the metabolite was tested [11]. This suggested the ability of nonylphenol to decrease the production and discharge of corticosteroid hormones [12]. Different byproducts of surfactants (like nonylphenol) have been found to have estrogen-mimicking effects and are responsible for human breast cancer and early puberty, with testicular cancer and low sperm count in men [13].

Surfactants have severe health implications on humans as well when they enter the human system through ingestion or drinking. They influence the endocrine system by reacting with proteins in the liver and serum. Nonylphenol has estrogenic effects on amphibians, mammals, and Pisces.

Linear alkylbenzene sulfonates can be aerobically degraded with a half-life of 7.33 days, but are persistent under anaerobic conditions. A concentration greater than 0.1 mg/L can lead to chronic toxicity due to anionic surfactants. One of the investigations revealed that the toxicity effects of surfactants were of the order cationic > anionic > nonionic surfactants. These surfactants were found to degrade the swimming capacity of fish after long exposure to the pollutants [14]. Salmon fish was found to be quite sensitive to surfactants (less than 1 ppm), which caused choking in the fish.

Surfactant in high concentrations in water introduces an unpleasant taste and odor. The toxicity of surfactants increases with increases in the level of hydrophobic nature brought about by an increase in the alkyl groups. This results in a disturbance in the endocrine system of several organisms. The 96 hours EC_{50}(half maximal effective concentration; ±95% CI) of a cationic surfactant for microalgae *Chlorella vulgaris* was found to be 145 ± 13.35 μg/L, while for invertebrates it was found to be 5.9 μg/L [15]. The growth pattern of microalgae was disturbed due to physico-chemical changes brought about by the presence of anionic surfactants. An anionic surfactant concentration of 10 mg/L decreased the chlorophyll and carotenoid content of algae, which retarded the growth of planktons. Surfactants have also been reported to cause discoloration of aquatic plants by completely breaking the chlorophyll protein complex, as was evident from a study with *Hydrilla verticillata*, which turned brown when exposed to a surfactant concentration of 10–30 mg/L [16]. Table 12.2 summarizes several toxicological impacts caused by surfactants to the biotic systems.

Surfactant poses an adverse impact on the flora and fauna existing in aquatic environment, along with their harmful effects on microbes and human beings. Health hazards associated with surfactants and their derivatives create an urgent need for their remediation from the aqueous environment. Thus, it is imperative to study and evolve efficient treatment technologies for surfactant wastewater to reduce the pollution load, improve the ecological environment, and also for economic benefits.

Table 12.2 Toxicological impact of surface active agents [16].

Aquatic species	Surface active agent	Species	Toxicological impact
Microflora and fauna	Linear alkyl sulfonate	Cryophytes	Linear alkyl sulfonate reduces photosynthetic activity
			It results in reduced growth
	Sodium dodecyl sulfate	*Dunaliella salina*	Leads to a decrease in the chlorophyll a content
			Reduces carotenoid levels
	Detergents	*Daphnia magna*	100% death of the female species at a concentration of 4 g/L in six days
	Shampoo	*Daphnia magna*	50% death of the female species at a concentration of 2.5 g/L in eight days
Aquatic plants	Sodium lauryl sulfate	*Azolla pinnata*	Reduced chlorophyll content at concentrations greater than 2 mg/L
	Sodium lauryl sulfate	*Hydrilla verticillata*	Complete browning of the leaves at a concentration range of 10–20 mg/L
	Sodium dodecyl sulfate	*Azolla pinnata*	Destruction of chlorophyll
			Reduced photosynthesis
			Reduced growth
		Hydrilla verticillata	Destruction of chlorophyll, complete discoloration, turning brown
			Delayed metabolism
			Reduced growth
	Linear alkyl sulfonate	*Elodea canadensis*	Reduced photosynthesis
			Reduced growth
	Linear alkyl sulfonate	*Myriophyllum spicatum*	Reduced photosynthesis
			It results in reduced growth.
Algae	Perfluorooctane sulfonate	*C. vulgaris*	About 50% of the species were affected at 96 μM concentration
	Alcohol ethoxylate	*Selenastrum*	About 10% of the species were destroyed at a 9.7 mg/L concentration

Aquatic species	Surface active agent	Species	Toxicological impact
Pisces	Linear alkyl sulfonate	Seabass fish larvae	LC_{50} of 96 h is observed to be 1.8 mg/L
			Vision-related problems
			Damage to gills
			Vascular degeneration of the liver
	Detergent	*Anabas testudineus*	Oxygen consumption capacity decreases
			Confusion with a slowed rate of swimming
			Lethal if the concentration exceeds 200 mg/L
	Anionic surfactant	Zebrafish *(Danio rerio)*	Macrocephaly
			Curvature in the tail
			Cortisol levels were found to be significantly higher
	Linear alkyl sulfonate	*Clarias gariepinus*	Histopathological impact on liver cells
			Diseased gills
			Damage to the intestinal walls, and mucosa
			Increased rate of mortality
	Perfluorooctane sulfonate	Silver carp	Muscular degeneracy at 3.21 ng/g
		Common carp fillets	Muscular degeneracy at 11–90 ng/g
Crustaceans	Benzalkonium chloride	*Daphnid*	LC_{50} was observed to be 0.1–1.0 mg/L

Environment and public health authorities have set regulatory limits for surfactants in water, yet the negative impacts and high concentration of surfactants in wastewater have raised concerns related to their removal, due to their non-biodegradable nature and related toxicity. Several methods have been recommended for the cleanup of surfactants from the environment.

The efficiency of a remediation technique is influenced by the characteristics of influent, time of treatment, and operational parameters like pH, contact time, temperature, dose of eliminating agent and the type of surfactant present in wastewater. Thus, it is essential to identify and understand the type of surfactant for deciding the best remedial strategy to be adopted.

12.4 Analytical Techniques for Surfactants

The type of technique chosen for analysis depends upon the composition of the wastewater and the nature of surfactant present. Usually, the process becomes complicated as wastewater is generally associated with more than one type of surfactant, organics and other contaminants. Sometimes, the concentration of surfactants is below the detectable limits of few analytical tools. However, the commonly used techniques for assessing the presence of surfactants in the environment include high-performance liquid chromatography (HPLC) and titrimetry. The direct influence of the surfactants on the living organism including the quantity as well as quality of the surfactants is carried out in the biological analysis of the wastewater sample. The analysis of the wastewater for the presence of surfactants can be carried out using the methods as depicted in Table 12.3 [15].

12.5 Treatment Techniques

The vast negative impacts of the surfactants on the environment and living organism reinforces the need to eliminate these from the aqueous environment. Figure 12.8 lists the various commonly used surfactant removal methods, which have shown good surfactant eradication potential.

12.5.1 Physical Treatment

1. **Adsorption:** Among the various physical methods of surfactant removal from wastewater, adsorption is the most common and efficient method. Activated carbon, nanomaterials, zeolites, clay, etc., have been utilized to remove this menace from the aqueous environment. In recent times, emphasis has been laid on the development of environmentally friendly, cost-effective, and efficient adsorbents, especially as alternatives to activated carbon, due to its high cost. The research community is focusing on the use of resourceful adsorbents from waste materials to accomplish the goal of sustainability. Reports suggest that at low concentrations of surfactants, adsorption takes place in the micropores, while if the concentration of the surfactants is high, adsorption proceeds by mesopores as well as macropores. Thus, controlling the pore size of the adsorbent material can help increase the performance of the process. In various experiments, the removal efficiency of the surfactants was

Table 12.3 Methods used for analysis of surfactants [15].

Method	Principle involved	Characteristics	Examples
Spectrophotometric analysis (UV–visible spectroscopy)	It is based on the detection of complex ions formed during interaction	1. It is a fast and easy method of analysis 2. Has low selectivity 3. Involves the development and extraction of surfactants with a suitable reagent 4. It generates toxic wastes 5. It involves the presence of a chromophoric group in the surfactants 6. The method is not feasible for complex mixture wastewater.	Alkylbenzene sulfonates, hydroxyl-oxo amides
Potentiometric analysis	The difference in the electromotive force forms the basis of analysis	1. Highly efficient method of analysis 2. The equipment is easy to handle 3. It is employed for ionic as well as nonionic surfactants 4. It involves ion-selective detection 5. It involves changes in the environmental mediums electrical property 6. Not an appropriate method for a mixture of surfactants	Disulfine blue active substances
Fourier transform infrared spectroscopy – attenuated total reflectance (FTIR-ATR)	Involves the use of infrared radiation to understand the response, in the form of an absorption pattern	1. It is an expensive but robust technique of analysis of surfactants 2. The method is capable of analyzing surfactants in complex wastewater 3. It involves absorbance in the infrared region 4. It is a reliable and quick technique 5. The technique is costly.	Sodium lauryl ether sulfate
Titrimetry	Includes conventional detection of the pollutants in the sample based on titration	1. It is a simple method of analysis. 2. It determines the critical micelle concentration of the surfactants.	Disodium cocoyl monoethanolamide sulfosuccinate

(Continued)

Table 12.3 (Continued)

Method	Principle involved	Characteristics	Examples
Gas chromatography – mass spectrometry (GC-MS)	This is mainly based on gaseous or volatile sample analysis	1. The technique is limited to specific analytes 2. It is a hybrid of the capillary column and flame ionization detector.	Linear alkylbenzene sulfonates
High-performance liquid chromatography (HPLC)	The partition or adsorption over stationary and mobile phases helps to identify the surfactants present	1. It is the most utilized technique for surfactant analysis 2. It does not require any derivative preparation 3. It has the prerequisite of isolation and extraction of the analyte 4. It has a low limit of surfactant detection 5. The technique can quantify the surfactants with high accuracy 6. The method has high analytical capacity	Octyl phenol ethoxylates, Lauryl alcohol ethoxylates
Supercritical fluid chromatography (EFC)	It is based on supercritical fluids	1. It is a recent method of analysis of surfactants 2. It involves the separation as well as the identification of surfactants 3. It has good efficiency and sensitivity	Pentaerythritol fatty acid esters
Voltammetry	Measures the electropotential of the surfactants	1. It is an electroanalytical method 2. It involves the quantification of the surfactants 3. It is less significant than other techniques 4. It has the limitation of being nonresponsive for a mixture of surfactants	Nonionic surfactants
Biological method (polymerase chain reaction technique)	It helps in studying the effect of surfactants on gene expression	1. It involves rapid, quick, and reliable analysis. 2. It involves immunochemistry, molecular, and biosensors for surfactants	Alkylphenols, ethoxylated surfactants

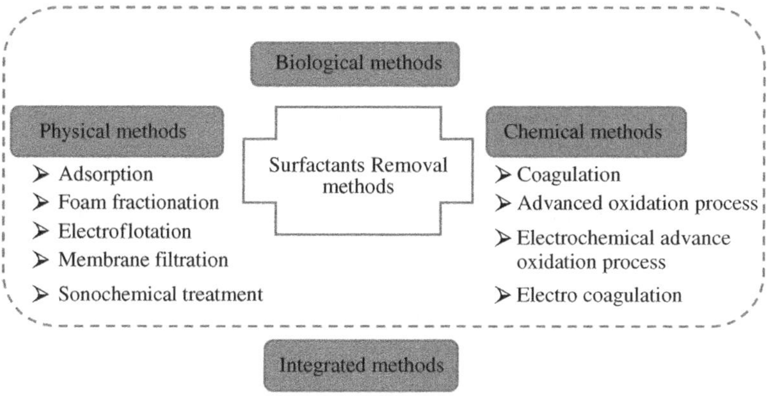

Figure 12.8 Surfactant removal techniques.

observed to be greater than 98%. The efficiency of the elimination is governed by the hydrophobic interaction between the surfactant molecules and the adsorbent surface. The adsorption process can only facilitate the sequestration of the surfactants from the wastewater, but it is not efficient enough to destroy the surfactant molecule completely. Thus, the desorbed surfactants require proper management and disposal. Nanomaterials have been found to eliminate a broad range of surfactants. Multiwalled carbon nanotubes, as well as modified nanotubes have been utilized for this purpose [17].

The evaluation of the adsorption potential of pine cone, an agricultural waste, was investigated for the elimination of sodium dodecyl sulfate from its aqueous solution by Sen and coworkers. The maximum adsorption potential was obtained as 95.75 mg/g at 20 °C. The extent of adsorption was influenced by initial surfactant concentration and contact time. The adsorption process was dominated by electrostatic forces and surfactant self-assembly. The adsorption process followed pseudo-second order kinetics. It involved rapid adsorption behavior of the surfactant over the adsorbent surface, followed by intraparticle diffusion [18].

Similarly, a marine biomass was employed for the removal of anionic and nonionic surfactants from aqueous solution, by the batch method. The anionic surfactant investigated was Sodium dodecyl benzene sulfonate, while the nonionic surfactant chosen was octylphenoxy polyethoxyethanol. In case of the anionic surfactant, the maximum adsorption capacity at 30 °C was reported to be 2.77 mg/g, while for the nonionic surfactant it was 1.81 mg/g. Both surfactants gave optimum performance at an acidic pH of 2. Adsorption occurred by homogenous distribution of the surfactant onto the biomass, which was well confirmed by the Langmuir isothermal model [19].

A composite adsorbent from egg shells and zeolite was prepared and investigated for the removal of linear alkylbenzene sulfonate, one of the most commonly used anionic surfactants for household hand washing products. The composite showed good adsorption characteristics due to the easy access of surfactant molecules onto the adsorbent with a large mesoporous surface area [20].

2. **Foam Fractionation:** Apart from the above-mentioned method, foam fractionation is a promising and interesting technique employed. Foam fractionation is based on adsorption at the gas-liquid interface. This technology is not only useful for wastewater treatment but is also adopted in metallurgy, biotechnology, pharmacy, and cosmetics. Foam fractionation separates various surfactants from wastewater by bubble-based foam generation. The water contaminated with surfactants is made to pass through a fractionation column, wherein foam is produced by air injection. The principle of separation is the mass concentration difference. The surfactant-rich foam floats at the top of the column while the aqueous-rich layer remains at the bottom. The process is influenced by temperature, nature of solute, acidity of the medium, flow rate in the column, column diameter, height, and the size of the bubbles. In this process, a controlled dispersion of the gas into the solution of the surfactant generates bubbles that expand the gas-liquid junction and enhance surfactant adsorption. The method has been utilized to eliminate perfluoroalkyl and polyfluoroalkyl substances from the wastewater matrix [21].

   Researchers have eliminated sodium dodecylbenzenesulfonate, an anionic surfactant, from an aqueous solution using this technique. The process was supported by an alcoholic reagent. Researchers were successful in recovering 77% of surfactant, while 95% of water was recovered in the process. The operative temperature of the process was 25 °C [22]. Similarly, Boonyasuwat et al. removed and recovered cationic surfactant cetylpyridinium chloride by multistage foam fractionation. This method demonstrated effectiveness in the removal of the surfactant [23].

   Investigators combined nanomaterials and foam fractionation to eliminate linear alkylbenzene sulfonates. The complete process was based on the utilization of nanoparticles for foam stabilization. The stability of the foam was governed by foam properties. Also, the rate of air input, critical micelle concentration (CMC), and power supply also significantly influenced foam fractionation. The method still requires more research for its commercialization and upscaling.

3. **Membrane Filtration:** Membrane filtration, like microfiltration, ultrafiltration, reverse osmosis, and nanofiltration, is a commonly employed membrane-based process in wastewater treatment plants to eradicate several contaminants, including microorganisms. These membranes are constituted

of thin membrane sheets capable of selectively allowing clear water through them and retaining the contaminants. The pores of the membranes are very small, and pressure is required to force the feed through the membranes. Membrane filtration is dependent upon temperature and pressure. Membranes can influence the removal of surfactants. Membranes like polysulfone and polyethersulfone have shown their potential in removing anionic surfactants. Nanofilter membranes efficiently eliminated 92% of surfactants from wastewater. This technique does not suffer from the issues related to permeability. Nanofilters are slightly affected by anionic and nonionic surfactants. Though surfactant removal can be carried out up to a maximum of 45 °C using nanofilters. Surfactant removal from wastewater by nanofilters is governed by the concentration of surfactants and the interaction between surfactants and the nanofilter membrane. Reports suggest that the fouling issue due to surfactants is difficult to get rid off due to the formation of protein surfactant complex having a complicated structure. This complex gets accumulated within the nanofilter membrane and thus causes problems. It is thus always required to replace the exhausted membrane. This also increases the cost and emphasizes the requirement of a ready stock of membrane filters. The process also has the limitation of generating of polluted water due to frequent backwashing [24].

4. **Sonochemical Treatment:** The sonochemical treatment involves the molecular reaction occurring in the presence of powerful ultrasound radiation in the range of 20 kHz–10 MHz. It operates on a cavitation mechanism. The low-frequency ultrasound generates cavitation bubbles. The collapse of these bubbles causes the easy breakdown of polymeric chains. It creates high local temperature and pressure, which leads to the decomposition of the organic matter pyrolytically. Also, the water vapor and oxygen present inside the bubble dissociate to produce hydroxyl radicals due to sonolysis. These primary hydroxyl radicals react with the organic or inorganic impurities present in the wastewater, leading to hydroxylation reaction, oxidation, and finally destruction of the contaminant species.

   In recent decades, sonochemical systems have evolved as an alternative for the effective elimination of surfactants from the wastewaters. These systems have no chemical requirement. Low byproduct formation and selectivity in degrading the contaminants from the aqueous solution are some advantages. Maximum production of radicals occurs between 200 and 600 kHz, as at this frequency the coalescence is less and the bubble surface is more unstable, favoring the generation of greater free radicals and ensuring greater surfactant removal.

   Anionic and cationic surfactants accumulate at the interfacial sites of cavitation bubbles due to their surface active properties. The sonolysis of different types of surfactants was investigated in a variety of environmental matrices, as the method can be applied to both homogenous and heterogeneous systems. The charge and properties of the surfactant polar head influence the cavitation bubble property, its growth, and stability [25].

The rate of anionic surfactant degradation, as successfully observed by using a sonochemical reactor, was proportional to the sonication time, but increased with the increase in the concentration of the surfactants. The potential of ultrasonic irradiation for the elimination of anionic surfactant linear alkyl sulfonate was found to be noteworthy in the presence of ferrous ion alone, or ferrous ion along with hydrogen peroxide. About a 93% reduction of linear alkyl sulfonate was observed when 20 kHz of ultrasound was subjected to the surfactant solution, having a pH of 2.5 and operating temperature of 40 °C. Similarly, a 125 W sonication for 120 minutes at 30 °C led to 80% degradation of linear alkyl sulfonate. The key findings based on sonochemical treatment for surfactant removal are listed in Table 12.4.

12.5.2 Chemical Treatment

1. **Coagulation:** Iron and aluminum compounds have been widely used as coagulants for the removal of surfactants. The choice of coagulants affects the removal process. The efficiency of the process increases in conjunction with the air flocculation and oxidation process [26].

   Researchers confirmed from their investigations that in addition to the coagulation process, the adsorptive micellar flocculation mechanism seems to contribute to the removal of surfactant and organic matter. The coagulant ions bind with the micelle and suppress the repulsion between the micelles, leading to their flocculation. This is followed by aggregation of the surfactant micelles for their effective removal. Coagulation and flocculation of wastewater were carried out using iron compounds, in a dosage of 60–1100 mg/L at pH 8. A 99% reduction in the level of surfactants was observed.
2. **Electrocoagulation:** In this method, the destabilization agents bring about the charge neutralization and remove the contaminants from the solution. This involves the pollutants present in the effluent to coagulate when an electrical charge is applied from the cathode and anode. The coagulants employed in the electrocoagulation method are mainly aluminum and iron salts. During the electrocoagulation process, the hydrolysis and agglomeration of aluminum hydroxide takes place at favorable hydrogen ion conditions, which forms various monomeric as well as several polymeric entities. The overall removal of surfactants by this method involves charge neutralization, facilitating coagulation, which is followed by sedimentation or precipitation of the coagulated matter. Iron-containing coagulants have been found

Table 12.4 Degradation of surfactants using sonochemical treatment [25].

Surface active agent	Initial concentration	Radiation energy	Radiation frequency	Other parameters	Key features
Polydisperse nonylphenol ethoxylate surfactant	0.01–0.11 mM	2 W/cm^2	363 kHz	Medium-argon saturation	1. Total organic content removal was 90%. 2. Removal was achieved after one day (24 h) of subjection to the sonication process. 3. The process helped achieve good mineralization output.
Sodium dodecylbenzene sulfonate (SDBS)	50–450 µM.	2.5 W/cm^2	362 kHz	–	1. The investigation resulted in a significant degradation. 2. The effect was more pronounced when the initial concentration was below the critical micelle concentration of sodium dodecylbenzene sulfonate (~150 µM)
Perfluorooctane sulfonate (PFOS) and perfluorooctanoic acid (PFOA)	10 and 100 mg/L	3 W/cm^2	200 kHz	Medium-air and argon saturation	1. Results reported effective removal of the surfactant. 2. The output was promising when the process was carried out under an argon-sonicated medium
Perfluorooctanoate (PFOA) and perfluorooctane sulfonate (PFOS)	10 µM	–	354 kHz (250 W/L), 500 kHz (150 W/L) and 618 kHz (250 W/L)	Medium-argon saturation	1. Effective degradation of both surfactants was achieved with the sonochemical method. 2. The results were good at lower concentrations

(Continued)

Table 12.4 (Continued)

Surface active agent	Initial concentration	Radiation energy	Radiation frequency	Other parameters	Key features
p-Octylbenzene sulfonate (LAS C_8), p-Nonylbenzene sulfonate (LAS C_9), and p-Dodecylbenzene sulfonate (LAS C_{12})	15–2000 μM	200 W power	200 kHz	–	1. Effective degradation of all linear alkyl sulfonates 2. The process was more effective, especially at a lower concentration level in the solution.
Lauryl pyridinium chloride (LPC)	0.1–0.6 mM,	18 W power	355 kHz	Temperature- $20 \pm 5\,°C$	1. Effective treatment of Lauryl pyridinium chloride solution was achieved. 2. A low confiscation of total organic content was seen. 3. The irradiation period of 20 hours was considered for investigation.
Octylbenzene sulfonate (OBS) anionic surfactant	1 mM	27 W power	616 and 205 kHz	pulse and continuous wave modes Temperature- 20 °C	1. The anionic surfactant degraded efficiently under the ultrasound sonolysis method. 2. Continuous wave sonolysis generated improved removal performance as compared to the continuous wave mode.
Perfluorooctanoic sulfonic acid	10 mg/L	40 W power	44 kHz, 400 kHz, 500 kHz, and 1000 kHz	pH-5.66 in an air atmosphere	1. The results indicated about 90% of the removal of the fluorosurfactant at 400, 500, and 1000 kHz. 2. The degradation of the surfactant was not observed when the frequency of the radiation was taken as 44 kHz.
Triton X 100, Nonionic surfactant	The investigation was carried out variable concentration	0.33 W/cm^2	358 kHz	–	1. Good elimination was reported for concentrations less than the critical micelle concentration of nonionic surfactant (250 μM).

to show a greater performance ability, which gets further enhanced on the addition of oxidizing agents like hydrogen peroxide. This generates hydroxyl radicals, which, along with ferrous ions of the coagulants, help easy degradation of surfactants like sodium dodecyl sulfonate [27].

The main mechanism involved is the oxidation of the ferrous ions produced at the anode. The ferric ion so formed in the solution combines with the hydroxyl ion produced at the cathode and results in the co-precipitation of the surfactant impurities. The pH of the solution is the deciding factor for the monomeric or polymeric state of iron. The ferric hydroxide formed is a rapidly settling floc containing the stabilized surface active agents. Electrocoagulation offers the advantage of high particulate removal efficiency, low cost, and complete automation quality. It is characterized by reduced sludge production and ease of operation.

Linear alkylbenzene sulfonate and sodium lauryl ether sulfate have been eliminated from the wastewater. Sodium sulfate (Na_2SO_4), iron sulfate ($Fe_3(SO)_4$), and hydrogen peroxide (H_2O_2) were added to increase the conductivity of wastewater, which acted as important factors in the coagulation of surfactants. They not only supported better coagulation but also decreased the energy requirement. In this regard, the energy consumption decreased about fivefold with the use of iron sulfate and hydrogen peroxide in the solution. A removal efficiency of 86% was reported in the investigation, with a 50 mg/L concentration of the aqueous solution, which involved an energy consumption of 0.1209 kWh/g. The efficiency of removal was found to increase with an increase in the current density, possibly owing to charge loading, which was proportional to the applied current. An increase in the concentration of the aqueous solution resulted in a decrease in the removal efficiency due to a lack of a sufficient amount of coagulant for sequestration of the surfactants. pH is a very important parameter in all coagulation and electrocoagulation processes. But, it was reported that good removal efficiency was obtained for both linear alkylbenzene sulfonate and sodium lauryl ether sulfate surfactants, at all pH values.

The efficiency of the process diminished due to the dissolution of the anode and cathode passivation. Thus, further research is needed to meet the required removal percentage for the targeted elimination [27].

Real wastewater from the Porsuk River was studied with electrocoagulation. The LAS concentration achieved was 2 mg/L. The energy consumption for the process was as low as 0.012 kWh/g. The study was carried out at an applied current density of 0.460 mA/cm^2. The results reflected an alteration in other parameters as well, including a change in the chemical oxygen demand from 91 to 36 mg/L, biological oxygen demand from 48 to 16 mg/L, and turbidity from 1.08 to 0.29, with pH remaining unaltered [28].

3. **Advanced Oxidation Process:** Recalcitrant surfactants can be removed from wastewater by using the advanced oxidation process. Techniques like the electrochemical method, electro-Fenton, and UV radiation have been employed for surfactant elimination. These processes do not pose the risk of generation of toxic byproducts or the possibility of release of pathogenic microbes.

   An anionic surfactant, sodium dodecyl sulfate, was eliminated using UV-based advanced oxidation process. The degradation percentage of surfactants was found to be 81% for a reaction time of 7 minutes and an H_2O_2 dose as 2 mol of H_2O_2/mol of surfactant. The surfactant degradation was less than 1% by direct UV irradiation, but the hydroxyl radicals generated by UV-H_2O_2 process enhanced the degradation process, as these radicals can attack the alkyl groups on all sites of the surfactant chain, resulting in the formation of aldehydes and carboxylic acids as the degradation products [29].
4. **Electrochemical Advanced Oxidation Process:** Utilization of the electrochemical advanced oxidation process involves transport of charge between the electrodes to remove anionic surfactants. An optimum power supply helps efficient removal of surfactants, while increased applied current density increases the energy consumption. Also, an electron beam based on irradiation was used for surfactant removal in the wastewater treatment plant. The species generated because of the irradiation degraded the aqueous contaminants. Similarly, the generation of free radicals was done by sonoreactor, which degraded the surfactants from the solution. This technique, however, requires a complete balance between the efficiency and cost involved in the energy input. Surfactant removal of 97% was achieved in laboratory scale by a team of workers by utilizing the Fenton process using an EDTA : iron : H_2O_2 ratio of 1 : 1 : 20 with an acidic pH of 6.4.
5. **Electroflotation:** Electroflotation comprises electrolysis of the solution and separation of contaminants by the flotation method, due to the creation of fine micro bubbles. The process is affected by treatment time, pH, and current density. Researchers have obtained an optimum efficiency value of 74.79%, with a current density of 8.42 mA/cm^2 [30].

12.5.3 Biological Treatment

The conventional physical and chemical methods adopted for surfactant removal from wastewater are not efficient in eliminating this pollutant. A high percentage of anionic surfactants remains unaltered and continuously becomes a part of the terrestrial or aqueous environment. The ecological behavior of the surfactants can be measured based on the biodegradation behavior of the surfactants. The molecular degradation process involves the conversion of the surfactant molecules to natural products like carbon dioxide and water, along with salts. Thus, biological treatment methods become a good option for the treatment of wastewater containing surfactants [27].

These methods are eco-friendly and sustainable for surfactant elimination. Algae, fungi, and bacterial populations break down the complex organics into simple compounds. Better results in the elimination and eradication of surfactants are observed, along with overcoming the possible shortcomings of the biological process, by combining the biological methods with physical and chemical processes. Several biological methods that can be used are discussed as under:

1. **Bioreactors:** Utilization of bioreactors for immobilization of surface active agents has received encouraging results.

   The cells of *Pseudomonas* over a glass bead support resulted in 85% sodium dodecyl sulfonate removal. The glass beads were used either unmodified or with a silanized surface. The anaerobic degradation of linear alkyl sulfonate was successfully performed by a microbial consortium comprising bacteroidetes, proteobacteria, and actinobacteria, over various inert support media, giving 99% removal efficiency for surfactants [31].

   Aerobic bioreactors and trickling filters were utilized for treating laundry wastewater, which depicted a stable decrease in the surfactant concentration by 96%, in a continuous four-month operational period. Aerobic treatment for surfactant removal sometimes reflects that the presence of a biological medium does not lead to an increase in the biodegradation ability. A decline in the degradation rate of microbial activity and binding of these with the surfactant molecules could be due to resistance posed by the presence of softeners, bleaching agents, and fragrances present in the detergents [27].

   In a conventional anaerobic batch reactor, linear alkyl sulfonate removal was achieved up to 99% in the presence of alcohol and nitrate, which promoted the population of the denitrifying microorganisms. An enhancement in the anaerobic degradation could also be achieved by means of supplements like iron. This supported the microbial population by acting as a terminal electron acceptor, providing energy to the microbes, like *Archaea* and bacteria. On the other hand, the presence of sulfide content inhibited the microbiotic activity [27].

   Other than the conventional biological process multichambered anaerobic biofilm reactor was investigated for household surfactant-containing wastewater. Coconut fibers were chosen as a biofilm support. Other components consisted of polyvinyl chloride spirals, which acted as a medium. The presence of a recirculation loop in the reactor enhanced the surfactant removal by enriching the microbes like proteobacteria (Gram-negative bacteria) and methanogenic bacteria (methane-producing anaerobic archaea).

2. **Biofilters:** A biofilter reactor was designed wherein microorganisms were grown on a media along with activated charcoal to lower the levels of organic material and smell. An alkaline pH of 7.3 and a temperature range of 25–37 °C served to be the optimum conditions. A reduction in the surfactant concentration significantly occurred to 94%, which was far below the prescribed standards [32]. Biofilters or artificial wetlands were developed by researchers to treat grey wastewater, which showed almost 77% removal of linear alkyl sulfonate, in 15 days of retention time. The aquatic plant investigated was *Schoenoplectus americanus*. Similarly, a horizontal flow type biofilter was installed for the removal of surfactant from grey water, collected from a commercial resort. It demonstrated efficient removal efficiency for COD, BOD, total suspended solids, and surfactant [27].
3. **Phytoremediation:** In an investigation, the phytoremediation ability of *Azolla filiculoides* was assessed for the removal of an anionic surfactant, sodium dodecylbenzene. This surfactant has a linear structure with twelve 12-carbon alkyl chain. The surfactant is widely used in household detergents due to its low cost and easy manufacturing. A water weed, *Azolla filiculoides* was chosen to remediate the surfactant from wastewater due to its rapid growth, high absorption capacity, and ability to survive even in contaminated water. The removal efficiency varied with the variation in concentration of the surfactant. The highest removal capacity was obtained as 77.79% in seven days, with medium having a surfactant concentration of 40 mg/L. It was also observed that with an increase in the biomass of *Azolla filiculoides*, the removal efficiency increased. This was attributed to a greater contact possibility of the pollutant and the biomass, and greater availability of the surface for adsorption of the surface active agent. They also reported that the endophytic microbes like *Anabaena, Azotobacter*, and *Arthrobacter* also contributed to the bioremediation process, but the complete mechanism of their action was still to be explored. *Azolla filiculoides* was found to have better tolerance against surfactants and thus acted as an efficient contaminant removal biomass.

An enhancement in the removal of an anionic and a nonionic surfactant was achieved from 94 to 99% and from 98 to 99%, respectively, by integrating physiobiological treatment processes. Similarly, an integrated phytoremediation and adsorption system led to 99.9% surfactant, 54% phosphates, and 77% chemical oxygen demand removal [33].

Membrane biological reactors are the most common method used for surfactant pollution elimination from wastewater. These have the benefits of compact design, but are costly in terms of energy input and operation.

Bacterial species like *Pseudomonas mendocina* and *Bacillus* were found to be dominating the degradation process of surfactants in actual domestic and commercial wastewater. A maximum removal efficiency of 66.63% was observed for sulfate dodecyl sodium (SDS) in around two days [27].

Several microbes and their consortia have been used for the degradation of surfactants from wastewaters. For example, removal of SDS has been successfully done using *Bacillus cereus* for detergent polluted pond, mixed consortia of *Acinetobacter calcoaceticus* and *Pantoea agglomerans* for wastewater from industry, *Klebsiella oxytoca*, *Pseudomonas aeruginosa*, *Pseudomonas mendocina*, *Pseudomonas stutzeri*, *Pseudomonas alcaligenes*, *P. pseudoalcaligenes*, *Pseudomonas putida*, and *Pseudomonas otitidis*, for detergent contaminated pond in Uttar Pradesh, India, *Delftia acidovorans* for river decontamination with 87% surfactant degradation rate after 11 days of incubation, *Pseudomonas plecoglossicida S5* for domestic laundry wastewater, mixed cultures of *Serratia* sp., *Pseudomonas beteli*, and *Acinetobacter johnsoni, Enterobacter cloacae, Alcaligenes faecalis*, for municipal sewage waste water and *Serratia marcescens consortia* for industrial wastewater. Similarly, *Alcaligenes odorans, Citrobacter diversus, Micrococcus luteus*, and *Pseudomonas putida* consortium were useful to eliminate linear alkyl sulfonate from surfactant-polluted streams [16].

12.5.4 Integrated Treatment Methods

A judicious choice of treatment method and its hybrids helps achieve maximum elimination efficiency, and at the same time helps mitigate the limitations of capital and energy input.

A hybrid of electrocoagulation and electroflotation cum electrolytic method was designed to treat detergent containing wastewater. The aluminum ion underwent hydrolysis and polymerization to form a cationic complex species, which increased the coagulation rate. During the electrolysis, the liberated oxygen and hydrogen gas led to an increased floc formation. This was investigated by Dimoglo et al., and it was reported that 90% surfactant removal, along with color and turbidity removal, took place. This was achieved with optimum parameters of pH 5.5 and contact time of five minutes. The process was executed at a current density of 5.26 mA/cm^2 [34].

Coagulation coupled with adsorption has also shown encouraging results. A household surfactant wastewater was treated by a group of researchers in which orange peel was taken as a source of coagulant, and the process was integrated with techniques like coagulation, flocculation, and adsorption over granulated activated carbon. A retention time of one day reflected excellent removal results due to the creation of large-sized flocs, possibly due to the active binding sites present in the pectin extract of orange peel. The combined technique eliminated about 92% of anionic surfactant, 54% of total suspended solids, and 86% of turbidity. Similarly, an integration of filtration, adsorption, and coagulation was successfully carried out for grey water treatment, in which an elimination of 85% of surfactant was registered [34].

Membrane filtration integrated with coagulation could abolish the disadvantage of membrane fouling and enhance the elimination of the organics. Pretreating wastewater with a cationic polymer coagulant, followed by membrane filtration, reduced fouling considerably. The capturing of small-sized impurities by the polymeric coagulants helped prevent high-resistance fouling and choking of the membrane pores [34].

Activated almond shell was utilized for the removal of sodium dodecylbenzene sulfonate. The activated adsorbent was prepared by oxidizing almond shells using hydrogen peroxide and nitric acid. The process was used in combination with the microbial degradation method to enhance the removal and degradation of the surfactant. The microbial population enhanced the hydrophobicity of the adsorbent surface. Also, due to availability of the support system for the bacteria in the form of adsorbents resulted in a greater contact time between the surfactant and degrading species. Thus, this study suggested that bioadsorption and biodegradation of the surfactants can be efficiently achieved using such adsorbents.

Thus, conventional coagulation process when coupled with techniques like filtration, flocculation, adsorption, membrane filtration, and flotation, resulted in an elevation in the surfactant removal efficiency. The disadvantage associated with using the combined techniques remained only in terms of the scalability of the integrated process, due to the high cost related to the use of chemicals. Research is being focused on finding substitutes that are sustainable and not based on the utilization of high-cost chemicals. In this regard utilization of biosurfactants can help solve the problem of environmental degradation associated with the use of conventional surfactants. Biotechnology has enabled the extraction of eco-friendly green surfactants manufactured from biomass resources. Among these, alkyl polyglucoside is found to be readily biodegradable and works without any side effects to the skin. Another surfactant, methyl ester sulfonate, is also popularly synthesized due to its biodegradability. Green surfactants like fatty acid methyl ester can be synthesized from renewable raw material like triglycerides via hydrogenation and trans-esterification. Similarly, thermal condensation of fatty amines present in cane molasses and corn syrup can form *N*-acetylglutamate surfactant. Biosurfactants are more effective, and their CMC which is 10–40 times lower than that of conventional surfactants. Large-scale production has not yet been executed due to low yield and high recovery, in addition to the purification cost.

The surfactant industry is constantly intensifying the production of a multitude of brands to meet the growing requirements of mankind. Statistical analysis has revealed the presence of surfactants in water bodies in concentrations greater than the required standards. Drastic environmental problems could be triggered due accumulation of these in the water bodies. Addressing the toxicity issues of surfactants, thus, becomes an aspect to be considered and managed carefully. Judicious utilization and suitable implementation of the techniques available for surfactant remediation can help trim down environmental vulnerability due to these chemicals.

12.6 Legislative Framework for Surfactant Discharge

Stringent guidelines have been framed to regulate the pollution caused by surfactants. In this regard, various public regulatory boards and governments have set limits for the discharge of surfactants or their presence in water.

The regulatory limit for anionic surfactants is 0.5 mg/L for drinking water, while up to 1.0 mg/L for nonionic surfactants. In France, the maximum surfactant value in drinking water is set as 0.2 mg/L for anionic surfactants, while 0.02 mg/L for nonionic surfactants. The Ministry of Health, Labor, and Welfare in Japan has set a surfactant limit of 0.2 mg/L and 0.02 mg/L, as a drinking water standard for anionic and nonionic surfactants, respectively [35]. According to the Croatian water regulatory criteria, the maximum allowed concentration of surfactant in wastewater effluents that can be released in a natural aquatic system is 1, 1, and 0.2 mg/L for anionic, nonionic, and cationic surfactants, respectively. The permissible limit for the discharge to the sewage system is set as 10 mg/L for anionic and nonionic surfactants, while 2 mg/L for cationic surfactants [35]. The concentration of one of the metabolites of surfactant, nonylphenol, in freshwater bodies should be less than 6.6 μg/L. Nonylphenol and octylphenol, the degradation products of nonylphenol ethoxylate and octylphenol ethoxylate, have been included in the list of hazardous substances by the European nations [9].

The legislative framework for surfactants has been specially formulated in the European nations, as these are manufactured and consumed on a large scale. Environmental European legislation is laid for the anionic and nonionic surfactants, but no imposed limit has been set for the cationic surfactants. In Europe, when European Regulation (EU) No. 648 came into existence, the need for control on cationic and other surfactants became evident. It was realized that detergents and cleansing products bring a quantitative and qualitative change in the environment both at manufacturing as well as application levels. EU No. 608/2004 mentions that each producer or importer of detergents should attest the biodegradability of surfactants in detergents. EU No. 907/2006 follows the assurance of biodegradability of surfactants and safety to human health. Another amendment of detergent regulation is referred as EU No. 259/2012, which standardized the use of phosphorus compounds in laundry and dishwashing detergents. The regulation also suggests that surfactants in detergents are permissible if these fulfill the aerobic degradability criteria. At an international level it is required that the producers are encouraged to obtain self-cleaning products and the complete life cycle of detergents should be secured from the beginning of the use of the raw materials, to the eliminating action. In Poland, the permissible concentration of anionic surfactant in surface water is between 0.2 and 1 mg/L while for nonionic surfactants the range is 0.5–2 mg/L. As per the national Regulation of the Slovak Government 269/2010, the permissible limit for surfactant in surface water discharges is 1 mg/L, while for water used in irrigation is 2 mg/L. Also, in the Ecuadorian legislation of environment quality, the anionic surfactant concentration limit is 0.5 mg/L, estimated as methylene blue active substances (MBAS) [9]. Nevertheless, strict actions to set the discharge limits for surfactants are still not implemented in a few developing nations.

12.7 Case Studies

12.7.1 Surfactant-loaded Wastewater – Lion Corporation, Thailand

Industrial wastewater containing soaps and detergents of Lion Corporation Thailand was taken to understand the surfactant removal efficiency by coagulation method. The untreated wastewater was stored at 4 °C for further analysis and treatment. Alum or aluminum sulfate was chosen as the coagulating agent. Wastewater on analysis revealed significant amount of turbidity, COD, presence of mixtures of surfactants like linear alkylbenzene sulfonates (LAS), alcohol ethoxysulfates (AES) and α-Olefin sulfonates (AOS), and other cationic and nonionic surfactants. The coagulation process was simulated to understand the feasibility of the same for surfactant removal using a test jar apparatus. The study was carried out at wide range of pH values ranging from 2 to 12. The optimum pH for removal was confirmed to be 10. A coagulant dose of 400 mg/L produced the removal efficiency of 95.4% for an anionic surfactant. Alum sludge from a consumer industry was reused for the process, to minimize the amount of the coagulants. A proficient removal of both the total active ingredient and surface active ingredients was successfully carried out. The analysis of total active ingredient and surface active ingredients was done as per the standard method of titrimetry, according to American Society for Testing and Materials (ASTM) [36].

12.7.2 River Tigris, Iraq – A Surfactant Loaded Freshwater Source

Tigris River in Baghdad city, Iraq, is one of the most convenient disposal sites for waste from various sources, including households. A 13-month study was carried at four different sampling sites to quantitatively and qualitatively analyze the presence of surfactants in water and their impact on environment. Various physicochemical parameters related to the water sample were investigated for one year time period, for dry and wet seasons. In dry season, high concentrations of nonionic and anionic surfactants were determined using high performance liquid chromatography. A surfactant deposition of 56.19 and 135.74 μg/g was recorded for nonionic and anionic surfactants, respectively. Lowest concentrations for both the type of surfactants were registered in the wet season. The adsorption of surfactant on the sediments was influenced by pH and charges present on surfactant and sediments. The electrical conductivity was found to vary between 580.50 and 1108.75 μs/cm in dry and wet seasons, respectively, which inferred the presence of higher levels of contaminants than the standards in the river. The adsorption of surfactant followed the order cationic surfactant > nonionic surfactant > anionic surfactant. The studies concluded that sediments served as a sink for the deposition of anionic and nonionic surfactants. Nonylphenol, and the degradation products of the nonionic surfactants, are crucial contaminants, owing to their toxic effects on the aquatic system in the dry season.

It is thus important that the wastewater from various sources must be treated for removal of surfactants, otherwise the quality of the river water would degrade with time [12].

References

1 Alessandra, Z., Stefania, F., Laura, B., et al. (2017). Embodied energy as key parameter for sustainable materials selection: the case of reusing coal fly ash for removing anionic surfactants. Journal of Cleaner Production, Volume 141, pp. 230–236. https://doi.org/10.1016/j.jclepro.2016.09.070.

2 Ciabattia, I., Cesaro, F., Faralli, L., et al. (2009). Demonstration of a treatment system for purification and reuse of laundry wastewater. Desalination, Volume 245, Issues 1-3, pp. 451–459. https://doi.org/10.1016/j.desal.2009.02.008.

3 Sostar-Turk, S., Petrinić, I., and Simonič, M. (2005). Laundry wastewater treatment using coagulation and membrane filtration. Resources, Conservation and Recycling, Volume 44, Issue 2, pp. 185–196. https://doi.org/10.1016/j.resconrec.2004.11.002.

4 Saxena, N., Islam, M.M., Baliyan, S., et al. (2023). A comprehensive review on removal of environmental pollutants using a surfactant based remediation process. RSC Sustainability, Volume 1, pp. 2148–2161. https://doi.org/10.1039/D2SU00069E.

5 Guerrero-Hernandez, L., Melendez-Ortiz, H.I., Cortez-Mazatan, G.Y., et al. (2022). Gemini and bicephalous surfactants: a review on their synthesis, micelle formation, and uses. International Journal of Molecular Science, Volume 4, Issue 23, p. 1798. https://doi.org/10.3390/ijms23031798.

6 Ahmer, A.S., Muhammad, R.S., Aaron, L., et al. (2020). A review on recent developments in the adsorption of surfactants from wastewater. Journal of

Environmental Management, Volume 254, p. 109797. https://doi.org/10.1016/j.jenvman.2019.109797.
7 Sutormin, O.S., Kolosova, E.M., Torgashina, I.G., et al. (2023). Toxicity of different types of surfactants via cellular and enzymatic assay systems. International Journal of Molecular Sciences, Volume 24, Issue 1, p. 515. https://doi.org/10.3390/ijms24010515.
8 Jahan, K., Balzer, S., and Mosto, P. (2008). Toxicity of nonionic surfactants. WIT Transactions on Ecology and the Environment, Volume 110, Issue 10, pp. 281–290. https://doi.org/10.2495/ETOX080301.
9 Gheorghe, S., Lucaciu, I., and Paun, I. (2013). Ecotoxicological behavior of some cationic and amphoteric surfactants (biodegradation, toxicity and risk assessment). Biodegradation – Life of Science, InTech. https://doi.org/10.5772/56199.
10 Jardak, K., Drogui, P., and Daghrir, R. (2016). Surfactants in aquatic and terrestrial environment: occurrence, behavior, and treatment processes. Environmental Science Pollution Research, Volume 23, pp. 3195–3216. https://doi.org/10.1007/s11356-015-5803-x.
11 Yuan, H., Xu, X., Sima, Y., et al. (2013). Reproductive toxicity effects of 4-nonylphenol with known endocrine disrupting effects and induction of vitellogenin gene expression in silkworm, *Bombyx mori*. Chemosphere, Volume 93, Issue 2, pp. 263–268. https://doi.org/10.1016/j.chemosphere.2013.04.075.
12 Al-Ani, R., Hassan, F.M., and Al-Obaidy, H.M.J. (2022). Environmental evaluation of surfactant: case study in sediment of Tigris River, Iraq. River deltas research – Recent advances, IntechOpen. https://doi.org/10.5772/intechopen.94324.
13 Badmus, S.O., Amusa, H.K., Oyehan, T.A., et al. (2021). Environmental risks and toxicity of surfactants: overview of analysis, assessment, and remediation techniques. Environmental Science and Pollution Research, Volume 28, pp. 62085–62104. https://doi.org/10.1007/s11356-021-16483-w.
14 Tripathi, S.K., Tyagi, R., and Nandi, B.K. (2013). Removal of residual surfactants from laundry wastewater: a review. Journal of Dispersion Science and Technology, Volume 34, Issue 11, pp. 1526–1534. https://doi.org/10.1080/01932691.2012.752328.
15 Motteran, F., Nascimento, R.F., Nadai, B.M., et al. (2019). Identification of anionic and nonionic surfactant and recalcitrants compounds in commercial laundry wastewater by GC-MS analysis after anaerobic fluidized bed reactor treatment. Water, Air and Soil Pollution, Volume 230, p. 301. https://doi.org/10.1007/s11270-019-4357-9.
16 Arora, J., Ranjan, A., Chauhan, A., et al. (2022). Surfactant pollution, an emerging threat to ecosystem: approaches for effective bacterial degradation.

Journal of Applied Microbiology, Volume 133, Issue 3, pp. 1229–1244. https://doi.org/10.1111/jam.15631.
17 Gao, Q., Wu, F., Hu, J., et al. (2020). Chemical composition-dependent removal of cationic surfactants by carbon nanotubes. Science of Total Environment, Volume 716, p. 137017. https://doi.org/10.1016/j.scitotenv.2020.137017.
18 Sen, T.K., Thi, M.T., Afroze, S., et al. (2012). Removal of anionic surfactant sodium dodecyl sulphate from aqueous solution by adsorption onto pine cone biomass of *Pinus Radiate*: equilibrium, thermodynamic, kinetics, mechanism and process design. Desalination and Water Treatment, Volume 45, Issue (1–3), pp. 263–275. https://doi.org/10.1080/19443994.2012.692036.
19 Ncibi, M.C., Mahjoub, B., and Seffen, M. (2008). Adsorptive removal of anionic and non-ionic surfactants from aqueous phase using *Posidonia oceanica* (L.) marine biomass. Journal of Chemical Technology and Biotechnology, Volume 83, Issue 1, pp. 77–83. https://doi.org/10.1002/jctb.1787.
20 Turyasingura, M., Wakatuntu, J., Lubwama, M., et al. (2023). Optimisation of eggshell-zeolite composite as a potential surfactant adsorbent for hand-washing wastewater. Case Studies in Chemical and Environmental Engineering, Volume 7, p. 100284. https://doi.org/10.1016/j.cscee.2022.100284.
21 Sochacki, M., Michorczyk, P., and Vogt, O. (2025). Foam fractionation as an efficient method for the separation and recovery of surfactants and surface-inactive agents: state of the art. ACS Omega, Volume 10, Issue 1, pp. 55–75. https://doi.org/10.1021/acsomega.4c08413.
22 Kumar, A.K., and Ghosh, P. (2022). Removal and recovery of an anionic surfactant in the presence of alcohol by foam fractionation. Industrial and Engineering Chemistry Research, Volume 61, Issue 21, pp. 7349–7360. https://doi.org/10.1021/acs.iecr.2c00372.
23 Boonyasuwat, S., Chavadej, S., Malakul, P., et al. (2005). Surfactant recovery from water using a multistage foam fractionator: part I effects of air flow rate, foam height, feed flow rate and number of stages. Separation Science and Technology, Volume 40, Issue 9, pp. 1835–1853. https://doi.org/10.1081/SS-200064595.
24 Krishnan, S., Chandran, K., and Sinnathambi, C.M. (2016). Wastewater treatment technologies used for the removal of different surfactants: a comparative review. International Journal of Applied Chemistry, Volume 12, Issue 4, pp. 727–739.
25 Merouani, S., Dehane, A., and Hamdaoui, O. (2024). Ultrasonic destruction of surfactants. Ultrasonics Sonochemistry, Volume 109, p. 107009. https://doi.org/10.1016/j.ultsonch.2024.107009.
26 Naumczyk, J., Marcinowski, P., and Bogacki Jan, B. (2017). Treatment of highly polluted cosmetic wastewater. Environment Protection Engineering, Volume 44, Issue 2, pp. 25–40. https://doi.org/10.5277/epe170203.

27 Ho, K.C., Teow, Y.H., Sum, J.Y., et al. (2021). Water pathways through the ages: integrated laundry wastewater treatment for pollution prevention. Science of the Total Environment, Volume 15, Issue 760, p. 143966. https://doi.org/10.1016/j.scitotenv.2020.143966.

28 Onder, E., Koparal, A.S., and Oğütveren, U.B. (2007). An alternative method for the removal of surfactants from water: electrochemical coagulation. Separation and Purification Technology, Volume 52, Issue 3, pp. 527–532. https://doi.org/10.1016/j.seppur.2006.06.006.

29 Mondal, B., Adak, A., and Datta, P. (2019). Degradation of anionic surfactant in municipal wastewater by UV-H_2O_2: process optimization using response surface methodology. Journal of Photochemistry and Photobiology A: Chemistry, Volume 375, pp. 237–243. https://doi.org/10.1016/j.jphotochem.2019.02.030.

30 Skender, A., Moulai-Mostefa, N., and Tir, M. (2010). Effects of operational parameters on the removal efficiency of non-ionic surfactant by electroflotation. Desalination and Water Treatment, Volume 13, Issue 1–3, pp. 213–216. https://doi.org/10.5004/dwt.2010.992.

31 Rebello, S., Asok, A.K., Mundayoor, S., et al. (2013). Surfactants: chemistry, toxicity and remediation. In: Lichtfouse E, Schwarzbauer J, Robert D (eds) Pollutant diseases, remediation and recycling. Environmental chemistry for a sustainable world, Volume 4. Springer, Cham. https://doi.org/10.1007/978-3-319-02387-8_5.

32 Suprijandani, S., Suryono, H., and Narwati, N. (2021). Detergent waste treatment through the modification of biofilter reactors. International Journal of Public Health Science, Volume 10, Issue 3, pp. 590–599. https://doi.org/10.11591/ijphs.v10i3.20843.

33 Siswoyo, E., Utari, A.W., and Mungkari, L.G.N. (2019). Adsorption combined phytoremediation system for treatment of laundry wastewater. MATEC Web Conferences, Volume 280. https://doi.org/10.1051/matecconf/201928005002.

34 Dimoglo, A., Sevim-Elibol, P., Dinç, O., et al. (2019). Electrocoagulation/electroflotation as a combined process for the laundry wastewater purification and reuse. Journal of Water Process Engineering, Volume 31, p. 100877. https://doi.org/10.1016/j.jwpe.2019.100877.

35 Kumar, M., Ahmed, S., Singh, S., et al. (2023). Assessment of anionic surfactant concentration in the Najafgarh drain: comparative analysis. Journal of Survey in Fisheries Sciences, Volume 10, Issue 3, pp. 539–546.

36 Jangkorn, S., Kuhakaew, S., Theantanoo, S., et al. (2011). Evaluation of reusing alum sludge for the coagulation of industrial wastewater containing mixed anionic surfactants. Journal of Environmental Sciences, Volume 23, Issue 4, pp. 587–594. https://doi.org/10.1016/S1001-0742(10)60451-2.

13

Fertilizers and Nutrients

Fertilizers are substances that contain nutrients helpful for plant growth. To cater to the global food demand of the growing population, there has been an extensive use of several fertilizers globally. Fertilizers aid in increasing the yield of the crops and improve their quality [1]. As per the data published (Figure 13.1) on a website "Our World in Data," shows a significant increase in a noteworthy amount of fertilizers being used and adopted worldwide.

There has been a substantial rise in the use of fertilizers (Figure 13.2) for the different income countries [1]. While the growing need for fertilizers is understood, this has also brought to notice several environmental problems that they can pose [2].

13.1 Fertilizers in Agriculture

Plants need a variety of nutrients such as nitrogen, phosphorus, calcium, magnesium, potassium, etc., for their growth and other biological processes. Due to repeated plantation of crops, there occurs a depletion of these nutrients from the soil, which needs to be replenished.

The use of fertilizers and nutrients plays a vital role in rejuvenating the soil nutrients, helping in the proper and healthy growth of plants [3]. Fertilizers also increase the tolerance of the plants toward several pests. These help in the proper development of plant roots, serving better water absorption and better growth [4]. The use of fertilizers helps boost food production. The nutrient value of the crops increases, leading to a healthy human life. Fertilizers also support sustainability, as the nutrient value of the soil is kept intact. With the greater yield of crops, the profitability for the farmers and their social and economic status also increases [5].

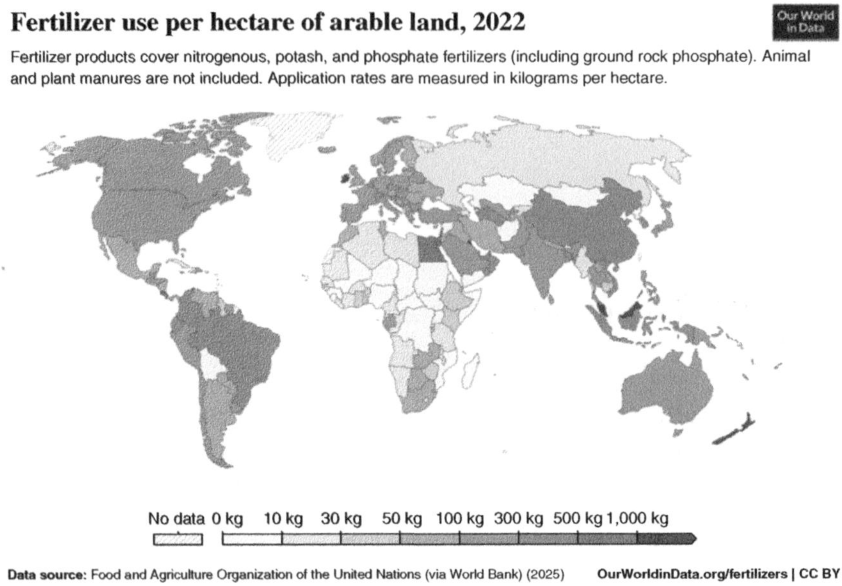

Figure 13.1 Fertilizer use per hectare of cropland Adapted from Food and Agriculture Organization of the United Nations (2025) – with major processing by Our World in Data.

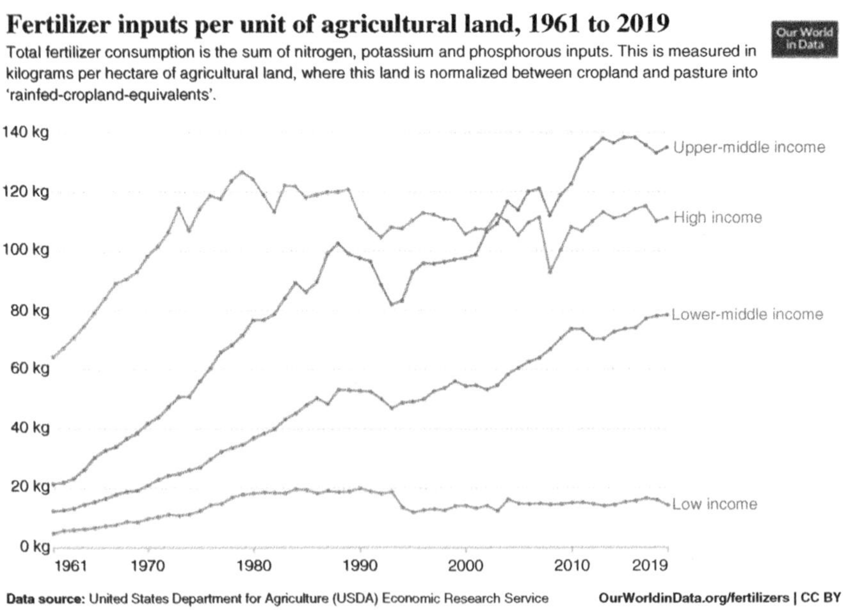

Figure 13.2 Amount of fertilizer input per unit of agricultural land Adapted from data published by the US Department of Agriculture (USDA) Economic Research Service. International Agricultural Productivity.

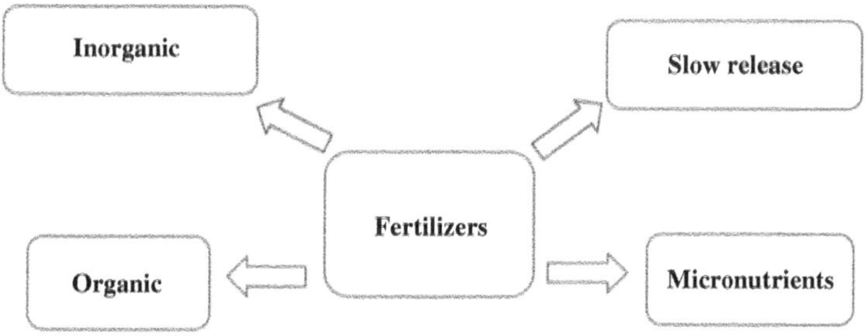

Figure 13.3 Common categories of fertilizers.

13.1.1 Types of Fertilizers

Depending upon the source or chemical composition of the fertilizers, they are divided into various types. Figure 13.3 portrays the different catcgories of fertilizers [6] being used to enhance the crop and soil quality.

13.1.1.1 Inorganic Fertilizers

Inorganic fertilizers are synthetic fertilizers that act almost immediately on the crops with a persistent nutrient supply. Synthetic fertilizers are quick in their action and provide precise nutrients to the soil [7]. These fertilizers can be nitrogen, phosphorus, or potassium-based. Nitrogen-based fertilizers have a significant impact on the yield of the crop [8]. Inorganic fertilizers can be chemically:

1. **Nitrogen (N)-based:** Nitrogen plays a vital role as a nutrient for plants due to the below [9]:
   a. Forms amino acids, which are the main components for the formation of proteins. These proteins form the structural components of the plants.
   b. Nitrogen is an essential component of adenosine triphosphate (ATP) which helps in the energy transfer, helping the cells to use and conserve energy.
   c. It is an essential component of nucleic acids such as deoxyribonucleic acid (DNA).

   There has been a rapid rise in the use of nitrogen-based fertilizers over the years [10]. Many of the nitrogen-based fertilizers contain nitrates. These are water-soluble and are easily available to plants. Nitrogen-based fertilizers can either be in ammoniacal form or urea [11]. These eventually get converted to nitrates for accumulation. Some important examples of nitrogen-based fertilizers are nitrate fertilizers (NO_3), ammonia (NH_3), ammonium (NH_4), urea (NH_2CONH_2), ammonium nitrate (NH_4NO_3), and ammonium sulfate ($(NH_4)_2SO_4$).

Accumulation of nitrogen based fertilizers can lead to certain negative impacts on the environment. One of the major mentions is their impact on the characteristics of water and wastewater. The wastewater containing nitrogenous substances can be a byproduct of the fertilizer manufacturing plants [12], leaching [13], or agricultural runoff.

2. **Phosphorus (P)-based:** These are the suppliers of another major nutrient for plant growth, phosphorus. Phosphorus helps the plant for the following reasons [14]:
   a. Helps in the root development.
   b. Helps in the seed and flower formation.
   c. Assists nitrogen fixation by roots.
   d. Enhances the quality of the crop.
   e. Improves the strength of the stem and stalk.
   f. Makes the plants disease resistant.
   g. Helps in the development of the plant

   Single super phosphate (SSP), double superphosphate (DSP), triple superphosphate (TSP), monoammonium phosphate (MAP), and diammonium phosphate (DAP) are some categories of phosphorous containing fertilizers [15].
3. **Potassium(K)-based:** Potassium is an important constituent present in the fertilizers and contributes to the plant growth in several ways, like [16]:
   a. It helps in enhancing the process of photosynthesis.
   b. Activates the enzymatic processes, including the synthesis of cellulose, vitamins, starch, and proteins.
   c. Enhances the flavor and color of fruits, including their size and growth.
   d. Increases crop yield and quality.
   e. Helps the plant roots fix nitrogen.
   f. Helps starch formation, especially in starch-storing crops such as potato, etc.
   g. Helps in the regulation of the opening and closure of the stomata.

   Important categories of potassium based fertilizers are sulfate of potash (SOP), sulfate of potash-magnesia (K_2SO_4–$MgSO_4$), potassium nitrate (KNO_3), and potassium hydroxide (KOH).

13.1.1.2 Organic Fertilizers

All the above categories of nutrient-containing fertilizers (N, P, K) discussed could also be organic. Organic fertilizers refer to the fertilizers that are of plant or animal origin and are available naturally, unlike synthetic fertilizers [17]. These include substances such as bone meal, plant material, manure, compost, etc. Organic fertilizers provide micro nutrients such as calcium, sulfur, magnesium, etc., to the plants, in addition to the essential macronutrients such as nitrogen, potassium, and phosphorus. The advantages offered by organic fertilizers are:

1. They provide organic matter to the soil, thereby improving the soil texture, increasing the moisture content.
2. Help in the growth of beneficial microorganisms surviving in the soil.
3. Improve the fertility of the soil.
4. Help in the gradual release of nutrients making them available for the plants for a longer period of time.
5. Help avoid the hazardous traits due to the use of toxic chemicals protecting the soil and the underground water table.
6. These are better than the synthetic fertilizers as they are not associated with the risk of overuse of fertilizers, because of their organic nature.

13.1.1.3 Slow-release Fertilizers

Most of the fertilizers release the nutrients quickly but that may not ensure the availability of the nutrient to the plants for longer duration. The main motive of the use of slow release fertilizers is that the nutrients are made available to the plants for prolonged period. Slow-release fertilizers (SRFs) release the nutrients gradually and in a steady manner and are thus better than others.

These fertilizers could be of the following two categories:

1. **Coated SRFs:** Coated SRFs possess a coating such as of polymers [18] on the fertilizers to ensure their slow release. The release of these nutrients depends upon the factors such as temperature, climate, moisture, and the coating material.
2. **Organic SRFs:** As these are of natural origin such as plant waste, bone meal, manure, etc., the release of nutrients from them is gradual owing to slow breakdown.

SRFs are preferred over fast-release release due to the following reasons:

1. Helps to prevent leaching of nutrients.
2. Supply nutrients consistently over a longer period.
3. As the nutrient supply is gradual, the need for resupply is less.
4. More environment friendly.

The SRFs make the nutrients available either by diffusion across the coated material due to a difference in the concentration gradient or osmosis, where migration of water molecules through the coating material is involved, leading to the dissolution of the nutrients inside, forcing them out of the coat. In certain cases, soil microbes also help the release of nutrients through the coat. Some examples of SRFs are urea-isobutyraldehyde/isobutylene diurea (IBDU), urea-acetaldehyde (UA)/cyclodiurea (CDU), and urea-formaldehyde (UF).

13.1.1.4 Micronutrient Fertilizers

Some fertilizers provide micronutrients such as zinc (Zn), boron (B), manganese (Mn), copper (Cu), iron (Fe), chlorine (Cl), and molybdenum (Mo), which are required lesser in quantities but are still essential for the plants. Several micronutrients are essential [19] in addition to the macronutrients because:

1. Micronutrients are involved in several physical and biological processes.
2. Help in different enzymatic reactions and plant processes such as photosynthesis, nitrogen metabolism, respiration, etc.
3. Help in structure formation and rigidity.
4. Help in the uptake of nitrogen.

13.2 Hazards of Fertilizers in Water/Wastewater

Several hazardous impacts of fertilizers as contaminants can be understood by first knowing the different pathways of their entry into water or wastewater sources, and then the effects caused by them.

13.2.1 Pathways of Contamination by Fertilizers

13.2.1.1 Leaching

Leaching refers to the movement of the fertilizers downwards along with the percolating water due to rain or irrigation [20]. The nutrients in such a case are not available to the plants as they move away from them. This would depend upon the water solubility of the fertilizer, the availability of soil water, and the soil characteristics such as porosity, soil water retention capacity, aeration, etc.

Leaching can take place at a rapid rate or a slow pace. In case a fertilizer is water soluble, it shall remain in the soil along with water in reasonable quantities. In case of an excess of supply of water in the form of rain, etc., this category of fertilizer will rapidly tend to move to surface or underground water. Certain nutrients show a greater mobility and therefore are more prone to leaching. Some common nutrients are nitrogen, phosphorus, potassium, and sulfur.

13.2.1.2 Surface Runoff

This refers to the flow of nutrients from the surface of the soil eventually into the water bodies. It refers to the portion of water that does not percolate into the soil. Nutrients can be lost either in the dissolved form or along with the sediments in the runoffs. At times when the nutrients are in excess and are not used by the plants, such nutrients get accumulated at the surface and are also carried away by the runoffs.

Snow melt or rainfall could be the possible factors responsible for nutrient runoff. The intensity of rainfall, soil water-holding capacity, soil structure, and slope are also important factors in determining the quantity of nutrients that can be taken away by runoffs. The transport of nutrients away from the soil is faster in higher rainfall intensity, steeper slopes, and improperly structured soil with less organic matter. The presence of organic matter positively helps in the retention of nutrients [21].

13.2.1.3 Agricultural Drainage

This refers to the artificial removal of excess water from the agricultural land. This removal of water is done either artificially or naturally by gravity [22].

The need to carry out agricultural drainage is to provide optimum aeration to poorly drained soils that are waterlogged. The use of agricultural drainage can vary from region to region, as shown in Figure 13.4 [22].

While the agricultural drainage has several advantages such as improved sanitation along with preservation of soil quality, etc., it is associated with some disadvantages also, like rapid flow which carries an excess of nutrients to the water bodies along with pollutants and also over drainage.

13.2.1.4 Acid Rain Due to Deposition in the Atmosphere

Fertilizers are also contributors to the formation of acids in the atmosphere, especially nitrogen-based fertilizers. Nitrogen-based fertilizers of synthetic nature are made readily available for plants, whereas in the case of organic nitrogen-based fertilizers, the nutrients are available to the plants gradually, which can lead to pollution. The production of nitrogenous fertilizers requires the use of petroleum or nitrogen gas, which results in the liberation of synthetic nitrogen into the air. The nitrogen that is not used up by the plants can be converted into harmless

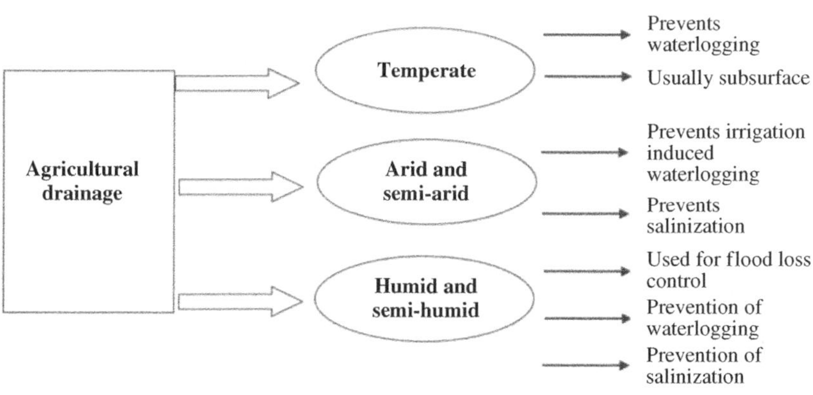

Figure 13.4 Agro-climatic zones for different categories of agricultural drainage.

nitrogen (N_2) or converted into oxide of nitrogen oxides (NO_x). These oxides of nitrogen combine with air and moisture in the presence of sunlight, leading to air pollution and acid rain [23].

Acid rain contains acids that have an adverse impact on the structures, soil, water bodies, and aquatic ecosystems. Acid rain contains nitrogen, which affects the water bodies, as a decline in the coastal fishes, shellfish fish, etc. [24]. In addition to the other harmful effects, acid rain leads to nitrogen deposition in the lakes, leading to algal bloom [25].

13.2.2 Impact on the Water Quality

Nutrients that pave their way into the aquatic ecosystem can lead to multiple effects, which have been discussed below:

13.2.2.1 Eutrophication

Eutrophication refers to an aquatic condition arising due to an excess of nutrients such as nitrogen and phosphorus. This condition gives rise to accelerated algal blooms in the water bodies, leading to a degradation of aquatic biodiversity and a decline in the water quality [26, 27]. Factors such as temperature, water retention, and the availability of ample sunlight govern the responses of the aquatic system in the presence of stimulating factors such as nutrients.

Eutrophication can be natural or anthropogenic. Anthropogenic eutrophication is human induced while natural eutrophication takes place due to natural processes such as erosion, decomposition or deposition of sediments [28].

Figure 13.5 shows the different factors responsible for natural and anthropogenic eutrophication.

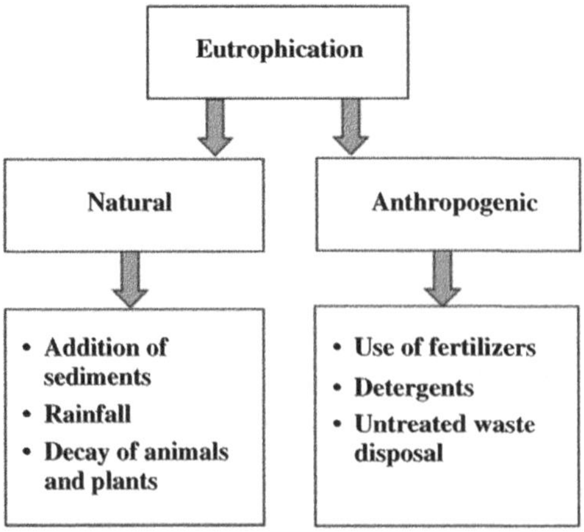

Figure 13.5 Factors responsible for natural and anthropogenic eutrophication.

13.2.2.2 Mechanism of Eutrophication

With the increase in the nutrient concentration, the plant biomass increases, leading to a decrease in the penetration of sunlight. Light becomes a limiting factor, and there is a shift in the aquatic system from a nutrient-rich to nutrient nutrient-saturated system. The old biotic species replaces the new one with changes in the population of zooplanktons, fish, etc. Large algal biomass production leads to the decay of the old species, which depletes the oxygen levels when degraded by bacteria. This situation gives rise to low oxygen levels in the water bodies, resulting in anoxia or hypoxia, which implies no oxygen or little oxygen content. The autotrophic algae contain bioplasm, which is formed in the presence of nutrients in the presence of sunlight [29, 30] (Equation 13.1).

$$106CO_2 + 16NO_3^- + HPO_4^{2-} + 122H_2O + 18H^+ \xrightarrow{\text{Energy+microelements}} \underbrace{C_{106}H_{263}O_{110}N_{16}P}_{\text{Bioplasm in algae}} + 138O_2 \tag{13.1}$$

13.3 Consequences of Eutrophication and Nutrients in Water

Owing to the excess of nutrients in water, the below-mentioned consequences are witnessed:

1. **Accelerated Growth of Algae:** Due to an increase in the nutrients there is an excessive growth of cyanobacteria, diatoms, phytoplankton, resulting in algal blooms [31]. The surface of the water bodies gets covered by the algal blooms, leading to a reduction in the light penetration and hampering the aesthetics of the water resource.
2. **Reduced Oxygen Levels in the Water Bodies:** The state of reduced oxygen concentration is known as hypoxia. After the accelerated growth, when these algal blooms die or sink to the bottom, in order to carry out their decomposition, there is an enhanced microbial activity, which increases the biological oxygen demand (BOD) of the water bodies. This leads to oxygen depletion. The stretches that have depleted oxygen levels are unable to support life and therefore are known as "Dead Zones." One such zone, which is considered to be the largest, is in the United States [32], which occurs in the Gulf of Mexico every spring.
3. **Harm to Biodiversity and Changes in Aquatic Species:** Due to eutrophication, there are sudden changes in the biodiversity of the lakes [33]. The structure of the aquatic community changes and is directly and indirectly affected [34]. At relatively low nutrient concentrations, the phytoplankton

and benthos (bottom) species productivity is high. With the rise in the nutrient level, the species at the benthic level can be more affected and can decrease due to a reduction in the light intensity reaching the bottom of the lake [35].

4. **Production of Toxins by the Algae:** Algal growth that leads to the production of toxins is termed harmful algal blooms (HABs). Several categories of algal blooms can grow as a result of over-nutrition, which include those that result in excessive biomass and those that lead to the production of toxins. Harmful algae can have an impact on the environment and human health. Dinoflagellates, diatoms, and cyanobacteria [36] are some examples. These HABs can be seen in the sea or freshwater, where they grow indefinitely, giving rise to the production of toxins. These also affect the aquatic organisms, such as fish, humans, who eat up a lot of oxygen and give rise to foul-smelling piles on the shores. In addition to this, they also have the ability to cause contamination of drinking water [37].

   HABs can occur in different aquatic environments such as marine water, brackish water, and freshwater bodies. Cyanobacteria, commonly known as blue-green algae, occur in freshwater, sometimes in marine water. Cyanobacteria are responsible for multiple toxin production, leading to harmful effects such as on nerves, skin, stomach functioning, etc. In marine brackish waters, species such as diatoms, dinoflagellates, and phytoplankton are observed. They impart different shades to the water bodies, ranging from red, brown to bioluminescent. Some of the species of dinoflagellates are capable of producing cysts, which can survive the harsh conditions and again produce algal blooms in favorable conditions. Several conditions, such as poisoning of the fish and respiratory irritation in human beings, and severe depletion of oxygen in the water bodies, can result from marine HABs [38].

5. **Degradation in the Water Quality:** With excessive nutrients, reduction in the oxygen content, and the toxins produced by the algal species and microorganisms, there is a severe degradation in the quality of water in the water bodies. The BOD levels accelerate exponentially, making the survival of useful aquatic organisms difficult. The quality of the water in freshwater sources also deteriorates, making it unfit for use by the human species.

   Nutrients get transferred to the underground water, leading to nitrogen pollution in the underground water. One such example that has been noticed is the United States, where such water is used as a drinking water source [39].

6. **Human Health Risks:** The hazardous effects of HAB are observed due to contact with phycotoxins through routes such as seafood contamination, inhalation, and direct contact with the skin. Generally, the impacts of algal blooms on human health have been observed after coming in contact with

recreational water, such as water skiing, wading, swimming, boating, etc. [40]. Eutrophication is a major cause leading to dermatological and gastrointestinal conditions such as nausea, vomiting, and diarrhea [41]. In the case of gastrointestinal disorders, *microcystis, aphanizomenon,* and *anabaena* have been identified from the stools and vomits of several patients [42]. Direct exposure to the toxins produced by the algae can lead to serious health effects such as respiratory problems, skin rashes, neurological disorders, and illnesses of the stomach and liver. Nitrates lead to eutrophication when they contaminate the drinking water and can lead to blue baby syndrome, characterized by blue-tinted skin and shortness of breath [43].

7. **Stimulation of Greenhouse Gas Release:** The increase in the HABs can lead to a substantial rise in the emission of greenhouse gases such as methane [44]. The HABs thrive in slow-moving water bodies where the water is warmer. The presence of carbon dioxide is necessary for the algal blooms. Therefore in they thrive in carbon dioxide (CO_2) rich environments. Methane (CH_4) production can be elevated due to the cyanobacterial blooms. Also, freshwater bodies significantly contribute to CH_4 production. The annual production of CH_4 by freshwater systems has been estimated to be 103 Tg, with a potential to increase even higher, if the HABs collapse [45].
8. **Economic and Social Consequences:** The effects of eutrophication rise beyond the level of aesthetics and hazardous implications to economic and social consequences. Eutrophication can lead to widespread social and economic impacts. One such example is the Gulf of Mexico, where there has been expansion of hypoxic zones has created a pressure on the fishing zones [46].

   In the Gulf of Mexico, the worth of the fishing industry annually is $680 billion, which is around 40% of the fishing yield of the United States. About $400 billion of the annual fishing is from shrimp. In a study by Duke University, the rise in the prices of the sold shrimps was linked to the rapid eutrophication in the Gulf [46]. Mass deaths of fish result due to inability to migrate from one region to another. This not only affects the fishing industry but also all those industries also which are dependent on them, such as the hotel industry.

   In the northern hemisphere, the demand for recreational activities in the water bodies has increased leading to a direct connection with eutrophication [47]. According to the US Environmental Protection Agency (EPA), there are seven major categories in which the impact due to eutrophication occurs. Some of which are cost of recreation, property value, cleanup costs, cost towards health management [48].

   Due to rise in the recreational activities, the availability of clean water surface becomes essential. The presence of algal blooms affects such activities leading to loss of economy. Further, the value of the properties close to the water bodies bearing algal blooms automatically deteriorates. An increased algal growth is an additional burden to the economy of any region due to the clean-up costs involved. Several hazards associated with HABs result in expenditures on health maintenance.

13.4 Control of Eutrophication and Nutrients in Water

Below are the measures used for the prevention of eutrophication:

13.4.1 Regulating Nutrient Inlet in the Water Bodies

As the reason for eutrophication is excessive nutrient content therefore reduction in the nutrient inlet in the water bodies can help to control the growth of algae. Treating wastewater and controlling nutrients in the agriculture runoff along with reduction in pollution of air can help to reduce the nutrient concentration in the water bodies to a substantial extent.

13.4.2 Improving Methods Adopted in Wastewater Treatment Plants

Increasing the efficiency of the wastewater treatment plants to eliminate excessive nutrients before the wastewater is let into the water bodies will be beneficial. An integration of technologies and the use of novel materials will help in this regard. The utilization of advanced treatment technologies can help solve the problem.

13.4.3 Management of Agricultural Practices

Agricultural practices which are adopted can substantially help the control of nutrient flow from the agricultural fields into the water bodies. Use of less quantity of fertilizers, especially reduction in the use of synthetic fertilizers can reduce the pollution load on water bodies. Water conservation tillage practices should also be adopted.

13.4.4 Public Contribution

Educating the general public, specifically the farmers, making them understand the causes of eutrophication for better management of runoffs from the fields will be supportive [49]. In Australia, there has been a substantial reduction in the nutrient load by educating the public to:

1. Reduce the use of fertilizers
2. Composting of domestic waste
3. Use of low-nutrient detergents
4. Use only full loads in the washing machines.

13.4.5 Regulatory Authorities

The role of the regulatory authorities in controlling the nutrients in the water bodies is vital. Strict discharge limits of the runoff water can control the concentration of nutrients in the water bodies.

13.4.6 Effluent Diversion

This is another means by which the nutrient load on the water bodies can be managed. Such a method has been tested in Lake Washington, which is the largest natural lake in Washington state, in the vicinity of Seattle, the United States of America, close to the sea. Cyanobacterial species *Oscillatoria rubescens* became dominant in 1955 due to the introduction of a large volume of sewage water from about 70000 people, increasing the nutrient levels. A diversion of sewage water from the lake was done into Puget Sound, a nearby inlet to the sea, which resulted in a decline in the bacterial population [49].

13.4.7 Phosphate Striping and Buffer Striping

In this method, the effluents are let into a tank containing a precipitator such as salts of iron, which lead to the precipitation of phosphorus in the water, precipitating at the bottom. The effectiveness of the process is high, with almost 90% nutrient removal [49].

Buffer striping involves the plantation of vegetation along the sides of the field along the water edge where the vegetation helps retain the nutrients getting drained off.

13.4.8 Treatment Using Constructed Wetlands

Constructed wetlands have proven their worth in removing the nutrients from wastewater before it enters the lakes and ponds. Constructed wetlands use a combination of processes and materials which are helpful in filtering off the excess of nutrients entering the water bodies.

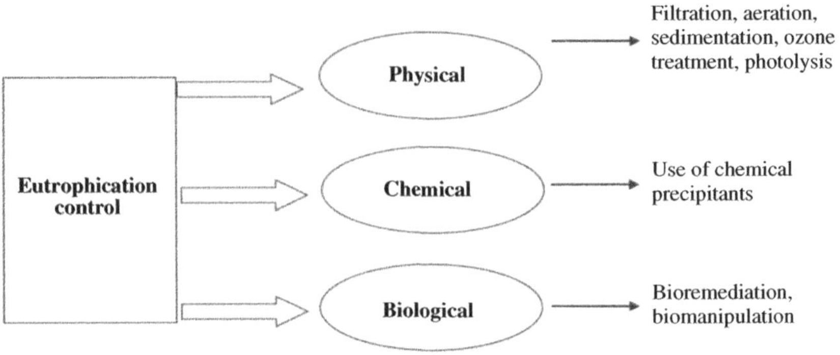

Figure 13.6 Methods to control eutrophication.

13.4.9 Other Methods

Several physical [50, 51], chemical [52], and biological/ecological methods [53] can be used for control of eutrophication. Figure 13.6 portrays the different methods that are a part of each of these categories:

Physical methods such as aeration, treatment with ozone, etc., can be used. Aeration increases the oxygen content in the water bodies. Methods such as ozonation have been used for the destruction of algal biomass. A chemical method, such as the use of precipitants, is generally considered to be nontoxic. This method has been generally used for the elimination of phosphorus, where the precipitator (iron) precipitates phosphorus to the bottom of the water body. This is then deactivated to prevent its release [52]. Biological/ecological methods for the control of eutrophication and the regulation of nutrients prove to be environment-friendly and cost-effective.

13.4.10 Ecological Approaches

Methods such as biomanipulation, bioremediation, and phytoremediation have been used for the effective management of nutrient flow into the water bodies.

Biomanipulation method has been tested as effective tools to combat eutrophication [53]. Biomanipulation is an intentional alteration in the aquatic ecosystem by addition or removal of species, to obtain the desired results (Table 13.1).

Phycoremediation involves the conversion of pollutants and nutrients in wastewater to biomass by the different metabolic processes in the micro or microalgae. This has been successfully employed in the elimination of nutrients leading to eutrophication. This technique has been successfully employed by several industries for wastewater treatment [62]. Optimum nitrogen and phosphorus concentration is required for the synthesis of nucleic acids, proteins, and phospholipids [63]. The phycoremediation method has proven its efficiency over other

Table 13.1 Different organisms used for biomanipulation along with the target organisms.

Organisms used	Target	References
Control of phytoplankton by the removal of zooplankton-eating fish	Increase in the zooplankton that feed on phytoplankton	[54, 55]
Control of phytoplanktons by the removal of benthic omnivorous fish	Decreasing internal nutrient cycle and sediment bioturbation	[56]
Stocking of predatory fish for controlling phytoplankton	Increase in the zooplankton feeding on phytoplankton	[57]
Stocking of pelagic herbivores for controlling phytoplankton	Removal of phytoplankton	[58]
Protection and transplantation of macrophytes	Maintenance of a high yield of macrophytes	[59]
Introducing mussels	Increasing the filtration of water to obtain clear water	[60]
Stocking herbivorous fish	Reduction in the growth of macrophytes	[61]

Source: Adapted from [53].

physicochemical methods in the removal of nutrients, with low operational cost, and is environmentally friendly. Also, after the removal of nutrients, the algal mass can be used as a fertilizer without any secondary pollution [64, 65].

Phytoremediation is yet another method for the reduction of nutrients and controlling eutrophication [66]. In this method, green plants are utilized for the elimination of environmental pollutants. Invasive and native species can be utilized to remove the pollutants due to their high accumulation capacities [67]. Nutrients are more efficiently removed with the use of native (invasive) species. Species such as water lettuce, duckweed, and water hyacinth effectively capture the nutrients from the water bodies [68, 69].

Bioremediation is also considered an effective way to deal with water pollution due to its high efficiency and low cost. Further, it does not lead to any secondary pollution [70]. Microbial remediation uses prokaryotes as biotic agents for the removal of pollutants from water. In addition to the removal of metals from water, bioremediation has also been successfully used for the removal of fertilizers and pesticides from water [71]. Besides the removal of pollutants from the water streams, the pollutants are converted into useful, recyclable, and reusable components. Factors such as pH and temperature affect the process of bioremediation. Bioremediation involves the eradication of pollutants, their degradation, detoxification, and immobilization [72].

13.5 The USEPA for Nutrient Pollution

EPA has been working continuously towards several measures for controlling nutrient pollution [73]. EPA is contributing in partnering and working in conjunction with different stakeholders. It is working with different agencies such as Water Research Foundation, US Department of Agriculture (USDA), National Tribal Water Council, USDA's Natural Resource Conservation Service (NRCS), etc. toward nutrient pollution control.

EPA also supports and helps the tribes, states, and territories in technical programs and workshops, and outreach programs for maintenance of water quality, awareness of toxins and the health effects of HABs, and helps in the nutrient requirement permits.

EPA support and help in reviewing and providing approval for water quality standards, developing total maximum daily loads (TMDLs), formulating guidelines for effluent discharge, working on the regulations to control nitrogen oxide emissions, which can lead to nitrogen pollution.

EPA is also instrumental in providing financial support in several nutrient reduction activities which include grants and funds for nonpoint source management, providing incentives for the use of different technologies, financing point and nonpoint pollution reduction strategies.

In addition to financial aid, collaborating with different agencies, setting up water quality standards, research, and development also is a key aspect of the efforts taken by EPA. It supports the national program to support research in various domains such as technological development for controlling nutrient pollution and HABs in water bodies, pathways and effects of pollutants, assessment of water quality standards, etc.

13.6 Food and Agriculture Organization of the United Nations

The Food and Agriculture Organization (FAO) of the United Nations has provided "The International Code of Conduct for the sustainable use and management of fertilizers" [74].

The document mentions the efficient use and management of fertilizers. It emphasizes the development of technologies that would lead to the production of fertilizers providing ample nutrients to the crops, and at the same time minimizing the harmful environmental impacts. The need for a holistic approach in the development and implementation of proper management practices, considering the crop productivity and environmental impacts be implemented. There should be proper information and dissemination related to the use of fertilizers and their impacts. Proper legislation should be in place with the application limits.

The use of nutrients by recycling to be encouraged with proper information about the side effects of reuse and recycled components, such as wastewater, compost, sewage, sludge, etc. In addition to the several important aspects linked with the use of fertilizers, the FAO also mentions about the composition of fertilizers, limits of application, and testing of the components in fertilizers. It mentions that the governments, along with several entities, national and international, should be responsible for the regulation of the components in the fertilizers, such as harmful microbes, toxic chemicals, additives, heavy metals, and other contaminants. The use of evidence-based safety standards, considering their pathways and impacts on humans and the environment.

13.7 Policies and Schemes for the Use of Fertilizers

13.7.1 The EU Nitrates Directive (1991)

This directive [75, 76] focuses on maintaining the water quality across Europe, preventing the pollution of surface and underground water bodies by preventing nitrogen contamination and the adoption of sustainable farming practices. These directives emphasize the reduction of pollution, specifically due to agriculture, restricting the use of fertilizers, pesticides, and other contaminants reaching the water bodies. The vulnerable water bodies are identified with a nitrogen concentration level above 50 mg/L or are susceptible to fall into this category. In deciding the nitrogen vulnerable zones (NVZ), the concentration of both phosphorus and nitrogen is considered. Measures mentioned below must be complied with by farmers for these zones such as limiting the use of nitrogenous fertilizers in these regions in specific durations to avoid losses to water bodies, consideration to the conditions of the application of fertilizers such as ground covered with snow, slope, etc., minimum capacity of storage for livestock manure, crop rotation and limitations to the quantity of livestock manure which can be applied to the soil.

Every four years, the state members are expected to report the concentration of nitrates in the surface and ground water, level of eutrophication in surface waters, impacts of application of the action plan, revision of plan as per the requirement, and future trend estimation of the quality of water.

13.7.2 The US Clean Water Act (1972)

The US Clean Water Act (CWA) [77] was an amendment in 1972 of the Federal Water Pollution Control Act (1948), which was the first major law in the

United States for controlling water pollution. The main aspects of the amendment are as below:

1. Establishment of primary structures to regulate the discharge of pollutants into the water.
2. Authority to the EPA for the implementation of pollution control programs and standards.
3. Maintenance of water quality standards.
4. Making the discharge of point source pollutants into the water bodies illegal unless a permit is obtained.
5. Funding sewage treatment plants.
6. Proper need of planning is needed for the management of nonpoint source pollutants.

There have been continuous amendments since then, leading to several modification such as inclusion of technology based standards, involving state partnerships and permitting discharges such as storm water etc., reduction of toxic pollutants in Great Lakes, consideration to the risks of diseases in recreational water bodies, provision of funds for surveys, etc.

13.7.3 New Soil Health Card Scheme (2015)

The New Soil Health Card Scheme by the Ministry of Agriculture and Farmers Welfare, India [78]. Information to the farmers on the nutrients in the soil is provided in the soil health card, along with the nutrient recommendations for improvement in the fertility and soil health. Several parameters, such as the presence of macro and micronutrients, pH, organic carbon, etc., are a part of the information. The recommendations for the use of fertilizers are also given based on the nutrients available in the soil.

13.8 Case Studies on Fertilizer Contamination

13.8.1 Dead Zone – Gulf of Mexico

The dead zone in the Gulf of Mexico is an area containing less than 2 ppm of dissolved oxygen at the border of the Mississippi River. The coverage of the dead zone is approximately 7000 square miles [79]. This is known as the "hypoxic zone" due to the lack of availability of oxygen. In August 2017, the largest dead zone was reported by the National Geographic News. This dead zone has been caused by the excessive pollution of the Mississippi River, specifically owing to the presence of phosphorus and nitrogen. The farming states in the river valley of the Mississippi, such as Illinois, Wisconsin, Mississippi, etc., contribute to the passage of nutrient contaminants into the river. The nutrients make their way into the river through the fertilizer runoff upstream, sewage, animal waste, and soil erosion. These nutrients, when let into the river along with the anthropogenic contribution to the water streams, lead to an accelerated algal growth. The food chains get altered, reducing the oxygen content in the water bodies substantially. The increase in the hypoxic zone depends on the weather conditions, hurricanes, floods, and farming practices.

Adoption of practices such as reduced use of fertilizers, their application time to avoid agricultural runoff, management of sewage inlet, and inlets from septic tanks, and reduced nutrient inlet are the practices which have been adopted for the zones such as the Gulf of Mexico.

13.8.2 Punjab – Groundwater Nitrate Pollution

Punjab, "India's bread-basket," where crops such as wheat, sugar cane, rice, have been a significant contributor of nutrient load [80]. There has been a remarkable rise in the population and the demand for food grains in the state of Punjab over the last three decades. An annual report in 2012 mentioned that there has been a 2% rise in the production of food grains against an 8.5% increase in the population [81].

The permissible limit of the nutrient nitrate is 50 mg/L in drinking water, as specified by the World Health Organization (WHO). The limit of nitrate set in India is 45 mg/L. In Punjab, because of the increase in the state crop productivity, high doses of chemical fertilizers were used in the state, with a consumption of 243 kg/year, significantly high in comparison to the other states in India [82, 83].

In a study, different water samples were taken from wells near villages and farms and were tested for nitrate concentrations, which were found to be significantly high [84]. In November 2023, an article published in "The Times of India" newspaper stated that a study carried out by the Indian Institute of Technology, Mandi, mentioned the severity of water pollution, owing to agricultural runoff, due to which several notices were served to different departments [85]. In a discussion paper on Water Global Practice in 2020 [86], several actions that can be taken to combat the situation were mentioned. Due emphasis on understanding the existing situation, taking proper action, and the involvement of institutions has been advocated.

13.8.3 China – Eutrophication of Lake Water

China has been facing the problem of eutrophication. In a study reported by Liang et al. in 2024 [87–89], the development in the urban areas of the Zhanjiang Bay

(ZJB) in China has led to a serious impact on the water bodies for decades, giving rise to nutrient accumulation and eutrophication.

There has been a direct discharge of sewage, agricultural runoff input, and the inlet of domestic effluents along the coast. A narrow channel connects the ZJB to the sea in the southeastern region. This narrow link has made the exchange of water and nutrients poor, impacting the water quality in the bay [89].

About 32% of the nitrogenous fertilizers are being consumed. This has led to the discharge of 14.5 ± 3.1 Tg nitrogen, per cent due to anthropogenic activities [90]. Coastal waters, around 20%, in China can be mentioned as "affected by eutrophication" [91]. The quality of coastal water in China has degraded in 2019, with a stretch of about 55340 km^2. Owing to the high usage of nitrogenous fertilizers, there is nutrient enrichment in the coastal waters. Several rivers, such as the Changjiang River, the Zhujiang River, the Huanghe River, etc., flow into the seas. A large amount of nutrient input and weak exchange capacity increases the retention time of the nutrients, leading to eutrophication. In an analysis of the Chinese coastal waters in 2019, there has been an increase in HAB growth every decade [92, 93].

Several government policies are in place by the government focusing on zero growth of fertilizer, water pollution control plan "Ten-Point Water Plan" and improving the recycling of manure [94–96]. Different environmental models have been explored for prevention of nutrient discharge [97, 98].

13.9 Efforts on Successful Remediation

13.9.1 Strategies Adopted in Denmark

Due emphasis has been given to the reduction of nutrients to improve the surface water quality in Denmark [99]. An action plan implemented in 1987 led to the reduction of nutrient load by almost 80%, corresponding to the reduction of nitrogen and phosphorus loads to a level of approximately 141600 tons of nitrogen and approximately 1820 tons of phosphorus.

The plan developed by the Danish Environment Protection Agency, the Ministry of Environment and Gender Equality, encompasses three major categories of water: agricultural wastewater, water from the municipal wastewater treatment plants, and discharges from the industries. Specific targets have been set under each category. Each plan at the onset included the adoption of several measures in a specific time frame, after which the situation was analyzed. In case of the results requiring further attention, these plans were further modified with an extended time frame. Some of the key features which were considered are as below:

1. Establishment and establishment of wet meadows converting nitrates to N_2.
2. Deciding durations for afforestation.
3. Financial support to farmers for the use of agricultural sensitive areas with a reduced use of fertilizers.
4. Changes in fodder and feeding practices, governing livestock density.
5. Promoting organic farming.
6. Reduction in the nitrogen application norm to 10%.

For municipal wastewater treatment plants, the discharges are regulated by the Environmental Protection Act, the Urban Wastewater Directive, along with official guidelines and derivative statutory orders. The year-wise nutrient load reduction plans were set, and almost 6600 tons per year of nitrogen reduction was achieved, and 1200 tons of phosphorus were achieved in the year 1996. The measures for industrial discharges are regulated by the EU Directive on Pollution Prevention Control (IPPC), the Environmental Control Act, and official guidelines. The pollution by major industrial sectors and integrated prevention is regulated by the IPPC Directives, which regulate industries such as energy, metal processing, chemical, waste management, etc. [99].

13.9.2 Initiatives for Organic Farming in Europe

The method of organic farming brings together the factors such as management of the farm in a Greener manner, along with food production. It works on high production standards along with natural resource preservation [100]. A target of conversion of 25% of the agricultural land in Europe is planned to be covered under organic farming by 2030, by the European Commission [101].

The objectives of the Green Deal's sustainability action plan are divided into Axes 1, 2, and 3. Out of the three the "Axes 3" deals with strengthening environmental sustainability with the use of organics. The farmers involved in organic farming are not permitted to use synthetic fertilizers, or organic pesticides.

13.9.3 Wetland Restoration Techniques

Constructed wetlands are cost-effective measures for regulating the flow of nutrients in the water bodies [102]. The USEPA [103] published several studies on the use of constructed wetlands for maintaining water quality standards, with the successful use of constructed wetlands. In addition to absorbing nutrients from wastewater, these also constitute a floral abundance serving as habitat and food for many species. The flow of the water slows done while passing through the wetlands. The suspended particles settle down, and organic matter and nutrients are assimilated.

13.10 Management of Fertilizer Use

The use of precision agriculture for fields has been helpful. In precision agriculture, the site-specific requirements of fertilizers are fulfilled with the use of GPS, soil mapping, and remote sensing. The soil samples in the field are analyzed for the availability of nutrients, after which a nutrient map is developed. The use of Integrated Nutrient Management (INM) deals with the incorporation of different types of nutrients (inorganic and organic), introducing compost, crop residues, and manure to make up for the nutritional requirements of the crops. Crop rotation also helps to maintain the fertility of the soil. Plantation of leguminous crops helps to make up for the nitrogen requirement of the soil [104].

The need for sustainable water management in agriculture is also to be emphasized in agriculture to balance the use of water, where appropriate amount of water is provided to the agricultural field. This would neither create a deficit of moisture in the soil nor would it lead to water running off from the fields into the water bodies, and also would keep nutrients in place [105].

13.11 Technologies and Methods for Nutrient Removal

In addition to the use of constructed wetlands discussed in the previous section, other technologies have also proved to be beneficial for nutrient removal. A shift from the conventional technologies used for nutrient treatment to advanced methods becomes crucial, considering the pace at which the nutrient inlet affects the ecological systems. Initiatives include the use of sequencing batch reactors (SBRs) and advanced membrane bioreactors (MBRs), integrated up-flow anaerobic sludge blanket (UASB), etc. [106].

SBR is an efficient technology for the treatment of wastewater as it is a simple process, with high efficiency of removal of suspended solids, BOD, and nutrients by aerobic and anoxic period alternation [107]. It is a kind of activated sludge system where water is added to a batch reactor. Modifications in SBRs can be used for the biological removal of nutrients biologically from wastewater [108]. This technology is also effective if any modification in case required. Factors affecting the performance include the rate of organic loading, dissolved oxygen, and solid content [109, 110]. In an investigation by Hu et al. in 2005, a cost-effective Taguchi technique for the design of experiments was investigated. Promising results were obtained with total phosphorus and total nitrogen removal efficiencies of 83.9% and 90.2%, respectively. Factors such as batch loading rate and aeration, and mixing ratio had a significant impact on the removal of nitrogen and phosphorus removal [111].

Advanced technologies offer high removal capacity of nutrients. Membrane bioreactor (MBRs) is one such category. In this technology, a combination of membrane bioreactor and biological treatment, along with filtration, is carried out using a membrane and system for nutrient recovery. The biggest advantage of the use of MBRs is that they not only are capable of nutrient removal but also help in their conversion into a reusable form [112]. In the present scenario, focus has been drawn to the Anammox process (anaerobic ammonium oxidation) [113]. Ammonia and nitrites are converted into nitrogen gas in the absence of any carbon-containing compound. Phosphorus and nitrogen are removed by a combination of aerobic and anaerobic conditions.

Another category of such advanced technologies is the UASB and short aerobic system (SAS) where the movement and flow of water is upwards through a blanket of granular sludge. Sludge granules and incoming wastewater mixes through upward flow with the sludge granules, leading to organic matter decomposition and removal of nutrients [114].

Use of single-celled photosynthetic organisms, microalgae, is a method for the treatment of agricultural, municipal, and industrial wastewater. These organisms can be autotrophs, heterotrophs, or mixotrophs, helping in the treatment of wastewater [115]. All microalgae have the same nutritional requirement hence helping in the removal of nutrients. They carry out biosorption of nutrients and heavy metals [116]. Microalgae can be introduced during the secondary or tertiary treatment of wastewater.

Granular biomass system, a much researched method, could be an alternative to activated sludge method [113, 117]. The organisms used in this method are capable of simultaneously removing nutrients such as nitrogen, phosphorous and COD with the use of a single reactor.

Advanced oxidation processes (AOPs) [118] can successfully be used for the removal of nutrients as these convert the available nitrogen and phosphorus into gaseous or less soluble forms. They involve the use of strong oxidizing agents. The use of photocatalysts such as TiO_2, ultraviolet radiation, and the use of ozone are some of the examples. AOPs can be combined with other conventional processes for enhancement in the nutrient removal efficiencies. The efficiency and functioning of AOPs are dependent upon wastewater composition, conditions of operation, the presence of the present inorganic and organic matter, and reactor design.

Use of strategies involving several chemicals has also proved its worth in the elimination and treatment of nutrients from wastewater [118]. Methods which involve the use of chemicals include precipitation, adsorption, the use of nanomaterials, ion exchangers, etc. Materials such as zeolites, biochar, clay minerals, and activated carbon have been efficiently used for the removal of nutrients [119]. Precipitating agents usually used for nutrient treatment are ferric chloride and alum, which help the precipitation of nutrients present in wastewater, which can be easily removed by filtration. The nutrient removal capacity can be enhanced by integrating with conventional methods. Conditions such as pH, residual concentration of metals in the treated water, and the optimization of chemical doses need careful consideration. Ion exchange resins exchange the ions of the nutrients, such as NH_4^+ and PO_4^{3-}, by ion exchange, due to high selectivity and easy operation, which can be regenerated. Cost considerations of ion exchange resins need careful consideration.

Membrane technologies are also a boon in the selective removal of nutrients from wastewater. Methods such as forward osmosis (FO), reverse osmosis (RO), and nanofiltration (NF) have proved to be successful in this regard [120]. The nutrient content can be substantially reduced from wastewater with the use of reverse osmosis, after which water can be safely released into the water bodies. Although the RO system demands the use of high energy inputs and is prone to fouling, it offers a compact footprint and operation flexibility [121]. NF can help the selective removal of nutrients, retaining the necessary ions. The energy consumption for NF is less, and they are more compatible for use in brackish waters, but these are also prone to fouling. FO, which utilizes the water movement across the semipermeable membrane from a lower to a higher concentration, retains the nutrients with a lesser consumption of energy. The tendency of fouling is minimum, but might require pretreatment for the removal of colloidal particles and suspended solids.

A comparison between some important technologies is presented in Table 13.2.

In the current scenario, there has been development in the use of novel materials, including nanomaterials, for the treatment of nutrients. These novel materials show high adsorption capacities for nutrients. Some such materials are metal–organic frameworks (MOFs), graphene based materials, nanoadsorbents, etc. A multifunctional treatment approach can be adopted with the promising use of nanomaterials with magnetic and photocatalytic properties [128].

Several challenges are encountered with the different removal technologies. While some technologies demand cost effectiveness, some technologies, such as wetlands, depend upon the weather conditions, with warmer climates more conducive for contaminant removal. The use of microorganisms depends on their successful harvesting. Technologies such as Anammox demand, denitrification, and carbon source [113]. Optimization and integration of techniques can help to achieve the maximum potential of nutrient removal [118]. Optimization can include the use of regulation of aeration rates, dosage of chemicals, use of artificial intelligence, machine learning, etc., which would help in overcoming the limitations of a single method and incorporating the advantages of beneficial methods. An integration of different types of methods can help in achieving the highest standards of nutrient removal.

Table 13.2 Advantages and disadvantages of some technologies for nutrient removal.

Method	Advantages	Disadvantages	References
Sequencing batch reactors (SBRs)	Flexibility of operation, economical	High maintenance, sophisticated	[122]
Advanced membrane bioreactors (MBR)	Easy to operate, less space, and less sludge production	High cost involvement, low sludge settling rate	[123]
Integrated up-flow anaerobic sludge blanket (UASB)	High loading rate, short retention time, high COD removal efficiency	Sensitivity of certain toxic substances, and the difficult-to-control granulation process	[124]
Microalgae	High removal efficiencies for nutrients, efficient, and cost-effective	Need for effective cultivation and efficient pre- and post-water treatment	[125]
Granular biomass system	Excellent performance at pilot scale	Instability of granular biomass	[126]
Wetland	Effective treatment in a passive manner, operation, and maintenance is less costly.	Large land area requirement, problem of mosquitoes, and insect vectors	[127]
Advanced oxidation processes (AOPs)	Efficient nutrient removal	Treatment optimization requirement, high energy requirement	[118]
Chemical methods	Rapid removal	Careful dose optimization, cost, and formation of byproducts	[118]

13.12 Recent Studies on Nutrient Removal

In a recent study by Hariz et al. in 2025, the potential of novel filamentous algae nutrient scrubbers (FANS) has been investigated, which can recover nutrients from wastewater. The performance of three categories of FANS, viz. standard, shallow, and deep, has been investigated. A comparison between the biomass productivity, photosynthetic efficiency, nutrient concentration in algal biomass, etc. was studied. In the studies, the shallow FANS were considered to be the most effective in the elimination of nutrients, which was 30% in summers and 40% in winters, endorsing a favorable nutrient assimilation [129].

Enhanced bioreactor systems have been explored for efficient nutrient removal. IMBMBR (integrated moving bed membrane bioreactor) was used for the study, which improved the effluent quality of municipal wastewater [130]. In addition to the COD and BOD, almost 24 ± 3.0% and 99 ± 0.2% removal was observed for PO_4^{3}-P and NH_4^+-N, respectively. Nitrite-oxidizing bacteria (NOB) and ammonia-oxidizing bacteria (AOB) together helped to enhance the removal of nutrients.

Li and coworkers [131] have recently investigated the potential of Duckweed for nutrient removal from pig farm slurry of varying concentrations. The effect on COD, BOD, total nitrogen and phosphorous along with ammonia nitrogen was studied. The removal of total nitrogen and phosphorous observed was 86.89% and 85.22% respectively, proving its efficiency.

Agro-industrial materials such as woodchip, zeolite, biochar, alum sludge, and flexible solid packing have been used as suspended fillers (SFs) in floating treatment wetlands (FTWs) for enhancing the removal of nutrients from wastewater [132]. Up to 20–57% total nitrogen and 23–63% total phosphorus were removed. Macrophytes showed accelerated growth and enhanced biomass production. In comparison to the five SFs, the use of a mixture of SFs in FTWs showed a considerable lead in the greater growth of the microbial community. The removal efficiencies on microcosm scale, mesocosm scale, and field scale, for total nitrogen and total phosphorus were 99.3%, 84.0%, and 15.0–73.7% for total nitrogen and 98.4%, 95.0%, and 31.5–77.1%, respectively. It was concluded that scaling up the method could be used for systems of eutrophic freshwater.

In a study, the nutrient removal capacity of microalgae such as *Coelastrella tenuitheca, Scenedesmus dimorphus, Parachlorella kessleri*, and *Chroococcus turgidus* was investigated with alteration in the concentrations of phosphate (PO_4-P) and nitrate (NO_3-N) and sewage water (SW) [133]. Species such as *Scenedesmus dimorphus* showed considerable nutrient removal capabilities towards different wastewaters. This species worked the best for nitrogen removal in comparison to the other species.

A review paper in 2024 elaborates on the discussion on the use of methods such as electrooxidation (EO) has also been explored to reduce the operational costs involved in nutrient removal [134]. Technologies such as electrooxidation and electrocoagulation have been examined for their efficiency and scalability.

EC methods can be electroflocculation (EFLOC), electroreduction (ER), electrophoresis (EP), electroflotation (EF), electrodeposition (ED), electrooxidation (EO), electrodialysis, and electrocoagulation [135–138]. EO can convert nitrogen present in the form of nitrate (NO_3^-) or ammonium ions (NH_4^+) into nitrogen (N_2) gas. Electro dialysis, on the other hand, is capable of removing ammonium ions selectively. Orthophosphates are converted into an insoluble form by electrocoagulation, and EC advanced oxidation processes (EAOPs) utilize the active oxygen for the removal of nitrogen and phosphorus from their compounds [139]. To enhance the nutrient removal capacity, optimization of operating conditions and the choice of the electrode play a vital role. Materials such as ruthenium dioxide (RuO_2), boron-doped diamond, lead oxide, and TiO_2 have been investigated. High nutrient removal rates are offered by EC processes in addition to being sustainable methods. High costs, by product formation, high energy demands, and electrode fouling, however, draw attention.

Some innovations in the technologies for nutrient removal have been discussed in a special issue on "Environmental Sustainability and Applications." Technologies such as the use of zerovalent metals [140] and struvite precipitation [141] for nitrogen removal have been used.

Zerovalent ions such as zinc (ZVZ), aluminum (ZVA), copper (ZVC), iron (ZVI), and magnesium (ZVM) have been studied for nitrate removal from water. The review paper presented [140] describes the studies on several zerovalent metals in different forms, for nutrient removal. Studies related to iron in the zerovalent form, microscopic (mZVI) and nanoscopic iron (nZVI), were considered. An evaluation of the different conditions, such as dissolved oxygen, temperature, and initial nitrate concentration, was conducted. Low pH values showed a positive impact on the process a similar effect of the increase in temperature. Several studies incorporated on the use of zerovalent ions endorse the developing interest of the scientific community in exploring the suitability of these materials for the abatement of nitrates.

Struvite ($MgNH_4PO_4{\cdot}6H_2O$) [141], which is a crystalline orthorhombic compound composed of equimolar quantities of NH_4^+, Mg^{2+}, and PO_4^{3-} can be used for the recovery of phosphates and ammonium from wastewater. It has the capacity to produce precipitates with nitrogen and phosphorus.

The use of microalgae-bacteria consortia for the removal of nutrients from municipal wastewater has also proven to be a sustainable approach. Oxygen is produced by the microalgae as a product of photosynthesis, utilizing carbon dioxide and sunlight. This produced oxygen is then consumed by bacteria to decompose the organic matter present. This self-sustaining economic cycle helps in the removal of nitrogen and phosphorus from the water bodies [142–144].

As the cost involved in the use of several technologies is high, future research should be amplified in the direction of integration of different methods for overcoming the limitation of individual techniques, proper optimization of conditions, and the possibility of scaling up play a significant role. Due consideration to exploring renewable measures while maintaining high nutrient standards should be the main focus of research in this direction. Attention to circular economy reduction in the impact on the environment should be given the necessary emphasis.

References

1 Hannah, R., Max, R., and Pablo, R. (2022). "Fertilizers" Published online at Our World in Data.org. https://ourworldindata.org/fertilizers

2 Penuelas, J., Coello, F., and Sardans, J. (2023). A better use of fertilizers is needed for global food security and environmental sustainability. Agriculture and Food Security, Volume 12, Issue 5. https://doi.org/10.1186/s40066-023-00409-5.

3 https://hefefertilizer.com/en/news/the-importance-of-the-use-of-fertilizers/ (Last accessed 05.02.25)

4 Alnaass, N.S., Agil, H.K., and Ibrahim, H.K. (2021). Use of fertilizers or importance of fertilizers in agriculture. International Journal of Advanced Academic Studies, Volume 3, Issue 2, pp. 52–57. https://doi.org/10.33545/27068919.2021.v3.i2a.770.

5 https://www.yara.com/crop-nutrition/crop-and-agronomy-knowledge/why-is-fertilizer-important/ (Last accessed 05.02.25)

6 https://eos.com/blog/types-of-fertilizers/ (Last accessed 05.02.25)

7 Shishir, S. (2015). Chapter 7: Synthetic fertilizers: role and hazards. In: *Fertilizer technology I synthesis*. Edition I, Studium Press LLC, USA, pp. 176–199. https://doi.org/10.13140/RG.2.1.4820.1688.

8 https://www.syfert.net/resources/nitrogen-fertilizers-types-benefits-and-application-tips.html (Last accessed 05.02.25)

9 https://www.cropnutrition.com/nutrient-management/nitrogen/ (Last accessed 09.02.25)

10 Ahmed, M., Rauf, M., Mukhtar, Z., et al. (2017). Excessive use of nitrogenous fertilizers: an unawareness causing serious threats to environment and human health. Environmental Science and Pollution Research, Volume 24, Issue 35, pp. 26983–26987. https://doi.org/10.1007/s11356-017-0589-7.

11 https://unacademy.com/content/nda/study-material/chemistry/artificial-fertilisers-and-its-types-in-chemistry/ (Last accessed 07.02.25)

12 Environmental, Health, and Safety Guidelines, for Nitrogenous Fertilizer Production. Report number 112800, World Bank Group, April 30, 2007. https://documents1.worldbank.org/curated/ru/824921487313555203/pdf/112800-WP-ENGLISH-Nitrogenous-Fertilizers-PUBLIC.pdf (Last accessed 07.02.25)

13 Yu, C., Huang, X., Chen, H., et al. (2019). Managing nitrogen to restore water quality in China. Nature, Volume 567, Issue 7749, pp. 516–520. https://doi.org/10.1038/s41586-019-1001-1.

14 https://www.cropnutrition.com/nutrient-management/phosphorus/ (Last accessed 07.02.25)

15 https://eos.com/blog/phosphorus-fertilizers/ (Last accessed 07.02.25)
16 https://icl-growingsolutions.com/agriculture/categories/potassium-based-fertilizers/ (Last accessed 07.02.25)
17 https://salsaagricola.es/en/what-are-organic-fertilizers-and-why-are-they-used/ (Last accessed 07.02.25)
18 Qiao, D., Liu, H., and Yu, L. (2016). Preparation and characterization of slow-release fertilizer encapsulated by starch-based superabsorbent polymer. Carbohydrate Polymers, Volume 147, pp. 146–154. https://doi.org/10.1016/j.carbpol.2016.04.010.
19 Tariq, A.K.R.H. (2020). Plant micronutrients: deficiency and toxicity management, Springer. https://doi.org/10.1007/978-3-030-49856-6.
20 https://agrinfobank.wordpress.com/2021/06/22/leaching-of-fertilizers-causes-effects-solution-and-extension-role/ (Last accessed 11.02.25)
21 https://www1.agriculture.alberta.ca/$department/deptdocs.nsf/all/epw11920/$FILE/8-1.pdf (Last accessed 12.02.25)
22 https://icid-ciid.org/Knowledge/basic_term/4/Irrigation (Last accessed 12.02.25)
23 https://beyondpesticides.org/dailynewsblog/2018/02/nitrogen-fertilizer-found-significant-source-air-pollution/ (Last accessed 13.02.25)
24 https://www.epa.gov/acidrain/effects-acid-rain (Last accessed 13.02.25)
25 https://science.howstuffworks.com/nature/climate-weather/atmospheric/acid-rain.htm (Last accessed 13.02.25)
26 Bali, M., and Gueddari, M. (2019). Removal of phosphorus from secondary effluents using infiltration–percolation process. Applied Water Science, Volume 9, Issue 54, pp. 1–8. https://doi.org/10.1007/s13201-019-0945-5.
27 Sonarghare, P.C., Masram, S.C., Sonparote, U.R., et al. (2020). Causes and effects of eutrophication on aquatic life (a review). International Journal for Environmental Rehabilitation and Conservation, Volume XI, Issue SP2, pp. 213–218. https://eoi.citefactor.org/10.11208/essence.20.11.SP2.147.
28 https://www.etl-technology.de/en/anthropogenic-eutrophication (Last accessed 14.02.25)
29 https://www.cnrs.fr/sites/default/files/download-file/Eutrophication_short-version_ENG_0.pdf (Last accessed 14.02.25)
30 Ahmed, T., Haque M.K., Haque, M.M., et al. (2024). Lake eutrophication mechanism and control: current status and future tendency. Life Science Studies, Volume 1, pp. 10–31.
31 Costa, J.A., Souza, J.P., Teixeira, A.P., et al. (2018). Eutrophication in aquatic ecosystems: a scientometric study. Acta Limnologica Brasiliensia, Volume 30, e2. https://doi.org/10.1590/S2179-975.3016.
32 https://oceanservice.noaa.gov/hazards/hypoxia/ (Last accessed 15.02.25)

33 Bunting, L., Leavitt, P., Simpson, G., et al. (2016). Increased variability and sudden ecosystem state change in Lake Winnipeg, Canada, caused by 20th century agriculture. Limnology and Oceanography, Volume 61, pp. 2090–2107. https://doi.org/10.1002/lno.10355.

34 Azevêdo, D., Barbosa, J., Gomes W., et al. (2015). Diversity measures in macroinvertebrate and zooplankton communities related to the trophic status of subtropical reservoirs: contradictory or complementary responses? Ecological Indicators, Volume 50, pp. 135–149. https://doi.org/10.1016/j.ecolind.2014.10.010.

35 Vadeboncoeur, Y., Jeppesen, E., Zanden, M.J.V., et al. (2003). From Greenland to green lakes: cultural eutrophication and the loss of benthic pathways in lakes. Limnology and Oceanography, Volume 48, pp. 1408–1418. https://doi.org/10.4319/lo.2003.48.4.1408.

36 https://www.epa.gov/habs/learn-about-harmful-algae-cyanobacteria-and-cyanotoxins (Last accessed 16.02.25)

37 https://www.noaa.gov/what-is-harmful-algal-bloom (Last accessed 16.02.25)

38 https://oceanservice.noaa.gov/hazards/hab/ (Last accessed 16.02.25)

39 https://www.epa.gov/nutrientpollution/problem (Last accessed 16.02.25)

40 Figgatt, M., Hyde, J., Dziewulski, D., et al. (2017). Harmful algal bloom-associated illnesses in humans and dogs identified through a pilot surveillance system – New York, 2015. MMWR Morbidity and Mortality Weekly Report, Volume 66, Issue 43, pp. 1182–1184. https://doi.org/10.15585/mmwr.mm6643a5.

41 Zhidkova, A.Y., Podberesnij, V.V., Zarubina, R.V., et al. (2020). The effect of eutrophication on human health on the example of the Gulf of Taganrog of the Sea of Azov. IOP Conference Series: Earth and Environmental Science, Volume 548, p. 052053. https://doi.org/10.1088/1755-1315/548/5/052053.

42 Scott, W.E., Steenderen, R.A.V., and Welch D.I. (1985). Health aspects of eutrophication, National Institute for Water Research.

43 https://us.envguide.com/eutrop/#:~:text=Drinking%2C%20accidentally%20swallowing%20or%20swimming,cause%20serious%20health%20problems%20including%3A&text=Rashes&text=Stomach%20or%20liver%20illness&text=Respiratory%20problems (Last accessed 17.02.25)

44 https://e360.yale.edu/digest/worsening-algae-blooms-could-significantly-increase-global-methane-emissions (Last accessed 17.02.25)

45 Xu, H., Liu, Yi., Tang, Z., et al. (2020). Methane production in harmful algal blooms collapsed water: the contribution of non-toxic microcystis aeruginosa outweighs that of the toxic variety. Journal of Cleaner Production, Volume 276, p. 124280. https://doi.org/10.1016/j.jclepro.2020.124280.

46 https://courses.lsa.umich.edu/healthy-oceans/group-3/economic-impacts-of-eutrophication/ (Last accessed 17.02.25)

47 Harper, D. (1992). The engineering, economic and social effects of eutrophication. In: Eutrophication of freshwaters. Springer, Dordrecht. https://doi.org/10.1007/978-94-011-3082-0_5.

48 U.S. Environmental Protection Agency. (2015). A compilation of cost data associated with the impacts and control of nutrient pollution. Technical Report.

49 https://www.open.edu/openlearn/nature-environment/environmental-studies/eutrophication/content-section-4.4.1 (Last accessed 18.02.25)

50 Zhang, Y., Luo, P., and Zhao, S. (2020). Control and remediation methods for eutrophic lakes in the past 30 years. Water Science and Technology, Volume 81, Issue 6, pp. 1099–1113.

51 Svrcek, C., and Smith, D.W. (2004). Cyanobacteria toxins and the current state of knowledge on water treatment options: a review. Journal of Environmental Engineering and Science, Volume 3, pp. 155–185. https://doi.org/10.1139/s04-010.

52 Cooke, G.D., Welch, E.B., Martin, A.B., et al. (1993). Effectiveness of Al, Ca, and Fe salts for control of internal phosphorus loading in shallow and deep lakes. Hydrobiology, Volume 253, pp. 323–335.

53 Jeppesen, E., Søndergaard, M., and Lauridsen, T.L. (2012). Biomanipulation as a restoration tool to combat eutrophication. Recent Advances and Future Challenges Advances in Ecological Research, Volume 47. https://doi.org/10.1016/B978-0-12-398315-2.00006-5.

54 McQueen, D.J. (1998). Freshwater food web biomanipulation: a powerful tool for water quality improvement, but maintenance is required. Lake and Reservoir Management, Volume 3, pp. 83–94. https://doi.org/10.1111/j.1440-1770.1998.tb00035.x.

55 Søndergaard, M., Liboriussen, L., Pedersen, A.R., et al. (2008). Lake restoration by fish removal: long-term effects in 36 Danish lakes. Ecosystems, Volume 1, pp. 1291–1305. https://doi.org/10.1007/s10021-008-9193-5.

56 Lazzaro, X., and Starling, F. (2005). Using biomanipulation to control eutrophication in a shallow tropical urban reservoir (Lago Paranoá, Brazil). In: Reddy MV (ed). Restoration and management of tropical eutrophic lakes. Oxford & IBH Publ. Co. Pvt. Ltd., New Delhi and Science Publishers Inc., New Hampshire, pp. 361–387.

57 Skov, C., and Nilsson, P.A. (2007). Evaluating stocking of YOY pike *Esox lucius* as a tool in the restoration of shallow lakes. Freshwater Biology, Volume 52, pp. 1834–1845. https://doi.org/10.1111/j.1365-2427.2007.01795.x.

58 Starling, F.L.R.M. (1993). Control of eutrophication by silver carp (*Hypophthalmichthys molitrix*) in the tropical Paranoá Reservoir (Brasília, Brazil): a mesocosm experiment. Hydrobiologia, Volume 257, pp. 143–152. https://doi.org/10.1007/BF00765007.

59 Lauridsen, T.L., Sandsten, H., and Møller, P.H. (2003). The restoration of a shallow lake by introducing *Potamogeton* spp. The impact of waterfowl grazing. Lakes and Reservoirs: Research and Management, Volume 8, pp. 177–187. https://doi.org/10.1111/j.1440-1770.2003.00224.x.

60 Roy, E.D., Martin, J.F., Irwin, E.G., et al. (2010). Transient social ecological stability: the effects of invasive species and ecosystem restoration on nutrient management compromise in Lake Erie, Ecology and Society, Volume 15, Issue 1. https://doi.org/10.5751/ES-03184-150120.

61 Hanlon, S.G., Hoyer, M.V., Cichra, C.E., et al. (2000). Evaluation of macrophyte control in 38 Florida lakes using triploid grass carp. Journal of Aquatic Plant Management, Volume 38, pp. 48–54.

62 Koul, B., Sharma, K., and Shah, M.P. (2022). Phycoremediation: a sustainable alternative in wastewater treatment (WWT) regime. Environmental Technology and Innovation, Volume 25, p. 102040. https://doi.org/10.1016/j.eti.2021.102040.

63 Apandi, N.M., Mohamed, R.M.S.R., Latiffi, N.A.A., et al. (2017). Protein and lipid content of microalgae Scenedesmus sp. biomass grown in wet market wastewater. MATEC Web Conference, Volume 103, p. 06011. https://doi.org/10.1051/matecconf/201710306011.

64 Liu, C., Subashchandrabose, S., Ming, H., et al. (2016). Phycoremediation of dairy and winery wastewater using *Diplosphaera* sp. MM1. Journal of Applied Phycology, Volume 28, Issue 6, pp. 3331–3341. https://doi.org/10.1007/s10811-016-0894-4.

65 Razak, S.B.A., and Sharip, Z. (2019). The potential of phycoremediation in controlling eutrophication in tropical lake and reservoir: a review. Desalination and Water Treatment, Volume 180, pp. 164–173. https://doi.org/10.5004/dwt.2020.25078.

66 Ansari, A.A., Trivedi, S., Khan, F.A., et al. (2015). Phytoremediation of eutrophic waters, phytoremediation. In: Ansari A, Gill S, Gill R, et al. (eds) Phytoremediation. Springer, pp. 41–50. https://doi.org/10.1007/978-3-319-10969-5_4.

67 Nahar, K., and Hoque, S. (2021). Phytoremediation to improve eutrophic ecosystem by the floating aquatic macrophyte, water lettuce (*Pistia stratiotes* L.) at lab scale. Egyptian Journal of Aquatic Research, Volume 47, Issue 2, pp. 231–237. https://doi.org/10.1016/j.ejar.2021.05.003.

68 Patel, H.A., and Sahoo, S. (2020). A review of water quality improvement with the help of aquatic macrophytes. Current World Environment, Volume 15, Issue 3. https://doi.org/10.12944/CWE.15.3.04.

69 Alam, A.R., and Hoque, S. (2017). Phytoremediation of industrial wastewater by culturing aquatic macrophytes, *Trapa natans* L. and *Salvinia cucullata* Roxb. Jahangirnagar University Journal of Biological Sciences, Volume 6, Issue 2, pp. 19–27. https://doi.org/10.3329/jujbs.v6i2.36587.

70 Shan, M., Wang, Y., and Xue, S. (2009). Study on bioremediation of eutrophic lake. Journal of Environmental Sciences, Volume 21, Supplement 1, pp. S16–S18. https://doi.org/10.1016/S1001-0742(09)60027-9.

71 https://ecampusontario.pressbooks.pub/microbio/chapter/10-4-bioremediation/ (Last accessed 18.02.25)

72 Kumara, U.M.A., Jayaprada, N.V.T., and Thiruchchelvan, N. (2023). Bioremediation of polluted water. In: Soni R, Suyal DC, Morales-Oyervides L, et al. (eds). Current status of fresh water microbiology. Springer, Singapore. https://doi.org/10.1007/978-981-99-5018-8_14.

73 https://www.epa.gov/nutrientpollution/epas-ongoing-efforts-reduce-nutrient-pollution (Last accessed 21.02.25)

74 FAO (2019). The International Code of Conduct for the Sustainable Use and Management of Fertilizers. Rome. Licence: CC BY-NC-SA 3.0 IGO. https://doi.org/10.4060/CA5253EN (Last accessed 21.02.25)

75 https://www.oecd.org/en/publications/ocean-policies_17318a16-en/european-union-nitrate-pollution-prevention-regulations_af94c88f-en.html (Last accessed 22.02.25)

76 Goodchild, R.G. (1998). EU policies for the reduction of nitrogen in water: the example of the nitrates directive. Environmental Pollution, Volume 102, Issue 1, pp. 737–740. https://doi.org/10.1016/S0269-7491 (98)80106-1.

77 https://www.epa.gov/laws-regulations/history-clean-water-act (Last accessed 22.02.25)

78 https://pib.gov.in/PressReleaseIframePage.aspx?PRID=1947891 (Last accessed 22.02.25)

79 https://serc.carleton.edu/microbelife/topics/deadzone/index.html (Last accessed 22.02.25)

80 Khajuria, A. (2016). Impact of nitrate consumption: case study of Punjab, India. Journal of Water Resource and Protection, Volume 8, pp. 211–216. https://doi.org/10.4236/jwarp.2016.82017.

81 Annual Report Pb (2012). Annual report from Punjab Environmental Centre. http://punenvis.nic.in/index2.aspx?slid=201&mid=1&langid=1&sublinkid=447

82 Virk, H.S. (2022). Groundwater contamination in Punjab due to high levels of nitrate (NO^{3-}) and its health hazards: a preliminary report. Research &Reviews: A Journal of Toxicology, Volume 12, Issue 3, pp. 18–26. https://doi.org/10.37591/RRJoT.

83 Chanda, I.K. (2013). Fertilizer consumption: sate wise pattern. Crops. Sustainable fertilizer. https://www.geographyandyou.com/sustainable-fertiliser/fertiliser-consumption-state-wise-pattern (Last accessed 25.02.25)

84 Singh, B., and Sekhon, G.S. (1976). Nitrate pollution of groundwater from nitrogen fertilizers and animal wastes in the Punjab, India. Agriculture and

Environment, Volume 3, Issue 1, pp. 57–67. https://doi.org/10.1016/0304-1131(76)90007-2.

85 https://timesofindia.indiatimes.com/city/chandigarh/ngt-notice-to-punjab-over-water-pollution/articleshow/105055290.cms (Last accessed 25.02.25)

86 World Bank (2020). Toward managing rural drinking water quality in the State of Punjab, India. World Bank, Washington, DC.

87 Liang, H., Zhang, J., Zhang, J., et al. (2024). Unveiling the eutrophication crisis: 20 years of nutrient development in Zhanjiang Bay, China. Frontiers in Marine Science, Volume 11. https://doi.org/10.3389/fmars.2024.1373716.

88 Li, X.Y., Li, B., and Sun, X.L. (2014). Effects of a coastal power plant thermal discharge on phytoplankton community structure in Zhanjiang Bay, China. Marine Pollution Bulletin, Volume 81, pp. 210–217. https://doi.org/10.1016/j.marpolbul.2013.08.006.

89 Lao, Q., Wu, J., Chen, F., et al. (2022). Increasing intrusion of high salinity water alters the mariculture activities in Zhanjiang Bay during the past two decades identified by dual water isotopes. Journal of Environmental Management, Volume 320, p. 115815. https://doi.org/10.1016/j.jenvman.2022.115815.

90 Yu, G., Jia, Y., He, N., et al. (2019). Stabilization of atmospheric nitrogen deposition in China over the past decade. Nature Geoscience, Volume 12, pp. 424–429. https://doi.org/10.1038/s41561-019-0352-4.

91 Ministry of Ecology and Environment of the People's Republic of China (MEE) (2020a). Bulletin of Marine Ecology and Environment Status of China in 2019 (in Chinese).

92 Wang, B., Ming, X., Wei, Q., et al. (2018). A historical overview of coastal eutrophication in the China Seas. Marine Pollution Bulletin, Volume 136, pp. 394–400. https://doi.org/10.1016/j.marpolbul.2018.09.044.

93 Wang, Y., Liu, D., Xiao, W., et al. (2021). Coastal eutrophication in China: trend, sources, and ecological effects. Harmful Algae, Volume 107, p. 102058. https://doi.org/10.1016/j.hal.2021.102058.

94 The State Council the People's Republic of China (2015).GOV, Action Plan for Prevention and Control of Water Pollution.

95 Ministry of Agriculture of the People's Republic of China (2015). MOA, Plan of Actions Aiming for Zero Growth in Synthetic Fertilizer Use from 2020 Onwards.

96 Production Department of Livestock, Ministry of Agriculture and Rural Affairs of the People's Republic of China (2017). MOA, Notice on Action Plan of Animal Manure Recycling from 2017–2020.

97 McCrackin, M.L., Gustafsson, B.G., Hong, B., et al. (2018). Opportunities to reduce nutrient inputs to the Baltic Sea by improving manure use efficiency in agriculture. Regional Environmental Change, Volume 18, pp. 1843–1854. https://doi.org/10.1007/s10113-018-1308-8.

98 Li, A., Strokal, M., Bai, Z., et al. (2019). How to avoid coastal eutrophication – a back-casting study for the North China Plain. Science of the Total Environment, Volume 692, pp. 676–690. https://doi.org/10.1016/j.scitotenv.2019.07.306.

99 https://www2.dmu.dk/1_viden/2_miljoe-tilstand/3_vand/4_eutrophication/nutrient_reduction.asp (Last accessed 28.02.25)

100 https://innovarum.es/en/horizon-europe/eu-action-plan-for-organic-production-what-is-it-about/ (Last accessed 28.02.25)

101 https://agriculture.ec.europa.eu/farming/organic-farming/organic-action-plan_en#organicsintheeu (Last accessed 28.02.25)

102 Kamilya, T., Majumder, A., Yadav, M.K., et al. (2022). Nutrient pollution and its remediation using constructed wetlands: insights into removal and recovery mechanisms, modifications and sustainable aspects. Journal of Environmental Chemical Engineering, Volume 10, Issue 3, p. 107444. https://doi.org/10.1016/j.jece.2022.107444.

103 https://www.epa.gov/wetlands/constructed-wetlands (Last accessed 28.02.25)

104 https://www.wastex.io/post/reduce-fertilizer-use (Last accessed 28.02.25)

105 https://www.wwfbalticfarmer.org/farming-practices/water-management/#:~:text=Sustainable%20water%20management%20is%20a%20balancing%20act&text=Too%20much%20water%20in%20agricultural,up%20nutrients%20from%20the%20soil (Last accessed 28.02.25)

106 Vidanage, P.W. (2024). State of art technologies for nutrient removal from wastewater. In: Yadav S, Ghangrekar MM, Yadava RN (eds) Emerging technologies in biological and hybrid wastewater treatment: lessons from developed to enhancing practices in developing countries. Earth and environmental sciences library, Springer, Cham. https://doi.org/10.1007/978-3-031-74515-7_7.

107 Rim, Y.T., Yang, H.J., Yoon, C.H., et al. (1997). A full-scale test of a biological nutrients removal system using the sequencing batch rector. Water Science and Technology, Volume 35, pp. 241–247.

108 Mahvi, A.H. (2008). Sequencing batch reactor: a promising technology in wastewater treatment. Iranian Journal of Environmental Health Science and Engineering, Volume 5, No. 2, pp. 79–90. https://www.bioline.org.br/request?se08017.

109 Hanaki, K., Wantawin, C., and Ohgaki, S. (1990). Nitrification at low level of dissolved oxygen with and without organic loading in a suspended-growth reactor. Water Research, Volume 12, pp. 297–302. https://doi.org/10.1016/0043-1354(90)90004-P.

110 Chang, C.H., and Hao, O.J. (1996). Sequencing batch reactor system for nutrient removal: ORP and pH profiles. Journal of Chemical Technology and Biotechnology, Volume 67, Issue 1, pp. 27–38. https://doi.org/10.1002/(SICI)1097-4660(199609)67:1<27::AID-JCTB430>3.0.CO;2-2.

111 Hu, J.Y., Ong, S.L., and Ng, W.J. (2005). Use of a sequencing batch reactor for nitrogen and phosphorus removal from municipal wastewater. Journal of Environmental Engineering, Volume 131, Issue 5. https://doi.org/10.1061/(ASCE)0733-9372(2005)131:5(734).

112 https://www.racoman.com/blog/nutrient-removal-wastewater-treatment-explained?srsltid=AfmBOopJ8kzCL66lSWN5gPuH8bCkuGexXaV0pLYS8df3eKW_eu49OzeF (Last accessed 28.02.25)

113 Cotton, M.C., Majdzadehmoghadam, F., and Mulligan, C. (2024). Advancements in nutrient removal technologies for wastewater. In: Yadav S, Ghangrekar MM, Yadava RN (eds) Emerging technologies in biological and hybrid wastewater treatment: lessons from developed to enhancing practices in developing countries. Earth and environmental sciences library, pp. 305–330. https://doi.org/10.1007/978-3-031-74515-7_16.

114 Mainardis, M., Buttazzoni, M., and Goi, D. (2020). Up-Flow anaerobic sludge blanket (UASB) technology for energy recovery: a review on state-of-the-art and recent technological advances. Bioengineering, Volume 7, p. 43. https://doi.org/10.3390/bioengineering7020043.

115 Díaz, V., Leyva-Díaz, J.C., Almécija, M.C., et al. (2022). Microalgae bioreactor for nutrient removal and resource recovery from wastewater in the paradigm of circular economy. Bioresource Technology, Volume 363, p. 127968. https://doi.org/10.1016/j.biortech.2022.127968.

116 Sukla, L.B., Subudhi, E., and Pradhan, D. (eds) (2019). The role of microalgae in wastewater treatment, Springer Singapore, Singapore.

117 Sarma, S.J., and Tay, J.H. (2018). Aerobic granulation for future wastewater treatment technology: challenges ahead. Environmental Science Water Research and Technology, Volume 4, pp. 9–15. https://doi.org/10.1039/C7EW00148G63.

118 Ugwuanyi, E.D., Nwokediegwu, Z.Q.S., Dada, M.A., et al. (2024). Review of emerging technologies for nutrient removal in wastewater treatment. World Journal of Advanced Research and Reviews, Volume 21, Issue 02, pp. 1737–1749.

119 Rathi, B.S., and Kumar, P.S. (2021). Application of adsorption process for effective removal of emerging contaminants from water and wastewater. Environmental Pollution, Volume 280, p. 116995. https://doi.org/10.1016/j.envpol.2021.116995.

120 OboteyEzugbe, E., and Rathilal, S. (2020). Membrane technologies in wastewater treatment: a review. Membranes, Volume 10, Issue 5, p. 89. https://doi.org/10.3390/membranes10050089.

121 Faroon, M.A., Alsaad, Z.A., Albadran, F.A., et al. (2023). Review on technology-based on reverse osmosis. Anbar Journal of Engineering Sciences. https://doi.org/10.37649/aengs.2023.139414.1047.

122 https://www3.epa.gov/npdes/pubs/sbr_new.pdf (Last accessed 01.03.25)

123 https://www.epa.gov/sites/default/files/2019-08/documents/membrane_bioreactor_fact_sheet_p100il7g.pdf (Last accessed 01.03.25)

124 https://pankajiitkgp.github.io/uasbr/assets/pdf/uasbr/uasbr3.pdf (Last accessed 01.03.25)

125 Maslaki, S.A., and Kouchaksaraei, A.H. (2023). Grand challenges in application of microalgae for nutrient recovery. Journal of Resource Recovery, Volume 1, p. 1019. http://doi.org/10.52547/jrr.2306.1015.

126 Rosa-Masegosa, A., Muñoz-Palazon, B., Gonzalez-Martinez, A., et al. (2021). New advances in aerobic granular sludge technology using continuous flow reactors: engineering and microbiological aspects. Water, Volume 13, Issue 13, p. 1792. https://doi.org/10.3390/w13131792.

127 https://www3.epa.gov/npdes/pubs/free_water_surface_wetlands.pdf (Last accessed 01.03.25)

128 Rajendran, S., Priya, A.K., Kumar, P.S., et al. (2022). A critical and recent developments on adsorption technique for removal of heavy metals from wastewater– a review. Chemosphere, Volume 303, p. 135146.

129 Hariz, H.B., Montemezzani, V., Shim, Y., et al. (2025). Nutrient removal performance of novel filamentous algae nutrient scrubber (FANS) configurations. Algal Research, Volume 86, p. 103937. https://doi.org/10.1016/j.algal.2025.103937.

130 Tari, K., Samarghandi, M.R., Shokoohi, R., et al. (2025). Nutrient removal performance and microbial composition analysis in hybrid membrane bioreactor for municipal wastewater treatment. Bioprocess and Biosystems Engineering, Volume 48, pp. 665–678. https://doi.org/10.1007/s00449-025-03135-3.

131 Li, J., Gao, Y., Xie, Y., et al. (2025). Optimal cultivation concentration of duckweed for pollutant removal from biogas slurry. Scientific Reports, Volume 15, p. 5193. https://doi.org/10.1038/s41598-025-89524-w.

132 Kumwimba, M.N., Huang, J., and Dzakpasu, M. (2023). Enhanced nutrient removal in agro-industrial wastes-amended hybrid floating treatment wetlands treating real sewage: laboratory microcosm to field-scale studies. Chemosphere, Volume 330, pp. 1–13. https://doi.org/10.1016/j.chemosphere.2023.138703.

133 Paul, T., Nath, P., Tapadar, S., et al. (2024). Growth potential, biochemical properties and nutrient removal efficiency of some freshwater microalgae and their consortia from wastewater. International Journal of Phytoremediation, Volume 27, Issue 2, pp. 145–155. https://doi.org/10.1080/15226514.2024.2405001.

134 Lee, J., Antonini, G., Al-Omari, A., et al. (2024). Electrochemical methods for nutrient removal in wastewater: a review of advanced electrode materials, processes, and applications. Sustainability, Volume 16, Issue 22, p. 9764. https://doi.org/10.3390/su16229764.

135 Merma, A.G., Santos, B.F., Rego, A.S.C., et al. (2020). Treatment of oily wastewater from mining industry using electrocoagulation: fundamentals and process optimization. Journal of Materials Research and Technology, Volume 9, pp. 15164–15176. https://doi.org/10.1016/j.jmrt.2020.10.107.

136 Ken, D.S., and Sinha, A. (2021). Dimensionally stable anode (Ti/RuO_2) mediated electro-oxidation and multi-response optimization study for remediation of coke-oven wastewater. Journal of Environmental Chemical Engineering, Volume 9, p. 105025. https://doi.org/10.1016/j.jece.2021.105025.

137 Wang, C., Li, T., Yu, G., et al. (2021). Removal of low concentrations of nickel ions in electroplating wastewater by combination of electrodialysis and electrodeposition. Chemosphere, Volume 263, p. 128208. https://doi.org/10.1016/j.chemosphere.2020.128208.

138 Yakamercan, E., Bhatt, P., Aygun, A., et al. (2023). Comprehensive understanding of electrochemical treatment systems combined with biological processes for wastewater remediation. Environmental Pollution, Volume 330, p. 121680. https://doi.org/10.1016/j.envpol.2023.121680.

139 Hu, Q., He, L., Lan, R., et al. (2023). Recent advances in phosphate removal from municipal wastewater by electrocoagulation process: a review. Separation and Purification Technology, Volume 308, p. 122944. https://doi.org/10.1016/j.seppur.2022.122944.

140 Curcio, G.M., Limonti, C., Siciliano, A., et al. (2022). Nitrate removal by zero-valent metals: a comprehensive review. Sustainability, Volume 14, Issue 8, p. 4500. https://doi.org/10.3390/su14084500.

141 Siciliano, A., Limonti, C., Curcio, G.M., et al. (2020). Advances in struvite precipitation technologies for nutrients removal and recovery from aqueous waste and wastewater. Sustainability, Volume 12, Issue 18, p. 7538. https://doi.org/10.3390/su12187538.

142 Razzak, S.A. (2024). Recent advances in sustainable biological nutrient removal from municipal wastewater. Cleaner Water, Volume 2, p. 100047. https://doi.org/10.1016/j.clwat.2024.100047.

143 Ramanan, R., Kim, B.H., Cho, D.H., et al. (2016). Algae-bacteria interactions: evolution, ecology and emerging applications. Biotechnological Advances, Volume 34, pp. 14–29. https://doi.org/10.1016/j.biotechadv.2015.12.003.

144 Zhou, Y., Zhu, Y., Zhu, J., et al. (2023). A comprehensive review on wastewater nitrogen removal and its recovery processes. International Journal of Environmental Research and Public Health, Volume 20, Issue 4, p. 3429. https://doi.org/10.3390/ijerph20043429.

14

Microplastics

After the discovery of Bakelite in 1889, many inexpensive methods for improving the production of thermosetting plastics and thermoplastics have been developed widely. These techniques have enhanced the properties like weight, the resilience, impermeability, and corrosion resistance, increasing their applicability and resourcefulness. Plastics are essential components in lucrative industries like packaging services, construction and real estate development, logistics, equipment for science and engineering, sustainable energy sources, healthcare, and many more. With this breakthrough, large-scale production of goods involving plastics began in the 1940s. As a result, in the years that followed, the manufacture of plastic products and their use globally increased rapidly. Plastic production mounted from half a billion tons in the 1950s to a whopping 335 million metric tons in 2016, entering almost every facet of the environment, including more than 10% in marine waters and freshwater bodies. More than 320 million tons of plastics are manufactured in industries each year, of which more than 40% are used as single-use packaging, contributing to plastic waste [1]. An estimated 5–13 million tons of plastic are thought to find their way into the ocean annually, most of it from resources such as landfills and the rest from other human-induced activities like fisheries [2]. Some potential sources of plastics in water bodies are from freshwater runoffs from drainage systems, tributaries, industrial spillages, and urban and industrial discharges. Recent studies indicate that around 95% of the more than 300 million tons of plastics generated yearly, worldwide, end up getting dumped in the oceans [3].

Until 2017, the yearly global plastic manufacturing was 348 million tons in total, which is significantly higher than in the early 1950s. The most frequently found polymers in seawater are acrylic, polyester, nylon, polyurethanes, polylactic acid, and polycarbonates. Other generally used polymers include polyethylene terephthalate (PET), polyvinyl chloride (PVC), low-density polyethylene (LDPE), high-density polyethylene (HDPE), polypropylene (PP), and polystyrene (PS) [4].

Analytical techniques like Fourier transform infrared spectroscopy (FTIR) have already confirmed their presence in marine and surface waters [5–7].

Long-chain polymers are also referred to as plastics. Plastics are used in various applications, such as packing, home and personal care goods, and industrial construction materials, owing to their lightweight, large strength-to-weight ratio, thermal, and biological degradation-resistant qualities [8]. Surrounding water reservoirs comprise of plastic garbage items like bottle caps, cosmetics tubes, clothes, styrofoam, syringes, along with the usual food wrappers, plastic shopping bags, and plastic films. Because of their excellent performance, low density, and many applications in the textile, automobile industry, construction, consumer goods, electric and electronic devices, pharma, food, and packaging industries, plastic materials are in extremely high demand on a global scale [9]. Polymers are ubiquitous contaminants in all aquatic ecosystems worldwide and are difficult to biodegrade. These break down into microplastics (MPs) and nanoplastics via an array of processes [9, 10]. General specifications of plastics in the surrounding environment with the potential of forming MPs are presented in Table 14.1.

14.1 Challenges in the Recycling of Plastics

Plastic pre-sorting before recycling is expensive, laborious, and energy-consuming. Production of low-grade polymers and many polymeric materials that cannot be managed via existing technologies is a major issue. The only extensively used technology for the extensive treatment of solid plastic waste is mechanical recycling. Thus, materials can only withstand a certain number of processing cycles before the ductility and molecular weight of the polymer are significantly reduced [13]. The process initially involves cleansing the organic residue, followed by shredding, melting, and remolding of the polymer, which is then mixed with a virgin polymer of the same type to produce an acçeptable material. Processing thermosensitive polymers, composites, and plastics that do not flow at high temperatures is not possible mechanically. Subsequently, only polyethylene terephthalate (PET) and polyethylene plastics, which account for 10–40% of the annual plastic production, are recovered and recycled using mechanical techniques. Because of energy expenditures, chemical recycling is not a standard method [14]. Fishing nets are one of the biggest sources of ocean plastic garbage, since their equipment typically has a limited period and casts off, is disposed off, or lost in seawater. Recycled resins of polyolefin from fishing nets, particularly those containing polyethylene (PE), have poor characteristics due to the development of impurities and considerable degradation. The usage of such recovered resins is thus constrained [15]. Lacking proper recycling methods is the major reason for the conversion of plastics to MPs.

Table 14.1 Plastics with the potential to form microplastics.

Sources of plastic	Type of plastic	Abbreviation	Specific gravity	Plastic to MPs conversion percentage
Soft drinks, water, juice, beer bottles	Polyester	PET	1.37	7%
Plastic film, bottles, cups, window cleaner, and dispensers of detergent, haircare products, cooking oil, food wrapping, medical kit, and boots	Polyvinyl chloride	PVC	1.38	18–19%
Single-use tableware, meat serving dishes, egg containers, carry-out containers, bottles	Polystyrene	PS	1.05	6%
Ropes, bottle lids, nets, car bumpers, flowerpots, folders	Polypropylene	PP	0.85–0.93	22–24%
Milk and juice jugs, household cleanser bottles, butter and yogurt containers	High-density polyethylene	HDPE	~0.94	17%
Plastic bags, rings, nets, drinking straws, wire shackles	Low-density polyethylene	LDPE	~0.9–0.93	21%
Cigarette filters, DVDs, goggles, iPod and computer gear, signs and displays, nylon, milk flasks for babies, bulletproof materials, netting and traps	Miscellaneous (nylon, cellulose acetate)	–	0.8–1.4	2–3%

Source: Adapted from [11, 12].

14.2 Plastics to Microplastics

Plastics can further disintegrate or fragment into tiny particles with diameters of millimeters, micrometers, and even nanometers when released into the environment by physicochemical and biological processes. Between 60 and 80% of plastic trash is in the form of particles. Moreover, industrial and consumer goods like paints, cosmetics, and body wash frequently contain micro- or nano-sized plastic microspheres that could release hazardous chemicals into the environment after usage. Eventually, finer plastic particles are already omnipresent in the environment [16]. Primarily, microplastics (MPs) are polymeric particles having a size less than 5 mm and are discharged into the environment directly. These particles also include pieces of fiber from laundry done at home. Scrubbing, bleaching, sanding, and other processes that degrade fibers and their interactions might exacerbate the fiber fragments during wet processes. Various mechanical and chemical textile finishing processes can be used to explain how different aspects of the fabric, such as type, weave or knitting pattern, type of detergent used, concentration, operating pH, temperature, spinning speed, and drying and washing time, affect the fiber fragment release [17]. Figure 14.1 depicts some of the sources of MPs contamination.

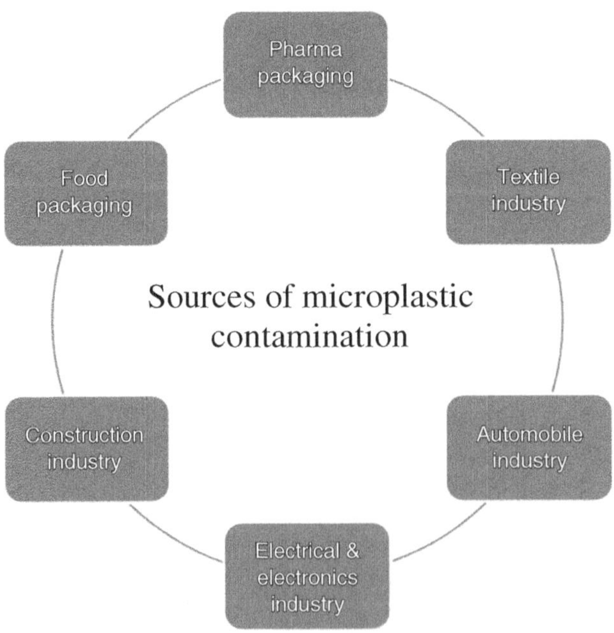

Figure 14.1 Sources of microplastic contamination.

Table 14.2 Microplastic distribution in the oceans of the world.

Ocean	Distribution (%)	Concentration	References
North East Atlantic Ocean	89	2.46 MPs/m^3	[20]
Arctic region polar waters	95	0–1.31 MPs/m^3	[21, 22]
The Laurentian Great Lake in Canada	20	~45000 MPs/km^2	[23]
Jade Bay north-east coast, Germany and the southern North Sea	70	1770 MPs/L	[24]
Northwest Atlantic	60	2500 MPs/km^2	[25]
Portuguese Coast	53	332–362 $/m^2$	[26]
Mediterranean Sea	74	0.90 ± 0.10 MPs/g	[27]
Yangtze estuary, including the East China Sea	90	0–144 MPs/m^3	[28]
Beaches of Guanabara Bay in southeast Brazil	56	12–1300 MPs/m^2	[29]

Source: [19]/with permission of Elsevier.

The distribution of MPs in several oceans of the world has been tabulated in Table 14.2. Marine plastic debris is primarily derived from terrestrial sources. This enters the oceans from land, and around 80% of it is human-made trash. The prominent sources of microplastics include sewage effluents, civic drainage systems, tire wear, poorly managed plastic waste on islands, or plastics being directly dumped on the coastline. The plastic pollution in the ocean is also a result of lost cargo shipping containers and marine aquaculture fishing fleets, which are again major sources of MPs [18]. The distribution of MPs in different oceans is shown in Table 14.2.

14.3 Toxicity Assessment of MPs

MPs are ingested by numerous aquatic species in both fresh and marine water, get incorporated in the energy chain, and can pose threats to human wellbeing, as contaminants in the wider and more diversified environment. Documentary evidences are available for the consumption of MPs by fish, turtles, seagulls, worms, and crabs. This includes MPs being present across the entire marine food chain. Polycarbonate and epoxy resins use bisphenol A (BPA) and are the best examples of leaching monomers. BPA, due to its estrogenic activity (EA), has demonstrated to having negative effects on humans, including numerous metabolic diseases, as well as behavioral and sexual issues. Poly (lactic-co-glycolic acid) MPs (0.2%), with a diameter of 3 μm were minimally consumed by humans and were present in colon mucosal tissue, according to an in-vitro investigation. When compared to healthy controls, the colon mucosal tissue of patients with severe illnesses, like inflammatory bowel disease (IBD), demonstrated faster transportation (0.45% versus 0.2%) and higher intestinal permeability. The preferred uptake pathway for MPs may be endocytosis or phagocytosis due to their tiny size. Using macrophages in the intestinal epithelium, phagocytosis may also happen when there are particles larger than 0.5 μm [4, 1]. Polystyrene particles that enter the human placenta are evidenced to move within the human body. Muscle cells contract and suffer cellular injury as a result of polystyrene particles. As a result of PS microsphere exposure, reactive oxygen species are produced, which affect the human brain and epithelial cells. Thus, MPs enter the human body mainly through fish and foods originating from the ocean [4]. Table salt is another significant aspect of marine origin that has been the subject of studies and may be impacted by marine pollution from plastic waste. Earlier this year, two studies assessed the amount of plastic waste in table salts. Plastic debris measurements were done on 17 salt varieties from 8 nations, and the results showed that MPs (170–520 m) were discovered in nearly all salts from all manufacturers, at amounts ranging from 1 to 10 MPs/kg of salt [30].

When animals, notably apex predators, consume these tiny plastic particles (MPs), these can have detrimental effects on aquatic ecosystems by increasing the concentration of other pollutants in the system and decreasing fitness and mortality [31]. Zooplankton is a vital food source for a large number of second-generation producers, including fish and aquatic mammals (cetaceans) that are important for the functioning of the economy. MPs have demonstrated to disrupt the feeding appendages of zooplanktons, restrict food intake, and even potentially harm or block their alimentary canals. Low egg production in copepods may result from a lack of nourishment. An energy deficit may result from a decrease in feeding behavior and a subsequent decrease in food intake. An obstructed or weakened digestive tract or insufficient nutrition resulting from reduced meals may cause a persistent loss of energy inputs and, ultimately, death [2]. Long-term exposure to ingested MPs in fish guts has the potential to cause malnutrition and a decline in fish numbers. Mackerels are more vulnerable because of their feeding environment, which consists mostly of the pelagic zone and surface waters, in which it is possible to consume resilient or neutrally floating MP particles. In laboratory settings, young fish that have consumed MPs die at a considerably higher rate than the control group. Benthic sediments include MPs, such as microfibers. Thus, it has also been noted that benthic invertebrates such as sea cucumbers, amphipods, blue mussels, and lugworms (*Arenicola marina*), among others, absorb MPs. Blue mussels exposed to MPs have revealed physical repercussions such as inflammation and oxidative stress, as evidenced by granulocytosis, instability in the membrane of lysosomes, increased phagocytic activity, and epithelial cell death. Additionally, lugworms that consumed MPs showed decreased eating, stamina, and weight loss. Across the world, seabirds are impacted by ingesting plastic. When compared to macroplastic ingestion, the effects of ingesting MPs do not always result in the rapid death of seabirds, because they forage on the surface. Species of seagulls, *Petrels*, seabirds, and *Fulmarus* consume MPs, but species that feed on zooplankton, including the tiny dovekie (*Allealle*), are more susceptible towards ingesting MPs. Dovekies cannot distinguish between zooplankton and MPs, which causes them to consume MPs directly. Sea turtle consumption of macroplastics has been documented in numerous studies. A case of mistaken identity may be to be blamed for sea turtles ingesting MPs. Invertebrate grazers and the microbial biofilm that grow on MPs may encourage sea turtles to eat these materials due to senses like taste and smell, in addition to the visual element. Partitioning allows MPs to absorb persistent organic pollutants (POPs) in seawater in minimum quantities. The hydrophobicity of POP makes it possible for them to accumulate in the litter of MPs more than in saltwater. A plausible pathway for POPs to permeate the marine food web is through the consumption of these contaminated plastics by marine creatures. MPs swallowed are never digested or absorbed by any of these organisms because no enzyme pathways exists to break down the synthetic polymers; thus, they are biologically inert. Consumption of MPs by the microbial community may result in issues such as the leaching of noxious additives used for compounding plastic or residual manufacturing monomers present in the plastics, and toxicity of some intermediates from partial degradation of plastics. Together with organic pollutants such as insect repellents, industrial chemicals, and pesticides entering the ocean via wastewater and effluents, POPs found in saltwater are gradually engrossed and condensed in MPs [11]. Prior research has revealed that one mechanism for contaminants to reach organisms is through the ingestion of plastics with sorbed pollutants [12]. This might be unfavorable for the growth and subsequent development to adulthood, for early larval stages. Ingestion of MPs can impair physical development, in addition to growth. A wide variety of marine creatures have been seen ingesting MP particles in oceanic places across the world. Majority of the times, marine creatures accidentally swallow MPs, frequently mistaking them for food. However, some organisms may target particular MPs for consumption. Marine species are harmed chemically and physically when they consume MPs. The ingestion of MPs by aquatic species may be due to circumstances like the polymers adhering to their body surfaces, hindering their mobility, and even blocking the gastro intestinal region. It may affect them through inflammation and hepatic stress, which leads to reduced growth. MPs can be consumed by a wide variety of aquatic creatures of different trophic levels, which includes fishes, fish-eating birds, turtles, amphipods,

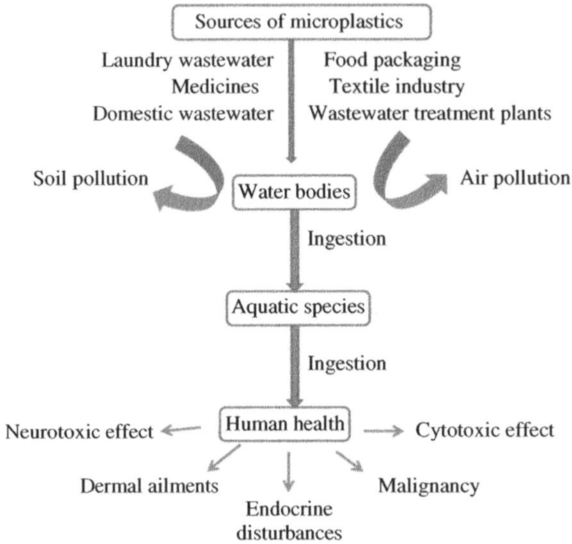

Figure 14.2 Sources, pathways, and effects of microplastics.

and mammals, invertebrates, especially lugworms, barnacles, mussels, sea cucumbers, and zooplankton. When MPs are swallowed by copepods, zooplanktons, and other lower trophic animals, the particles have the potential to travel up the food chain, upsetting the balance of life on Earth [19]. It has been discovered that POPs absorbed onto MPs are metabolized by marine microorganisms. POPs and heavy metal-coated MPs are transported across oceans and thus pollute the adjacent ecosystems. The marine biota, which moves up the food chain, may consume these plastic particles. These harmful contaminants can cause birth deformities, cancer, endocrine disruption, immune system disorders, and child development problems, for every organism in the food chain (Figure 14.2).

In reaction to high polylactic acid (PLA) MP dosages, *Ostrea edulis* respiration rates increased, indicating stress in the oysters. The biomass and diversity of related benthic organisms, such as peppery furrow shell clams (*Scrobicularia plana*), isopods (*Idotea balthica*), and periwinkles (*Littorina* sp.), also decreased. The decrease was related to reduced nutrition as well as reduced reproduction output and morbidity, brought about by MP intake [19]. Bivalves and mussels are popular marine meals in Europe. Due to their feeding mechanics, mollusks can expel bigger particles as pseudo-feces before ingesting them [32]. MPs and microbeads have been discovered to be lodged in the tissues and organs of mussels, increasing the likelihood of bioaccumulation and reaching higher trophic levels, including human beings. After intracellular MP absorption, oxidative stress and lysosomal disruptions were seen in bivalve species. According to a different study on coastal dolphins, organisms connected to the ocean bottom are recognized to consume more MP waste compared to those connected to the surface. These studies show the existence of MPs in the intestines of marine mammals; however, they do not discuss any potential effects on these creatures. Studies on the effects of MPs on these creatures are not possible due to their longer lifespan and larger bodies [4]. Table 14.3 gives the various categories of microplastics that have been associated with different types of microbes. Microplastic ingestion by marine organisms and allied effects are mentioned in Table 14.3.

14.4 Legislation for MP Management

MPs, as described by the Joint Group of Experts on the Scientific Aspects of Maritime Environmental Protection (GESAMP), are "plastic particles <5 mm in diameter, which include particles in the nano-size range (1 nm)" [72]. In a study, MP concentrations were calculated in laundry and personal care products, along with potential routes and sources of transmission into the southern region of latitude 60 °S. The estimations revealed that the quantity of microplastic pollution in this region from ships and scientific research stations was likely to be minimal at the size of the Southern Ocean, but potentially noteworthy on a local scale. The Antarctic Treaty System supervises the southern region of latitude 60 °S. The Protocol of the Antarctic Treaty on Environmental Protection came into force in 1998, which included particular annexes on the management and disposal of marine waste (Annex III) and the inhibition of marine pollution (Annex IV). It is illegal for ships to discharge wastewater within twelve nautical miles of the coast, however, organizations are not obliged to clean up wastewater generated from their scientific research stations. Environmental Impact Assessments (EIAs) must be conducted for all activities inside the Antarctic Treaty region, as required by Annex 1 of EIAs. Although MP pollution is although not particularly mentioned in any international maritime regulation, microplastic-related issues still need to be addressed [73]. In 1990, stringent laws and regulations prohibiting the disposal of marine waste and strengthened raw material recycling management in the plastics industry stabilized, and even decreased the amount of plastic pellets entering terrestrial and marine habitats. Still, massive amounts of MPs have been discovered on an almost daily basis due to the deterioration of already-existing plastic waste in the environment [74]. Reducing the quantity of MPs entering the water environment requires identifying the main sources and types of plastics and MPs that invade the marine environment. Additionally, a major step in this direction will be education about MPs in the public, private, and governmental domains. Several organizations have created management rules as a result of concern over MPs. The Expert Panel of the United Nations Environmental Programme (UNEP) demanded swift measures to remove MPs from the oceans back in 2014. UNEP has launched an initiative involving over 40 million people across more than 120 nations to foster comprehension and strive

Table 14.3 Microplastics and marine organisms [1, 19].

Organism	Type of microplastic	Concentration/size	Effects	References
Carcinus maenas	Polystyrene (PS)	107 microspheres/L	Breathing disorders	[33]
Mytilus edulis, *Mytilus trossulus,* *Crassostrea gigas,* *Macoma balthica*	Polyethylene; PS	250 beads mg/L	Accumulates in soft tissues	[34–36]
Artemia nauplii, *Pomatoschistus microps,* *Danio rerio,* *Oryzias latipes*	Polyethylene; PS	1.2 × 106 particles/mg; 12 mg/L	Pathological stress, liver swelling, oxidative stress, and the build-up of lipids in the liver	[37–40]
Mesoplodon mirus, *Balaenoptera physalus,* *Megaptera novaeangliae*	PE PVC Terephthalate Nylon Polypropylene	0.5 mg–2.5 MPs/mg	Increase in toxicological symptoms	[21, 22] [41, 42]
Flounder, cod, dab, herring, pelagic fish, and mackerel	PE	54 particles/mg	Digestive disorders	[43]
Daphnia magna, *Centropages typicus*	PS beads	4000 mg/L and 400 mg/L	Immobilization due to decreased feeding	[44, 45]
Crangon	MPs	0.68 ± 0.55/gw.w (200–1000 μm)	Difficulty in food swallowing	[46]

Allorchestes compressa, Mytilus edilus	PE, PS, microbeads	2.5 g/L (0–80 μm), 0.51 g/L (2 μm and 9 μm)	Granulocytoma formation; destabilization of lysosomal membranes; vector for accumulation of POPs	[36, 47–49]
Ostrea edulis	Polylactic acid	80 μg/L and 0.8 μg/L	Increase in respiration rates	[50]
Arenicola marina	PS, polyethylene, polyvinyl chloride	1.4–707 and 250 μm, 2.5–316 and 112.9 μm, 8.7–478 and 143.5 μm	Metabolic rate increases production of fewer fecal casts affects fitness	[51, 52]
C. helgolandicus, C. cristatus, Euphausia pacifica	PS	75 MPs/mL	Decreased reproduction, reduced feeding, and effect on egg production	[53, 54]
Chlorella spp., *Scenedesmus* spp.	Nanosized plastic beads	–	Induction of the generation of reactive organic species	[55]
Scenedesmus, Obliquus	Nano PS particles	103 and 0.22 mg/L	Decreased chlorophyll content, reduced growth rate	[56]
Dunaliella tertiolecta, Chlorella vulgaris, Thalassiosira pseudonana	Polystyrene particles	0.05 and 6 μm	Significant reduction in photosynthesis but no marked change in algal growth	[57]
Chlamydomonas reinhardtii	HDPE, polypropylene particles	–	Significant growth inhibition	[58]
Protozoan Paramecium spp.	PS microparticles	1 μm size	Ingestion of polystyrene MPs	[59]
Cnidarians, Hydra attenuata	Polyethylene flakes (< 400 μm)	0.01–0.08 g/mL	Significant changes in the morphology due to the Ingestion of microplastic particles, but no lethal impact	[60]

(Continued)

Table 14.3 (Continued)

Organism	Type of microplastic	Concentration/size	Effects	References
Crustaceans, Daphnia magna (Cladoceran)	Carboxylated PS MPs	20–100 nm size	Build-up in gut epithelial layer and quicker release of larger beads	[61]
Daphnia magna	Polymethyl methacrylate, Micro and nanoplastics	29.5 ± 26 μm size, 1 μm size, 50–10 μm	Significant bioaccumulation, dissipation. and transformation are reported, reduced feeding and reproduction at high microplastic levels, immobilization	[45, 59, 62]
Oryzias latipes (Medaka fish)	LDPE	3 mm	Alteration in immune responsive system, modification in metabolic functions, and growth of tumor cells	[40]
Daphnia pulex	PS	75 nm, 0.1 mg/L, and 1 mg/L	Biochemical response including homeostasis and in vivo oxidation	[63]
Crassostrea gigas	PS-COOH PS-NH_2	100 nm and 0.1–100 mg/L	Reactive oxygen species production	[64]
Caenorhabditis elegans	PS and TiO_2 nanoparticle	108.2 ± 4.5 nm and 10 ± 2 nm at 0.01–1 mg/L	Alteration in the molecular basis of oxidative pressure	[65]
Brachionus plicatilis	PS-COOH PS-NH_2	40–50 nm and 0.5–50 mg/L	Acute toxicity in rotifers	[66]
Mytilus galloprovincialis	Polystyrene	110 ± 6.9, 0.5–50 mg/l carbamazepine, and 6.3 mg/L	Decrease in enzymatic activity, induced effects on neurotransmission	[67]
Dicentrarchus labrax	Polymethyl-methacrylate (PMMA)	45 nm at 0–20 mg/L	Altered molecular signaling pathways and interference in lipid metabolism mechanisms	[68]
Danio rerio	PS	50 nm at 1 mg/L	Significant inhibition in acetylcholinesterase activity	[69]
Amphibalanus amphitrite	PMMA	45 nm and 5–25 ppm	Nanoplastics could persist in the cells throughout the growth stages	[70]
Mytilus galloprovincialis	PS-NH_2	50 nm, 1–50 mg/L	Decrease lysosomal membrane stabilization	[71]

to decrease plastic usage, avoid MP formation, promote recycling, and evaluate disposal facilities [19]. The bulk of MPs in the garbage may be removed by the primary treatment zones through the scanning and subsiding processes. Reducing the quantity of MPs from the wastewater discharge is possible by applying microbial degradation to sewage wastewater treatment facilities. MP polymers could also be biodegraded by microbes. The International Maritime Organization (IMO) created a "Marine Environmental Awareness" program to educate people about the relevance of shipping regulations and procedures, as well as the sustainability of the marine environment from MPs (IMO, 2015). These initiatives can help maritime crews become more conscious and encourage healthy behavioral and mental adjustments [8]. The Marine Strategy Framework Directive (MSFD) of the European Union aims to achieve "Good Environmental Status" (GES) for the marine environment throughout Europe, which targets plastic pollutants along with MPs in the marine litter (GES, Subgroup and Litter, 2011). The Convention for the Protection of the Marine Environment of the North-East Atlantic, or OSPAR Commission, also focuses on the issue of marine trash, especially including plastics and MPs, to safeguard and preserve the environment of the North East Atlantic region and its resources (OSPAR, 2014) [2]. Earlier, the Ciénaga Grande de Santa Marta (CGSM) lagoon complexes selected stations of the OSPAR (2010), where the methodology for monitoring marine litter with plastics and MPs was modified to account for access, structure, and mangrove conditions [75].

The European Union uses 11 descriptors to characterize the ecological condition of a region, MPs constituting one of them. The European Union incorporated MPs in the water framework regulation (WFD-2008/ 56/EMSFD-EU, 2008), after realizing the necessity for additional study on their existence and impacts on the environment. Efforts are being made for sediment sampling methodologies [76]. Studies have drawn a lot of interest from the scientific community, and public health organizations around the world are prioritizing the problem of human exposure to MPs through drinking water. As a result, securing access to safe drinking water is a top political priority, with Sustainable Development Goal (SDGs) 6, including safe and cheap drinking water as a specific goal (WHO and UNICEF 2017) [77].

14.5 Treatment of Microplastic-containing Wastewater

Among the techniques explored for the removal of MP from wastewater are the conventional physical, chemical, and biological processes (Figure 14.3). All the methods suffer some disadvantages of difficulty in wide scale applicability, as well as cost effectiveness.

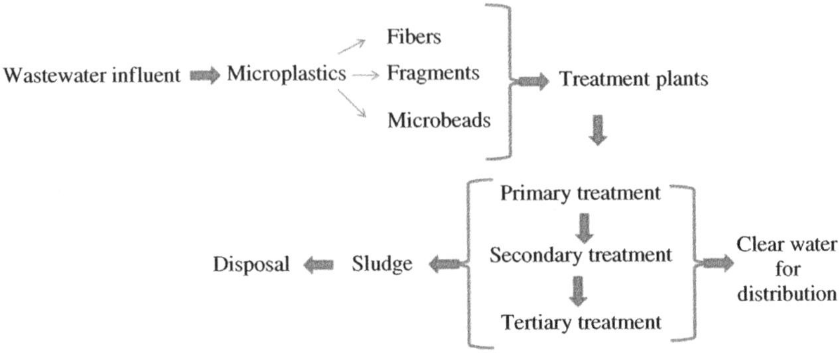

Figure 14.3 Treatment process for microplastic elimination.

14.5.1 Primary Treatment

Primary treatment is done to settle down the insoluble solids from the waste stream using screening, grit removal, skimming, and sedimentation process. Approximately 60% of the MPs can be eliminated during the preliminary treatment. The principle involved in the process is mainly sedimentation of the entrapped MPs by gravity. It is not possible to remove MPs with a wide range of sizes during the primary process.

In studies carried out, MP removal up to 36% was achieved by employing aluminum salts for coagulation, mainly for particles less than 0.5 μm. Aluminum-based salts provided better zeta potential and rendered easy charge neutralization and surface area. The accumulated MPs were eliminated by ultrafiltration membranes. The destabilization of MPs could also be achieved by electrocoagulation, where the cost of chemicals is reduced and coagulants are produced in an eco-friendly manner. Flocs interact with the microplastics via hydrogen bonding, Van der Waals, and electrostatic interactions. The plastic settling technology removes the settleable part from the suspended MP. Lack of identification of the degradation products, their properties, and related toxicity are the areas to be addressed. Investigations have reported that MPs of size varying from 0.1 μm to 2 nm can be removed using microfiltration, ultrafiltration, and nanofiltration [78, 79]. Sand filter systems with integrated biochar showed better removal efficiency, up to 95%, while with granular activated filters, up to 60.9%. These, however, were, clogged, and suffered drawbacks such as high costs involved and membrane regeneration requirement.

14.5.2 Secondary Treatment

Secondary treatment usually involves degradation using microbial species by either suspended growth system or attached growth system. The secondary treatment can bring about 2–55% of the microplastic removal, but on specific examination investigations activated sludge process can help in achieving around 43% of the removal [80]. The efficiency of the removal of plastics using these alone does not give encouraging results. Sludge obtained during the process is loaded with plastics, further giving rise to disposal issues, though modification can lead to better removal. Sludge containing MP is removed in subsequent secondary processes. The conventional anoxic/oxic bioreactor system activated sludge involves the use of bacterial strains like *Rhodococcus* and *Ideonella* to degrade MPs. However, a retention time of 14 hours is not sufficient to degrade MPs.

14.5.3 Tertiary Treatment

Tertiary treatment methods involve processes like membrane filtration, disinfection, reverse osmosis, ozonation, etc.

Chemical oxidation treatment involves mineralization, wherein decomposition of MPs occurs into small molecules. Chlorination is one of the most commonly used chemical oxidation and disinfection process, and all MPs are not resistant to chlorine. The oxidizing nature of the chlorine leads to a change in the chemical and physical properties of MPs. The modified chemical entity leads to better accumulation and adsorption of MPs. It is reported that tertiary treatment can help to achieve 97% of the MP removal. None of the conventional methods target the degradation of the MPs. It is essential to modify the existing treatment systems instead of developing the newer treatment technologies to develop infrastructure and equipment familiarity with the operation.

Among the emerging techniques are the electrocoagulation-electroflotation, membrane bioreactor, algal masses, bioinspired molecules, metal–organic frameworks, photocatalytic method, advanced oxidation process, and the use of hybrid technologies.

14.5.4 Newer Treatment Methods

Several novel methods have been investigated for the treatment of MPs in wastewater, which have been discussed as under:

14.5.4.1 Biomass-based Treatment

Granular activated charcoal derived from coconut shells was reported to show a maximum adsorption capacity of 6.33 mg/g with 98% removal efficiency for PS MPs. Biomass-derived activated charcoal provides a sustainable approach for the removal of MPs from wastewater. Granular activated charcoal treatment offered a good removal efficiency for PE and PP, which have a lower gravity than water. Biochar, which is a highly porous and carbon-rich material, has shown good potential in adsorbing and removing MPs from the aqueous environment [81]. PS removal efficiency of 70% was observed using magnetic biochar. Also, iron-modified biochar showed easy and quick removal of MPs from water bodies. Corncob biochar showed a removal efficiency of 80% [82]. Biochar obtained from jujube biomass reflected removal efficiency greater than 80% for nylons and PE. The utilization of biochar for the eradication of MPs from wastewater is still in its early stages, but so far, investigations have demonstrated its potential for practical scalability. Chemically modified biochars can have greater affinity for MPs, encouraging greater exclusion of MPs from wastewater.

A myriad of factors govern the efficiency of MPs removal from aqueous environment using bio adsorbents. These include the nature of the bioadsorbents, the quality of the adsorbent, surface chemistry, and modifications of the bioadsorbents, MPs concentration, pH, and zeta potential. The characteristics of bioadsorbents depend on the methods adopted for their preparation, used raw material and the surface modification carried out for bioadsorbent preparation, along with the technology used for ageing.

The type of method employed for the preparation of adsorbents governs the specific surface area availability for adsorption. Investigations have shown that supercritical water gasification produces better surface area and pore volume as compared to pyrolysis. Temperature also influenced the removal efficiency of the adsorbents. The adsorption capacity increases with the adsorbent concentration due to the creation of plenty of binding sites for the adsorption of the pollutants. Although an excessive amount of adsorbents leads to clumping together of the adsorbent and decreases the number of accessible sites. Thus, an optimum amount of the adsorbents plays a crucial role in the optimization of the adsorption efficiency.

Tailored materials obtained from natural sources play an important role in addressing the process of contaminant remediation from wastewaters. Magnetic biochar modified using magnesium and zinc was effective in the removal of the MPs. The removal efficiency obtained was of the order zinc-modified biochar > magnesium-modified biochar > magnetic biochar [83].

In a noteworthy study, ball milling was utilized to prepare the Fe_3O_4 nanoparticle biochar for PS removal, and it was found that the removal efficiency increased from 43.67 to 82.73%. This indicates that the surface chemistry involved in the bioadsorbent use is an important factor for the elimination of MPs from aqueous ecosystem [84]. Initial MPs concentration and increase in the contact time elevate the adsorption capacity, which may be due to an increase in the concentration gradient of MPs, which improves mass driving force effect, promoting the diffusion of MPs to the adsorbent surface. It was reported that during the initial phase of process adsorbent site availability facilitates adsorption, which gradually reaches saturation with time. Thus, with boost in the concentration of the pollutant, the effectiveness declined due to the saturation of the adsorbent sites. It is thus necessary to optimize the initial MP concentration for remediation strategy implementation.

It is imperative to note that pH impacts the surface properties of both the adsorbent and the adsorbate. The surface charge depends upon the type of biomass material employed. Variation in pH is related to different rates of MPs adsorption.

Metal–organic frameworks are excellent adsorbents for MPs due to their large surface area, easy preparation, tuning flexibility, and high interaction ability with MPs. Metal–organic frameworks can interact with MPs through hydrogen bonding or by Vander Waals forces. The metal–organic frameworks have positively charged defects that have a great affinity for the negatively charged MPs.

Researchers successfully reported metal–organic frameworks-based foam to eradicate polyvinylidene fluoride (PVDF), PMMA, and PS MPs from simulated water. A removal rate of 95.5% was reported, which was influenced by particle size and zeta potential of the MPs [85]. A metal–organic framework called MIL-100/Fe was used for extracting MPs from textile wastewater. The metal–organic framework is comprised of mesoporous and microporous structures. The same when blended with polysulfone matrix provided 10.3-fold increase in the clear water output. An alkaline pH of 9.0 showed an optimum performance due to the electrostatic repulsion phenomenon. The findings also demonstrated the reversibility of metal–organic frameworks. Though further research is needed to improve framework design and operating conditions, to increase the performance of the process. Mohana and team used metal–organic framework based ED-MIL101/Cr ultrafiltration membranes, which showed excellent water permeability and 90% microplastic and nanoplastic removal from wastewater [86]. MPs are found to interact with other contaminants present in wastewater, which may change their toxicity and behavior making their removal more crucial. Metal–organic frameworks have been also utilized to trap MPs due to their high surface area, porous characteristic, versatile functionality, and durability. Metal–organic framework loaded with melamine was employed to remove MPs like PVDF, PMMA, and PS. These metal–organic frameworks not only efficiently remove MPs but also show good recyclability.

14.5.4.2 Membrane Technology

Conventional wastewater treatment plants and potable water treatment plants cannot eliminate MP from water. Membrane technology offers a good option for making water suitable for humans and the environment. Membranes employed in wastewater treatment are either polymeric or inorganic membranes. Polymeric membranes include polyether sulfones, polyamides, cellulose acetate, polycarbonate, PTFE, while inorganic membranes include alumina, silica, silicon carbides, zeolite, etc. [87].

Inorganic membranes are usually employed in wastewater treatment due to their better thermal, mechanical, and chemical characteristics. These also offer the advantage of high flux, better separation, good antimicrobial property and long durability. But, due to their high fabrication cost they are less employed in water treatment than polymeric membranes. Microfiltration membranes have pore size of 0.1–10 μm, and are pressure driven membranes. Investigations have revealed that although the size of microfiltration membrane is smaller than the MPs, 100% removal is not achieved due to low membrane strength. Ultrafiltration provides superior removal of MPs due to smaller pore size of ultrafiltration membranes. Hundred percent removal was achieved by PVDF flat sheet ultrafiltration membranes, having pore sizes of 30 nm. In another investigation, MP removal efficiency was found to be 99.2%, with the use of ultrafiltration membranes. Also, PES ultrafiltration membranes removed MP with effectiveness greater than 96% [87]. A very limited research has been carried out for understanding nanofiltration separation for MPs. 99% of MP removal from a landfill leachate treatment plant was obtained by Kara et al. [88]. However, MP removal from wastewater by nanofiltration membranes requires more investigation and future research. The separation process involved in the reverse osmosis membrane operates based on a sorption diffusion mechanism. Reverse osmosis technology exhibits better removal than other membrane techniques. However, MP can even be present in the permeate of RO membranes. Dalmau-Soeler et al. reported about 60% fibrous and 40% fragment-type MP in the permeate [89]. Similar reports were obtained by Sun et al., who reported that MP less than 50 μm in size were obtained in permeate after the RO unit [90].

Microfiltration in combination with biological treatment was employed for MP elimination from aqueous solution. Hundred percent removal efficiency of MPs was reported using a flat sheet membrane in primary wastewater treatment with a membrane biological reactor. In another study, conducted in Finland, 99.4% removal efficiency was observed for wastewater treatment through grit chamber, primary sedimentation and membrane biological reactor [87].

Cai and team reported that MPs removal could be achieved up to 98% in the wastewater treatment plants using primary, secondary, membrane biological reactor, and reverse osmosis. They also reported that non fibrous MPs greater than 0.5 mm were completely eliminated from the effluent. The combination of ultrafiltration and reverse osmosis, demonstrated consistent performance and contribution to fewer MPs in the treatment plant effluent [91].

Lares and team noticed membrane biological reactor technology to deduce the concentration of MPs by 99.3%, in comparison to traditional wastewater treatment methods. However, the removal efficiency was influenced by structure and morphological properties of MPs and presence of other contaminants. Removal is a function of particle size, and is the most common investigation is carried out for

MPs particles of 100–200 μm, in the wastewater treatment plant. It is quite effective in the elimination of MPs of all size fractions and all shapes, from wastewater [80]. A membrane biological reactor having 20 ultrafiltration membranes and flat sheets of 0.4 μm pore size, with surface area of 8 m^2, efficiently removed almost all MPs and showed 99% removal [92].

Membrane bioreactors (MBR) show around 99% efficiency in removing the MPs. These are an integration of membrane separation and a conventional sludge treatment method. The main operative mechanism in membrane bioreactors is biofiltration and absorption. The biofilm supported on the inert media retains the MPs, which are removed by backwashing. An in-depth study is still needed to explore the exact mechanism involved in the microplastic removal through microbial action, as well as the influence of MPs on the function of the microbes.

14.5.4.3 Advanced Oxidation Process

Advanced oxidation process involves use of strong oxidizing agents to break down MP contaminants. Type and size of MP, and water quality parameters influence the effectiveness of advanced oxidation process. Advanced oxidation processes are effective against a variety of MP.

Advanced oxidation processes are based on physicochemical changes involving molecular disassembly, breaking of chemical structures, and mineralization. These involve the generation of hydroxyl, superoxide, chlorine, and sulphate radicals. Among these, photocatalysis demonstrates a high degradation rate for MPs [93]. Electro-Fenton process was carried out by using TiO_2/graphite cathode to degrade PVC. Cathodic reduction and hydroxyl radical oxidation can degrade PVC with mean removal of 75%. But the process suffered from the limitation of existence of short chain organic compounds in post-treatment solution. Also, MPs may adhere on electrodes and hinder mass transfer by blocking the reactive sites on the catalyst, reducing the removal efficiency [94]. Electrooxidation technology decomposed PS MPs with a removal of 58% in 60 minutes of contact. It suffered from high energy requirements and complex reaction devices [94].

Photocatalysis involves the generation of hydroxyl radicals by electron excitation of a semiconductor material by radiation. UV light could degrade 98.4% of PS using TiO_2 nanoparticles within 12 hours. Photocatalysis results in complete neutralization of MPs and has emerged as a green and cost-effective approach for MP pollution control. In another study, zinc oxide nanorods degraded 65% of PP after visible light irradiation for 2 weeks [94].

Ultraviolet oxidation modifies the surface morphology of MPs and alters its chemical characteristics. Polyethene, polypropene, and polystyrene have been reported to show surface change by formation of cracks and flakes, which further breaks down the MPs to smaller sizes and oxidation results in the formation of free radicals species, with peroxy linkages, which may form crosslinked structures.

However, the intermediates and toxicity of ultraviolet oxidation treated MPs still needs to be explored further. MPs larger than the size of membranes used for filtration, have been efficiently removed from wastewater. Reports suggest interactions between MPs and the pores and surface of the membranes, leading to the permeation of MPs across the membranes. The creation of a polarization layer on to the surface of membrane results in fouling of the membranes, whereby the performance of membrane filtration decreases. Further research and efficient membrane cleaning procedures are crucial in determining the influence of MPs on membranes. Dynamic membrane technology is one of the promising techniques for the treatment of microplastics from contaminated aqueous environments. The efficiency of the membrane technology relies on the MPs flux concentration and particle size retention time. The factors crucial in the elimination of MPs include the shape of the plastics [95]. It is reported that membrane technology in conjunction with biological technology removes MPs up to 99% [96].

A combination of catalytic oxidation and hydrothermal hydrolysis process using carbon nanotubes has degraded MPs exceptionally. The removal was influenced by initial concentration, catalyst dose, operating temperature, and temperature [97]. Microbes like bacteria and fungi have also degraded the MPs successfully. Caruso reported one of the fungal strains, with the ability to degrade PE by 43% in 14 days of exposure time [98]. Similarly, PP was degraded using a bacterial strain, with complete degradation of PET using *Ideonella sakaiensis* [99]. Though a large-scale evaluation of the biodegradation of MPs is not well explored, it can still be seen as one of the promising remediation technologies for the future. Factors like temperature, pH, concentration of the plastics, and bacterial and fungal availability govern the overall degradation process. In a similar exploration, a large amount of MPs have been found to be present in sludge produced from wastewater treatment plants (WWTP), at various stages. This sludge has generally been used in agricultural fields as a fertilizer, causing environmental pollution. MBR-combined biological and membrane separation technology saves space and limits sludge production. In a study by Talvitie et al., most of the MPs were retained after going through the MBR system [100]. Very promising results were obtained by Lares et al. for the removal efficiency of polyvinyl chloride gel, with particle size less than 5 μm, at a temperature around 19.1 °C and pH 7.5 [80]. Sludge treatment and fouling are the major drawbacks associated with MBR. The fouling mechanism is still unexplored and depends on the chemistry, stability, surface properties and size of the MPs.

14.5.4.4 Electrocoagulation

Electrocoagulation is used to improve the exclusion of MPs and provides a rapid, simple, and cost-effective solution for wastewater treatment. It involves electrochemical dissolution of metal electrodes, especially aluminum and iron, which

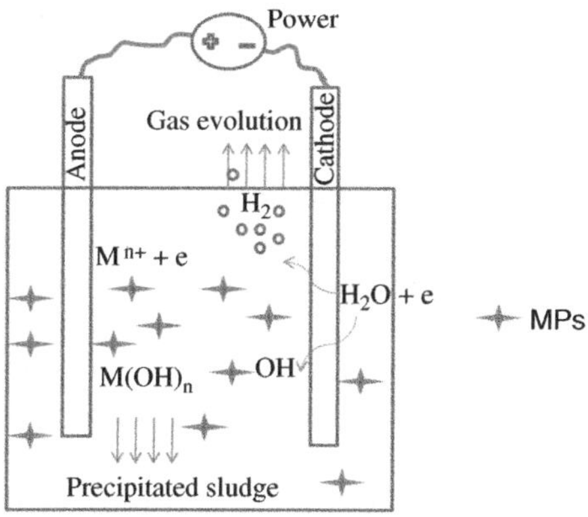

Figure 14.4 Electrocoagulation for microplastics removal.

generates metal ions. These ions generated at the anodic surface react with the hydroxide ion to form the corresponding metal hydroxide, which further forms flocs and absorbs the contaminants present in the wastewater. Electrocoagulation is usually operated at a pH range of 4–8.5. Parameters like intensity of current, time of operation, and composition of wastewater influence the performance of electrocoagulation (Figure 14.4).

The technique was employed for the removal of PE MP beads, and it was observed that with the addition of sodium chloride as an electrolyte solution and the use of surfactant, led to the exclusion of 99% MPs from synthetic wastewater. Electrocoagulation was found to have the potential to remove MPs from real wastewater effluents. In an investigation, a removal efficiency of 96.5% was obtained for MPs in real wastewater, along with 92.2% of COD and 88.8% of bacterial population elimination, by employing the electrocoagulation technique. The process executed on the synthetic wastewater showed 99% removal efficiency, with current densities of 1.92–8.07 mA/cm^2, at acidic pH within 90 minutes [101].

An investigation was performed to remove PS MP from synthetic wastewater. It was revealed that the plastic can be removed using the electrocoagulation process in a short period with an optimum current density of 16.3 A/m^2. The removal efficiency was dependent upon the size of MPs and electrolyte concentration [102]. Findings demonstrated that electrocoagulation treatment affected MPs by flaking and charged surface conditions, along with chemical modification, which was confirmed by changes in the vibrational energies of the contaminants. PA, PE, PET, and PP fibers and fragments were effectively removed by the electrocoagulation method [103].

Literature suggests that electrocoagulation with an aluminum electrode best eliminates MPs from wastewater, including surfactants. An investigation was done for the elimination of MPs from laundry wastewater using electrocoagulation. The chemical composition analysis and fiber count were done by using optical microscopy and the FTIR technique. The percentage removal of 97.9% was obtained for MPs along with a reduction in COD and surfactants at an operating time of 25 minutes and current density of 300 A/m^2. The total operational cost was also not huge [104].

One of the studies meticulously examined various electrode combinations of stainless steel and aluminum and discovered that aluminum–aluminum coupling demonstrated the highest efficiency (95.54%) for MPs elimination. The optimal electrolyte concentration was determined by decision making tool called TOPSIS, and was obtained as 0.5 g/L. The optimum time taken was 100 min and input concentration of 2 g/L removed the MPs of different sizes with a removal efficiency of 89.80–94.08% [105].

PS was reduced using electrocoagulation from an effluent obtained from the wastewater treatment plant of Dengkil, Malaysia in a duration of 60–120 min. About 82% removal was achieved employing the technique [106].

A technique employed aluminum and iron, which served as excellent sacrificial anodes due to their good coagulation ability. At the cathode, gas generation takes place, which helps flocs to rise up to the upper surface of the water and enhance the elimination of contaminants or MPs. Electrocoagulation offers several advantages, like efficiency (Table 14.4), successful separation of organic compounds, pH level adjustment, low cost, and simple automation. It helps prevent the degradation of the pollutants and stabilizes them [107].

Table 14.4 Efficiency of electrocoagulation method for elimination of MPs from wastewater.

Plastic	Source of wastewater	Electrode employed	pH	Current density/ voltage	Time	Percentage efficacy
PE, PVC	Domestic	Al-Fe	7	20	10 min	100
PA	Synthetic	Fe-Al	7	10 V	120 min	96.82
PE, PP, PVC	Synthetic	Al-Al	4	2.88	90 min	98.5
PE	Secondary treatment effluent	Al-Al	6	12	20 min	97.5
PE, PP, PVC	Municipal	Al-Al	4	2.88	90 min	96.5
PET	Laundry	Fe-steel	8.38	15 V	–	89–91
PA	Laundry	Fe-steel	8.38	15 V	–	70–75
PS	Synthetic	Fe-Al	7	10 V	80 min	82.5
PET, LDPE	Synthetic	Fe-steel	8.38	15 V	–	94.91

Source: [107]/EDP Sciences/CC BY 4.0.

The biggest disadvantage associated is the need for frequent replacement of the sacrificial anodes, post-treatment, sludge treatment issues, cathode passivation, and related cost of elevated energy expenditure.

Two polymeric plastics were investigated for their removal from wastewater using the electrocoagulation and flotation methods. A reaction time of 10 minutes at neutral pH with aluminum-iron electrodes helped reach 100% removal efficiency. The method was found to be more favorable in comparison to secondary and conventional tertiary processes [108, 109].

14.5.4.5 Photocatalysis

The technique involves focusing on contaminants that form microbubbles and combine to large masses that can be efficiently removed without any specific chemical and biological reagents. This is also one of the energy-efficient methods of plastic removal. In a similar attempt, 3D solar evaporator technique is also seen as a promising method for the elimination of plastic contaminants. It is a low-cost biomass-derived photothermal platform for the removal of plastics [110]. Photocatalytic degradation of polypropene particles by visible light of zinc oxide nano rods resulted in greater than 60% reduction in the particle volume.

Organic compounds, hydrocarbons, pesticides and dyes have been well removed by photocatalysis process. Figure 14.5 illustrates the mechanism involved in the process. Photocatalysis has been tried for the removal of MPs in recent years. Photocatalytic degradation of PS was carried out under UV irradiation source. Similarly, PS microspheres and PE were degraded photocatalytically over titanium oxide nanoparticles in the UV irradiation. About 98.40% of degradation occurred in 12 hours for PS. Other than UV irradiation, sunlight was also successfully utilized to degrade PE using a composite of polypyrrole and titanium oxide.

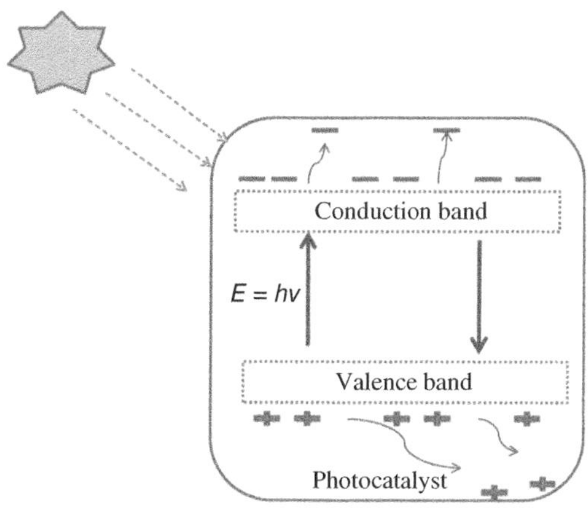

Figure 14.5 Photocatalytic mechanism for microplastics removal.

The degradation was the result of strong interaction between the plastic and photocatalyst surface, which was confirmed by FTIR study. HDPE MPs isolated from cosmetic wastewater was broken down using titanium oxide a photocatalyst in both solid and aqueous mediums. Microfibers in wastewater treatment plants can be effectively removed using photocatalyst like titanium oxide under UV radiation [93]. In a study, zinc oxide coated with glass fiber substance removed low density polypropylene (LDPP) from wastewater, with recorded decrease in the particle volume by 65%, when PP was irradiated for about two weeks with visible light. The photocatalytic degradation products were harmless products. The plastic litter could be cost effectively removed from wastewater using the photocatalysis approach [111]. The performance of photocatalysis depends on operational mode, hydrodynamics, and reaction mixture composition. The radiation sources are the most important parts of the process, as photocatalytic oxidation is initiated by photons. A Fenton reaction in combination with photocatalysis also contributed to a novel wastewater treatment approach [111].

14.5.4.6 Plasma Technology

Plasma treatment subjects the MPs to a plasma discharge. The plasma leads to the generation of reactive species that can degrade MPs. Plasma treatment for 30 min can reduce MP weight by 11.3%. The mechanism involved is oxidation and hydrolysis, which was assessed by photochemical analysis. It is nonthermal plasma comprising of partially ionized gas having electrons, UV photons, and reactive species (Figure 14.6). It has been used widely for the decomposition of inorganic pollutants in water, including pesticides, dyes, drugs, etc. PP, LDPE MPs were degraded with oxygen, nitrogen and their mixture with plasma technology. PP had a larger weight reduction than LDPE [112, 113].

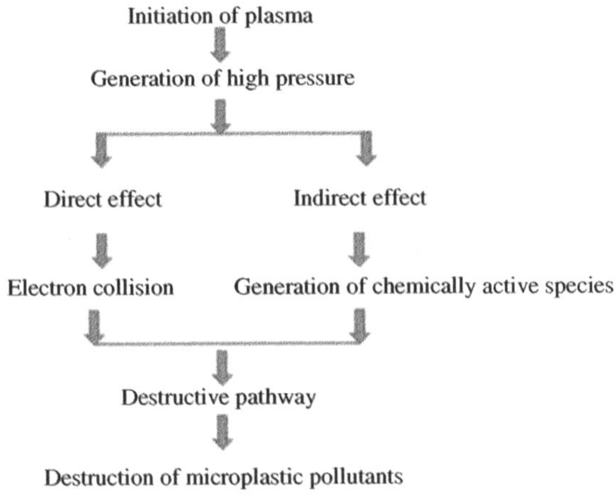

Figure 14.6 Plasma technology for microplastic removal.

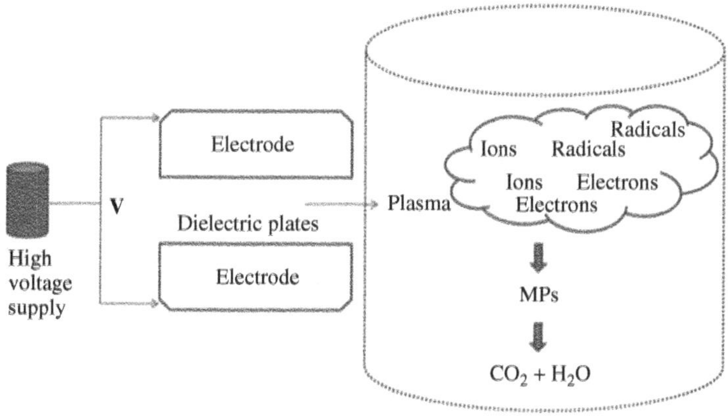

Figure 14.7 Assembly involved in plasma technology.

Similarly, PVC was removed by employing this technique, and 85.9% of removal efficiency was obtained. PVC removal was affected by discharge voltage and pollution load. The —CH group was the most important reaction site of PVC, and removal was facilitated by hydroxyl and hydrogen radical species [114].

In the plasma method, the reactive species are created in situ, reducing the expense involved in transportation and storage (Figure 14.7).

It is advantageous in terms of the removal of MPs as it is a simple decomposition and eliminates the possibility of leakage. The dielectric barrier discharge generates plasma, which produces uniform and stable degrading species. A cold plasma treatment in combination with microfiltration was demonstrated for the removal of MPs, with a 99.9% removal rate. This was attributed to an increased coagulation due to an increase in ionic strength of the solution and a reduction in the thickness of the electrical double layer. The process avoided the liberation of secondary contaminants to the environment [115].

14.5.4.7 Phytoremediation

Phytoremediation is yet another technique employed for wastewater treatment, but there is insufficient understanding of the interaction between plastics and biomass, as the interaction between the two has not been investigated systematically. An investigation into the PE MPs and a floating aquatic species named *Lemna minor* was carried out. The results showed that the maximum number of MPs adhered to the plant in one day. The increase in the biomass of the plant further increased the adherence capacity for MPs. It was reported that 15 cycles of phytoremediation could help remove 75% of the MPs, from the aqueous solution, under test conditions [116]. The investigations helped in understanding the removal efficiency under static conditions, as well as with moving water, at different concentration of the biomass. No significant change in MPs adsorption characteristics and efficiency was observed with movement in water. Thus, the technique can be used as an environmental friendly method for in situ removal of MPs from wastewater.

In a similar investigation carried out by Yin and co-workers, the capacity of *E. crassipes* (water hyacinth) was tested to eliminate MPs from contaminated water. They reported that 0.5, 1, and 2 μm PS particles were removed with a removal efficiency of 55.3%, 69.1%, and 68.8%, respectively, within a 48-hour time of contact. The roots of the plants were the principal sites for the entrapment of the MPs. The same was confirmed by fluorescence microscopy (FM). Also, the vascular system present in the stem of the plant prevented the translocation of the contaminants to other tissues. Thus, phytoremediation strategy could be an important contributor to mitigate the MP pollution [117].

Few investigations have reported nanoparticle elimination by phytoremediation. PS nanoparticles (NPs) were sequestered using water hyacinth, and 44.7% elimination was obtained for 200 nm PS-NPs. The plants used for phytoremediation have shown high tolerance to plastic exposure.

The fastening of MPs to aquatic vegetation may be due to the existence of hydrophobic attraction between MPs and the anionic polysaccharides, oozing out of the vegetation. This helps to stabilize the MPs on the plant surface and minimizes their interaction with the other organisms (Figure 14.8) [118].

14.5.4.8 Biotechnological Methods

Emerging genetic engineering has enabled the utilization of biological species for contaminant elimination from wastewaters. Several successful attempts have been made to carry out the biodegradation of the pollutants from the

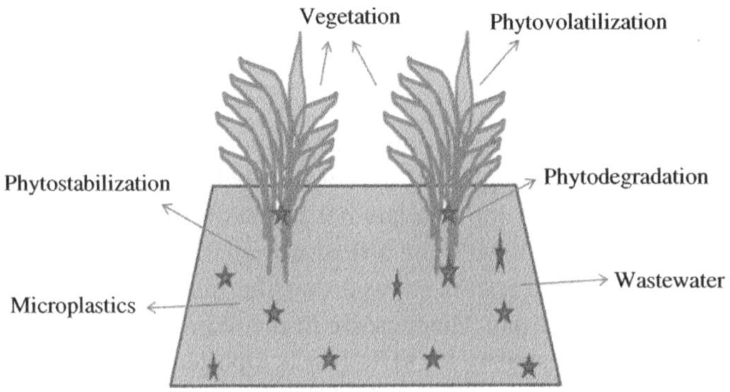

Figure 14.8 Phytoremediation for microplastics removal.

wastewaters. A myriad of factors influence the biodegradability of the pollutants, which include [119]:

1. **Microplastic Particle Morphology:** It includes particle size, particle shape, and surface roughness.
2. **Chemical Composition of MPs:** It includes factors like surface hydrophobicity, surface electrostatic interaction, and presence of additives, including the chemical structure of MPs.
3. **Operating Conditions:** pH, temperature, salinity, humidity, agitation speed, optical density, and addition of biostimulants affect the process.

The MP removal can be facilitated by following biological species, a summary of which is given in Figure 14.9 [119]:

Bioinspired Molecules: Bioinspired molecules are prepared by the combination of organic and inorganic molecular blocks, especially for the removal of hydrophobic MPs. These molecules basically comprise of an inclusion unit component and a capturing unit part. When the MPs are captured, the displacement of the embedded water from the inclusion compound takes place.

1. **Use of Algal Species:** Microalgae are reported to colonize MPs, altering their buoyancy, which facilitates their increased settlement. The algal surface electrostatically interacts with the MPs. The positively charged center of MPs sorbs onto the algal surfaces due to the existence of anionic polysaccharides

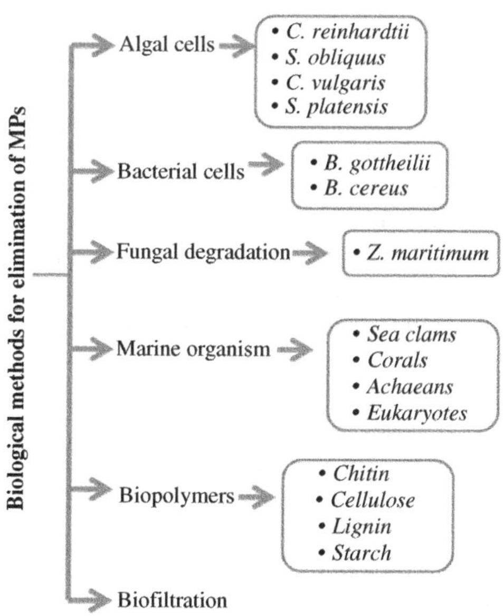

Figure 14.9 Biological methods for the removal of microplastics from wastewater.

in the algal cell wall. *F. vesiculosus* has been used to eliminate MPs via translocation in the algal tissues, as the algal cells have narrow channels that help in the capture of the MPs. A microalgae named *P. tricornutum* was isolated, which has the potential to secrete polyethylene terephthalate enzymes at a temperature of 21 °C. It can degrade PET polymer into 2-hydroxyethyl terephthalate and terephthalic acid [99].

One of the investigations revealed an 81% elimination rate of PS MPs using *S. platensis*, while 85% of polyethene MPs using *C. vulgaris*. The species, *C. vulgaris*, has a unique composition and surface properties, due to which it has a high biosorption capacity. It is a fast-growing green algae with a special cell wall rich in polysaccharides and protein content. Removal of PS MPs was carried out at pH 7.5 with *C. Vulgaris* (274.05 mg/L), taking a prepared solution of PS as 789.37 mg/L. A maximum removal of 73.01% was achieved for PS, within 31.90 minutes. The use of algae, therefore, is an eco-friendly and sustainable methodology to address MP contamination in water bodies [120].

Genetically engineered green algae *C. reinhardtii* and diatom, *P. tricornutum*, degraded PET. These demonstrated degradation and a catalytic activity against the polymer, increasing the possibility of developing an eco-friendly solution to degrade MPs from wastewaters [120].

The accumulation of MPs in marine water is more alarming than any other contaminants due to bioaccumulation in the living system. Algal cell releases alginate substance from their cell walls, which increases the adherence of MPs. This suggests that the charge of MPs and chemical features of algal cells play an important role in determining the removal efficiency [121]. *S. obliquus* is reported to adsorb nanoplastics and MPs, which eventually decrease its growth, photosynthesis and chlorophyll generation. *P. subcapitata* was observed to adsorb PS particles of size 20–500 nm over it. The positively charged PS is firmly attached to the algal cell wall, which is composed of ionic polysaccharides [121].

2. **Use of Fungi:** Fungi are largely saprophytic or obligate. They have good adaptability and can grow under adverse environmental conditions. They can produce a diverse range of extracellular enzymes that can degrade complex molecules, including MPs. Fungi like *Z. maritimum, A. niger, Cladosporium, and P. simplicissimum* use MPs as a carbon source and degrade them. Degradation of PU was studied using *A. fumigata, A. tubingensis*, and *P. microspora* by several researchers. Serine hydrolase played an important role in the degradation of PU. Pre-treatment of MPs can increase the possibility of degradation of MPs by fungi. *Z. maritimum* exhibited degradation of PE MPs. A removal efficiency of 43% was observed on subjecting the MPs to the fungi for 14 days. This was possibly due to the existence of biological contents on the MP's surface, which

helped in the degradation of the pollutants. PE MPs were removed by oxidative degradation, which was confirmed by FTIR analysis [122].

3. **Use of Bacteria**: *B. gottheilii* and *B. cereus* were investigated for PS, PE, and PET removal from wastewater. Appreciable loss in plastic mass was observed by both types of bacteria. *B. gottheilii* showed a removal potential of 3%, 5.8%, and 6.0% for PET, PS, and PE, respectively. Similarly, bacterial enzymes like lipases, cutinases, and esterases converted PET into nontoxic monomeric molecules. These enzymes present in *I. sakaiensis* hydrolyzed PET at 30 °C [120].

MPs have proved to be a menace to the pollution control system, as no specific or 100% efficient technology has been formulated yet to get rid of the problem. Current treatment methods rely on single treatment setup or a hybrid of various techniques, but still, the performance of the process is not very encouraging, as far as real wastewater is concerned. Removal of contaminants depends upon their concentration and the time of retention of the wastewater in the specific treatment unit. The research community is still struggling to find a way out for the alarming and increasing pollution load, the MPs are creating. The remediation of the problem is not limited to the treatment but also considers the usage and handling of the plastics and their disposal strategies. Legislative execution, awareness, willingness, and adaptability for eco-friendly materials are essential to stop the contaminants from entering the environment. Inclination, environmental concern, and research for green technology solutions, along with green materials, are the demand of today's sustainable approach. Legislation alone cannot help combat the problem. Strategies to implement an efficient eco-friendly treatment technology on real wastewater need to be developed to fight against the negative impact of MPs on the ecosystem and human health.

14.6 Case Studies

14.6.1 MP Removal Efficiency of Wastewater Treatment Plants in Zhengzhou, China

A study on the removal efficiency of MPs from wastewater treatment plants situated in Zhengzhou, China, was carried out. Detection of MPs was done by using Raman spectroscopy. The removal efficiency was found to be in the range of 76.2–91.2%. The plants were equipped with a tertiary treatment process. The three plants under investigation had the treatment scale of 600000, 200000, and 100000 m^3/day. The influent, secondary treatment process outlet, and effluent were the sampling points in all three plants under study.

Zhengzhou, the capital of Henan (Province) in China, is a significant textile production center and contributes to the presence of fibrous MPs in wastewater. Also, domestic laundry generates a substantial amount of fibers. The treatment plants receive sewage wastewater via long-distance pipelines, whose abrasion increases the percentage of small sized plastics in the influent. The results inferred that the percentage of MPs at the secondary treatment outlet and effluent outlet was relatively high, suggesting the inefficacy of the existing wastewater treatment units in eradicating the fibrous MPs.

Raman spectroscopy-based spectrometer suggested the presence of ten different types of MPs like PET, PP, PVC, polyphenylene oxide, PTFE, PE, PS, polyformaldehyde, PA, and PMMA. The PP amount was found to be the highest in all three treatment plants. This could be due to use of PP predominately in catering and medical sectors. MPs derived from PET were also found to be noteworthy contributors to the occurrence of MPs within all wastewater treatment plants. The effluents were also investigated for types and color of the MPs, and it was found that fibers dominated in the wastewater, which ranged from 10 to 100 μm. MPs of several colors were seen, but transparent MPs were prominent. The sludge of the wastewater treatment plants was also found to accumulate the MPs from wastewater bodies, especially PET (42.4%). It was suggested that gravity settling and activated sludge treatments played an important role in the removal of MPs during the secondary treatment method. The tertiary treatment was devoid of the sedimentation tank, which clearly demonstrated the lowered overall efficiency of all the wastewater treatment plants under investigation. The results showed that there is immense prospective for increasing the efficiency of plastics exclusion within the plants in China. It is needed to efficiently regulate the operational parameters of the existing wastewater treatment plants, to ensure better MPs elimination [123].

14.6.2 Groundwater Contamination by MPs

The groundwater is getting contaminated due to MPs because of various anthropogenic activities including improper plastic disposal and discharge of wastewater. The impact of groundwater pollution on human health is significant.

The narrated investigation is a study focused on MPs present in the groundwater in West Jakarta and South Jakarta. Jakarta is one of the highly dense metro cities of Indonesia. High levels of anthropogenic influences have resulted in groundwater contamination. A sampling was carried out in August'23 from 56 points in West Jakarta and 65 points in South Jakarta. The investigations revealed the abundance of MPs, which ranged between 28 and 236 particles, per liter. West Jakarta water had higher average MPs abundance than South Jakarta. MPs presence was found to be of various shapes and colors. West Jakarta and

South Jakarta groundwater was found to contain manly fragments, fibers, microbeads, and films. The prominent type was fragments as per result which was also confirmed by their frequent occurrence in the wastewater treatment plants. Other than fragments, fibers were also found in noticeable amounts in the samples. This was due to the entry of synthetic fibers in the groundwater through wastewater from laundry and washing facilities. Microbeads occurrence was very less in the samples. The groundwater has MPs of black, red, blue, yellow, green, and transparent color. The sampling sites had dominance of black (85–93%). Colors play an important role in the light absorption which facilitates photodegradation of the plastics. The dominance of black color was attributed to the raw plastic utilization in food packaging, automobile components and toys manufacturing etc.

The study also measured the size of the MPs through glass ocular lenses. FTIR helped confirmation and identification of the plastics like PE, PP, PA resin, and silicone rubbers in groundwater. The most common was PE. The results confirmed high level of contamination of both the sites, may be due to onsite sewage treatment plant. The findings alarmed the need for improved waste management practice to reduce the MPs pollution [124].

References

1. Sana, S.S., Dogiparthi, L.K., Gangadhar, L., et al. (2020). Effects of microplastics and nanoplastics on marine environment and human health. Environmental Science and Pollution Research, Volume 27, Issue 36, pp. 44743–44756. https://doi.org/10.1007/s11356-020-10573-x.
2. Botterell, Z.L.R., Beaumont, N., Dorrington, T., et al. (2019). Bioavailability and effects of microplastics on marine zooplankton: a review. Environmental Pollution, Volume 245, pp. 98–110. https://doi.org/10.1016/j.envpol.2018.10.065.
3. Vianello, A., Boldrin, A., Guerriero, P., et al. (2013). Microplastic particles in sediments of lagoon of Venice, Italy: first observations on occurrence, spatial patterns and identification. Estuarine, Coastal and Shelf Science, Volume 130, pp. 54–61. https://doi.org/10.1016/j.ecss.2013.03.022.
4. Ajith, N., Arumugam, S., Parthasarathy, S., et al. (2020). Global distribution of microplastics and its impact on marine environment – a review. Environmental Science and Pollution Research, Volume 27, Issue 21, pp. 25970–25986. https://doi.org/10.1007/s11356-020-09015-5.
5. Mintenig, S.M., Löder, M.G.J., Primpke, S., et al. (2019). Low numbers of microplastics detected in drinking water from ground water sources. Science of the Total Environment, Volume 648, pp. 631–635. https://doi.org/10.1016/j.scitotenv.2018.08.178.

6 Selvam, S., Manisha, A., Venkatramanan, S., et al. (2020). Microplastic presence in commercial marine sea salts: a baseline study along Tuticorin Coastal salt pan stations, Gulf of Mannar, South India. Marine Pollution Bulletin, Volume 150. https://doi.org/10.1016/j.marpolbul.2019.110675.
7 Talvitie, J., Mikola, A., Koistinen, A., et al. (2017). Solutions to microplastic pollution – removal of microplastics from wastewater effluent with advanced wastewater treatment technologies. Water Research, Volume 123, pp. 401–407. https://doi.org/10.1016/j.watres.2017.07.005.
8 Li, W.C. (2018). The occurrence, fate, and effects of microplastics in the marine environment. In: Zeng EY (ed) Microplastic contamination in aquatic environments: an emerging matter of environmental urgency. Elsevier, pp. 133–173. https://doi.org/10.1016/B978-0-12-813747-5.00005-9.
9 Vaughan, R., Turner, S.D., and Rose, N.L. (2017). Microplastics in the sediments of a UK urban lake. Environmental Pollution, Volume 229, pp. 10–18. https://doi.org/10.1016/j.envpol.2017.05.057.
10 Alimba, C.G., and Faggio, C. (2019). Microplastics in the marine environment: current trends in environmental pollution and mechanisms of toxicological profile. Environmental Toxicology and Pharmacology, Volume 68, pp. 61–74. https://doi.org/10.1016/j.etap.2019.03.001.
11 Andrady, A.L. (2011). Microplastics in the marine environment. Marine Pollution Bulletin, Volume 62, Issue 8, pp. 1596–1605. https://doi.org/10.1016/j.marpolbul.2011.05.030.
12 Wang, J., Tan, Z., and Peng, J. (2016). The behaviors of microplastics in the marine environment. Marine Environmental Research, Volume 113, pp. 7–17. https://doi.org/10.1016/j.marenvres.2015.10.014.
13 Perkins, S. (2015). Nearly every seabird may be eating plastic by 2050. Science. https://doi.org/10.1126/science.aad1694.
14 Garcia, J.M., and Robertson, M.L. (2017). The future of plastics recycling. In Science. American Association for the Advancement of Science, Volume 358, Issue 6365, pp. 870–872. https://doi.org/10.1126/science.aaq0324.
15 Juan, R., Domínguez, C., Robledo, N., et al. (2021). Challenges and opportunities for recycled polyethylene fishing nets: towards a circular economy. Polymers, Volume 13, Issue 18. https://doi.org/10.3390/polym13183155.
16 Wang, J., Liu, X., Liu, G., et al. (2019). Size effect of polystyrene microplastics on sorption of phenanthrene and nitrobenzene. Ecotoxicology and Environmental Safety, Volume 173, pp. 331–338. https://doi.org/10.1016/j.ecoenv.2019.02.037.
17 Periyasamy, A.P., and Tehrani-Bagha, A. (2022). A review on microplastic emission from textile materials and its reduction techniques. Polymer Degradation and Stability, Volume 199. https://doi.org/10.1016/j.polymdegradstab.2022.109901.

18 Khalid, N., Aqeel, M., Noman, A., et al. (2021). Linking effects of microplastics to ecological impacts in marine environments. Chemosphere, Volume 264. https://doi.org/10.1016/j.chemosphere.2020.128541.

19 Auta, H.S., Emenike, C.U., and Fauziah, S.H. (2017). Distribution and importance of microplastics in the marine environment: a review of the sources, fate, effects, and potential solutions. Environment International, Volume 102, pp. 165–176. https://doi.org/10.1016/j.envint.2017.02.013.

20 Lusher, A.L., Burke, A., O'Connor, I., et al. (2014). Microplastic pollution in the North East Atlantic Ocean: validated and opportunistic sampling. Marine Pollution Bulletin, Volume 88, pp. 325–333.

21 Lusher, A.L., Tirelli, V., O'Connor, I., et al. (2015a). Microplastics in arctic polar waters: the first reported values of particles in surface and sub-surface samples. Scientific Reports, Volume 5, p. 14947. https://doi.org/10.1038/srep/14947.

22 Lusher, A.L., Harnandez-Milian, G., O'Brien, J., et al. (2015b). Microplastic and macroplastic ingestion by a deep diving, oceanic cetacean: the true's beaked whale *Mesoplodon mirus*. Environmental Pollution, Volume 199, pp. 185–191.

23 Eriksen, M., Mason, S., Wilson, S., et al. (2013). Microplastic pollution in the surface waters of the Laurentian Great Lakes. Marine Pollution Bulletin, Volume 7, pp. 177–182. https://doi.org/10.1016/j.marpolbul.2013.10.007.

24 Dubaish, F., and Liebezeit, G. (2013). Suspended microplastics and black carbon particles in the jade system, southern North Sea. Water, Air and Soil Pollution, Volume 224, p. 1352. https://doi.org/10.1007/s11270-012-1352-9.

25 Law, K.L., and Thompson, R.C. (2014). Microplastics in the seas. Science, Volume 345, Issue 6193, pp. 144–145. https://doi.org/10.1126/science.1254065.

26 Antunes, J.C., Frias, J.G.L., Micaelo, A.C., et al. (2013). Resin pellets from beaches of the Portuguese coast and adsorbed persistent organic pollutants. Estuarine, Coastal and Shelf Science, Volume 130, pp. 62–69.

27 Alomar, C., Estarellas, F., and Deudero, S. (2016). Microplastics in the Mediterranean Sea: deposition in coastal shallow sediments, spatial variation and preferential grain size. Marine Environmental Research, Volume 115, pp. 1–10.

28 Zhao, S., Zhu, L., Wang, T., et al. (2014). Suspended microplastics in the surface water of the Yangtze estuary system, China: first observations on occurrence, distribution. Marine Pollution Bulletin, Volume 15, Issue 86 (1–2), pp. 562–568. https://doi.org/10.1016/j.marpolbul.2014.06.032.

29 de Carvalho, D.G., and Neto, J.A.D. (2016). Microplastic pollution of the beaches of Guanabara Bay, Southeast Brazil. Ocean and Coastal Management, Volume 128, pp. 10–17.

30 Renzi, M., and Blašković, A. (2018). Litter and microplastics features in table salts from marine origin: Italian versus Croatian brands. Marine Pollution Bulletin, Volume 135, pp. 62–68. https://doi.org/10.1016/j.marpolbul.2018.06.065.

31 Vermaire, J.C., Pomeroy, C., Herczegh, S.M., et al. (2017). Microplastic abundance and distribution in the open water and sediment of the Ottawa river, Canada, and its tributaries. Facets, Volume 2, Issue 1, pp. 301–314. https://doi.org/10.1139/facets-2016-0070.

32 Scott, N., Porter, A., Santillo, D., et al. (2019). Particle characteristics of microplastics contaminating the mussel Mytilus edulis and their surrounding environments. Marine Pollution Bulletin, Volume 146, pp. 125–133. https://doi.org/10.1016/j.marpolbul.2019.05.041.

33 Watts, A.J.R., Lewis, C., Goodhead, R.M., et al. (2014). Uptake and retention of microplastics by the shore crab *Carcinus maenas*. Environmental Science and Technology, Volume 24, pp. 8823–8830.

34 Van Cauwenberghe, L., and Janssen, C.R. (2014). Microplastics in bivalves cultured for human consumption. Environmental Pollution, Volume 193, pp. 65–70.

35 Setala, O., Norkko, J., and Lehtiniemi, M. (2016). Feeding type affects microplastic ingestion in a coastal invertebrate community. Marine Pollution Bulletin, Volume 102, pp. 95–101.

36 von Moos, N., Burkhardt-Holm, P., and Köhler, A. (2012). Uptake and effects of microplastics on cells and tissues of the blue mussel *Mytillus edilus* L. after experimental exposure. Environmental Science and Technology, Volume 46, pp. 11327–11335.

37 Ferreira, P., Fonte, E., Soares, M.E., et al. (2016). Effects of multi-stressors on juveniles of the marine fish Pomatoschistus microps: gold nanoparticles, microplastics and temperature. Aquatic Toxicology, Volume 170, pp. 89–103.

38 Batel, A., Linti, F., Scherer, M., et al. (2016). The transfer of benzo(a)pyrene from microplastics to *Artemia* nauplii and further to zebrafish via trophic food web experiment – CYP1A induction and visual tracking of persistent organic pollutants. Environmental Toxicology and Chemistry, Volume 35, Issue 7, pp. 1656–1666. https://doi.org/10.1002/etc.3361.

39 Lu, Y., Zhang, Y., Deng, Y., et al. (2016). Uptake and accumulation of polystyrene microplastics in zebra fish (Danio rerio) and toxic effects in liver. Environmental Science and Technology, Volume 50, pp. 4054–4060.

40 Rochman, C.M., Hoh, E., Kurobe, T., et al. (2013). Ingested plastic transfers hazardous chemicals to fish and induces hepatic stress. Scientific Reports, Volume 3, p. 3263. https://doi.org/10.1038/srep03263.

41 Fossi, M.C., Marsili, L., Baini, M., et al. (2016). Fin whales and microplastics: the Mediterranean Sea and the sea of Cortez scenarios. Environmental Pollution, Volume 209, pp. 68–78.

42 Besseling, E., Foekema, E.M., Van Franeker, J.A., et al. (2015). Microplastics in a macro filter feeder: humpback whale *Megaptera novaeangliae*. Marine Pollution Bulletin, Volume 95, pp. 248–252. https://doi.org/10.1016/j.marpolbul.2015.04.007.

43 Rummel, C.D., Löder, M.G.J., Fricke, N.F., et al. (2016). Plastic ingestion by pelagic and demersal fish from the North Sea and Baltic Sea. Marine Pollution Bulletin. https://doi.org/10.1016/j.marpolbul.2015.11.043.

44 Cole, M., Lindeque, P., Fileman, E., et al. (2013). Microplastic ingestion by zooplankton. Environmental Science and Technology, Volume 47, pp. 6646–6655. https://doi.org/10.1021/es400663f.

45 Rehse, S., Kloas, W., and Zarfi, C. (2016). Short-term exposure with high concentrations of pristine microplastic particles leads to immobilization of Daphnia magna. Chemosphere, Volume 153, pp. 91–99.

46 Devriese, L.I., van der Meulen, M.D., Maes, T., et al. (2015). Microplastic contamination in brown shrimp (*Crangon crangon*, Linnaeus 1758) from coastal waters of the southern North Sea and channel area. Marine Pollution Bulletin, Volume 98, pp. 179–187.

47 Avio, C.G., Gorbi, S., Milan, M., et al. (2016). Pollutants bioavailability and toxicological risk from microplastics to mussels. Environmental Pollution, Volume 198, pp 211–222.

48 Browne, M.A., Dissanayake, A., Galloway, T.S., et al. (2008). Ingested microscopic plastic translocate to the circulatory system of the mussel, *Mytilus edulis* (L). Environmental Science and Technology, Volume 42, pp. 5026–5031.

49 Chua, E.M., Shimeta, J., Nugegoda, D., et al. (2014). Assimilation of polybrominated diphenyl ethers from microplastics by the marine amphipod, *Allorchestes compressa*. Environmental Science and Technology, Volume 48, Issue 14, pp. 8127–8134. https://doi.org/10.1021/es405717z.

50 Green, D.S. (2016). Effects of microplastics on European flat oysters, Ostrea edulis and their associated benthic communities. Environmental Pollution, Volume 216, pp. 95–103.

51 Green, D.S., Boots, B., Sigwart, J., et al. (2016). Effects of conventional and biodegradable microplastics on a marine ecosystem engineer (*Arenicola marina*) and sediment nutrient cycling. Environmental Pollution, Volume 208, pp. 426–434.

52 Besseling, E., Wang, B., Lürling, M., et al. (2014). Nanoplastic affects growth of *S. obliquus* and reproduction of *D. magna*. Environmental Science and Technology, Volume 48, Issue 20, pp. 12336–12343.

53 Cole, M., Lindeque, P.K., Fileman, E., et al. (2016). Microplastics alter the properties and sinking rates of zooplankton faecal pellets. Environmental Science and Technology, Volume 50, pp. 3239–3246. https://doi.org/10.1021/acs.est.5b05905.

54 Desforges, J.W., Galbraith, M., and Ross, P.S. (2015). Ingestion of microplastics by zooplankton in the Northeast Pacific Ocean. Archives of Environmental Contamination and Toxicology, Volume 69, pp 320–330. https://doi.org/10.1007/s00244-015-0172-5.

55 Bhattacharya, P., Lin, S., Turner, J.P., et al. (2010). Physical adsorption of charged plastic nanoparticles affects algal photosynthesis. Journal of Physical Chemistry C, Volume 114, pp. 16556–16561.

56 Besseling, E., Wegner, A., Foekema, E.M., et al. (2012). Effects of microplastics on fitness and PCB bioaccumulation by the lugworm *Arenicola marina* (L). Environmental Science and Technology, Volume 47, Issue 1, pp. 593–600. https://doi.org/10.1021/es302763x.

57 Sjollema, S.B., Redondo-Hasselerharm, P., Leslie H.A., et al. (2016). Do plastic particles affect microalgal photosynthesis and growth? Aquatic Toxicology, Volume 170, pp. 259–261.

58 Lagarde, F., Olivier, O., Zanella, M., et al. (2016). Microplastic interactions with freshwater microalgae: hetero-aggregation and changes in plastic density appear strongly dependent on polymer type. Environmental Pollution, Volume 215, pp. 331–339.

59 Holm, P., Schulz, G., and Athanasopulu, K. (2013). Meeresverschmutzung der neuen Art: Mikroplastik—ein unsichtbarer Störenfried. Biologie in Unserer Zeit, Volume 43, Issue 1, pp. 27–33. https://doi.org/10.1002/biuz.201310497.

60 Murphy, F., and Quinn, B. (2018). The effects of microplastic on freshwater *Hydra attenuata* feeding, morphology and reproduction. Environmental Pollution, Volume 234, pp. 487–494.

61 Rosenkranz, P., Chaudhary, Q., Stone, V., et al. (2009). A comparison of nanoparticle and fine particle uptake by Daphnia magna. Environmental Toxicology and Chemistry, Volume 28, pp. 2142–2149.

62 Imhof, H.K., Ivleva, N.P., Schmid, J., et al. (2013). Contamination of beach sediments of a sub alpine lake with microplastic particles. Current Biology, Volume 23, pp. R867–R868.

63 Liu, Z., Cai, M., Yu, P., et al. (2018). Age-dependent survival, stress defense, and AMPK in *Daphnia pulex* after short-term exposure to a polystyrene nanoplastic. Aquatic Toxicology, Volume 204, pp. 1–8.

64 Gonzalezfernandez, C., Tallec, K., Le, N.G., et al. (2018). Cellular responses of Pacific oyster (*Crassostrea gigas*) gametes exposed in vitro to polystyrene nanoparticles. Chemosphere, Volume 208, pp. 764–772.

65 Ding, J., Zhang, S., Razanajatovo, R.M., et al. (2018). Accumulation, tissue distribution, and biochemical effects of polystyrene microplastics in the freshwater fish red tilapia (*Oreochromis niloticus*). Environmental Pollution, Volume 238, pp. 1–9.

66 Machado, A.A.S., Valyi, K., and Rillig, M.C. (2017). Potential environmental impacts of an “underground revolution”: a response to Bender et al. Trends in Ecology and Evaluation, Volume 32, Issue 1, pp. 8–10.

67 Brandts, I., Teles, M., Gonçalves, A.P., et al. (2018). Effects of nanoplastics on *Mytilus galloprovincialis* after individual and combined exposure with carbamazepine. Science of the Total Environment, Volume 643, pp. 775–784.

68 Barboza, L.G.A., Vieira, L.R., Branco, V., et al. (2018). Microplastics cause neurotoxicity, oxidative damage and energy-related changes and interact with the bioaccumulation of mercury in the European seabass, *Dicentrarchus labrax* (Linnaeus, 1758). Aquatic Toxicology, Volume 195, pp. 49–57.

69 Chen, Q., Yin, D., Jia, Y., et al. (2017). Enhanced uptake of BPA in the presence of nanoplastics can lead to neurotoxic effects in adult zebrafish. Science of the Total Environment, Volume 609, pp. 1312–1321.

70 Bhargava, S., Lee, S.S.C., Ying, L.S.M., et al. (2018). Fate of nanoplastics in marine larvae: a case study using barnacles, *Amphibalanus amphitrite*. ACS Sustainable Chemistry and Engineering, Volume 6, pp. 6932–6940.

71 Canesi, L., Ciacci, C., Fabbri, R., et al. (2016). Interactions of cationic polystyrene nanoparticles with marine bivalve hemocytes in a physiological environment: role of soluble hemolymph proteins. Environmental Research, Volume 150, pp. 73–81.

72 Frias, J.P.G.L., and Nash, R. (2019). Microplastics: finding a consensus on the definition. Marine Pollution Bulletin, Volume 138, pp. 145–147. https://doi.org/10.1016/j.marpolbul.2018.11.022.

73 Waller, C.L., Griffiths, H.J., Waluda, C.M., et al. (2017). Microplastics in the Antarctic marine system: an emerging area of research. Science of the Total Environment, Volume 598, pp. 220–227. https://doi.org/10.1016/j.scitotenv.2017.03.283.

74 Wu, P., Huang, J., Zheng, Y., et al. (2019). Environmental occurrences, fate, and impacts of microplastics. Ecotoxicology and Environmental Safety, Volume 184. https://doi.org/10.1016/j.ecoenv.2019.109612.

75 Garcés-Ordóñez, O., Castillo-Olaya, V.A., Granados-Briceño, A.F., et al. (2019). Marine litter and microplastic pollution on mangrove soils of the Ciénaga Grande de Santa Marta, Colombian Caribbean. Marine Pollution Bulletin, Volume 145, pp. 455–462. https://doi.org/10.1016/j.marpolbul.2019.06.058.

76 Piperagkas, O., Papageorgiou, N., and Karakassis, I. (2019). Qualitative and quantitative assessment of microplastics in three sandy Mediterranean beaches, including different methodological approaches. Estuarine, Coastal and Shelf Science, Volume 219, pp. 169–175. https://doi.org/10.1016/j.ecss.2019.02.016.

77 Koelmans, A.A., Mohamed Nor, N.H., Hermsen, E., et al. (2019). Microplastics in freshwaters and drinking water: critical review and assessment of data quality. Water Research, Volume 155, pp. 410–422. https://doi.org/10.1016/j.watres.2019.02.054.

78 Perren, W., Wojtasik, A., and Cai, Q. (2018). Removal of microbeads from wastewater using electrocoagulation. ACS Omega, Volume 3, Issue 3, pp. 3357–3364.

79 Ma, B., Xue, W., Hu, C., et al. (2019). Characteristics of microplastic removal via coagulation and ultra-filtration during drinking water treatment. Chemical Engineering Journal, Volume 359, pp. 159–167.

80 Lares, M., Ncibi, M.C., Sillanpää, M., et al. (2018). Occurrence, identification and removal of microplastic particles and fibers in conventional activated sludge process and advanced MBR technology. Water Research, Volume 133, pp. 236–246.

81 Mulindwa, P., Kasule, J.S., Nantaba, F., et al. (2024). Bioadsorbents for removal of microplastics from water ecosystems: a review. International Journal of Sustainable Engineering, Volume 17, Issue 1, pp. 582–599. https://doi.org/10.1080/19397038.2024.2374003.

82 Ahmad, M., Lubis, N.M.A., Usama, M., et al. (2023). Scavenging microplastics and heavy metals from water using jujube waste-derived biochar in fixed-bed column trials. Environmental Pollution, Volume 335, p. 122319. https://doi.org/10.1016/j.envpol.2023.122319.

83 Wang, J., Sun, C., Huang, Q.X., et al. (2021). Adsorption and thermal degradation of microplastics from aqueous solutions by Mg/Zn modified magnetic biochars. Journal of Hazardous Materials, Volume 419, p. 126486. https://doi.org/10.1016/j.jhazmat.2021.126486.

84 Shi, Y., Du, J., Zhao, T., et al. (2023). Removal of nanoplastics from aqueous solution by aggregation using reusable magnetic biochar modified with cetyltrimethylammonium bromide. Environmental Pollution, Volume 318, p. 120897. https://doi.org/10.1016/j.envpol.2022.120897.

85 Honarmandrad, Z., Kaykhaii, M., and Gębicki, J. (2023). Microplastics removal from aqueous environment by metal organic frameworks. BMC Chemistry, Volume 17, p. 122. https://doi.org/10.1186/s13065-023-01032-y.

86 Mohana, A.A., Rahman, M., Sarker, S.K., et al. (2022). Nano/microplastics: fragmentation, interaction with co-existing pollutants and their removal from wastewater using membrane processes. Chemosphere, Volume 309, p. 136682. https://doi.org/10.1016/j.chemosphere.2022.136682.

87 Acarer, S. (2023). A review of microplastic removal from water and wastewater by membrane technologies. Water Science Technology, Volume 88, Issue 1, pp. 199–219. https://doi.org/10.2166/wst.2023.186.

88 Kara, N., Sari, E.H., and Onkal, E.G. (2023). Characterization and removal of microplastics in landfill leachate treatment plants in Istanbul, Turkey. Analytical Letters, Volume 56, Issue 9, pp. 1535–1548. https://doi.org/10.1080/00032719.2022.2137808.

89 Dalmau-Soler, J., Ballesteros-Cano, R., Boleda, M. et al. (2021). Microplastics from headwaters to tap water: occurrence and removal in a drinking water treatment plant in Barcelona metropolitan area (Catalonia, NE Spain). Environmental Science and Pollution Research, Volume 28, pp. 59462–59472. https://doi.org/10.1007/s11356-021-13220-1.

90 Sun, J., Zhu, Z.R., Li, W.H., et al. (2021). Revisiting microplastics in landfill leachate: unnoticed tiny microplastics and their fate in treatment works. Water Research, Volume 190, p. 116784. https://doi.org/10.1016/j.watres.2020.116784.

91 Cai, Y., Wu, J., Lu, J., et al. (2022). Fate of microplastics in a coastal wastewater treatment plant: microfibers could partially break through the integrated membrane system. Frontiers of Environmental Science and Engineering, Volume 16, Article 96. https://doi.org/10.1007/s11783-021-1517-0.

92 Bodzek, M., and Bodzek, P. (2025). Remediation of micro- and nanoplastics by membrane technologies. Membranes, Volume 15, Issue 3, p. 82. https://doi.org/10.3390/membranes15030082.

93 Hamd, W., Daher, E.A., Tofa, T.S., et al. (2022). Recent advances in photocatalytic removal of microplastics: mechanisms, kinetic degradation, and reactor design. Frontiers in Marine Science, Volume 9, p. 885614. https://doi.org/10.3389/fmars.2022.885614.

94 Gao, W., Zhang, Y., Mo, A., et al. (2022). Removal of microplastics in water: technology progress and green strategies. Green Analytical Chemistry, Volume 3, p. 100042. https://doi.org/10.1016/j.greeac.2022.100042.

95 Li, L., Liu, D., Song, K., et al. (2020). Performance evaluation of MBR in treating microplastics polyvinylchloride contaminated polluted surface water. Marine Pollution Bulletin, Volume 150, pp. 110–724.

96 Padervand, M., Lichtfouse, E., Robert, D., et al. (2020). Removal of microplastics from the environment: a review. Environmental Chemistry Letters, Volume 18, pp. 807–828.

97 Kang, J., Zhou, L., Duan, X., et al. (2019). Degradation of cosmetic microplastics via functionalized carbon nanosprings. Matter, Volume 1, pp. 745–758.

98 Caruso, G. (2015). Microplastics in marine environments: possible interactions with the microbial assemblage. Journal of Pollution Effects and Control, Volume 3, Issue 2, pp. 1–11.

99 Dey, T.K., Uddin, M.E., and Jamal, M. (2021). Detection and removal of microplastics in wastewater: evolution and impact. Environmental Science and Pollution Research, Volume 28, pp. 16925–16947.

100 Talvitie, J., Mikola, A., Setal, O., et al. (2016). How well is microlitter purified from wastewater? A detailed study on the stepwise removal of microlitter in a tertiary level wastewater treatment plant. Water Research, Volume 109, pp. 164–172.

101 Dounia, E., Vinka, O., and Elvis, C. (2021). Electrocoagulation applied for the removal of microplastics from wastewater treatment facilities. Separation and Purification Technology, Volume 276, p. 118877. https://doi.org/10.1016/j.seppur.2021.118877.

102 Sara Mateo, S., Zhang, A., Piedra, A., et al. (2024). Electrocoagulation assessment to remove micropolystyrene particles in wastewater. ACS EST Water, Volume 4, Issue 7, pp. 3049–3058. https://doi.org/10.1021/acsestwater.4c00312.

103 Kala, S., Raji, K., Logeshwaran, P., et al. (2023). Fate and transformation of microplastics due to electrocoagulation treatment: impacts of polymer type and shape. Environmental Pollution, Volume 334, p. 122159. https://doi.org/10.1016/j.envpol.2023.122159.

104 Yaranal, N.A., Kuchibhotla, S.A., Subbiah, S., et al. (2024). Identification, removal of microplastics and surfactants from laundry wastewater using electrocoagulation method. Water Emerging Contaminants and Nanoplastics, Volume 3, Issue 1, pp. 1–15. https://doi.org/10.20517/wecn.2023.46.

105 Jain, P., Yadav, K.D., and Dholakiya, B.Z. (2025). Assessing the efficacy of electrocoagulation process for polypropylene microplastics removal from wastewater: optimization through TOPSIS approach. Environmental Science and Pollution Research, Volume 32, Issue 19, pp. 12066–12085. https://doi.org/10.1007/s11356-025-36410-7.

106 Kasmuri, N., Rosli M.S., Zaini, N., et al. (2024). Reduction of microplastic in wastewater via electrocoagulation process. IOP Conference Series: Earth and Environmental Science, Volume 1303, p. 012020. https://doi.org/10.1088/1755-1315/1303/1/012020.

107 Esskifati, M., Hammoudani, Y.E., Haboubi, K., et al. (2024). Electrocoagulation as a possible treatment for wastewater contaminated with microplastics – a review. Bio Web Conference, Volume 109, p. 9. https://doi.org/10.1051/bioconf/202410901011.

108 Ceyhun, A., Halil, K., Ahmet, E.K., et al. (2021). Removal of microplastics from wastewater through electrocoagulation-electroflotation and membrane filtration processes. Water Science Technology, Volume 84, Issue 7, pp. 1648–1662. https://doi.org/10.2166/wst.2021.356.

109 Rahman, N.A., Jol, C.J., Linus, A.A., et al. (2021). Emerging application of electrocoagulation for tropical peat water treatment: a review. Chemical Engineering Process, Volume 165, p. 108449.

110 Meng, X., Peng, X., Xue, J., et al. (2021). A biomass-derived, all-day-round solar evaporation platform for harvesting clean water from microplastic pollution. Journal of Materials Chemistry A. Volume 9, Issue 17, pp. 11013–11024.

111 Uheida, A., Mejía, H.G., Abdel-Rehim, M., et al. (2020). Visible light photocatalytic degradation of polypropylene microplastics in a continuous water flow system. Journal of Hazardous Materials, Volume 406, p. 124299. https://doi.org/10.1016/j.jhazmat.2020.124299.

112 Sacco, N.A., Zoppas, F.M., Devard, A., et al. (2023). Recent advances in microplastics removal from water with special attention given to photocatalytic degradation: review of scientific research. Microplastics, Volume 2, Issue 3, pp. 278–303. https://doi.org/10.3390/microplastics2030023.

113 Velasquez, J.A., Bao, Y., and Huang, J. (2024). Atmospheric cold plasma as a novel approach to remediating microplastics pollution in water. Environmental Pollution, Volume 356, p. 124390. https://doi.org/10.1016/j.envpol.2024.124390.

114 Ren, J., Li, J., Zhen, Y., et al. (2022). Removal of polyvinyl chloride microplastic by dielectric barrier discharge plasma. Separation and Purification Technology, Volume 290, p. 120832. https://doi.org/10.1016/j.seppur.2022.120832.

115 Kim, H., Lee, U., Lee, I., et al. (2023). Enhancement of polystyrene microplastic removal by near dissolved organic matter microfiltration (NDOM MF) coupled with cold plasma treatment. Journal of Water Process Engineering, Volume 54, p. 103901. https://doi.org/10.1016/j.jwpe.2023.103901.

116 Rozman, U., Blažič, A., and Kalčíková, G. (2023). Phytoremediation: a promising approach to remove microplastics from the aquatic environment. Environmental Pollution, Volume 338, p. 122690. https://doi.org/10.1016/j.envpol.2023.122690.

117 Yin, J., Zhu, T., Li, X., et al. (2025). Phytoremediation of microplastics by water hyacinth. Environmental Science and Ecotechnology, Volume 24, p. 100540. https://doi.org/10.1016/j.ese.2025.100540.

118 Yuan, W., Xu, E.G., Shabaka, S., et al. (2024). The power of green: harnessing phytoremediation to combat micro/nanoplastics. Eco-Environment and Health, Volume 3, Issue 3, pp. 260–265. https://doi.org/10.1016/j.eehl.2024.04.001.

119 Singh, S., Kalyanasundaram, M., and Diwan, V. (2021). Removal of microplastics from wastewater: available techniques and way forward. Water Science and Technology, Volume 84, Issue 12, pp. 3689–3704. https://doi.org/10.2166/wst.2021.472.

120 Lotfigolsefidi, F., Davoudi, M., Sarkhosh, M., et al. (2025). Removal of microplastics by algal biomass from aqueous solutions: performance, optimization, and modeling. Scientific Reports, Volume 15, Issue 1, p. 501. https://doi.org/10.1038/s41598-024-84114-8.

121 Amjad, M., Intisar, A., Afzal, A., et al. (2023). Chapter Four – biological methods for the removal of microplastics from water. In: Ferreira LFR, Kumar A, Bilal M (eds) Advances in chemical pollution, Environmental management and protection Volume 9, Elsevier, pp. 65–78. https://doi.org/10.1016/bs.apmp.2022.10.003.

122 Anand, U., Dey, S., Bontempi, E., et al. (2023). Biotechnological methods to remove microplastics: a review. Environmental Chemistry Letters, Volume 21, Issue 3, pp. 1787–1810. https://doi.org/10.1007/s10311-022-01552-4.

123 Li, Y., Qin, T., Bai, X., et al. (2024). Occurrence and removal of microplastics in tertiary wastewater treatment plants: a case study of three plants in Zhengzhou, China. Processes, Volume 12, Issue 4, p. 803. https://doi.org/10.3390/pr12040803.

124 Wilyalodia, H.C., Moersidik, S.S., and Pratama, M.A. (2024). Microplastic contamination in urban groundwater: a case study from Jakarta Megacity, Indonesia. The Scientific World Journal. https://doi.org/10.1155/tswj/1870706.

15

Wastewater Treatment Technologies: Performance and Mechanism

Several methods have been proposed and investigated to eliminate various contaminants from wastewater. No single method is sufficient to remove the pollutants from aqueous solutions completely. The appropriateness of the method is decided based on the nature and chemistry of the contaminant and the pathway of its elimination. The mechanism of elimination thus becomes crucial to be understood so as to ascertain the feasibility of the technique for the removal of pollutants from wastewater. The present chapter gives an insight into the mechanistic routes involved in various treatment methods employed for contaminant removal.

Though the mechanism of removal differs from one pollutant to another, the general approach can be important to understand. An example is the mass transfer mechanism in some processes, such as adsorption, ion exchange, and membrane filtration [1].

15.1 Mechanisms of Different Processes

The mechanisms operating in several processes have been discussed in the sections below:

15.1.1 Adsorption

Adsorption refers to the transport of contaminants onto the surface of the adsorbent, which holds them either through physical or chemical intramolecular interactions. During the process, forces like Van der Waals forces, hydrophobic interactions, and hydrogen bonding may exist along with processes like surface diffusion and intraparticle pore diffusion. The porous nature of the adsorbent and the total exposed area of the adsorbent facilitate efficient removal of the pollutants from wastewater. The process involves the formation of inner sphere and outer sphere coordination compounds on the adsorbent surface. The adsorption of metal ions involving the emplacement of water molecules results in the formation of outer sphere complexes, while with no water molecule, inner complexes are formed. The process involves coordinate linkage formation on the surface functional group existing on the adsorbent. In a few cases, electrostatic interactions can be predominant [2]. Biosorption involves chelation along with the ion exchange process. Weak Van der Waals forces constitute the primary removal mechanism in physisorption.

Remediation of contaminants by adsorbents is carried out by numerous physical and chemical interactions involving the formation of outer and inner sphere complexes, electrostatic forces, exchange of ions, and precipitation at the surfaces of the adsorbents. The interior structure of the adsorbent, like biochar, is made up of a blend of macropores, mesopores, and micropores, which helps retain the contaminants. Alteration in the functionality takes place during their association, which is confirmed by the Fourier transform infrared (FTIR) technique. pH and redox potential are important in influencing the adsorption characteristics. Cross-linked structure of adsorbents like coal facilitates complex formation with contaminants, which are further confirmed by the developed metal carboxylate, chemical composition, solubility, structural arrangement, particle size, and surface chemistry, which are the deciding factors in the performance of absorbents. High surface area to volume ratio greatly influences the matrix properties of certain nanoparticle-based adsorbent materials. The adsorption kinetics of metal ions over fly ash reveals diffusion as the key mechanism involved [3].

The adsorption process has the advantage of reproducibility via desorption. The adsorption of adsorbate onto solid adsorbents involves three steps:

1. Transportation of contaminant to the adsorbent surface,
2. Association or linkage with the adsorbent surface, and
3. Transportation within the adsorbent particles.

Electrostatic interaction dominates in removal via adsorption. Adsorption can proceed either physically or chemically. Physical interaction mainly involves nonspecific Van der Waals forces, whereas chemisorption involves bonding through a chemical reaction, providing greater attractive strength. This makes chemical adsorption irreversible. The adsorptive removal of contaminants over biosorbents like bacteria, fungi, industrial and agricultural waste involves intraparticle diffusion, hydrogen bond, redox reaction, complexation, precipitation, pore adsorption, etc. Adsorbent material having a biological origin usually contains hydroxyl, carboxyl, carbonyl, and ammonia groups and is quite porous with a large number of cavities and binding sites. Additionally, hydrophobic and hydrogen bond interactions also contribute to the contaminant removal (Figure 15.1). Complex

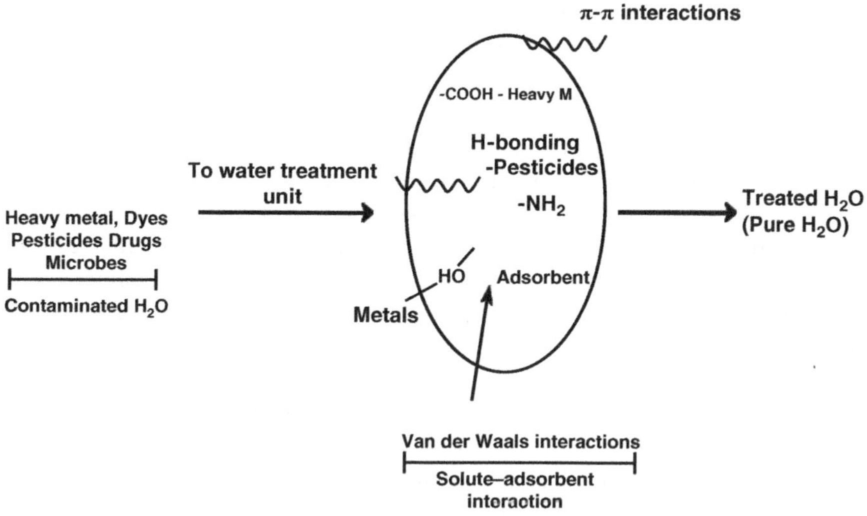

Figure 15.1 Adsorption mechanism.

formation occurs when a metal pollutant is surrounded by ligands, while the development of rings occurs in chelation when pollutant is present on the surface of the adsorbents [4].

Adsorbent biochar is produced by pyrolysis in an oxygen-free environment, whereby small-sized biomass material of dimension nanometers or centimeters can shape up. Heavy metal removal over biochar is a synergistic effect of various mechanisms, including surface complexation and ion exchange. Adsorbent biochar has shown the potential to eliminate inorganic salts present in wastewater via surface functional groups, electrostatic interaction, surface co-precipitation, and complexation, along with hydrogen bonding. The rich microporous structure and large specific surface area of biochar provide active size for inorganic ions like phosphate. Physical deposition occurs due to the diffusion of the inorganic ions onto the surface of biochar. The adsorbent is equipped with various functional groups like hydroxyl, carboxyl, carbonyl, amine, and nitro, which can readily combine with the organic ion. Sometimes, chemical adsorption occurs in the presence of metal oxides and metal hydroxides on the adsorbent surface. The removal mechanism of dyes or surfactants depends on the chemistry of the contaminants, but is mainly governed by π–π interaction and multilayer mechanism like pore filling [5].

The fabrication of adsorbents for wastewater treatment had a breakthrough with the application of nanotechnology, which allowed the synthesis of adsorbents with better surface area, adjustable pore size, structure, and functionality.

Also, adsorbents can be modified to improve their adsorption characteristics and performance. Methods like thermal treatment, acid–base cleaning, irradiation, impregnation, and grafting can be employed. These activation processes alter the surface area, porosity, electrical nature, and hydrophobicity of the adsorbents. The selectivity of the adsorbents to adsorb hydrophobic molecules from an aqueous environment is possible by exploiting hydrophobic and hydrophilic interactions. For example, an oil spill can be remediated from the water surface using adsorption. Another fabrication method involves enriching the adsorbent with electrons or protons by modification in the surface chemistry using acid–base interaction. Adsorbents are designed to facilitate specific interaction with the target molecule [6]. The alteration in the adsorptive properties of the adsorbent can be done by using steam activation, thermal air treatment, acid modification, base modification, hydrothermal methods, ball milling, impregnation, pyrolysis, etc.

Extensive research has been conducted related to metal–organic frameworks (MOF), as promising and versatile contenders for the remediation of a variety of pollutants from wastewater. They have proven to be better than traditional options for the utilization of adsorbents like activated charcoal and zeolites for wastewater treatment. A MOF effectively adsorbs contaminants from wastewater due to its high surface area, selectivity, sustainable properties, and functional tunability.

Water-soluble MOFs have been employed in the wastewater treatment process. Such MOFs can be

1. Hyper valent metal ions incorporated with metal carboxylate frameworks.
2. Nitrogen donors incorporated into a metal azo framework or
3. Functionalized metal ions.

Post synthetic modifications can improve the pore reactivity of MOFs.

MOFs are promising, efficient, and viable options for emerging contaminants from wastewater in the future. Some of the synthesis methods of MOFs are summarized below [7]:

1. **Conventional Synthesis:** This method of synthesizing MOFs includes solvothermal and nonsolvothermal methods. In the solvothermal process, MOFs are synthesized through electric heating in NMR tubes, using certain solvents like ethanol, methanol, formamides, acetones, etc.

   This technique was employed for the fabrication of titanium-based MOFs for heavy metal removal from wastewater. Similarly, microwave-assisted solvothermal preparation was done for bimetallic MOFs, which showed 96% elimination of colorants from wastewaters. The nonsolvothermal approach involves methods like nanoprecipitation and emulsification. This method is not commercially applicable, as the MOFs yield is in the form of a fine powder.

Solvothermal synthesis is one of the most widely used ways of fabricating MOFs. The synthesis involves highly solubilizing organic solvents like dimethyl formamide, methanol, acetonitrile, etc. The MOFs acquire a high surface area at low temperature and a slow reaction time. Some solvothermal syntheses involve the use of a coordination complex of copper, which greatly affects the size as well as the shape of the MOFs. Such MOFs are equipped with mesoporous and macroporous cavities. The choice of solvent is crucial during the solvothermal synthesis of MOFs, as it influences the morphology and structure of the framework. This synthesis method suffers from the limitations of long reaction periods, cost-intensive nature, and solvent-related toxicity.

2. **Slow Diffusion Method:** MOF, like MOF-5, has been synthesized using the slow diffusion of triethylamine. Researchers have synthesized MOF-5 with different degrees of crystallinity, BET surface area (481.1 m^2/g), and pore volume (0.19 cm^3/g). But this occurs in the presence of triethylamine and the specific molar ratio of solvents, in a reaction time of two days. In a similar synthesis, MOF was incorporated into the pores of hyper-cross-linked polymers. This was possible due to the formation of a complex during the synthesis. Plate and rod-like MOFs were synthesized at high temperature using the slow diffusion method, which reflected different particle sizes.
3. **Sonochemical Method:** This method involves acoustic cavitation for the synthesis of MOF like MOF-5 with a size of 5–25 μm. The preparation occurs at 155 °C, where 1-methyl-2-pyrrolidine was used as the approach solvent. MOF-based zirconium-fumaric acid was prepared under sonochemical conditions by employing fumaric acid at 180 °C. The synthesized MOF displayed a high specific surface area with a good potential for the removal of contaminants from wastewaters. A few fabrications were conducted at room temperature as well. For example, a small particle size containing zinc MOF was fabricated using the sonochemical method by employing a modulator, with half an hour of ultrasound irradiation. Modulators like benzoic acid and sodium acetate can influence the crystallinity, particle size, and time of synthesis. Sonochemical method enhances the crystallization and growth rate of the MOFs.
4. **Mechanochemistry:** This is yet another synthesis method of fabricating the MOF at room temperature. This is also observed as a green synthesis method for MOF preparation. The technique has an advantage of reduced reaction time and solvent requirements. The method is based on grinding or ball milling. The process employs three pathways namely neat grinding, liquid-assisted grinding, and ion-liquid–assisted grinding. Julien and coworkers synthesized zinc-based MOF using zinc oxide and 2,5-dihydroxy terephthalic acid, via liquid-assisted grinding method [8]. Mechanosynthesis results in

the formation of metal MOF with a high surface area. The synthesis process involves solvation, crystallization, formation of monoclinic and polymorphic structures, and finally, MOF formation having a highly crystalline nature.

5. **Electrochemistry:** This method has the advantage of mild reaction conditions, low preparation time, and controllable conditions. A MOF-5 was prepared by an in situ electrochemical synthesis method using an electrolyte under an electric field. A highly porous, pure, and high surface area containing MOF was prepared, showing π–π interactions and coordination linkages. Here, the preparation of MOFs was influenced by solvent, electric pressure, and reaction time.

   Anodic dissolution and cathodic electro-synthesis are the two types of electrochemical methods used for the synthesis of MOFs. In anodic dissolution, current is applied to an electrode immersed in an organic ligand and electrolyte. On oxidation, metal and metal ions are liberated to the solution of organic linkers, which forms the layer of MOF over the electrode surface. In the case of cathodic dissolution, MOFs are formed over the cathodic electrode.
6. **Microwave-assisted Method:** This method yields narrow narrow-sized crystal of MOF which can be controlled. It requires less time for reaction than another conventional method. The synthesis involves keeping a mix of substrate and suitable solvent in a sealed tank, which is subjected to a microwave system at a particular temperature for a fixed time.
7. **Continuous Flow Spray Drying Synthesis:** The method is a hybrid of continuous flow and spray drying techniques. A large-scale production of MOFs can be done using this method. MOFs are prepared and developed as nanoparticles, composites, or spherical structures. MOFs like Fe-BTS/MIL-100, UiO-66, and Zr-fumarate beads are synthesized by this technique [9]. The general procedure involved in this process is illustrated in Figure 15.2.

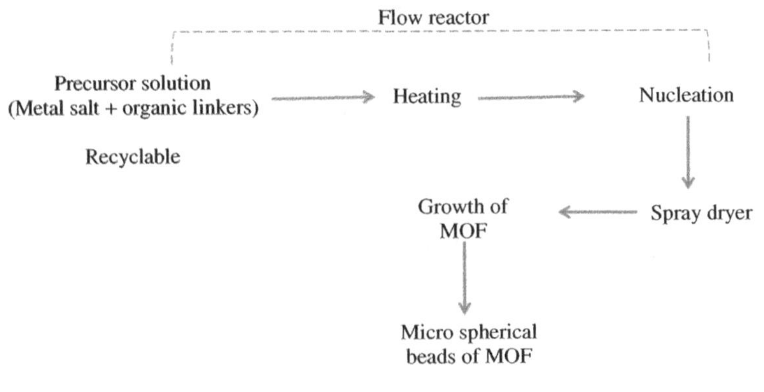

Figure 15.2 Basic steps involved in the continuous flow spray drying method.

8. **Ionothermal Synthesis:** This fabrication technique involves ionic liquids as a solvent, and a template which helps provide zero vapor pressure, solvating properties, and high thermal stability during synthesis. The cationic part of the ionic liquids attaches to the framework to form a neutral composite. The method suffers from the disadvantage of a strong interactive effect between the liquid ions and the MOF, which can influence the characteristics of the ionic liquid when incorporated with the MOF. Sustainable application of MOF and economic viability is achievable only by the regeneration of MOF. Also, environmental toxicity due to linkers and solvents used during the synthesis poses a major concern. Low mass productivity and a high temperature requirement in synthesis restricts large-scale production of MOFs, which limits their utilization for wastewater treatment [10].
9. **Carbonization:** The carbonization process results in the development of MOFs with metal and metal oxide particles of various sizes, structures, and oxidation characteristics. The thermal decomposition releases volatile compounds and forms carbon carbon-rich structure within the MOF. This structural makeup remediates the contaminants from wastewater.

MOF is a unique three-dimensional blend of organics and inorganics, with high porosity, and having designable pores, metal ions, and linkers. MOFs have high flexibility and selectivity due to pore structure and efficient design.

Synthesis of MOFs is always preceded by purification and activation to increase the performance.

The mechanism of using MOFs for contaminant elimination from wastewater involves physical as well as chemical interactions. MOFs undergo π–π interaction, ion exchange, H-bonding, acid–base exchanges, and electrostatic interactions with the contaminants (Figure 15.3). The interactions counterbalance the charges and involve deprotonation and protonation, which favor the electrostatic exchanges between the MOF and the pollutants. Higher surface area of the adsorbents facilitates the adsorption of dyes, heavy metals, pesticides, drugs, and the dominant interactions between adsorbate and adsorbent significantly influence the mechanism [11].

The physical interaction of the pollutant with the MOF is mainly the Van der Waals force. On the other hand, chemisorption involves ionic and covalent bonding, and hydrogen bonding. Chemisorption needs a long contact time between the MOF and adsorbate, while a short contact time can help in the interaction of the pollutant via physisorption. Properly incorporated functionalized groups help entice contaminant molecules through ionic interactions.

The MOFs employ π–π interaction, ion exchange, H-bonding, acid–base interaction, and ion exchange process for the removal of toxic contaminants like dyes, pharmaceutics, surfactants, micro pollutants, etc., from wastewaters. The net

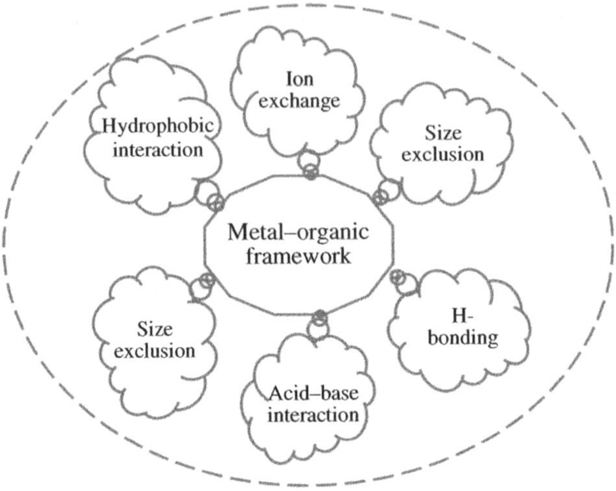

Figure 15.3 MOFs and operative interaction in wastewater remediation.

surface charges involve protonation and deprotonation, which favor electrostatic interaction between the MOF and contaminants. MOFs contain hydroxyl groups, which can effectively attach adsorbates via hydrogen bonds. Sometimes acid acid–base interaction is involved in the adsorptive removal of pollutants. A high surface area of the MOF facilitates the adsorption of contaminants. The overall mechanism is complex [12]. A closer contact between the MOF and contaminants enhances the reaction between them.

MOF serves both as an adsorbent and a catalyst in complex water matrices for separating pollutants. The toxic element elimination primarily involves π–π interaction, ion exchange, and electrostatic interactions. The catalytic action of MOFs degrades various pollutants in combination with reactive species like sulfate and hydrogen peroxide for the oxidation of pollutants. In an investigation, MOF, represented as NbCo–MOF, demonstrated a removal efficiency of 99.7% for pharmaceutical drug tetracycline in wastewater within half an hour. Sulfur and ammonia-functionalized MOFs have shown good capacity for eliminating heavy metals from wastewater with rapid kinetics. Cationic dyes are removed via adsorption over MOF at pH 6 or greater due to the deprotonation action of the MOF surface, while positively charged MOF effectively eradicates anionic dyes from wastewaters [10].

New and innovative approaches are still required to improve the efficiency of MOFs and reduction in the environmental implications so that it can be applied on a broad scale and contribute to greener and sustainable water sources.

15.1.2 Ion Exchange

Ion exchange is a sorption process, which comprises of exchange resins, which are operated on a cyclic basis. The exchange resins are either cationic resins or anionic resins. The resin attracts the undesirable ions present in wastewater and exchanges them with the required ions, purifying the wastewater. Ion exchange resins were conventionally developed using copolymerization of styrene and cross-linking with divinylbenzene. These resins have been modified to meet the requirements for the removal of specific contaminants from wastewater. A very important advancement involves the development of macroreticular resins. These are resins with discrete pores within cross-linked organic polymer. These have higher physical strength, more resistance to oxidizing agents, good resistance to thermal degradation, and better resistance against fouling.

The functional ability of the resin depends on the ionizable group attached to the resin bed. These can be strong acid cationic resin (—SO_3H^+), weak acid cationic resin (—COOH), strong base anionic resin (quaternary ammonium group) and weak base anionic resin (primary, secondary and tertiary amines or Cl^- or OH^-) [13].

Ion exchange resins or reactive polymers are insoluble cross-linked polyelectrolytes available in the form of beads. The classification of ion exchange resins is based on the charge carried by the cationic or anionic exchanger and the ionic strength of the exchanging unit. Ion exchange resins, especially designed for specific containment, contribute to decreasing the chemical footprint of wastewater treatment. The selection of ion exchangers for wastewater treatment depends on the internal structure of the resins, high surface area, and greater affinity to larger ions for exchange, irrespective of the nature of the solvent. Microporous resins have no discrete pore availability, thus, solute contaminants need to diffuse through the particle to interrelate with the exchange spot. These are less fragile and have high loading capacity.

Langmuir and Freundlich isotherms provide insights into the sorption mechanism along with surface properties and affinities of the exchanger. The ion exchange process involves interactions, linking the ion exchange resins placed in a packed column and the contaminants present in the influent. The ion exchange mechanism depends upon the characteristic groups present on the resin beads and the chemical nature of the pollutant molecules. Figure 15.4 illustrates the exchange involved between the wastewater and the ion exchange resins [5].

Sodium alumino silicate or zeolites have shown excellent ion-exchanging capacity due to the presence of reversible exchangeable ions. Zeolite is formed by linking several silica tetrahedrals, with each oxygen of the tetrahedron shared with the adjacent one. The Silicon ions (Si^{4+}) are isomorphously replaced by

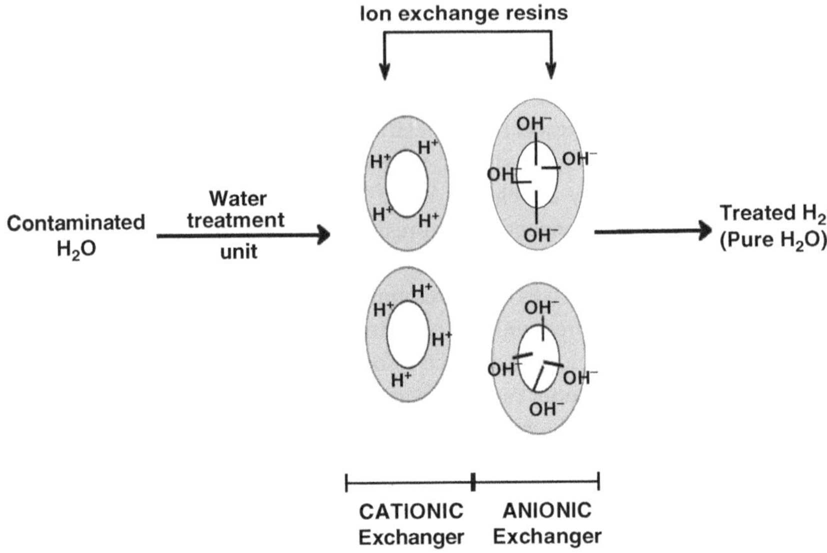

Figure 15.4 Ion exchange mechanisms.

aluminum ions (Al^{3+}), and the charge is balanced by an extra cation (Na^+/K^+). The linking of tetrahedron results in open structure having cavities. This porous nature of the structure helps permit ion exchange.

Electrochemical ion exchange is a newer technique, which is a combination of ion exchange and electrochemical reaction. In this system, electric current propels the ions toward the polymeric resin, which facilitates better removal of wastewater contaminants.

Ion exchange resins are in particular utilized for demineralization. When the resins become saturated with the pollutants these have the advantage of getting regenerated. The regeneration process mainly comprises of backwashing, regenerating solution, and rinsing. The regeneration mechanism ensures that ion exchange is a reliable, cost effective, eco-friendly and sustainable method of wastewater treatment [13].

15.1.3 Membrane Filtration

Membrane refers to a boundary which keeps two phases apart by restricting the movement of its constituents through it. Membranes can either be isotropic or anisotropic. Membranes having uniform composition and structure are termed isotropic, while those with nonuniform area and variable structural layer are termed anisotropic membranes. Isotropic membranes can either be microporous or dense. They are widely employed for microfiltration. Anisotropic membranes are used in reverse osmosis (RO), and they have a thin selective layer supported by a broad permeable layer. Membranes are chemically organic or inorganic. Usually, synthetic organic polymers are employed for pressure-driven applications like polythene (PE), polytetrafluoroethene (PTFE), polypropylene (PP), etc. Inorganic membranes are chemically stable and made up of substances like ceramics and metals, etc., and are used in ultrafiltration and microfiltration. Organic membranes are not suitable for a harsh aqueous environment.

The working of membranes is based on equilibrium and nonequilibrium conditions, or pressure-driven or nonpressure-driven processes. Pressure-driven processes rely on hydraulic pressure for the separation of contaminants. These include microfiltration, ultrafiltration, nanofiltration, and reverse osmosis, which differ from each other in their pore sizes. Several combinations of these processes have been used in wastewater treatment to facilitate better removal efficiencies. Forward osmosis follows the natural osmosis process to eradicate the impure molecules from aqueous solution. This involves a concentrated solution which provides a concentration gradient to draw water from the feed solution. The unit recovers freshwater. This process operates in the absence of any external pressure, reducing the energy consumption. The process has the advantage of easy fouling reversal, flexibility of choosing draw solution, regeneration and reuse of draw solution. However, it suffers from the drawbacks of concentration polarization and separation of the drawn solute from the treated water. Selectivity and productivity of the process decide the feasibility of membrane filtration for wastewater treatment. Selectivity is expressed as a retention factor, while productivity is a flow parameter, named flux. Microfiltration and ultrafiltration are low-pressure membrane systems which operate between 10 and 30 psi, while high-pressure membrane systems are nanofiltration and reverse osmosis, which require a high pressure of 75–250 psi. Thus, variation in the pressure conditions facilitates the removal of a number of toxic contaminants from wastewater. Reports have shown that contaminants like pharmaceutically active compounds are eliminated via hydrophobic adsorption and size exclusion mechanisms, using ultrafiltration and nanofiltration membranes [14].

Other than the pressure given membrane electrical potential is also utilized to desalinate water. Electro dialysis (ED) or reverse electro dialysis (RED) involves such a principle. Here, the direct current induces transfer of cations across the cationic membrane to the negatively charged electrode while restrained at the anionic interface. The opposite phenomenon happens with the anionic impurities. This results in obtaining a unit of dilute water and another section of concentrated water. The process is, however, associated with a large cost of membrane systems [15].

Microfiltration utilizes membranes having pore sizes ranging from 0.1 to 10 μm for water purification, while for nanofiltration, the size varies from 0.5 to 0.2 nm. Size exclusion and electrostatic interaction govern the removal process [16].

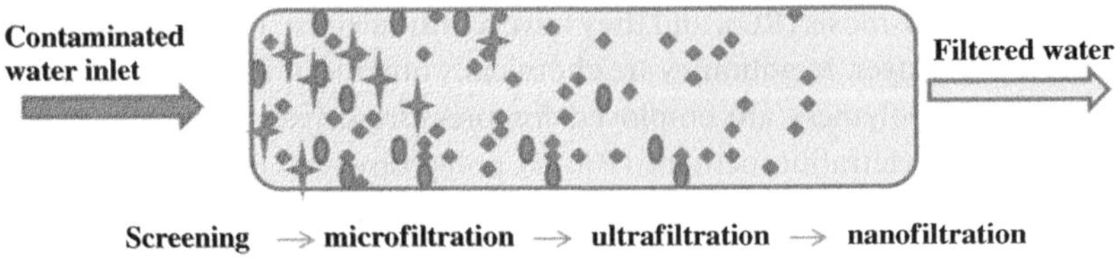

Figure 15.5 Membrane filtration.

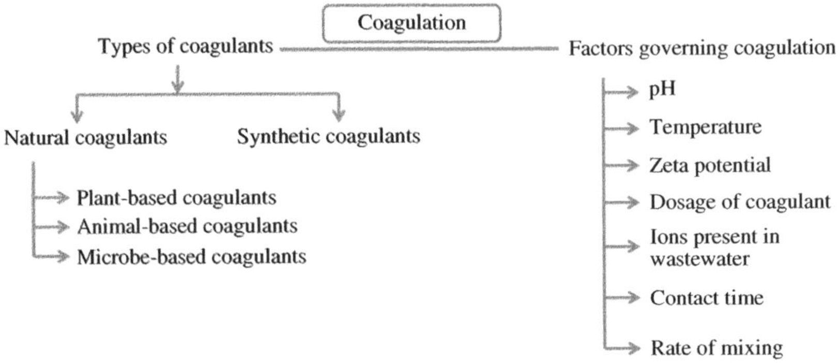

Figure 15.6 Coagulation and important parameters.

Figure 15.5 illustrates the mechanism involved. The incoming contaminated water passes through various types of filters in a sequence wherein the pollutants are eliminated by size exclusion, and finally, clean water is obtained.

15.1.4 Coagulation

Coagulation is observed as a crucial pre-treatment process in wastewater treatment. The mechanism of coagulation is explained by the Schultz-Hardy rules and the particle collision function given by Smoluchowski. Later, the theory given by Langelier and Ludwig was used to remove colloidal impurities. The mechanism involves a double-layer compression process and precipitate enmeshment, which were termed as coagulation and flocculation [17].

The properties of the coagulant material influence the performance of the wastewater treatment. Factors like dosage, pH, speed, time of contact, rate of mixing and settling time, etc., influence the process (Figure 15.6). The effectiveness of the process is influenced by the flocculation speed as well. However, excessive chemicals lead to large sludge formation.

Most of the suspended contaminants or particles are charged species when associated with an aqueous medium. The presence of similar charge obstructs the agglomeration and quick settlement of these suspended particles. Thus, destabilization of the particles using the process of coagulation can facilitate the removal of such substances from wastewater. Destabilization is achieved by electrostatic double layer pressure, absorption phenomenon, and neutralization of the charged particles. Aluminum-based metal salts involve three major steps during coagulation, namely dissociation, hydrolysis and polymerization.

Polymerization of a hydroxo complex occurs to form hydroxometal polymers like $Al_6(OH)_{15}$, $Al_7(OH)_{17}^{4+}$, $Al_{13}(OH)_{34}^{5+}$, etc. Sodium aluminate is one of the most commonly used coagulants in wastewater treatment due to its alkaline nature [18]. Its reaction can be represented in Equation 15.1:

$$NaAlO_2 + CO_2 + H_2O \rightarrow 2Al(OH)_3 + Na_2CO_3 \quad (15.1)$$

Polymeric coagulants can facilitate excellent coagulation due to appreciable ability of destabilization of particles by interparticle bridging and neutralization. The anionic polymers are capable of destabilizing negatively charged particles through their bridging effect, while the cationic polymers have the capacity to coagulate the contaminant, both by neutralization and interparticle bridging.

Natural coagulants have gained recognition in recent years due to their biodegradable nature. The mode of action of natural coagulants is same as that of polyelectrolytes. Plant-based coagulants include starch, gum, tannin, seeds of *M. oleifera, S. potatorum*, leaf extract of *H. rosasinesis*, etc. Animal-based coagulants like carboxyl methyl chitosan decontaminate water by a mechanism of charge neutralization, sweep flocculation, double layer compression, and bridging mechanism. Microbe-based coagulants are derived from fungi, bacteria, and yeasts. Their mode of action is dependent on their shape and ability to produce extracellular polysaccharides. Both chemical and natural coagulants show good pollutant [18].

15.1.5 Advanced Oxidation Processes

These processes seem to be the future of the wastewater treatment sector. Rapid development in several advanced oxidation processes (AOPs) includes photocatalysis, electrocatalysis, Fenton method, and the integration of these individual techniques. These are the treatment processes involving the creation of hydroxyl radicals and are characterized by a substantial potential for oxidation. These convert pollutants into harmless nontoxic oxidized forms. The primary stage involved is the generation of the hydroxyl radicals, and the interaction between radical and targeted pollutant occurs in the secondary stages [19, 20]. The hydroxyl radicals interact with each other to form oxidant hydrogen peroxide as (Equation 15.2):

$$OH^{\cdot} + OH^{\cdot} \rightarrow H_2O_2 \quad (15.2)$$

The formation of unwanted oxidation by-products delays the mineralization and leads to their accumulation. Removal is greatly influenced by pH, acid–base equilibrium, and thus, these are critical parameters to be taken care of during the process.

15.1.6 Fenton Process

Fenton introduced this process in 1894, which involved the evaluation of the oxidation ability of ferrous ions. A mixture of hydrogen peroxide and ferrous ions is termed a Fenton reagent and is successfully employed for remediating a wide variety of waste streams. Fenton treatment involves four steps. The first step is the modification in the pH to the lower range, followed by an oxidation reaction in the pH range of 3–5 [21]. The next step is neutralization using sodium hydroxide and lime, followed by precipitation of the impurity. Hydroxyl radicals formed due to the decomposition of water at acidic pH react with the pollutant. The major disadvantage associated with the process is a lower rate of reaction and high concentration of sludge, along with the accumulation of ferric ions as inhibitors.

The rate of degradation of the pollutants increases with the UV radiation or light. This process is also favored at low pH conditions. Photo Fenton decreases the color and chemical oxygen demand more than the Fenton process, and also reduces the amount of sludge production. It is governed by pH, hydrogen peroxide concentration, UV radiation duration, contact time, iron sulfate dosage, etc.

This is a catalytic advanced oxidation process involving the formation of hydroxyl radicals takes place as (Equation 15.3):

$$Fe^{2+} + H_2O_2 \rightarrow Fe^{3+} + OH^- + OH^{\cdot} \quad (15.3)$$

Dyes and pesticides have been successfully eliminated during the process [22].

15.1.7 Electro-Fenton Process

The process (Figure 15.7) is an emerging technique where hydrogen peroxide is electro-generated in situ at the carbon cathode. An iron catalyst is generated at the cathode continuously to facilitate the settling process [23].

15.1.8 Ozone Treatment

Ozone has a high oxidation potential. It can degrade the pollutants in two ways. First is a direct pathway where ozone directly interacts with the pollutants and oxidizes them. The indirect method involves decomposition of ozone to structure-free radicals, which react with the pollutants, to form harmless water and carbon dioxide. The rate of ozonation depends on the presence of hydroxyl radicals formed. Ozone has an outstanding ability for inactivation of microbial species

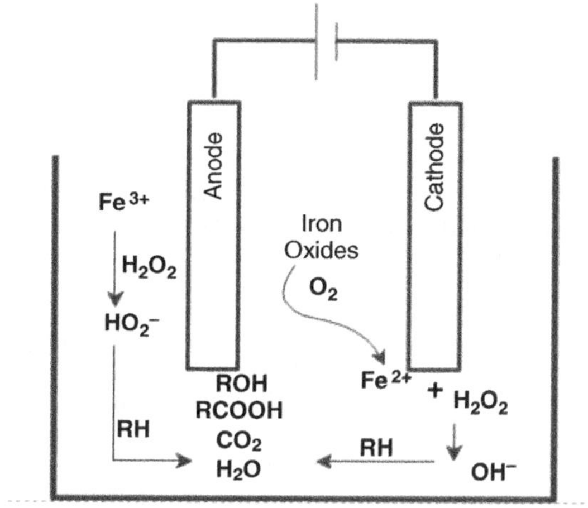

Figure 15.7 Electro-Fenton process.

like viruses, bacteria and protozoa. Ozone decomposes by a complex mechanism and involves the formation of hydroxyl ion and several oxidizing species. The formation of these radicals is enhanced in the presence of a solid catalyst. Catalytic ozonation is a technology which mineralizes pollutants, using free radicals generated by ozone degradation. It can be executed either using homogeneous or heterogeneous catalysis. Heterogeneous catalytic ozonation is a promising technique as its removal efficiency is high, recovery of the catalyst is easy and quick, and the cost of wastewater treatment is low. The catalysts employed in heterogeneous ozonation are metal oxides and metal oxide/noble metals combinations, immobilized over a substrate [24]. Metal oxides are characterized by the presence of Lewis acid sites, which allow the water molecules to attach to their surface for efficient remediation of the pollutant. Ozone catalytic oxidation is capable of reducing various by-products formed during the degradation process.

15.1.9 Electrochemical Oxidation

The method involves direct anodic oxidation, where oxygen or hydroxyl radicals are generated, while in certain processes, oxidation is caused due to mediating substances like chlorine, hypochlorous acid, hydrogen peroxide, etc. The hypochlorite ion is formed as below (Equations 15.4–15.6) and helps in the oxidation process [25].

$$Cl^- \rightarrow Cl_2 + 2e \tag{15.4}$$

$$Cl_2 + H_2O \rightarrow HOCl + H^+ + Cl^- \tag{15.5}$$

$$HOCl \rightarrow H^+ + OCl^- \tag{15.6}$$

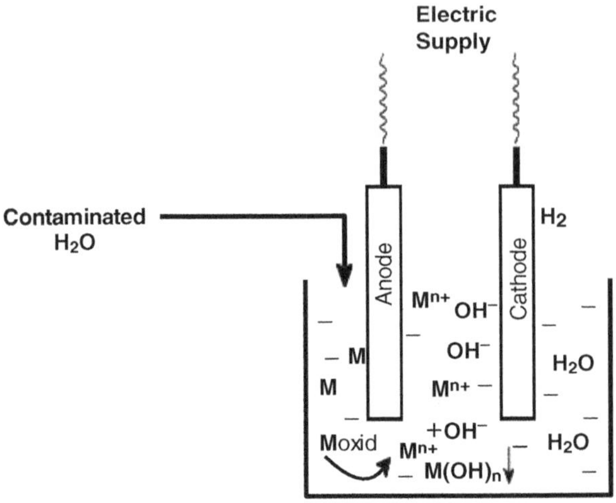

Figure 15.8 Electrocoagulation.

Organic contaminant removal is possible by combining oxidation and coagulation, along with electro Fenton process. Electrocoagulation involves the use of two electrodes for in situ coagulant generation and power supply. Iron metal acts as a speed enhancer for the reaction, as well as helps in the generation of a coagulant internally, while the cathode is the site for hydrogen gas production [26]. The process involves the production of coagulants due to the oxidation at the anode, which forms hydroxyl radical species that help to counterbalance the electrostatic charge of pollutants, enhancing the cluster formation. The process is followed by destabilization of the pollutants and formation of flocs and their settlement. The mechanism of electrochemical oxidation, along with coagulation, is illustrated in Figure 15.8. Equations (15.7–15.10) suggest the formation of iron hydroxide precipitates during the process.

$$Fe \rightarrow Fe^{2+}(aq.) + 2e \tag{15.7}$$

$$2H_2O(l) + 2e \rightarrow H_2(g) + 2OH^-(aq.) \tag{15.8}$$

$$Fe^{2+}(aq.) + 2OH^- \rightarrow Fe(OH)_2(s) \tag{15.9}$$

$$Fe(s) + 2H_2O(l) \rightarrow Fe(OH)_2(s) + H_2 \tag{15.10}$$

Electrochemical oxidation depends upon the concentration of salt solution, pH, temperature, size, types of anode electrodes and voltage applied. The anode discharges the water, resulting in hydroxyl radicals. These are adsorbed onto the active sites of the electrode surface and finally oxidize the organic matter.

The involved process or mechanism is electron transfer and oxidation of the reactive species. Electrochemical anoxic oxidation successfully eliminated ibuprofen using a Ti/Pt/PbO_2 anode in half an hour, with reduction in total organic content and chemical oxygen demand greater than 90% [24].

15.1.10 Photocatalytic Degradation

It is a form of AOP, which is dependent on the usage of light from different natural and artificial sources for pollutant degradation. This process is particularly suitable for contaminants which are complex, recalcitrant, higher in concentration, and difficult to remove by other common modes of remediation. It is based on the principle of incorporating a light source to target efficient removal of pollutants. Degradation of pollutants occurs due to the generation of electron and hole pairs, which carry out the reduction and oxidation processes, respectively (Figure 15.9). These form several radicals due to subsequent reactions, which further combine with the pollutants and convert these into simple and harmless products.

The condition for the occurrence of the reaction is that the intensity of the light must be equal to or larger than the energy bandgap of the photocatalysts. Different ranges of light which work successfully in this reaction, as the light source, are ultraviolet (UV), infrared (IR), and visible (VIS).

Photo catalytic degradation is an advanced oxidation process employed for the degradation of pollutants with low biodegradability. It harnesses the ability of

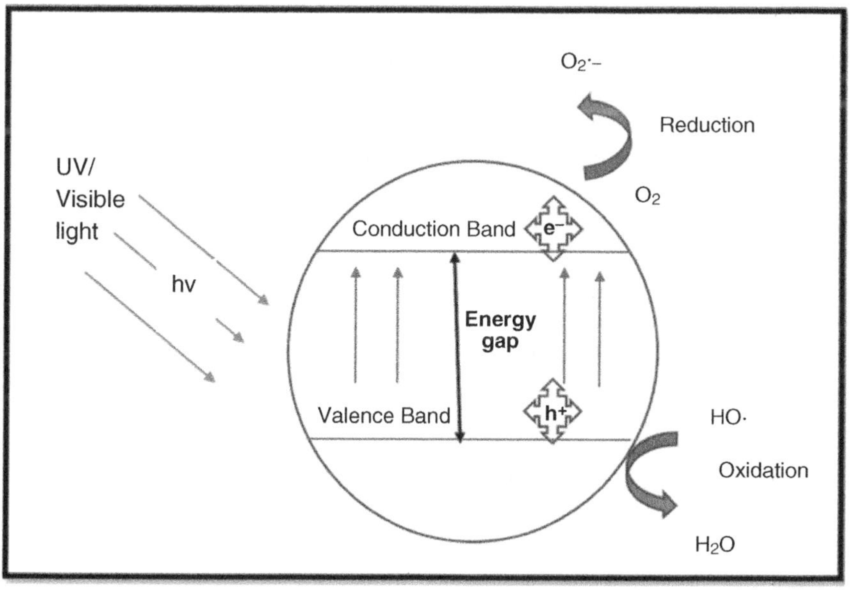

Figure 15.9 Photocatalytic degradation.

the catalyst under UV–VIS radiation to generate highly reactive hydroxyl radicals and exhibits several advantages over the conventional wastewater treatment processes. Photo catalysts degrade the organic pollutants rapidly under light exposure. The process operates under mild temperature and pressure conditions, which minimizes energy consumption and operational costs. Such processes are scalable and adaptable to various industrial wastewater matrices.

The general photo catalytic degradation mechanism involves [27]:

1. Electron and hole generation by absorption of photons
2. Redox reaction with adsorbed species on the surface of the catalyst
3. Degradation of the contaminants to form smaller, less harmful by-products.

The valence band has a significant oxidizing nature, and the conduction band has reducing character. The excited electrons and holes have the potential to perform redox reactions with adsorbed species. The photo catalysts employed are affordable, nonhazardous and have a regeneration potential.

The photo catalytic degradation mechanism can be understood using two pathways [27], direct and indirect.

Direct Method: The direct photo catalytic degradation mechanism is elaborated by using Langmuir-Hinshelwood theory, which states that both electrons and holes produced by photo excitation of the catalyst degrade the pollutants through the formation of active radical species.

Another approach by Eley-Rideal suggests the trapping of holes on the surface of the catalyst, where the photo catalyst species reacts with the pollutant to form the intermediate, and finally forms degraded products. The direct mechanism is operative in the presence of visible light. The pollutant molecules are excited under visible radiation from the ground state to a higher energy state. The excited state contaminant is further converted into semi-oxidized radical species. This happens by electron entry into the conduction band of the photo catalyst. Further, the trapped electron and dissolved oxygen of the system form superoxide radical, which in turn forms hydroxyl radical. This radical is responsible for the oxidation of contaminants. The direct mechanism is believed to be slower than the indirect mechanism, which is generally prevalent.

Indirect Method: Indirect photo catalytic degradation involves the processes mentioned below [28]:

1. **Photo Excitation:** When light strikes a semiconductor, a photo catalytic reaction is initiated due to the promotion of photo electrons from the filled valence band to the empty conduction band. During the excitation, a hole is created in the valence band. Thus, during photo excitation creation of an electron and a hole occurs.
2. **Ionization of Water:** The produced holes in the valence band react with water to form a hydroxyl radical. This is a very strong oxidizing agent. The

radical attacks the pollutant near the catalytic surface, leading to mineralization, depending upon their structure and stability. The radical can decontaminate wastewater from microbes as well.

3. **Oxygen Ionosorption:** In a parallel reaction to the above, oxygen takes up the electron of the conduction band to generate anionic superoxide radical. The superoxide ion maintains the electron neutrality within the catalyst molecule and prevents electron–hole recombination.
4. **Protonation of Superoxide Radical:** The superoxide radical gets protonated and ultimately forms the hydroxyl radical. The surface of the photoexcited semiconductor acts as a site for degradation of the pollutant, where the product of oxidation is formed.

The mechanism of photocatalytic degradation is influenced by pH, time, pollutant strength, photo catalytic load, intensity of light, initial contaminant concentration, effect of inorganic salts, and presence of additional oxidizing species.

Photocatalytic degradation has been observed as an environmentally friendly and sustainable approach for contaminant remediation from wastewater. Materials like zinc sulfide, titanium oxide, silver chloride, and zinc oxide are often employed in the process, which can absorb visible and UV radiation. Photo catalysts like metal, semiconductors, metal oxides, dendrimers, and MOFs have been employed for the removal of pollutants from wastewater. Photo catalytic materials are categorized into three types [29]:

1. Single-component materials like oxides and nitrides
2. Multicomponent material like $WO_3/NiWO_4$ and C_3N_4/Ag_3VO_4
3. Immobilized photocatalyst like steel/TiO_2-WO_3 and glass/P-YiO_2.

The photo catalytic semiconductor approach requires two components, viz. catalytic surface, and a source of photons. Semiconductor material is characterized by having a valence band and a conduction band, separated by a band gap.

The process of photo catalytic degradation also results in the formation of certain intermediate products. For example, photo catalytic degradation of pesticide acephate using $Fe_3O_4/SiO_2/TiO_2$ resulted in the formation of several intermediates. The ability of the contaminant to bind with the surface of the catalyst by adsorption is also important in determining the degradation of the pollutants. Establishment of the interfacial contact determines the speed of the degradation process. Researchers have reported that organic contaminants are degraded to appreciable levels when the association of the pollutant with the catalyst surface occurs via adsorption.

Several photo-catalytic materials are used for eradicating the products from wastewater. These includes carbon waste materials (polythiophene, polypyrrole cellulose), metal sulfides (copper iron sulfide, molybdenum disulfide), iron based magnetite materials (Ilmenite, spinel ferrite nanoparticles), metal oxide (zinc oxide, titanium oxide, tungsten oxide, cerium oxide), MOFs (MOF-199), layer double hydroxide and its derivative.

pH is a critical operating factor and affects the charge on the surface of the photo catalyst and pollutant, and also the position of the valence band and the conduction band. It also regulates the electrostatic influence between the pollutant, the catalytic surface, and degradation radicals. For example, photo catalytic degradation of malachite green decreased at acidic pH due to the strong repulsive force between the reactants. However, alkaline conditions favor degradation, due to strong adsorption on the surface. Photo catalytic degradation depends upon light, and thus no temperature control is necessary. Though the increase in temperature increases the mobility of hydroxyl ions toward the pollutant, but not significantly.

An increase in the radiation power increases the pollutant removal tendency. Some reports suggest the removal process of the dyes gets doubled with an increase of light intensity from 10 to 70 mW/cm^2. The removal process involves the generation of reactive oxygen species, which are instrumental in the breakdown of pollutants. But inorganic anions like carbonates, chlorides, nitrates, and sulfates can scavenge hydroxyl radicals and reduce their concentration and efficiency. Additionally, oxidizing agents like hydrogen peroxide, ammonium persulfate, etc., enhance degradation. Monitoring of these intermediate products is very crucial during photo-catalytic degradation. Catalysts with a high surface area to volume ratio have enhanced photo catalytic activity as they have greater availability of the active sites for adsorption and subsequent reactions. The porosity, crystalline nature, and specific particle size can facilitate better degradation kinetics. The drawback of kinetics related to the electron–hole pair recombination can be eliminated by the incorporation of carbon nanotubes.

The mechanism of photo catalysis is also divided into two types:

1. Homogeneous photo catalysis, where the photo catalyst and the reactants are in the same phase. The reactive species involved in homogeneous photocatalysis is the hydroxyl. The advantage of homogeneous photo catalyst is that they allows for a direct and rapid interaction between the catalyst and reactant, in the same phase. The practical utility of homogeneous catalysts is limited due to their unstable nature. Sometimes, the retrieval of the homogeneous catalyst is not achievable.
2. Heterogeneous photo catalytic degradation, in which sunlight is used, in the presence of a photo catalytic semiconductor, to remediate environmental contaminants. The nature of the catalyst is a crucial factor in the type of radiations involved in the process. Heterogeneous catalyst is an efficient technique where catalyst, product and reacting species exist in distinct phases. The use

of transition metal oxide and semiconductors, as heterogeneous photo catalysts, is prevalent in wastewater treatment.

Heterogeneous photo catalysis has been found to be superior because [29]:

1. Recovery and reuse of photo catalyst is possible
2. The process becomes cost-effective and environmentally sustainable
3. Process has an enhanced selectivity
4. It has higher degradation products
5. It has fewer side reactions
6. Solid heterogeneous photo catalyst can withstand harsh conditions
7. Catalysts have a longer lifetime
8. Heterogeneous systems are scalable and can be industrially commercialized for wastewater treatment.

15.1.11 Sonochemical Method

The sonochemical method works on the utilization of ultrasound, which differs from other light, heat or radiations in various aspects. The principal operative is acoustic cavitation. It involves ultrasound to act as a source of concentration of sound energy in liquids. Such phenomena occurs due to the generation of high energy radical species from solvents at high temperature and pressure in the cavitating bubble. The mechanism of degradation of a contaminant depends upon whether the chemical reaction occurs in the cavitation bubble, in the solution, or at the interface of the bubble-solution. The degradation pathway followed is either a thermal pathway or an oxidative pathway, induced by a radical species.

It is proposed that compounds with a high vapor pressure enter the cavitation bubble gas phase rapidly and undergo degradation quickly. A nonvolatile and hydrophilic compound remains in the bulk, liquid during the sonochemical treatment. Thus, at a low concentration liquid medium and at a high concentration bubble-liquid interface, effective destruction of contaminants by oxidative degradation occurs by the radicals, due to cavity collapse. Research reports that vapor pressure is an important parameter, influencing the sonochemical degradation kinetics, and efficiency [30].

Sonochemical degradation has also been found to effectively disinfectant microbes from water. Disinfection is enhanced due to the formation of radicals by gas bubbles, initiated due to ultrasound. It is a widely used alternative for the direct thermal activation of radical species. Easy procedure and nonproduction of by-products are the prominent advantages associated with the sonochemical method.

Ultrasonic technique results in better degradation due to induced cavitation. Bubble formation occurs due to a difference in the acoustic pressure. The bubble expands and reaches a critical size, and collapses. This compresses the gas in the bubbles, and a high temperature develops. This causes the creation of free radicals, and an increased concentration of radicals in the aqueous environment leads to the degradation of contaminants by a series of chemical reactions. An ultrasonic wave effectively decontaminates the water from microbial species. The cavitation bubbles reach high temperature and pressure, stress, and resultant shock wave energy, stress and temperature kill the microorganism. High energy ultrasound is more effective in water disinfection in comparison to low frequency ultrasound, with the best outcome in the range given 100–500 KHz [31].

The chemical reaction involves a combination of hydroxyl groups to form hydrogen peroxide, while hydrogen radicals react with oxygen to generate peroxide radicals. These radicals participate further in the reaction with hydrogen radicals or water, forming hydrogen peroxide. These radicals, being strong oxidizing agents, react with the contaminants of wastewater, and break them down into simpler, less harmful, and nontoxic substances. The optimization of parameters like frequency and power regulates the efficiency of the sonochemical process.

The process is not quite effective for high concentration of pollutants. The energy cost is another hurdle in its large scale implementation. Thus, further research and technological advancement is required to make the process more realistic and reasonable [32].

15.1.12 Biologically Assisted Degradation

This is a promising treatment technology, especially in terms of its eco-friendliness. Biodegradation is based on biomaterial contaminant interactions, which involve physical and chemical association with the biomass or biological species. These interactions include adsorption, ion exchange, complexation, chelation, precipitation, and degradation by enzymes present in the biosorbent, biomass, or biological species. The efficiency of the process depends on the microbial activity, which is governed by the concentration of the biological species, pH, temperature, contact time, and nature of the contaminants. Bacterial degradation is possible either by the entire bacterial cell (*Pseudomonas, Staphylococcus*, and *Clostridium*), or enzymes present in the bacterial cells like azoreductase, laccase, peroxidase, or by a bacterial consortium, where two or more bacterial species are mixed for degradation of pollutants. Bacterial-assisted degradation of pollutants is usually carried out by azo reductase under aerobic and anaerobic conditions, and involves the cleavage of azo bond linkages. Strong electrostatic interactions between the pollutants and associated characteristic groups like carboxyl, hydroxyl, phosphates, and other charged moieties present in the microbial body govern the biosorption competence, along with pH, temperature, type of species, concentration of pollutant, etc. The process is supported by coenzymes like Flavin mononucleotide (FMN) and NADPH-dependent reductases. The degradation

Figure 15.10 Biologically assisted degradation.

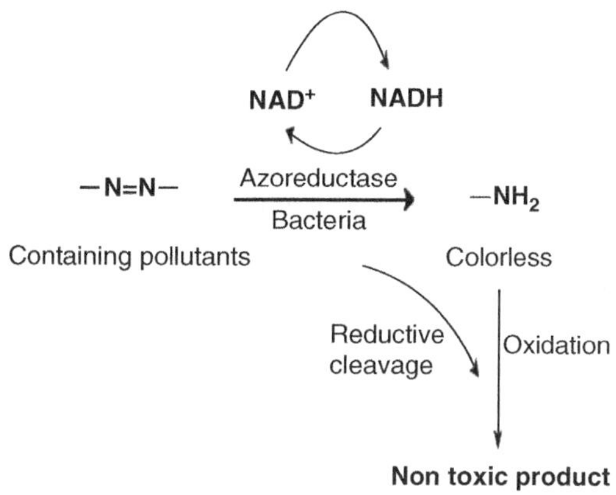

results in the formation of colorless amino acids. The anaerobic process results in the production of toxic by-products, thus necessitating an aerobic stage in the process (Figure 15.10). Bacterial-assisted degradation has the ease of cultivation of the bacterial population, both isolated or in consortium, with a high specific growth rate and versatile characteristics of mineralization [33, 34].

Enzyme-assisted degradation involves enzymes such as peroxidase, oxidoreductase, laccases, and azoreductases for the conversion of complex organics into inorganics via the flocculation method. It suffers from the disadvantages of biocatalyst deactivation due to the denaturation process, but the method results in better performance, if associated with immobilization of pollutants, using cross-linking by various means, including charcoal, biochar pellets, etc. [35, 36].

Yeasts, namely *Saccharomyces cerevisiae*, *Saccharomyces lipolytic*, *Kluyveromyces marxianus*, and *Saccharomyces uvarum*, have successfully degraded several pollutants by azo bond cleavage, and mostly by the use of azoreductase enzymes. This degradation can be carried out even at extremely low pH, as yeast has shown good tolerance under such conditions [37]. Fungi produce lignin, degrading enzymes, and facilitate the decomposition and transformation of various complex, chemically tolerant substances [38]. White rot fungus, *Aspergillus fumigatus*, *Aspergillus flavus*, and *Alternaria solani* have been widely used in the decolorization of dye-loaded wastewater, and for degradation of other contaminants from wastewater. The most widely used fungi are ligninolytic fungi of the Basidiomycetes family [39, 40].

Algae are considered important biocatalysts and are seen as a future of wastewater treatment options. These can act as biocoagulants during the removal of complex contaminants from wastewater. The efficiency of the species is attributed to the nature of their cell wall, greater surface area, and thus a high tendency for binding with the pollutants [41–43].

15.1.13 Membrane Bioreactor

Membrane bioreactor is a hybrid of conventional biological and physical treatment systems, which involves the use of a membrane filter as a single unit membrane. Bioreactors offer advantages over other conventional activated sludge processes in terms of high-quality effluent, shorter water retention period, but longer solid retention, and high loading rate. This technology also eliminates the need for secondary clarifiers, which helps set up the treatment unit in a small area only [44]. It is a biochemical engineering technology involving the use of suspended growth bioreactors, which undergo reactions like fermentation, biological oxidation, nitrification, and denitrification. The structure and symphony of bioreactor microbial species are variable, as they depend upon the source of the wastewater and its composition. In bioreactor groups like bacteria (90%), protozoa, metazoan, and fungi are employed [45]. It operates on the principle of biodegradation and membrane filtration. The biological degradation employs microbial species in the degradation of organic matter present in wastewater. This component of the membrane bioreactor is the same as the conventional activated sludge process system. The biological process effectively decreases the chemical oxygen demand and biological oxygen demand, which indicate the organic content of wastewater. The maintenance of optimal conditions for bioreactors, like temperature, pH, and nutrients, is crucial for good performance of the system.

The membrane unit of the system separates the treated water from micro contaminants. Membrane bioreactors are configured as submerged and side-stream types [46]. But the technology is associated with limitations of greater energy input, periodic maintenance of membranes, and a very crucial fouling issue. Fouling reduces the life of the membranes and their performance. Figure 15.11 illustrates how the membrane fouling leads to a retarded process.

Different foulants of inorganic natures like polysaccharides, proteins, calcium, magnesium, sulfate, phosphate, hydroxide, etc., decrease the efficiency of the

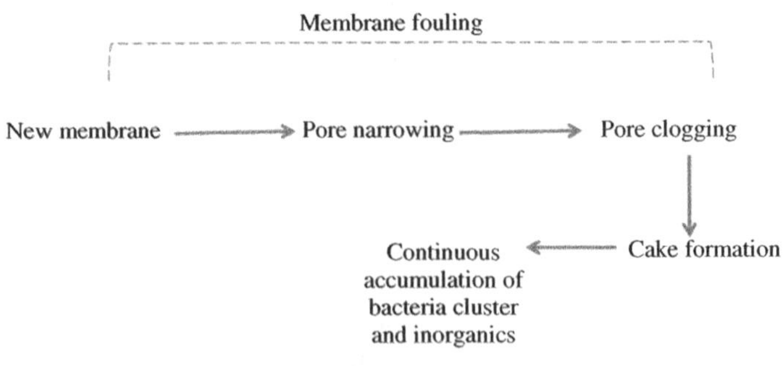

Figure 15.11 Mechanism of membrane fouling.

membranes. The fouling is influenced by factors like characteristics of the membrane, operational conditions, feed characteristics, and type and activity of biomass involved [44].

In membrane bioreactors, membranes are used as ceramic membranes, polymeric membranes, and composite membranes. Ceramic membranes have hydrophilic characteristics, which increase their performance and ease of cleaning. They are prone to less fouling drawback, but their fragile nature makes them uneconomical and restricts their large-scale implementation for wastewater treatment. On the other hand, polymeric membranes have good resistance but are hydrophobic, which increases their fouling tendencies. These include polyvinylidene fluoride, polysulfone, polyethene, cellulose acetate, etc. To culminate the disadvantages associated with the above membranes and the disadvantage of fouling, a hybrid of two or more materials is fabricated, where a hydrophilic polymer is coated over a hydrophobic membrane [44].

Membrane bioreactors use microfiltration and ultrafiltration for separating sludge obtained from the biological treatment. This ensures better separation. It can also be employed as an anaerobic membrane bioreactor, which produces high-quality effluent and has a lower chemical oxygen demand than the conventional methods [47].

15.1.14 Constructed Wetlands

Constructed wetlands are specially designed wastewater treatment systems which utilize natural processes and microbial assemblages to convert the contaminants into nontoxic products, especially involving the removal of nitrogen and ammonia. Constructed wetlands are associated with the emission of greenhouse gases, like methane and nitrous oxide, due to the methanogenesis process, and due to de-nitrification of nitrogen by microbial populations. Aerenchyma wetland plants play a crucial role in the process.

Constructed wetlands comprise of three major processes [48]:

1. **Filtration:** The vegetation and soil matrix acts as natural filter, removing the suspended solids and other pollutants from wastewater.
2. **Sedimentation:** The sedimentation of the heavier particles to the bottom makes supernatant less loaded with pollutants.
3. **Microbial Action:** The microbial population surviving in the wetland soil and plants break down the organic matter to form less harmful substances.

Nitrogen removal occurs via denitrification, while phosphorus removal occurs by ligand exchange process, in which phosphates are displaced by water or hydroxides of iron and aluminum. The plant provides substrate for the growth of attached microorganisms, distribution of nutrients from roots to the rhizosphere, and insulation of the microbial population in variable degree of climatic conditions.

In a constructed wetland with vertical subsurface flow, the feed is introduced to the wetland, which percolates down the sand medium. The next wastewater input is done only after complete filtration, which enables diffusion of air properly through the bed. Here nitrification occurs but not denitrification.

Recently, tidal constructed wetlands are designed which involve percolation of the wastewater until the surface of the wetland is flooded. After a fixed set time wastewater is drained downward which results into the completion of the treatment cycle. The microbial degradation in most of the constructed wetlands is aerobic, while in horizontal flow constructed wetland, anoxic and anaerobic conditions exist. In sub-surface flow constructed wetlands the amputation means are proposed to be adsorption and precipitation. Volatilization is also an important route of nitrogen removal in constructed wetlands, especially with acidic conditions and daytime.

The mechanism involves complex physical, chemical, and biological processes for contaminant confiscation. The removal percentage is influenced by wastewater feed rate, hydrologic regime, detention time, mode of operation, and type of plant species. Nutrient uptake by plants and microbial transformation of the contaminants are influenced by climatic factors like solar radiation, temperature, and microbial activity.

Wetlands have a large majority of vascular plants, which slow down the water flow and create micro microenvironment and sorption sites for contaminants and microbes. With the death of the plant, an additional layer of attachment site and exchange site is created. A microbial consortium performs the transformation of the pollutant. Every microbe has a specific degradation pathway. Constructed wetlands are capable of removing 100% nitrogen and phosphorus from petroleum-contaminated wastewater [49].

The existence of various sorption sites in soil and the substrates influences the adsorption and the decontamination ability of the biomass. The technique still demands understanding of the capabilities of available microbes, to judge the complex mechanism operative in the degradation of the contaminants. The degradation involves processes like ammonification, nitrification, denitrification, volatilization, and adsorption. Also, various autotrophic and bacterial species contribute to the degradation of pollutants. Aerobically, organics serve as electron donors. Oxygen gains electrons and the pollutants are broken down to carbon dioxide and water (Equation 15.11), while in anaerobic activity, pollutant acts as electron donors, and carbon dioxide, nitrates, sulfates and carbonates serve as electron acceptors. The resultants involved are acid or alcohol along with carbon dioxide (Equation 15.12).

$$\text{Organics} + \text{Oxygen} \rightarrow CO_2 + H_2O \quad (15.11)$$

$$\text{Organics} \rightarrow \text{Lactic acid or Ethanol} + CO_2 \quad (15.12)$$

The removal route is dependent on pH, temperature (20–30 °C), oxygen availability, concentration of organic load, feeding mode, retention rate, plant harvesting, etc. Phosphorus removal involves immobilization through chemical precipitation, bacterial action, physico-chemical pathways, and plant intake [50, 51].

15.1.15 Phytoremediation

The phytoremediation technique is an environmentally conscious method that is used to convert the pollutants into simpler and nontoxic forms. It incorporates the use of different parts of plants to carry out the remediation process [52]. Figure 15.12 shows the involvement of different plant parts in phytoremediation.

Based on the particular part of the plant involved in remediation, different mechanisms like phytodegradation, phytoextraction (uptake by biomass), phytostabilization (reduction in mobilization of contaminants), phytovolatilization (conversion of toxic components into nontoxic and their release from leaves), and filtration using roots, that is, rhizofiltration, take place [53].

Figure 15.12 Diverse approaches in phytoremediation.

Phytoremediation is primarily a wastewater treatment approach based on using plants for contaminant remediation. Phytoremediation is the application of compatible plant species to decompose the pollutants in the water resources. It involves the plant activity and the activity of its microbial community for the degradation and inactivation of toxic pollutants in wastewater. Phytoremediation is an innovative in situ contaminant elimination method.

Plant components like leaves, peels, and roots are efficient in removing the pollutants. It is seen as an emerging green technology to degrade and remove several pollutants, including dyes, drugs, surfactants, microplastics, etc. Plant species take off the toxic component from the environment and detoxify their impact. It utilizes the symbiotic relationship between the rhizospheric organisms and physiology of the plant. The process of absorption and transfer of a large volume of groundwater in phytoremediation is called hydraulic monitoring. This manages the movement and leaching of contaminants. The removal of contaminants depends on exposure duration, concentration of pollutants, environmental temperature, pressure, and the type of plant species involved. This method has an advantage over other traditional methods in terms of environmental friendliness.

Phytoremediation depends on diverse mechanisms. Phytoremediation includes a number of approaches like phytostabilization, rhizodegradation, rhizofiltration, phytoextraction, photodegradation, phytovolatilization, phytoaccumulation, etc. (Figure 15.12) [54].

The initiation step of phytoremediation functions as a natural filter whereby the plant selectively absorbs the pollutants from wastewater. Plants undergo various biochemical transformation reactions for the detoxification of pollutants. The degradation is supported by the microbial population surviving in the zone of pollutant accumulation. Exploration of factors affecting the uptake by the plants is crucial here. Various approaches used in phytoremediation are:

1. **Phytostabilization:** It involves the adsorption and accumulation of toxic contaminants on the root or rhizosphere. It involves processes like precipitation, sorption, and complexation. This prevents the transfer and dispersion of pollutants from one place to another through precipitation runoff. This inhibits their entry into groundwater and migration to the soil. Plants having extensive root growth, like legumes and grasses, are frequently used in immobilizing the contaminants. However, the utilization of phytostabilization is not a promising option with highly contaminated wastewater. The choice of suitable plant species for tolerating the extreme toxic load is very crucial for the effectiveness of this technique. Advances in genetic engineering have increased the efficiency of phytostabilization in the long term for remediating wastewater. Phytostabilization reduces the mobilization of pollutants.

2. **Rhizodegradation:** In this, the pollutants are degraded and broken down into nontoxic products. Heavy metals are specially degraded by this mechanism. The microbial population associated facilitates the process.
3. **Rhizofiltration:** Here, pollutant adsorption occurs in the roots, and enzymes present in the plant root help in the degradation process. This is also called phytotransformation. It involves absorption, adsorption, precipitation, along with chelation, ion exchange, and chemical precipitation through root exudates. Translocation does not occur in rhizofiltration, and the pollutants accumulate in the root tissues. After mobilization, these get stored in vacuoles and apoplast, and get bonded with the cell wall, due to chelation with phytochelatins and metallothioneins. Aquatic plant species like *E. crassipes, L. minor, E. canadensis*, and *Typha* have been widely used in rhizofiltration [55].
4. **Phytovolatilization:** In this, the plants absorb the toxic wastewater and convert the toxic components into nontoxic volatile forms, followed by their release into the atmosphere, via leaves. Various organic contaminants can be eliminated in this way. The mechanism involved is direct uptake of pollutants in the tissues of plant, where they get accumulated and are metabolized. The volatile degradation products are released into the atmosphere by transpiration process. This triggers the activity of microbes associated with the plant like mycorrhizal fungi and various microbial consortia which degrade the accumulated pollutants. Phytovolatilization is specially employed when removing organic contaminate like trichloroethane, benzene, atrazine, phenols, along with mercury, selenium and arsenic from wastewaters. Under ideal conditions, plants like mustard and poplar can volatilize 70–80% of the pollutants in their nontoxic forms. But this process also has a limited applicability due to its potential to release volatile chemicals into the atmosphere, causing air pollution. Factors like soil type, temperature, humidity, and plant species influence the effectiveness of the process. Plant breeding with genetic tools has developed the possibility of increasing stronger volatilization capacity in the plant with quick growth rates. This makes the approach viable and an efficient option for wastewater treatment.
5. **Phytoaccumulation:** When the waste is stored in plants parts like roots, shoots or leaves, it is known as phytoaccumulation. Accumulation is a multistep complex process involving transportation through plasma membrane of cells of roots, loading of ions into the root xylem, translocation, sequestration and detoxification at the cellular level.
6. **Phytodegradation:** It utilizes the enzymatic action of the plant and microbes flourishing in their body for converting organic pollutants into inorganic compounds. The approach is a combined effect of interactions between soil, water, microbes and plant. This is a popular method of treating wastewater, but several limitations restrict its large-scale applicability. The prime

hindrance is the heterogeneity in the plant's absorption capacities. Other factors include climatic factors, temperature, nutrient availability, pH, etc. The overall process is quite slow. Genetic engineering has helped reduce the problem to some extent. For example, the introduction of the gene encoding for laccase enzyme has shown an increased efficiency for the degradation of pollutants like pesticides. The symbiotic association of plants and microbes has helped enhance degradation ability. Enzymes act as biocatalysts for degradation in plant tissue [56].

Plant microbial interactions are essential to achieve the degradation of organic pollutants. Uptake of the pollutants occurs in roots, which involves apoplastic and symplastic pathways. Organic pollutants move to the leaves through the xylem sap and finally metabolized to less toxic components by the plant enzymes.

Phytoremediation has several benefits which include its autotrophic nature, ability to produce plenty of biomass, which is quite acceptable by society, due to the aesthetic appeal, and associated remediation benefits. The mechanism of phytoremediation includes alteration in root exudates, biodegradation in root tissues, abiotic losses and microbial degradation. Algal species like *S. fusiforme, E. prolifera*, along with *P. stratiotes, H. umbellata, S. acutus, C. zizanioides L, H. annuus, I. aquatic*, and *S. biloba,* have been successfully employed in phytoremediation [52]. The remediation capacity of the process depends on molecular and physiological mechanism. Phytoremediation has been successfully used for the eradication of heavy metals from contaminated wastewater, and is under constant research for mitigating the majority categories of pollutants. This method can remove the contaminants when they are present in lower to average concentrations, which is also a limitation of this method. Also, phytoremediation is still in its early stages of optimization and requires breakthroughs in making the technique applicable on a large scale.

References

1 Qasem, N.A.A., Mohammed, R.H., and Lawal, D.U. (2021). Removal of heavy metal ions from wastewater: a comprehensive and critical review. Clean Water, Volume 4, Issue 36. https://doi.org/10.1038/s41545-021-00127-0.

2 Mudhoo, A., Ramasamy, D.L., Bhatnagar, A., et al. (2020). An analysis of the versatility and effectiveness of composts for sequestering heavy metal ions, dyes and xenobiotics from soils and aqueous milieus. Ecotoxicology and Environmental Safety, Volume 197, p. 110587.

3 Younas, F., Mustafa, A., Farooqi, Z.U.R., et al. (2021). Current and emerging adsorbent technologies for wastewater treatment: trends, limitations, and

environmental implications. Water, Volume 13, Issue 2, p. 215. https://doi.org/10.3390/w13020215.

4 Raji, Z., Karim, A., Karam, A., et al. (2023). Adsorption of heavy metals: mechanisms, kinetics, and applications of various adsorbents in wastewater remediation– a review. Waste, Volume 1, Issue 3, pp. 775–805. https://doi.org/10.3390/waste1030046.

5 Wang, Y., Chen, L., Zhu, Y., et al. (2024). Research status, trends, and mechanisms of biochar adsorption for wastewater treatment: a scientometric review. Environmental Science Europe, Volume 36, p. 25. https://doi.org/10.1186/s12302-024-00859-z.

6 Satyam, S., and Sanjukta, P. (2024). Innovations and challenges in adsorption-based wastewater remediation: a comprehensive review. Heliyon, Volume 10, Issue 9, pp. e29573. https://doi.org/10.1016/j.heliyon.2024.e29573.

7 Essalmia, S., Lotfi, S., BaQais, A., et al. (2024). Design and application of metal organic frameworks for heavy metals adsorption in water: a review. RSC Advance, Volume 14, pp. 9365–9390. https://doi.org/10.1039/d3ra08815d.

8 Julien, P.A., Užarević, K., Katsenis, A.D., et al. (2016). In situ monitoring and mechanism of the mechanochemical formation of a microporous MOF-74 framework. Journal of the American Chemical Society, Volume 138, Issue 9, pp. 2929–2932. https://doi.org/10.1021/jacs.5b13038.

9 Kaur, H., Devi, N., Siwal, S.S., et al. (2023). Metal-organic framework-based materials for wastewater treatment: superior adsorbent materials for the removal of hazardous pollutants. ACS Omega, Volume 8, Issue 10, pp. 9004–9030. https://doi.org/10.1021/acsomega.2c07719.

10 Singh, S., Sivaram, N., Nath, B., et al. (2024). Metal organic frameworks for wastewater treatment, renewable energy and circular economy contributions. npj Clean Water, Volume 7, p. 124. https://doi.org/10.1038/s41545-024-00408-4.

11 Rojas, S., and Horcajada, P. (2020). Metal-organic frameworks for the removal of emerging organic contaminants in water. Chemical Reviews, Volume 120, Issue 16, pp. 8378–8415. https://doi.org/10.1021/acs.chemrev.9b00797.

12 Salahshoori, I., Vaziri, A., Jahanmardi, R., et al. (2024). Molecular simulation studies of pharmaceutical pollutant removal (Rosuvastatin and Simvastatin) using novel modified-MOF nanostructures (UIO-66, UIO-66/Chitosan, and UIO-66/Oxidized Chitosan).ACS Applied Materials and Interfaces, Volume 16, Issue 20, pp. 26685–26712. https://doi.org/10.1021/acsami.4c01365.

13 Wawrzkiewicz, M., and Hubicki, Z. (2015). Anion exchange resins as effective sorbents for removal of acid, reactive, and direct dyes from textile wastewaters. Ion exchange – studies and applications, InTech. https://doi.org/10.5772/60952.

14 Obotey Ezugbe, E., and Rathilal, S. (2020). Membrane technologies in wastewater treatment: a review. Membranes, Volume 10, Issue 5, p. 89. https://doi.org/10.3390/membranes10050089.

15 Gurreri, L., Tamburini, A., Cipollina, A., et al. (2020). Electrodialysis applications in wastewater treatment for environmental protection and resources recovery: a systematic review on progress and perspectives. Membranes, Volume 10, Issue 7, p. 146. https://doi.org/10.3390/membranes10070146.
16 Akpomie, K.G., and Conradie, J. (2020). Advances in application of cotton-based adsorbents for heavy metals trapping, surface modifications and future perspectives. Ecotoxicology and Environmental Safety, Volume 201, p. 110825.
17 Jia-Qian, J. (2015). The role of coagulation in water treatment. Current Opinion in Chemical Engineering, Volume 8, pp. 36–44. https://doi.org/10.1016/j.coche.2015.01.008.
18 Reem, M.E.l., Nora, M., Khlood, A., et al. (2023). A review of coagulation explaining its definition, mechanism, coagulant types, and optimization models; RSM, and ANN. Current Research in Green and Sustainable Chemistry, Volume 6, p. 100358. https://doi.org/10.1016/j.crgsc.2023.100358.
19 Behera, M., Nayak, J., Banerjee, S., et al. (2021). A review on the treatment of textile industry waste effluents towards the development of efficient mitigation strategy: an integrated system design approach. Journal of Environmental Chemical Engineering, Volume 9, Issue 4, p. 105277.
20 He, H., Xin, X., Qiu, W., et al. (2021). Waste sludge disintegration, methanogenesis and final disposal via various pretreatments: comparison of performance and effectiveness. Environmental Science and Ecotechnology, Volume 8, p. 100132. https://doi.org/10.1016/j.ese.2021.100132.
21 Pandis, P.K., Kalogirou, C., Kanellou, E., et al. (2022). Key points of advanced oxidation processes (AOPs) for wastewater, organic pollutants and pharmaceutical waste treatment: a mini review. Chemical Engineering, Volume 6, Issue 1, p. 8. https://doi.org/10.3390/chemengineering6010008.
22 Rahmani, A.R., Mousavi-Tashar, A., Masoumi, Z., et al. (2019). Integrated advanced oxidation process, sono-Fenton treatment, for mineralization and volume reduction of activated sludge. Ecotoxicology and Environmental Safety, Volume 168, pp. 120–126.
23 Saratale, R.G., Sivapathan, S., Saratale, G.D., et al. (2019). Hydroxamic acid mediated heterogeneous Fenton-like catalysts for the efficient removal of Acid Red 88, textile wastewater and their phytotoxicity studies. Ecotoxicology and Environmental Safety, Volume 167, pp. 385–395.
24 Li, X., Fu, L., Chen, F., et al. (2023). Application of heterogeneous catalytic ozonation in wastewater treatment: an overview. Catalysts, Volume 13, Issue 2, p. 342. https://doi.org/10.3390/catal13020342.
25 Mojiri, A., Ohashi, A., Ozaki, N., et al. (2018). Pollutants removal from synthetic wastewater by the combined electrochemical, adsorption and sequencing batch

reactor (SBR). Ecotoxicology and Environmental Safety, Volume 161, pp. 137–144.

26 Montanes, M.T., García-Gabaldon, M., Roca-Perez, L., et al. (2020). Analysis of norfloxacin ecotoxicity and the relation with its degradation by means of electrochemical oxidation using different anodes. Ecotoxicology and Environmental Safety, Volume 188, p. 109923.

27 Khan, S., Noor, T., Iqbal, N., et al. (2024). Photocatalytic dye degradation from textile wastewater: a review. ACS Omega, Volume 9, Issue 20, pp. 21751–21767. https://doi.org/10.1021/acsomega.4c00887.

28 Ajmal, A., Imran, M., Riffat, N.M., et al. (2014). Principles and mechanisms of photocatalytic dye degradation on TiO_2 based photocatalysts: a comparative overview. RSC Advances, Volume 4, pp. 37003–37026. https://doi.org/10.1039/C4RA06658H.

29 Jayaraj, I., Baburao, G., Gnanasekaran, R., et al. (2024). Critical review on wastewater treatment using photo catalytic advanced oxidation process: role of photocatalytic materials, reactor design and kinetics. Case Studies in Chemical and Environmental Engineering, Volume 9, p. 100599. https://doi.org/10.1016/j.cscee.2023.100599.

30 González-García, J., Sáez, V., Tudela, I., et al. (2010). Sonochemical treatment of water polluted by chlorinated organocompounds. A review. Water, Volume 2, Issue 1, pp. 28–74. https://doi.org/10.3390/w2010028.

31 Dehghani, M.H., Karri, R.R., Koduru, J.R., et al. (2023). Recent trends in the applications of sonochemical reactors as an advanced oxidation process for the remediation of microbial hazards associated with water and wastewater: a critical review. Ultrasonics Sonochemistry, Volume 94, p. 106302. https://doi.org/10.1016/j.ultsonch.2023.106302.

32 Bao, J., Guo, S., Fan, D., et al. (2023). Sonoactivated nanomaterials: a potent armament for wastewater treatment. Ultrasonics Sonochemistry, Volume 99, p. 106569. https://doi.org/10.1016/j.ultsonch.2023.106569.

33 Al-Tohamy, R., Ali, S.S., Li, F., et al. (2022). A critical review on the treatment of dye-containing wastewater: ecotoxicological and health concerns of textile dyes and possible remediation approaches for environmental safety. Ecotoxicology and Environmental Safety, Volume 231, p. 113160. https://doi.org/10.1016/j.ecoenv.2021.113160.

34 Brüschweiler, B.J., and Merlot, C. (2017). Azo dyes in clothing textiles can be cleaved into a series of mutagenic aromatic amines which are not regulated yet. Regulatory Toxicology and Pharmacology, Volume 88, pp. 214–226.

35 Guo, G., Hao, J., Tian, F., et al. (2020). Decolorization and detoxification of azo dye by halo-alkaliphilic bacterial consortium: systematic investigations of performance, pathway and metagenome. Ecotoxicology and Environmental Safety, Volume 204, p. 111073.

36 Samsami, S., Mohamadizaniani, M., Sarrafzadeh, M.H., et al. (2020). Recent advances in the treatment of dye-containing wastewater from textile industries: overview and perspectives. Process Safety and Environmental Protection, Volume 143, pp. 138–163.

37 Buscio, V., García-Jimenez, M., Vilaseca, M., et al. (2016). Reuse of textile dyeing effluents treated with coupled nanofiltration and electrochemical processes. Materials, Volume 9, Issue 6, p. 490.

38 Ruscasso, F., Bezus, B., Garmendia, G., et al. (2021). *Debaryomyces hansenii* F39A as biosorbent for textile dye removal. Revista Argentina de Microbiología, Volume 53, pp. 257–265.

39 Kathing, C., and Saini, G. (2022). A review of various treatment methods for the removal of dyes from textile effluent. Recent Progress in Materials, Volume 4, Issue 4, p. 21. https://doi.org/10.21926/rpm.2204028.

40 Abd Ellatif, S., El-Sheekh, M.M., and Senousy, H.H. (2021). Role of microalgal ligninolytic enzymes in industrial dye decolorization. International Journal of Phytoremediation, Volume 23, pp. 41–52.

41 Yesilada, O., Birhanli, E., and Geckil, H. (2018). Bioremediation and decolorization of textile dyes by white rot fungi and laccase enzymes. In: Prasad R (ed) Mycoremediation and environmental sustainability. Springer. pp. 121–153.

42 Doria, A.R., Pupo, M., Santos, G.D.O.S., et al. (2020). Electrochemical oxidation of indanthrene blue dye in a filter-press flow reactor and toxicity analyses with *Raphidocelis subcapitata* and *Lactuca sativa*. Ecotoxicology and Environmental Safety, Volume 198, p. 110659.

43 An, X., Chen, Y., Chen, G., et al. (2020). Integrated metagenomic and metaproteomic analyses reveal potential degradation mechanism of azo dye-Direct Black G by thermophilic microflora. Ecotoxicology and Environmental Safety, Volume 196, p. 110557.

44 Iorhemen, O.T., Hamza, R.A., and Tay, J.H. (2016). Membrane bioreactor (MBR) technology for wastewater treatment and reclamation: membrane fouling. Membranes, Volume 6, Issue 2, p. 33. https://doi.org/10.3390/membranes6020033.

45 Sameer, A., Marzieh, B., and Ahmed, A. (2021). Membrane bioreactor for wastewater treatment: a review. Case Studies in Chemical and Environmental Engineering, Volume 4, p. 100109. https://doi.org/10.1016/j.cscee.2021.100109.

46 Fazal, S., Zhang, B., Zhong, Z., et al. (2015). Industrial wastewater treatment by using MBR (membrane bioreactor) review study. Journal of Environmental Protection, Volume 6, pp. 584–598. https://doi.org/10.4236/jep.2015.66053.

47 Rahman, T.U., Roy, H., Islam, M.R., et al. (2023). The advancement in membrane bioreactor (MBR) technology toward sustainable industrial wastewater management. Membranes, Volume 13, Issue 2, p. 181. https://doi.org/10.3390/membranes13020181.

48 Ahmad, A., Mohd-Setapar, S.H., Chuong, C.S., et al. (2015). Recent advances in new generation dye removal technologies: novel search for approaches to reprocess wastewater. RSC Advances, Volume 5, Issue 39, pp. 30801–30818.

49 Hassan, I., Chowdhury, S.R., Prihartato, P.K., et al. (2021). Wastewater treatment using constructed wetland: current trends and future potential. Processes, Volume 9, p. 1917. https://doi.org/10.3390/pr9111917.

50 Ji, Z., Tang, W., and Pei, Y. (2021). Constructed wetland substrates: a review on development, function mechanisms, and application in contaminants removal. Chemosphere, Volume 286, p. 131564.

51 Takkar, S., Shandilya, C., Agrahari, R., et al. (2022). Chapter 17 - Green technology: phytoremediation for pesticide pollution. In: Phytoremediation technology for the removal of heavy metals and other contaminants from soil and water. pp. 353–375. https://doi.org/10.1016/B978-0-323-85763-5.00008-8.

52 Diaconu, L.I., Covaliu-Mierlă, C.I., Păunescu, O., et al. (2023). Phytoremediation of wastewater containing lead and manganese ions using algae. Biology, Volume 12, Issue 6, p. 773. https://doi.org/10.3390/biology12060773.

53 DalCorso, G., Fasani, E., Manara, A., et al. (2019). Heavy metal pollutions: state of the art and innovation in phytoremediation. International Journal of Molecular Science, Volume 20, p. 3412. https://doi.org/10.3390/ijms20143412.

54 Khan, A.U., Khan, A.N., Waris, A., et al. (2022). Phytoremediation of pollutants from wastewater: a concise review. Open Life Sciences, Volume 17, Issue 1, pp. 488–496. https://doi.org/10.1515/biol-2022-0056.

55 Delgado-González, C.R., Madariaga-Navarrete, A., Fernández-Cortés, J.M., et al. (2021). Advances and applications of water phytoremediation: a potential biotechnological approach for the treatment of heavy metals from contaminated water. International Journal of Environmental Research and Public Health, Volume 18, Issue 10, p. 5215. https://doi.org/10.3390/ijerph18105215.

56 Sharma, M., Rawat, S., and Rautela, A. (2024). Phytoremediation in sustainable wastewater management: an eco-friendly review of current techniques and future prospects. AQUA – Water Infrastructure, Ecosystems and Society, Volume 73, Issue 9, pp. 1946–1975. https://doi.org/1.2166/aqua.2024.427.

Conclusions and Recommendations

This chapter gives a summary of the entire findings and chapters presented in this book. The authors have made an attempt to bring together the details of all the prominent contaminants present in wastewater.

An overview highlighting the problem of wastewater generation, mentioning SDGs, the global scenario of water pollution, and the need to address it has been provided. It becomes evident that there is an urgent need to deal with the wastewater contaminants, enabling the possible reuse of wastewater to cater to the arising water scarcity. Several factors were found to be responsible for contributing to wastewater generation, for example, sewage, industries, oil spills, etc. These contaminants contribute to the degradation in the quality of wastewater with varied implications depending on the sources of generation, region, and the type of surrounding inhabitants, flora, and fauna.

Having understood the need to treat the contaminants and address the issue of the toxic effects of these contaminants on the living organism and the ecosystem, the contaminants were discussed in detail in the subsequent chapters, mentioning the different methods used for their treatment.

Sewage water has been considered as one of the several categories of wastewater contaminants. Methods such as screening and sedimentation, followed by further treatment, have been used as methods for treatment. Industrial contaminants include an array of several contaminants, depending on the type of industry being referred to. Wastewater generated from industries has been considered a major contributor to water pollution. For the treatment of industrial contaminants, various processes ranging from simple processes like coagulation to complex methods such as advanced oxidation processes have been successfully used. It has also been emphasized that an integration of different methods and technologies has an edge over the solitary methods for industrial contaminants. Constructed wetlands have also been widely used nowadays for the elimination of excess nutrients and contaminants. Heavy metals also pose several hazardous

Hazards and Treatment of Wastewater Contaminants, First Edition. Dipika Jaspal and Arti Malviya. Published 2026 by John Wiley & Sons, Inc.

effects. For the treatment of heavy metals in wastewater, several conventional, novel, and hybrid methods have been used for efficient eradication. The practical applicability of these methods is still being explored as the studies and research on the actual wastewater are scarce. Also, the treatment in the presence of a matrix of heavy metals in the wastewater might sometimes become challenging. Though several method act in conjunction for efficient treatment, still these methods are associated with some advantages and disadvantages, which need to be carefully looked into before converging on a specific method. Certain novel materials are also been efficiently utilized, in addition to some novel methods like nanotechnology and its integration with other methods.

Dyes have also been identified as very toxic and hazardous. Several physical, chemical, and biological techniques have proved to be fruitful for efficient treatment. Methods such as adsorption have proved to be promising for their eradication from wastewater. Several materials have proved their worth for the elimination of a wide variety of dyes. Methods for the treatment of oil as a contaminant range from the use of some physical and simple methods, such as oil absorption pads, to methods such as bioremediation, where different organisms, such as bacteria, fungi, microalgae, etc., are successfully being used for the treatment. With the increase in population, there has been an increase in the use of chemicals such as pesticides, which are toxic. When these leach into the water bodies, they pose a threat to the aquatic life and humans. Different methods for the treatment of pesticides were presented in this book. Several toxic effects that these pesticides pose have been brought to light. Methods such as advanced oxidation processes, chlorination, photochemical degradation, adsorption, membrane filtration, and biological treatments have been mentioned as efficient methods along with the pros and cons of each of these methods. A mention of the several studies with the use of different processes on the removal of different pesticides, along with the results of each of these studies, has been elaborated to get a fair idea of how the different methods work for separate classes of pesticides. Radioactive contaminants were also thought to be an important class of contaminants, and hence, a discussion on the toxic effects and treatment methods was carried out at length.

Mentioning the contaminants in wastewater, it becomes crucial to make mention of the microbes that are responsible for several pathogenic effects on humans. The effects of these microbes have been elaborated, and it has been mentioned that the processes, such as primary, secondary, and tertiary, are important for the complete and efficient treatment of these contaminants. Day-to-day use of pharmaceutical and pharma care products has led to the inclusion of these chemicals in the water bodies. A detailed description of the pharmaceutical wastewater composition, category of such wastes, toxicity, and different categories and subcategories of the treatment methods has been given. A discussion has also been carried out on the use of novel methods being adopted at an accelerated rate for the treatment of pharmaceutical contaminants.

Surfactants are one of the emerging contaminants, and are accountable for several toxic effects, depending on their structure and properties. Water matrices are usually found to have greater amounts of cationic and nonionic surfactants. These compounds need immediate inclusion in the list of priority hazardous substances due to their toxicological impacts on various organisms. It is required to formulate strict regulations for their discharge limits in the environment. Future research related to investigations on the persistence of surfactants and their degradation ability needs to be carried out. Also, potential investigations regarding the impacts of the different surfactant constituents in the real water are crucial.

Proper storage, handling, and application are essential to protect the water sources from chemical fertilizers. It is required that laws governing nutrients should be more stringently enforced. To alleviate the pollution due to fertilizers, improved management, the use of biodegradable, and environmentally friendly fertilizers are recommended.

Microplastics are an important type of contaminant gaining attention in recent times. Different sources of microplastics, their toxicity, and primary, secondary, and tertiary treatments have been discussed. In addition, novel methods reflecting better performance have also been incorporated.

It was thought worthwhile to understand the mechanisms behind the processes for contaminant removal, and hence, a separate chapter constituting the mechanisms involved in some prominent processes of treatment was also made a part of the content.

Several contaminants of emerging nature, arising from the conventional, identified contaminants, are being constantly explored and are under the radar of the research community. More focus is to be kept on such contaminants also, which might become concerning later.

The authors have attempted to categorize and put all the important information related to the several contaminants in wastewater, their toxicity, and treatment methods, all in one place, to help the research community get all the necessary information related to the pollutants in wastewater. The matter presented will help in designing the possible treatment methods and deciding on the efficient removal technique, integrating the conventional methods. It will be helpful to guide policymakers and decision-makers in regulating the required action on a priority basis.

Index

Note: Page numbers in *italics* and **bold** refers to figures and tables respectively.

a

abiotic environment 388–389
acid dyes 157
acid mine drainage (AMD) 78
acid rain 421–422
activated sludge process 359–360
  biological treatment process 61, *61*
  conventional activated sludge process 58–59
  modified activated sludge process 59–62, *60*
adsorption 69, 89, **89**, 355–357, 394, 397–398
  biochar 496
  carbonization process 501
  continuous flow spray drying method 500, *500*
  conventional synthesis 498–499
  cross-linked structure 496
  electrochemistry 500
  electrostatic interaction 496
  ionothermal synthesis 501
  mechanism 496, *497*
  mechanochemistry 499–500
  microwave-assisted method 500
  slow diffusion method 499
  sonochemical method 499
advanced oxidation processes (AOPs) 91, **93**, 404, 507–508
  Japan 32
aerobic biological treatment
  activated sludge process 57–62, *58*, *60*
  anaerobic treatment method 62, *63*
  biological action 54
  constructed wetlands 65–67, *66*
  electrolytic action 56
  mechanical straining 54
  oxidation ditch 64–65, *65*
  oxidation pond 62–64
  sedimentation 54
  treatment process 54
  trickling filters 56–57, *57*
  types of **55**
agricultural wastewater 7
agriculture fertilizers
  acid rain 421–422
  case studies 432–434
  consequences of eutrophication and nutrients 423–426

Hazards and Treatment of Wastewater Contaminants, First Edition. Dipika Jaspal and Arti Malviya. Published 2026 by John Wiley & Sons, Inc.

agriculture fertilizers (*Cont.*)
control of eutrophication and nutrients 426–429
EU nitrates directive 431
eutrophication 422–426
fertilizer use management 436
Food and Agriculture Organization 430–431
improving methods 426
inorganic fertilizers 417–418
leaching 420
micronutrient fertilizers 420
New Soil Health Card Scheme 432
nutrient removal technologies 436–438, **439**
organic fertilizers 418–419
remediation 434–435
slow-release fertilizers 419
surface runoff 420–421
US Clean Water Act (CWA) 431–432
USEPA for nutrient pollution 430
alpha spectrometry 275–276, *275*
anionic surfactants 383–384
autoclaving 350
azo dyes 157

b

basic dyes 157
beach washing 218
beta counting 276
bicephalous surfactants 385
bioaugmentation 180
biofilters 330–332, 406
biologically assisted degradation 516–517, *517*
biological oxygen demand (BOD) 48
biological stains 159
biological treatment, pesticide wastewater
aerobic and anaerobic biological treatment 253–254
bioaugmented activated sludge 253
ligninolytic fungi 252
membrane bioreactor for pesticide removal 254, 257
microbial consortium 253
pressurized activated sludge 253
biomass-based treatment 467–469
biopesticides 233–234
bioreactors 405
for wastewater treatment (Saudi Arabia) 33
bioremediation 213–214, **214**, 218, 284–287, **285–286**
biosurfactants 385
biotic environment 389–394

c

capacitive deionization 133
carbamates 232
carbonization process 501
carbon nanotubes 134–135
cationic surfactants 384
Cauvery river, Tamil Nadu 141–142
Central Pollution Control Board (CPCB) 180
Ceratophyllum demersum 174
chemical oxygen demand (COD) 48, 77
chemical precipitation 277–279
chemical treatment, pesticide wastewater
advanced oxidation process 248–249
chlorination 247–248
electrocoagulation 250–251
Fenton process 250
iron-enhanced sand filters 247
ozonation 249–250
photochemical degradation 249
chitin poison 234
chlorination 324
Cienaga Grande de Santa Marta (CGSM) 465
Clean Water Act (CWA) 206

coagulation 277–279, 351, 400, 506–507
coagulation-flocculation 68–69
coastal ecosystems 196–197
Code of Federal Regulations 192
constructed wetlands (CWs) 93–95, *94*, 326–328, 358–359, 519–521, 531
  Sweden 32
contact pesticides 234
contact stabilization activated sludge process 60, *60*
contaminants
  aesthetics effect 18
  categories of 17–18, *18*
  challenges 36
  detecting contaminants 28–29, *28*
  economic implications 18
  education and public awareness 33
  elimination of 31–33
  environment effect 18
  incidents of 30–31
  nonsewage wastewater 20–28
  policies and collaboration 35–36
  polluted water body restoration 34–35
  prevention of pollution 33
  removal of wastewater 29
  research and development 34
  risk to human and aquatic life 18
  River Ganga, India 31
  sewage wastewater 19–20
  use of technology 34
  wastewater management 34
contaminants elimination
  advanced oxidation processes (AOPs) (Japan) 32
  bioreactors for wastewater treatment (Saudi Arabia) 33
  constructed wetlands (Sweden) 32
  Tuas water reclamation plant (Singapore) 31–32
  wastewater reclamation (Taiwan) 32
continuous flow spray drying method 500, *500*
conventional disinfection 323–324
cosmetics and drug industries 159

d

de-chlorination 68
detritus tanks 51
direct dyes 157
disinfection process 67–68
disperse dyes 157–158
dissolved oxygen (DO) 47
domestic/municipal wastewater 6–7, *7*
dye wastewater 532
  application of 159
  biological treatment 171–175, *174*, **176–178**
  characteristics of 155, **155**
  chemical treatment 166–171
  classification of 156–161, *156*, **160–161**
  legislative guidelines 180–181
  natural sources 153
  novel/hybrid treatment methods 175, 178, *179*, **179**, 180
  organic molecules 153
  physical treatment 165–166
  properties of 159, 161
  removal techniques 165, *165*, **172–173**
  textile industry 154
  toxicity effects on living organisms 163–164, **164**
  toxicity in aqueous environment 162–163

e

Eichhornia crassipes 174
electrochemical advanced oxidation process 404
electrochemical method 132–133, *132*
electrochemical oxidation 509–511

electrochemical removal 354
electrocoagulation 132, 400, 403, 472–475, *473*, **474**
electrodeposition 133
electrodialysis 132–133
electro-Fenton process 508, *509*
electroflotation 132, 404
electrokinetic remediation (EKR) 287–289
electro-technologies 91
elimination of contaminants
  advanced oxidation processes (AOPs) (Japan) 32
  bioreactors for wastewater treatment (Saudi Arabia) 33
  constructed wetlands (Sweden) 32
  Tuas water reclamation plant (Singapore) 31–32
  wastewater reclamation (Taiwan) 32
Elk River chemical spill, West Virginia 95
emerging contaminants
  categories of 25, **26**
  microplastics 26–27
  pharmaceuticals 27–28
  surfactants 27
EU nitrates directive 431
eutrophication 422–426
evaporation 352

f
Facility Response Plan (FRP) rule 220
Fenton process 508
fertilizer use management 436
filamentous algae nutrient scrubbers (FANS) 439
Flavin mononucleotide (FMN) 516
Flint, Michigan water crisis 30
foam fractionation 398
Food and Agriculture Organization (FAO) 237, 430–431
food dyes 159
fumigants 234
fungicides 233

g
gamma ray spectrometry 271–273
Gemini surfactants 384
global pesticide consumption 237
graphene nanomaterials 135
grit chamber 51, *51*
groundwater contamination 482–483
groundwater pollution, Southeast Vietnam 95

h
Haematoxylum campechianum 153
heavy metals
  adsorption by novel adsorbents 125
  domestic sources 114
  effect on aquatic organisms 117–118
  electrochemical method 132–133, *132*
  harmful effects and sources **121–122**
  human health effects 118–120
  industrial sources 114
  nanotechnology for metal removal 133–137, *134*
  nonpoint/diffuse sources 114–116, *115*, *116*
  overview 113
  photocatalysis 128–130, **131**
  phytoremediation method 125–128, **128**
  plant effects 118
  toxicity assessment 116–117
  treatment methods 123–124, **124**
  in water bodies 137–139, **139**
herbicides 233
high-rate algal pond 366–367
Hydrilla verticillata 174
hydro environment, pesticides
  adsorption 244–246, **246–247**

advanced oxidation process 248–249
biological treatment methods 252–257
chlorination 247–248
classification of 232–234, **235–236**
country-wise consumption *237*
electrocoagulation 250–251
evaporation beds 247
Fenton process 250
global pesticide consumption 237
hybrid technologies 257–259
iron-enhanced sand filters 247
legislative framework 259–263
membrane filtration 243
mode of action 234
overview 231
ozonation 249–250
pentachlorophenol removal 263–265
photochemical degradation 249
plasma process 243
removal from secondary wastewater 265–266
toxicity assessment 238–241
ultrasonication 243
use of 238

i

Indigofera tinctoria 153
industrial wastewater 7
adoption of green alternatives 97–98
agricultural industry 78–79
categories of 82–83, *82*
challenges and research directions 99
circular economy 98–99
fate of contaminants 84–86
food industry 80, **81**
iron and steel industry 79
mining industry 78
monitoring and detection techniques 83, **83–84**
nuclear industry 79
organic chemicals industry 81–82, **81**
overview 77–78, *78*
pulp, paper, and textile industry 79–80
recreation industry 82
regulatory compliance 98
toxicity 86–87, **88**
treatment technologies 88–95, 97
zero liquid discharge (ZLD) 97
inorganic fertilizers 417–418
insecticides 233
integrated moving bed membrane bioreactor (IMBMBR) 439
integrated techniques 92, 93
International Maritime Organization (IMO) 206
International Petroleum Industry Environmental Conservation Association (IPIECA) 206
ion exchange 69, 350
mechanisms 503–504
iron nanoparticles 135–136
Isatis tinctoria 153

k

Kala Shah Kaku industrial complex, Pakistan 96
Kutubdia channel, Bangladesh 140–141

l

landfills 341
leaching 420
liquid scintillation counting (LSC) 273

m

marine ecosystems 196
marine organisms **462–464**
mass spectrometry (MS) 273–275, *274*
membrane bioreactor 180, 254, 329–330, 362–363, 518–519, *518*

membrane filtration 398–399, 504–506, *506*
membrane separation technology 69, 89–90, **90**, 360–361
metal complex dyes 158
metal oxide nanoparticles 136–137, *137*
microbes as contaminants
  advanced oxidation process 332
  biofilters 330–332
  chlorination 324
  coagulation-flocculation 322–323
  constructed wetlands 326–328
  conventional disinfection 323–324
  filtration 323
  membrane bioreactor 329–330
  ozonation 325–326
  ultraviolet (UV) light 324–325
  waste stabilization ponds (WSPs) 326
microbial fuel cells 178
microfiltration 505
micronutrient fertilizers 420
microplastics 533
  advanced oxidation process 471–472
  biomass-based treatment 467–469
  biotechnological methods 478–481
  distribution **457**
  electrocoagulation 472–475, *473*, **474**
  groundwater contamination 482–483
  legislation 461, 465
  marine organisms **462–464**
  membrane technology 469–471
  photocatalysis 475–476, *475*
  phytoremediation 477–478, *478*
  plasma technology 476–477, *476*, *477*
  primary treatment 466
  recycling challenges 454, **455**
  removal efficiency 481–482
  secondary treatment 466–467
  sources of *456*, *460*
  tertiary treatment 467
  toxicity assessment 457–461
microwave-assisted method 500
Minamata bay disaster 31
Minamata disease (Kumamoto Prefecture) 96
mordant dyes 158

n

NADPH-dependent reductases 516
nanoparticle-based bioremediation 180
National Oceanic and Atmospheric Administration (NOAA) 206
National Pollutant Discharge Elimination System (NPDES) 180
natural dyes 153
nematicides 233
neonictinoids 233
nerve poison 234
New Soil Health Card Scheme 432
nonionic surfactants 383
nonsewage wastewater
  agricultural runoffs 21–22
  emerging contaminants 25–28, **26**
  industrial contaminants 20–21, **21**, *22*
  microbial contaminants 25
  oil spills 22–23, **23**
  radioactive contaminants 23–24, **24**
Nopal cochenillifera 175
nutrient removal technologies 436–438, **439**

o

Occupational Safety and Health Administration (OSHA) 206
Oil Pollution Act (OPA) 206
oil spills
  in aquatic environment 193–195, *194*
  bioremediation 213–214, **214**
  booms 212–213, *212*
  burning in situ (thermal remediation) 208–209
  characteristics 191–192
  cleanup activities 206–207
  Deepwater Horizon oil rig 214–215

detection of 198–200
dispersants 211–212
disposal and treatment of wastes 216–219, *216*, *217*, **217**, **218**, *219*
economic impacts 197–198
environment impacts 196–197
Exxon Valdez 215
Gulf of Mexico 30
heavy oils 192
hot water and high pressure washing 211
human impacts 197
management and control 219–221
manual labor 213
medium oils 192
Montara 215–216
nonpersistent light oils 192
oil-absorbing pads 208
persistent light oils 192
Prestige 215
prevention and mitigation 205–206, 219
properties 192–193, *193*
sinking oils 192
skimmers, use of 207–208
sorbents 209–210, **210**
technologies and materials 200–205
UK regulations 220–221
US regulations 220
Okhla industrial area, New Delhi 74
organic fertilizers 418–419
organochlorines 232
oxidation ditch 64–65, *65*
oxidation pond 62–64
ozonation 325–326, 354–355
ozone treatment 508–509

p

personal care products 340
pesticide degradation
bioaugmented activated sludge 253
ligninolytic fungi 252
microbial consortium 253
pesticides, hydro environment
adsorption 244–246, **246–247**
advanced oxidation process 248–249
biological treatment methods 252–257
chlorination 247–248
classification of 232–234, **235–236**
country-wise consumption *237*
electrocoagulation 250–251
evaporation beds 247
Fenton process 250
global pesticide consumption 237
hybrid technologies 257–259
iron-enhanced sand filters 247
legislative framework 259–263
membrane filtration 243
mode of action 234
overview 231
ozonation 249–250
pentachlorophenol removal 263–265
photochemical degradation 249
plasma process 243
removal from secondary wastewater 265–266
toxicity assessment 238–241
ultrasonication 243
use of 238
pesticides-loaded wastewater 239
pharmaceutical manufacturing facilities 340
pharmaceutical waste
activated sludge process 359–360
adsorption 355–357
advanced oxidation process 353
anaerobic treatment 359
autoclaving 350
biological treatment 358–360
case studies 365–367
coagulation 351
composition of 343, **344**

pharmaceutical waste (*Cont.*)
 constructed wetlands 358–359
 electrochemical removal 354
 evaporation 352
 floatation process 351
 healthcare facilities 340
 ion exchange 350
 landfills 341
 legislative framework 363–365
 membrane bioreactor system 362–363
 membrane separation technology 360–361
 metal–organic frameworks 357–358
 ozonation 354–355
 pathways of *342*
 personal care products 340
 pharmaceutical manufacturing facilities 340
 photocatalytic degradation 353–354
 physio-chemical methods 353–358
 precipitation 350
 sewerage treatment plant 365–366
 solvent extraction process 352
 sources of 339–340, *340*
 toxicity assessment 344–349
 treatment using high-rate algal pond 366–367
 types of 341–344, *343*
 veterinary institutions 341
 wastewater treatment plants 341
photocatalysis 128–130, **131**, 475–476, *475*
photocatalytic degradation 353–354
 direct method 512
 heterogeneous photo catalytic degradation 514–515
 homogeneous photo catalysis 514
 indirect method 512–513
 ionization of water 512–513
 oxygen ionosorption 513
 photo excitation 512
 superoxide radical 513
 UV–VIS radiation 512
photo-induced/AOP Processes 281, 284
Phragmites australis 174, 175
physical treatment, pesticide wastewater
 adsorption 244–247
 evaporation beds 247
 membrane filtration 243–244
 plasma process 244
 ultrasonication 243
physio-chemical methods 353–358
phytoaccumulation 523
phytodegradation 523–524
phytoremediation 125–128, **128**, 406–407, 477–478, *478*, 521–524, *521*
phytostabilization 522
phytovolatilization 523
Pistia species 174
polycyclic aromatic hydrocarbons (PAHs) 191
polymer-based nanomaterials 136
precipitation 350
preliminary treatment
 detritus tanks 51
 grit chamber 51, *51*
 screening 50–51
 skimming tanks 51–52
pressurized activated sludge 253
printing inks 159
protoplasmic poison 234
pyrethroids 232–233

q

Quercus tinctoria 153

r

radioactive contaminants
 case studies 293–296
 challenges and future needs 301–302
 detection of 271–276
 regulatory authorities 300–301

technologies 297–298
treatment of 276–292, **293**
wastewater remediation materials 298–300
reactive dyes 158
repellents 234
respiratory poison 234
reverse electro dialysis (RED) 505
Rheum rhabarbarum 175
rhizodegradation 523
rhizofiltration 523
rodenticides 233

S

Salvinia species 174
sand washing 218
Sargassum latifolium 175
screening 50–51
sedimentation tanks 52–54
sewage wastewater 19–20, 531
analysis of 44
biological features 48–49
chemical features 46–48
physical characteristics 45–46
preliminary treatment 50–52
primary treatment 50, 52–54
secondary treatment 50, 54–67
solid categories 45–46, **46**
tertiary treatment 50, 67–69, **70–72**, 73
treatment plant 73–74
sewerage treatment plant 365–366
silicon-based nanomaterials 135
silver and gold nanoparticles 135–136
skimming tanks 51–52
slow diffusion method 499
slow-release fertilizers 419
solvent extraction process 352
sonochemical method 499, 515–516
sonochemical treatment 399–400
sorption 280–281
sources of 382
Spill Prevention, Control, and Countermeasure (SPCC) rule 220
stomach poisons 234
sulfur dyes 158
surface runoff 420–421
surfactants 533
abiotic environment 388–389
adsorption 394, 397–398
advanced oxidation process 404
analytical techniques for 394
anionic surfactants 383–384
bicephalous surfactants 385
biofilter 406
biological treatment 404–409
bioreactors 405
biosurfactants 385
biotic environment 389–394
case studies 410–411
cationic surfactants 384
chemical treatment 400–404
coagulation 400
electrochemical advanced oxidation process 404
electrocoagulation 400, 403
electroflotation 404
foam fractionation 398
Gemini surfactants 384
integrated treatment methods 407–409
legislative framework 409–410
membrane filtration 398–399
nonionic surfactants 383
physical treatment 394–400
phytoremediation 406–407
sonochemical treatment 399–400
sources of 382
toxicity assessment of 385, 388–394
types of 383–385, **386–388**
Zwitterion surfactants 384
systemic pesticides 234

t

textile dyes 159
textile wastewater treatment plant
  Arequipa, Peru 181–183, **182**
  Foshan, China 183–184
toxicity, industrial wastewater
  biodiversity and ecosystem 87
  contaminants 87, **88**
  human health 87
  soil and groundwater 87
  waterbodies 86
Tuas water reclamation plant (Singapore) 31–32
Typha angustifolia 174

u

ultraviolet (UV) light 324–325
US Clean Water Act (CWA) 431–432
USEPA for nutrient pollution 430

v

vat dyes 158
volatile organic compounds (VOCs) 82

w

waste stabilization ponds (WSPs) 326
wastewater
  agricultural 7
  aquatic ecosystem 11
  characteristics 8–11
  color 8, **8**, *9*
  contaminants 12
  domestic/municipal 6–7, *7*
  humans 11–12
  industrial 7
  nutrients 10
  odors 8, *9*
  storm water/surface runoff 6
  surfactants 10, *11*
  treatment 12
  treatment plants 341
  turbidity 9, *10*
  types of 5, *6*
wastewater dye
  application of 159
  biological treatment 171–175, *174*, **176–178**
  characteristics of 155, **155**
  chemical treatment 166–171
  classification of 156–161, *156*, **160–161**
  legislative guidelines 180–181
  natural sources 153
  novel/hybrid treatment methods 175, 178, *179*, **179**, 180
  organic molecules 153
  physical treatment 165–166
  properties of 159, 161
  removal techniques 165, *165*, **172–173**
  textile industry 154
  toxicity effects on living organisms 163–164, **164**
  toxicity in aqueous environment 162–163
wastewater reclamation (Taiwan) 32
water bodies contamination 197
water pollution
  climate change 3–5, *4*
  sustainable development goals 2–3
  wastewater 5–7
  water stress 3–5, *5*
West Virginia chemical spill 30
World Trade Organization (WTO) 237

y

yeasts 517

z

Zwitterion surfactants 384

Printed and bound by CPI Group (UK) Ltd, Croydon, CR0 4YY
17/09/2026
14964311-0008